Springer Finance

Springer Finance

Springer Finance is a programme of books addressing students, academics and practitioners working on increasingly technical approaches to the analysis of financial markets. It aims to cover a variety of topics, not only mathematical finance but foreign exchanges, term structure, risk management, portfolio theory, equity derivatives, and financial economics.

Ammann M., Credit Risk Valuation: Methods, Models, and Application (2001)
Back K., A Course in Derivative Securities: Introduction to Theory and Computation (2005)
Barucci E., Financial Markets Theory. Equilibrium, Efficiency and Information (2003)
Bielecki T.R. and Rutkowski M., Credit Risk: Modeling, Valuation and Hedging (2002)
Bingham N.H. and Kiesel R., Risk-Neutral Valuation: Pricing and Hedging of Financial Derivatives (1998, 2nd ed. 2004)
Brigo D. and Mercurio F., Interest Rate Models: Theory and Practice (2001, 2nd ed. 2006)
Buff R., Uncertain Volatility Models – Theory and Application (2002)
Carmona R.A. and Tehranchi M.R., Interest Rate Models: An Infinite Dimensional Stochastic Analysis Perspective (2006)
Dana R.-A. and Jeanblanc M., Financial Markets in Continuous Time (2003)
Deboeck G. and Kohonen T. (Editors), Visual Explorations in Finance with Self-Organizing Maps (1998)
Delbaen F. and Schachermayer W., The Mathematics of Arbitrage (2005)
Elliott R.J. and Kopp P.E., Mathematics of Financial Markets (1999, 2nd ed. 2005)
Fengler M.R., Semiparametric Modeling of Implied Volatility (2005)
Filipovic D., Term-Structure Models (2008 forthcoming)
Fusai G. and Roncoroni A., Implementing Models in Quantitative Finance (2008)
Geman H., Madan D., Pliska S.R. and Vorst T. (Editors), Mathematical Finance – Bachelier Congress 2000 (2001)
Gundlach M., Lehrbass F. (Editors), CreditRisk$^+$ in the Banking Industry (2004)
Jondeau E., Financial Modeling Under Non-Gaussian Distributions (2007)
Kabanov, Y.A. and Safarian M., Markets with Transaction Costs (2008 forthcoming)
Kellerhals B.P., Asset Pricing (2004)
Külpmann M., Irrational Exuberance Reconsidered (2004)
Kwok Y.-K., Mathematical Models of Financial Derivatives (1998, 2nd ed. 2008)
Malliavin P. and Thalmaier A., Stochastic Calculus of Variations in Mathematical Finance (2005)
Meucci A., Risk and Asset Allocation (2005)
Pelsser A., Efficient Methods for Valuing Interest Rate Derivatives (2000)
Prigent J.-L., Weak Convergence of Financial Markets (2003)
Schmid B., Credit Risk Pricing Models (2004)
Shreve S.E., Stochastic Calculus for Finance I (2004)
Shreve S.E., Stochastic Calculus for Finance II (2004)
Yor M., Exponential Functionals of Brownian Motion and Related Processes (2001)
Zagst R., Interest-Rate Management (2002)
Zhu Y.-L., Wu X., Chern I.-L., Derivative Securities and Difference Methods (2004)
Ziegler A., Incomplete Information and Heterogeneous Beliefs in Continuous-time Finance (2003)
Ziegler A., A Game Theory Analysis of Options (2004)

Attilio Meucci

Risk and Asset Allocation

Reprint of the 2007 edition

Springer

Attilio Meucci
Bloomberg L.P.
Head of Portfolio Research
731 Lexington Avenue
New York, NY 10022, USA
attilio_meucci@symmys.com

ISSN 1616-0533
ISBN 978-3-540-22213-2 (hardcover) e-ISBN 978-3-540-27904-4
ISBN 978-3-642-00964-8 (softcover)
DOI 10.1007/978-3-540-27904-4
Springer Dordrecht Heidelberg London New York

Library of Congress Control Number: 2009926061

Mathematics Subject Classification (2000): 15-xx, 46-xx, 62-xx, 65-xx, 90-xx
JEL Classification: C1, C3, C4, C5, C6, C8, G0, G1

Cover illustration: courtesy of Linda Gaylord
Cover design: WMX Design GmbH, Heidelberg

Printed on acid-free paper

Springer is part of Springer Science+Business Media (www.springer.com)

alla mia Capretta

Contents

Part II Classical asset allocation

Part III Accounting for estimation risk

Preface

In an asset allocation problem the investor, who can be the trader, or the fund manager, or the private investor, seeks the combination of securities that best suit their needs in an uncertain environment. In order to determine the optimum allocation, the investor needs to model, estimate, assess and manage uncertainty.

The most popular approach to asset allocation is the mean-variance framework pioneered by Markowitz, where the investor aims at maximizing the portfolio's expected return for a given level of variance and a given set of investment constraints. Under a few assumptions it is possible to estimate the market parameters that feed the model and then solve the ensuing optimization problem.

More recently, measures of risk such as the value at risk or the expected shortfall have found supporters in the financial community. These measures emphasize the potential downside of an allocation more than its potential benefits. Therefore, they are better suited to handle asset allocation in modern, highly asymmetrical markets.

All of the above approaches are highly intuitive. Paradoxically, this can be a drawback, in that one is tempted to rush to conclusions or implementations, without pondering the underlying assumptions.

For instance, the term "mean-variance" hints at the identification of the expected value with its sample counterpart, the mean. Sample estimates make sense only if the quantities to estimate are market invariants, i.e. if they display the same statistical behavior independently across different periods. In equity-like securities the returns are approximately market invariants: this is why the mean-variance approach is usually set in terms of returns. Consider instead an investment in a zero-coupon bond that expires, say, in one month. The time series of the past monthly returns of this bond is not useful in estimating the expected value and the variance after one month, which are known with certainty: the returns are not market invariants.

Similarly, when an allocation decision is based on the value at risk or on the expected shortfall, the problem is typically set in terms of the portfolio's profit-and-loss, because the "P&L" is approximately an invariant.

In general, the investor focuses on a function of his portfolio's value at the end of the investment horizon. For instance, the portfolio's return or profit-and-loss are two such functions which, under very specific circumstances, also happen to be market invariants. In more general settings, the investor needs to separate the definition of his objectives, which depend on the portfolio value at a given future horizon, from the estimation of the distribution of these objectives, which relies on the identification and estimation of some underlying market invariants.

To summarize, in order to solve a generic asset allocation problem we need to go through the following steps.

Detecting invariance
In this phase we detect the market invariants, namely those quantities that display the same behavior through time, allowing us to learn from the past. For equities the invariants are the returns; for bonds the invariants are the changes in yield to maturity; for vanilla derivatives the invariants are changes in at-the-money-forward implied volatility; etc.

Estimating the market
In this step we estimate the distribution of the market invariants from a time series of observations by means of nonparametric estimators, parametric estimators, shrinkage estimators, robust estimators, etc.

Modeling the market
In this phase we map the distribution of the invariants into the distribution of the market at a generic time in the future, i.e. into the distribution of the prices of the securities for the given investment horizon. This is achieved by suitable generalizations of the "square-root-rule" of volatility propagation. The distribution of the prices at the horizon in turn determines the distribution of the investor's objective, such as final wealth, or profit and loss, etc.

Defining optimality
In this step we analyze the investor's profile. We ascertain the features of a potential allocation that are more valuable for a specific investor, such as the trade-off between the expected value and the variance of his objective, or the value at risk of his objective, etc.; and we determine the investor's constraints, such as budget constraints, reluctance to invest in certain assets, etc.

Only after performing separately the above steps can we proceed toward the final goal:

Computing the optimal allocation
At this stage we determine exactly or in good approximation the allocation that best suits the investor, namely the allocation that maximizes the valuable features of the investor's objective(s) given his constraints.

Nevertheless, the approach outlined above is sub-optimal: two additional steps are needed.

Accounting for estimation risk

It is not clear from the above that an allocation based on one month of data is less reliable than an allocation based on two years of data. Nevertheless, the effect of estimation errors on the allocation's performance is dramatic. Therefore we need to account for estimation risk in the optimization process.

Including experience

The most valuable tool for a successful investor is experience, or a-priori knowledge of the market. We need to include the investor's experience in the optimization process by means of a sound statistical framework.

Purpose of this book is to provide a comprehensive treatment of all the above steps. In order to discuss these steps in full generality and consistently from the first to the last one we focus on one-period asset allocation.

Audience and style

A few years ago I started teaching computer-based graduate courses on asset allocation and risk management with emphasis on estimation and modeling because I realized the utmost importance of these aspects in my experience as a practitioner in the financial industry. While teaching, I felt the need to provide the students with an accessible, yet detailed and self-contained, reference for the theory behind the above applications. Since I could not find such a reference in the literature, I set out to write lecture notes, which over the years and after many adjustments have turned into this book.

In an effort to make the reader capable of innovating rather than just following, I sought to analyze the first principles that lead to any given recipe, in addition to explaining how to implement that recipe in practice. Once those first principles have been isolated, the discussion is kept as general as possible: the many applications detailed throughout the text arise as specific instances of the general discussion.

I have tried wherever possible to support intuition with geometrical arguments and practical examples. Heuristic arguments are favored over mathematical rigor. The mathematical formalism is used only up to (and not beyond) the point where it eases the comprehension of the subject. The MATLAB® applications downloadable from `symmys.com` allow the reader to further visualize the theory and to understand the practical issues behind the applications.

A reader with basic notions of probability and univariate statistics could learn faster from the book, although this is not a prerequisite. Simple concepts of functional analysis are used heuristically throughout the text, but the reader is introduced to them from scratch and absolutely no previous knowledge of the subject is assumed. Nevertheless the reader must be familiar with multivariate calculus and linear algebra.

For the above reasons, this book targets graduate and advanced undergraduate students in economics and finance as well as the new breed of practitioners with a background in physics, mathematics, engineering, finance or economics who ever increasingly populate the financial districts worldwide. For the students this is a textbook that introduces the problems of the financial industry in a format that is more familiar to them. For the practitioners, this is a comprehensive reference for the theory and the principles underlying the recipes they implement on a daily basis.

Any feedback on the book is greatly appreciated. Please refer to the website `symmys.com` to contact me.

Structure of the work

This work consists of the printed text and of complementary online resources.

- **Printed text**

  The printed text is divided in four parts.

  Part I

  In the first part we present the statistics of asset allocation, namely the tools necessary to model the market prices at the investment horizon. Chapters 1 and 2 introduce the reader to the formalism of financial risk, namely univariate and multivariate statistics respectively. In Chapter 3 we discuss how to detect the market invariants and how to map their distribution into the distribution of the market prices at the investment horizon.

  Part II

  In the second part we discuss the classical approach to asset allocation. In Chapter 4 we show how to estimate the distribution of the market invariants. In Chapter 5 we define optimality criteria to assess the advantages and disadvantages of a given allocation, once the distribution of the market is known. In Chapter 6 we set and solve allocation problems, by maximizing the advantages of an allocation given the investment constraints.

  Part III

  In the third part we present the modern approach to asset allocation, which accounts for estimation risk and includes the investor's experience in the decision process. In Chapter 7 we introduce the Bayesian approach to parameter estimation. In Chapter 8 we update the optimality criteria to assess the advantages and disadvantages of an allocation when the distribution of the market is only known with some approximation. In Chapter 9 we pursue optimal allocations in the presence of estimation risk, by maximizing their advantages according to the newly defined optimality criteria.

  Part IV

  The fourth part consists of two mathematical appendices. In Appendix A we review some results from linear algebra, geometry and matrix calculus.

In Appendix B we hinge on the analogies with linear algebra to introduce heuristically the simple tools of functional analysis that recur throughout the main text.

- **Online resources**

The online resources consist of software applications and ready-to-print material. They can be downloaded freely from the website `symmys.com`.

Software applications

The software applications are in the form of MATLAB programs. These programs were used to generate the case studies, simulations and figures in the printed text.

Exercise book

The exercise book documents the above MATLAB programs and discusses new applications.

Technical appendices

In order to make the book self-contained, the proofs to almost all the technical results that appear in the printed text are collected in the form of end-of-chapter appendices. These appendices are not essential to follow the discussion. However, they are fundamental to a true understanding of the subjects to which they refer. Nevertheless, if included in the printed text, these appendices would have made the size of the book unmanageable.

The notation in the printed text, say, "Appendix www.2.4" refers to the technical appendix to Chapter 2, Section 4, which is located on the internet. On the other hand the notation, say, "Appendix B.3" refers to the mathematical Appendix B, Section 3, at the end of the book.

A guided tour by means of a simplistic example

To better clarify the content of each chapter in the main text we present a more detailed overview, supported by an oversimplified example which, we stress, does not represent a real model.

Part I

A portfolio at a given future horizon is modeled as a random variable and is represented by a univariate distribution: in Chapter 1 we review univariate statistics. We introduce the representations of the distribution of a generic random variable X, i.e. the probability density function, the cumulative distribution function, the characteristic function and the quantile, and we discuss expected value, variance and other parameters of shape. We present a graphical interpretation of the location and dispersion properties of a univariate distribution and we discuss a few parametric distributions useful in applications.

For example, we learn what it means that a variable X is normally distributed:

$$X \sim \mathrm{N}\left(\mu, \sigma^2\right), \tag{0.1}$$

where μ is the expected value and σ^2 is the variance.

The market consists of securities, whose prices at a given future horizon can be modeled as a multivariate random variable: in Chapter 2 we discuss multivariate statistics. We introduce the representations of the distribution of a multivariate random variable $\mathbf{X}$, namely the joint probability density function, the cumulative distribution function and the characteristic function. We analyze the relationships between the different entries of $\mathbf{X}$: the marginal-copula factorization, as well as the concepts of dependence and of conditional distribution.

We discuss expected value, mode and other multivariate parameters of location; and covariance, modal dispersion and other multivariate parameters of dispersion. We present a graphical interpretation of location and dispersion in terms of ellipsoids and the link between this interpretation and principal component analysis.

We discuss parameters that summarize the co-movements of one entry of $\mathbf{X}$ with another: we introduce the concept of correlation, as well as alternative measures of concordance. We analyze the multivariate generalization of the distributions presented in Chapter 1, including the Wishart and the matrix-variate Student t distributions, useful in Bayesian analysis, as well as very general log-distributions, useful to model prices. Finally we discuss special classes of distributions that play special roles in applications.

For example, we learn what it means that two variables $\mathbf{X} \equiv (X_1, X_2)'$ are normally distributed:

$$\mathbf{X} \sim \mathrm{N}\left(\boldsymbol{\mu}, \boldsymbol{\Sigma}\right), \tag{0.2}$$

where $\boldsymbol{\mu} \equiv (\mu_1, \mu_2)'$ is the vector of the expected values and where the covariance matrix is the identity matrix, i.e. $\boldsymbol{\Sigma} \equiv \mathbf{I}$. We represent this variable as a unit circle centered in $\boldsymbol{\mu}$: the radius represents the two eigenvalues and the reference axes represent the two eigenvectors. As it turns out, the normal distribution (0.2) belongs to the special elliptical, stable and infinitely divisible classes.

In Chapter 3 we model the market. The market is represented by a set of securities that at time t trade at the price $\mathbf{P}_t$. The investment decision is made at the time T and the investor is interested in the distribution of the prices $\mathbf{P}_{T+\tau}$ at a determined future investment horizon τ. Modeling the market consists of three steps. First we need to identify the invariants hidden behind the market data, i.e. those random variables $\mathbf{X}$ that are distributed identically and independently across time.

For example suppose that we detect as invariants the changes in price:

$$\mathbf{X}_{t,\widetilde{\tau}} \equiv \mathbf{P}_t - \mathbf{P}_{t-\widetilde{\tau}}, \tag{0.3}$$

where the estimation horizon $\widetilde{\tau}$ is one week.

Secondly, we have to associate a meaningful parametric distribution to these invariants

For example suppose that the normal distribution (0.2) with the identity as covariance is a suitable parametric model for the weekly changes in prices:

$$\mathbf{X}_{t,\widetilde{\tau}} \sim \mathrm{N}(\boldsymbol{\mu}, \mathbf{I}). \tag{0.4}$$

In this case the market parameters, still to be determined, are the entries of $\boldsymbol{\mu}$.

Finally, we have to work out the distribution of the market, i.e. the prices $\mathbf{P}_{T+\tau}$ at the generic horizon τ, given the distribution of the invariants $\mathbf{X}_{t,\widetilde{\tau}}$ at the specific horizon $\widetilde{\tau}$. This step is fundamental when we first estimate parameters at a given horizon and then solve allocation problems at a different horizon.

For example, suppose that the current market prices of all the securities are normalized to one unit of currency, i.e. $\mathbf{P}_T \equiv \mathbf{1}$, and that the investment horizon is one month, i.e. four weeks. Then, from (0.3) and (0.4) the distribution of the market is normal with the following parameters:

$$\mathbf{P}_{T+\tau} \sim \mathrm{N}(\mathbf{m}, 4\mathbf{I}), \tag{0.5}$$

where

$$\mathbf{m} \equiv \mathbf{1} + 4\boldsymbol{\mu}. \tag{0.6}$$

In a market of many securities the actual dimension of risk in the market is often much lower than the number of securities: therefore we discuss dimension-reduction techniques such as regression analysis and principal component analysis and their geometrical interpretation in terms of the location-dispersion ellipsoid. We conclude with a detailed case study, which covers all the steps involved in modeling the swap market: the detection of the invariants; the "level-slope-hump" PCA approach to dimension reduction of the swap curve invariants, along with its continuum-limit interpretation in terms of frequencies; and the roll-down, duration and convexity approximation of the swap market.

Part II

In the first part of the book we set the statistical background necessary to formalize allocation problems. In the second part we discuss the classical approach to solve these problems, which consists of three steps: estimating the market distribution, evaluating potential portfolios of securities and optimizing those portfolios according to the previously introduced evaluation criteria.

In Chapter 4 we estimate from empirical observations the distribution of the market invariants. An estimator is a function that associates a number, the estimate, with the information i_T that is available when the investment decision in made. This information is typically represented by the time series of the past observations of the market invariants.

For example, we can estimate the value of the market parameter $\boldsymbol{\mu}$ in (0.4) by means of the sample mean:

$$i_T \equiv \{\mathbf{x}_1, \ldots, \mathbf{x}_T\} \mapsto \widehat{\boldsymbol{\mu}} \equiv \frac{1}{T} \sum_{t=1}^{T} \mathbf{x}_t, \tag{0.7}$$

where we dropped the estimation interval from the notation.

We discuss general rules to evaluate the quality of an estimator. The most important feature of an estimator is its replicability, which guarantees that a successful estimation does not occur by chance. An estimator's replicability is measured by the distribution of its loss and is summarized by error, bias and inefficiency. Then we introduce different estimators for different situations: nonparametric estimators, suitable in the case of a very large number of observations; maximum likelihood estimators under quite general non-normal assumptions, suitable when the parametric shape of the invariants' distribution is known; shrinkage estimators, which perform better when the amount of data available is limited; robust estimators, which the statistician should use when he is not comfortable with a given parametric specification of the market invariants. Throughout the analysis we provide the geometrical interpretation of the above estimators. We conclude with practical tips to deal, among other problems, with outliers detection and missing values in the time series.

In Chapter 5 we show how to evaluate an allocation. The investor can allocate his money in the market to form a portfolio of securities. Therefore, the allocation decision is defined by a vector $\boldsymbol{\alpha}$ whose entries determine the number of units (e.g. shares) of the respective security that are being purchased at the investment time T. The investor focuses on his primary objective, a random variable whose distribution depends on the allocation and the market parameters: different objectives corresponds to different investment priorities, such as benchmark allocation, daily trading (profits and losses), financial planning, etc.

For example, assume that the investor's objective is final wealth. If the market is distributed as in (0.5) the objective is normally distributed:

$$\Psi \equiv \boldsymbol{\alpha}' \mathbf{P}_{T+\tau} \sim \mathrm{N}\left(\boldsymbol{\alpha}' \mathbf{m}, \sigma^2\right), \tag{0.8}$$

where $\mathbf{m}$ is given in (0.6) and σ^2 is a simple function of the allocation.

Evaluating an allocation corresponds to assessing the advantages and disadvantages of the distribution of the respective objective. We start considering stochastic dominance, a criterion to compare distributions globally: nevertheless stochastic dominance does not necessarily give rise to a ranking of the potential allocations. Therefore we define indices of satisfaction, i.e. functions of the allocation and the market parameters that measure the extent to which an investor appreciates the objective ensuing from a given allocation.

For example, satisfaction can be measured by the expected value of final wealth: a portfolio with high expected value elicits a high level of satisfaction. In this case from (0.6) and (0.8) the index of satisfaction is the following function of the allocation and of the market parameters:

$$(\boldsymbol{\alpha}, \boldsymbol{\mu}) \mapsto \mathrm{E}\{\Psi\} = \boldsymbol{\alpha}'(\mathbf{1} + 4\boldsymbol{\mu}). \tag{0.9}$$

We discuss the general properties that indices of satisfaction can or should display. Then we focus on three broad classes of such indices: the certainty-equivalent, related to expected utility and prospect theory; the quantile of the objective, closely related to the concept of value at risk; and coherent and spectral measures of satisfaction, closely related to the concept of expected shortfall. We discuss how to build these indices and we analyze their dependence on the underlying allocation. We tackle a few computational issues, such as the Arrow-Pratt approximation, the gamma approximation, the Cornish-Fisher approximation, and the extreme value theory approximation.

In Chapter 6 we pursue the optimal allocation for a generic investor. Formally, this corresponds to maximizing the investor's satisfaction while keeping into account his constraints. We discuss the allocation problems that can be solved efficiently at least numerically, namely convex programming and in particular semidefinite and second-order cone programing problems.

For example, suppose that transaction costs are zero and that the investor has a budget constraint of one unit of currency and can purchase only positive amounts of any security. Assume that the market consists of only two securities. Given the current market prices, from (0.9) the investor's optimization problem reads:

$$\boldsymbol{\alpha}^* \equiv \underset{\substack{\alpha_1+\alpha_2=1 \\ \boldsymbol{\alpha}\geq \mathbf{0}}}{\operatorname{argmax}} \, \boldsymbol{\alpha}'(\mathbf{1} + 4\widehat{\boldsymbol{\mu}}), \tag{0.10}$$

where $\widehat{\boldsymbol{\mu}}$ are the estimated market parameters (0.7). This is a linear programming problem, a special case of cone programing. The solution is a 100% investment in the security with the largest estimated expected value. Assuming for instance that this is the first security, we obtain:

$$\widehat{\mu}_1 > \widehat{\mu}_2 \quad \Rightarrow \quad \alpha_1^* \equiv 1, \; \alpha_2^* \equiv 0. \tag{0.11}$$

In real problems it not possible to compute the exact solution to an allocation optimization. Nevertheless it is possible to obtain a good approximate solution by means of a two-step approach. The core of this approach is the mean-variance optimization, which we present in a general context in terms market prices, instead of the more common, yet more restrictive, representation in terms of returns. Under fairly standard hypotheses, the computation of the mean-variance frontier is a quadratic programming problem. In special cases we can even compute analytical solutions, which provide insight into the effect of the market on the allocation in more general contexts: for example, we prove wrong the common belief that uncorrelated markets provide better investment opportunities than highly correlated markets. We analyze thoroughly the problem of managing assets against a benchmark, which is the explicit task of a fund manager and, as it turns out, the implicit objective of all investors. We discuss the pitfalls of a superficial approach to the mean-variance problem, such as the confusion between compounded returns and linear returns which gives rise to distortions in the final allocation. Finally, we present a case study that reviews all the steps that lead to the optimal allocation.

Part III

In the classical approach to asset allocation discussed in the second part we implicitly assumed that the distribution of the market, once estimated, is known. Nevertheless, such distribution is estimated with some error. As a result, any allocation implemented cannot be truly optimal and the truly optimal allocation cannot be implemented. More importantly, since the optimization process is extremely sensitive to the input parameters, the sub-optimality due to estimation risk can be dramatic.

The parameter $\widehat{\mu}$ in the optimization (0.10) is only an estimate of the true parameter that defines the distribution of the market (0.4). The true expected value of the second security could be larger than the first one, as opposed to what stated in (0.11). In this case the truly optimal allocation would read:

$$\mu_1 < \mu_2 \quad \Rightarrow \quad \overline{\alpha}_1 \equiv 0, \; \overline{\alpha}_2 \equiv 1. \tag{0.12}$$

This allocation is dramatically different from the allocation (0.11), which was implemented.

As a consequence, portfolio managers, traders and professional investors in a broader sense mistrust the "optimal" allocations ensuing from the classical approach and prefer to resort to their experience. In the third part of the book we present a systematic approach to tackle estimation risk, which also includes within a sound statistical framework the investor's experience or models.

Following the guidelines of the classical approach, in order to determine the optimal allocation in the presence of estimation risk we need to introduce a new approach to estimate the market distribution, update the evaluation criteria for potential portfolios of securities and optimize those portfolios according to the newly introduced evaluation criteria.

In Chapter 7 we introduce the Bayesian approach to estimation. In this context, estimators are not numbers: instead, they are random variables modeled by their posterior distribution, which includes the investor's experience or prior beliefs. A Bayesian estimator defines naturally a classical-equivalent estimator and an uncertainty region.

For example, the Bayesian posterior, counterpart of the classical estimator (0.7), could be a normal random variable:

$$\widehat{\boldsymbol{\mu}}_B \sim \mathrm{N}\left(\frac{1}{2}\widehat{\boldsymbol{\mu}} + \frac{1}{2}\boldsymbol{\mu}_0, \mathbf{I}\right), \tag{0.13}$$

where $\boldsymbol{\mu}_0$ is the price change that the investor expects to take place. Then the classical-equivalent estimator is an average of the prior and the sample estimator:

$$\widehat{\boldsymbol{\mu}}_{\text{ce}} \equiv \frac{1}{2}\widehat{\boldsymbol{\mu}} + \frac{1}{2}\boldsymbol{\mu}_0; \tag{0.14}$$

and the uncertainty region is a unit circle centered in $\widehat{\boldsymbol{\mu}}_{\text{ce}}$:

$$\mathcal{E} \equiv \left\{\boldsymbol{\mu} \text{ such that } (\boldsymbol{\mu} - \widehat{\boldsymbol{\mu}}_{\text{ce}})' (\boldsymbol{\mu} - \widehat{\boldsymbol{\mu}}_{\text{ce}}) \leq 1\right\}. \tag{0.15}$$

Since it is difficult for the investor to input prior beliefs directly in the model, we discuss how to input them implicitly in terms of ideal allocations.

In Chapter 8 we introduce criteria to evaluate the sub-optimality of a generic allocation. This process parallels the evaluation of an estimator. The estimator's loss becomes in this context the given allocation's opportunity cost, i.e. a positive random variable which represents the difference between the satisfaction provided by the true, yet unattainable, optimal allocation and the satisfaction provided by the given allocation.

In our example, from (0.9) the opportunity cost of the sub-optimal allocation (0.10) reads:

$$\mathrm{OC}(\boldsymbol{\alpha}^*, \boldsymbol{\mu}) = (\mathbf{1} + 4\boldsymbol{\mu})' \overline{\boldsymbol{\alpha}} - (\mathbf{1} + 4\widehat{\boldsymbol{\mu}})' \boldsymbol{\alpha}^*, \tag{0.16}$$

where $\overline{\boldsymbol{\alpha}}$ is the truly optimal allocation (0.12).

We analyze the opportunity cost of two extreme approaches to allocation: at one extreme the prior allocation, which completely disregards any information from the market, relying only on prior beliefs; at the other extreme the sample-based allocation, where the unknown market parameters are replaced by naive estimates.

In Chapter 9 we pursue optimal allocations in the presence of estimation risk, namely allocations whose opportunity cost is minimal. We present allocations based on Bayes' rule, such as the classical-equivalent allocation and the Black-Litterman approach. Next we present the resampling technique by Michaud. Then we discuss robust allocations, which aim at minimizing the maximum possible opportunity cost over a given set of potential market parameters. Finally, we present robust Bayesian allocations, where the set of potential market parameters is defined naturally in terms of the uncertainty set of the posterior distribution.

In our example, the sub-optimal allocation (0.10) is replaced by the following robust Bayesian allocation:

$$\boldsymbol{\alpha}^* \equiv \underset{\substack{\alpha_1+\alpha_2=1 \\ \boldsymbol{\alpha}\geq\mathbf{0}}}{\operatorname{argmin}} \left\{ \max_{\boldsymbol{\mu}\in\mathcal{E}} \mathrm{OC}\left(\boldsymbol{\alpha},\boldsymbol{\mu}\right) \right\}, \tag{0.17}$$

where the opportunity cost is defined in (0.16) and the uncertainty set is defined in (0.15). The solution to this problem is a balanced allocation where, unlike in (0.11), both securities are present in positive amounts.

In general it is not possible to compute exactly the optimal allocations. Therefore, as in the classical approach to asset allocation, we resort to the two-step mean-variance setting to solve real problems.

Acknowledgments

I wish to thank Carlo Favero, Carlo Giannini, John Hulpke, Alfredo Pastor and Eduardo Rossi, who invited me to teach finance at their institutions, thereby motivating me to write lecture notes for my courses.

A few people provided precious feedback on different parts of the draft at different stages in its development, in particular Davide DiGennaro, Luca Dona', Alberto Elices, Silverio Foresi, Davide Guzzetti, Philip Stark, Dirk Tasche, Kostas Tryantafyllapoulos and an anonymous referee. Francesco Corielli and Gianluca Fusai furnished insightful comments and suggested new material for the book during many pleasant conversations throughout the last few years.

I am indebted to Catriona Byrne, Susanne Denskus and Stefanie Zoeller at Springer for their active support; to EDV-Beratung Frank Herweg for carefully correcting the proofs; and to George Pearson and John MacKendrick at MacKichan Software, Inc. for helping me discover the capabilities of Scientific WorkPlace®, which I used to write this book.

A special thank is due to Jenifer Shiu, for her support during the last year of writing.

Greenwich, January 2005,

Attilio Meucci

Part I

The statistics of asset allocation

1

Univariate statistics

In this chapter we review the basics of univariate statistics. For more on this subject see Mood, Graybill, and Boes (1974) and Casella and Berger (2001).

In Section 1.1 we introduce the definition of random variable and the concept of distribution, as well as four equivalent ways to represent a distribution: the most intuitive, i.e. the probability density function, and three equivalent representations, namely the cumulative distribution function, the characteristic function and the quantile. Depending on the applications, all of the above representations prove useful.

In Section 1.2 we discuss the parameters that summarize the main features of a distribution, such as the location, the dispersion, the degree of symmetry and the thickness of the tails. Then we present the graphical representation of these properties.

In Section 1.3 we introduce a few distributions that are useful to model and solve asset allocation problems.

1.1 Building blocks

A *random variable* X is the number that corresponds to a measurement that has yet to take place. The measurement can assume a range of values on the real axis $\mathbb{R}$, each with a specific probability.

For example, consider a stock that trades today on the exchange at the following price (e.g. in dollars):

$$\widetilde{x} \equiv 100. \tag{1.1}$$

Tomorrow's price X for this stock is a random variable. Something about this measurement is known: for example we might argue that tomorrow's measurement is more likely to be in the neighborhood of today's value (1.1) than in the neighborhood of, say, $x \equiv 10$.

A. Meucci, *Risk and Asset Allocation*, Springer Finance,

The stochastic features of the different possible measurements of a random variable X can be described in terms of a *distribution*. A distribution is characterized by a *space of events* $\mathfrak{E}$ and a *probability* $\mathbb{P}$.

The unknown outcome x of the measurement of X corresponds to one specific event $\mathfrak{e}$ among many that can take place in the space of events $\mathfrak{E}$. Therefore, a random variable is a function from the space of events to the range of measurements on the real line $\mathbb{R}$: if a specific event $\mathfrak{e}$ takes place, the measurement will take on the value $x \equiv X(\mathfrak{e})$. In a different universe, a different event $\mathfrak{e}'$ might have taken place and thus the measurement would have been a different value $x' \equiv X(\mathfrak{e}')$.

The likelihood of different possible events is described by a probability $\mathbb{P}$, which is a measure on the space of events. The following notation stands for the probability of all the events $\mathfrak{e}$ in the space of events $\mathfrak{E}$ that give rise to a measurement of X in a given interval $[\underline{x}, \overline{x}]$:

$$\mathbb{P}\{X \in [\underline{x}, \overline{x}]\} \equiv \mathbb{P}\{\mathfrak{e} \in \mathfrak{E} \text{ such that } X(\mathfrak{e}) \in [\underline{x}, \overline{x}]\}. \tag{1.2}$$

A distribution can be represented in three equivalent ways.

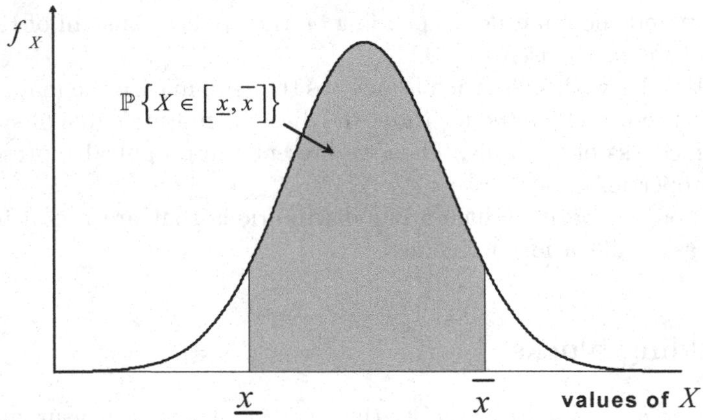

Fig. 1.1. Probability density function

The most intuitive way to represent the distribution of the random variable X is by means of the *probability density function (pdf)* f_X. Intuitively, the pdf shows a peak where the outcome of the measurement of X is more likely to occur. More formally, the probability density function is defined in such a way that the probability $\mathbb{P}$ that a measurement takes place in a generic interval $[\underline{x}, \overline{x}]$ is the area comprised the interval and the density, see Figure 1.1:

$$\mathbb{P}\{X \in [\underline{x}, \overline{x}]\} \equiv \int_{\underline{x}}^{\overline{x}} f_X(x)\, dx. \tag{1.3}$$

In particular, we notice that, since a probability is non-negative, the probability density function is non-negative:

$$f_X(x) \geq 0. \tag{1.4}$$

Furthermore, since the measurement of X must assume a value on the real axis, the following normalization must hold:

$$\int_{-\infty}^{+\infty} f_X(x)\, dx = 1. \tag{1.5}$$

For example the function

$$f_X(x) \equiv \frac{1}{\sqrt{\pi}} e^{-(x-\widetilde{x})^2}, \tag{1.6}$$

which we plot in Figure 1.1, has a bell shape which is peaked around the current price (1.1). We show in a more general context in Section 1.3.2 that (1.6) satisfies (1.4) and (1.5). Therefore it is a probability density function which could model tomorrow's price for the stock.

To introduce the second equivalent way to describe a distribution we notice from (1.3) that, in order to compute probabilities, we always need to integrate the probability density function f_X over some interval. The *cumulative distribution function (cdf)* F_X is defined as the probability that the measurement be less than a generic value x, see Figure 1.2. In formulas:

$$F_X(x) \equiv \mathbb{P}\{X \leq x\} = \int_{-\infty}^{x} f_X(u)\, du. \tag{1.7}$$

In other words, the cumulative distribution function is obtained from the probability density function by applying (B.27), the integration operator:

$$F_X = \mathcal{I}[f_X]. \tag{1.8}$$

This means that the probability density function can be recovered from the cumulative distribution function by applying the derivative operator (B.25), which is the inverse of the integration operator:

$$f_X = \mathcal{D}[F_X]. \tag{1.9}$$

Therefore the two representations are equivalent.

Given the properties (1.4) and (1.5) of the probability density function, it is easy to check that the cumulative distribution function is non-decreasing and satisfies the following normalization conditions:

$$F_X(-\infty) = 0, \quad F_X(+\infty) = 1. \tag{1.10}$$

On the other hand, any function with the above properties defines a cumulative distribution function.

We plot in Figure 1.2 the cumulative distribution function that corresponds to the density (1.6). This cumulative distribution function can be expressed in terms of the error function (B.75) as follows:

$$F_X(x) = \frac{1}{2}\left(1 + \operatorname{erf}\left(x - \widetilde{x}\right)\right), \tag{1.11}$$

where $\widetilde{x} = 100$ is today's price (1.1) of the stock. This is a specific instance of a more general result, see Section 1.3.2.

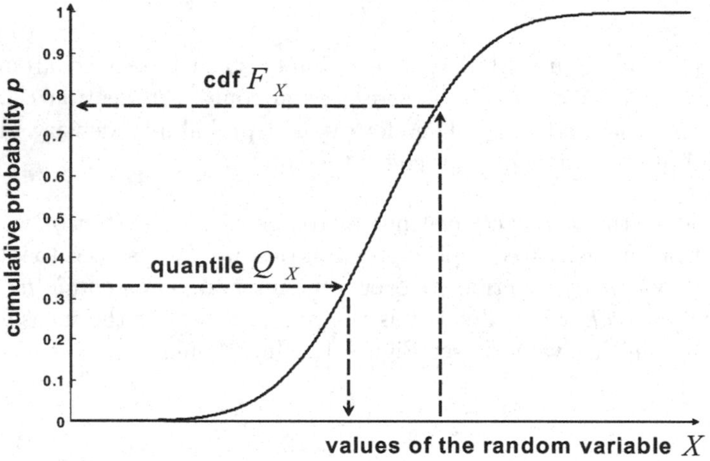

Fig. 1.2. Cumulative distribution function and quantile

A third way to describe the properties of a distribution is through the *characteristic function (cf)* ϕ_X, defined in terms of the expectation operator (B.56) as follows:

$$\phi_X(\omega) \equiv \mathrm{E}\left\{e^{i\omega X}\right\}, \tag{1.12}$$

where $i \equiv \sqrt{-1}$ is the imaginary unit. The characteristic function can assume values in the complex plane.

It is not straightforward to determine the properties of a generic characteristic function implied by the properties (1.4) and (1.5) of the probability density function. Nevertheless, a set of sufficient conditions is provided by

Polya's theorem, which states that function ϕ is a characteristic function of a distribution if it is real-valued, even, convex on the positive real axis, and if it satisfies:

$$\phi(0) \equiv 1, \qquad \lim_{\omega \to \infty} \phi(\omega) \equiv 0, \tag{1.13}$$

see Cuppens (1975).

A comparison of (1.12) with (B.34) and (B.56) shows that the characteristic function is the Fourier transform of the probability density function:

$$\phi_X = \mathcal{F}[f_X]. \tag{1.14}$$

Therefore the probability density function can be recovered from the characteristic function by means of (B.40), i.e. the inverse Fourier transform:

$$f_X = \mathcal{F}^{-1}[\phi_X]. \tag{1.15}$$

At times, the characteristic function proves to be the easiest way to describe a distribution.

The characteristic function of the distribution in the example (1.6) reads:

$$\phi_X(\omega) = e^{i\widetilde{x}\omega - \frac{1}{4}\omega^2}, \tag{1.16}$$

where $\widetilde{x} = 100$ is today's price (1.1) of the stock. This is a specific instance of a more general result, see Section 1.3.2.

We stress that the probability density function f_X, the cumulative distribution function F_X and the characteristic function ϕ_X are three equivalent ways to represent the distribution of the random variable X. We summarize in Figure 1.3 the mutual relationships among these representations.

We also discuss a fourth, fully equivalent way to describe all the properties of a random variable which is very important in financial applications, see Section 5.5. The *quantile* Q_X of the random variable X is the inverse of the cumulative distribution function:

$$Q_X(p) \equiv F_X^{-1}(p), \tag{1.17}$$

where $p \in [0,1]$ denotes a specific value of cumulative probability, see Figure 1.2. By definition, the quantile associates with a cumulative probability p the number x such that the probability that X be less than x is p. In other words, the quantile is defined implicitly by the following equation:

$$\mathbb{P}\{X \leq Q_X(p)\} = p. \tag{1.18}$$

Since the quantile is equivalent to the cumulative distribution function, it is equivalent to any of the above representations of the distribution of X.

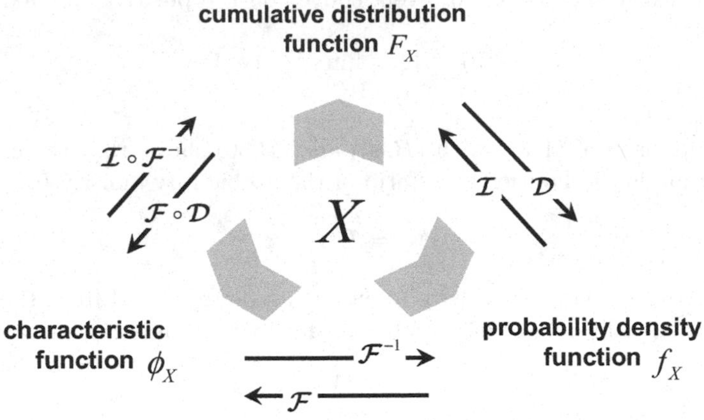

Fig. 1.3. Equivalent representations of a univariate distribution

The quantile of the distribution of our example (1.6) reads in terms of the inverse of the error function (B.75) as follows:

$$Q_X(p) = \widetilde{x} + \operatorname{erf}^{-1}(2p - 1), \tag{1.19}$$

where $\widetilde{x} = 100$ is today's price (1.1) of the stock. This is a specific instance of a more general result, see Section 1.3.2.

In the above discussion we have made the implicit assumption that the probability density function f_X is smooth and positive. This is not always the case.

For instance, the definition of quantile provided in (1.17) only makes sense if the cumulative distribution function is *strictly* increasing, because only in this case with each point on the vertical axis of the cumulative function is associated one and only one point on the horizontal axis, see Figure 1.2. In order for the cumulative distribution function to be strictly increasing, the probability density function must be strictly positive. Indeed, the cumulative distribution function is flat in those regions where the probability density function is null.

To handle situations such as the above example we have two options: either we build a more sophisticated mathematical framework that does not rely on the assumptions of smoothness and positivity for probability density function, or we make the above hypotheses legitimate by regularizing the probability density function as in Appendix B.4. We choose throughout the book the second approach, for practical as well as "philosophical" reasons, see $(B.54)$ and comments thereafter.

To handle the above example, Since the regularized probability density function $f_{X;\epsilon}$ obtained with $(B.54)$ is strictly positive, the respective regularized cumulative distribution function $F_{X;\epsilon}$ is strictly increasing and thus invertible. Therefore we can properly define the regularized quantile as in (1.17) as the inverse of the cumulative distribution function:

$$Q_{X;\epsilon} \equiv F_{X;\epsilon}^{-1}. \tag{1.20}$$

The exact quantile is recovered as the limit of the regularized quantile when the bandwidth ϵ tends to zero, if this limit exists. Otherwise, we simply work with the approximate quantile.

1.2 Summary statistics

In this section we discuss a few parameters that summarize the most information about the properties of a distribution.

1.2.1 Location

Suppose that we need to summarize all the information regarding the random variable X in only one number, the one value that best represents the whole range of possible outcomes. We are looking for a *location* parameter $\mathrm{Loc}\{X\}$ that provides a fair indication of where on the real axis the random variable X will end up taking its value.

A location parameter should enjoy a few intuitive features. In the first place, if the distribution is peaked around a specific value, the location parameter should be close to that peak. In particular, a constant a can be seen as an infinitely peaked random variable, see $(B.22)$ and comments thereafter. Thus the location of a constant should be the constant itself:

$$\mathrm{Loc}\{a\} = a. \tag{1.21}$$

More in general, the location parameter should track any affine transformation of the random variable:

$$\mathrm{Loc}\{a + bX\} = a + b\,\mathrm{Loc}\{X\}, \tag{1.22}$$

where a and $b > 0$ are the constants that define the affine transformation.

Property (1.22) is called the *affine equivariance* of the location parameter.

To understand this property, imagine that the variable X is the price of a stock in cents and that we are interested in the value of our portfolio, which consists of that stock and an extra dollar in cash. Assume that we believe that tomorrow the stock price will be located in a neighborhood of, say, the following value in cents:

$$\operatorname{Loc}\{X\} = 298c. \tag{1.23}$$

Then the whole portfolio should be located around the following value in dollars:

$$\operatorname{Loc}\left\{1 + \frac{X}{100}\right\} = 3.98\$ = 1 + \frac{\operatorname{Loc}\{X\}}{100}. \tag{1.24}$$

An immediate choice for the location parameter is the center of mass of the distribution, i.e. the weighted average of each possible outcome, where the weight of each outcome is provided by its respective probability. This corresponds to computing the *expected value* (B.56) of the random variable:

$$\mathrm{E}\{X\} \equiv \int_{-\infty}^{+\infty} x f_X(x)\, dx. \tag{1.25}$$

As we prove in Appendix www.1.4, the expected value is affine equivariant, i.e. it satisfies (1.22). Therefore the expected value of a random variable is a sensible parameter of location, when the integral that defines it converges.

Whenever the characteristic function (1.12) of X is known and analytical, i.e. it can be recovered entirely from its Taylor series expansion, computing the expected value is easy, as we show in Appendix www.1.6.

An alternative choice for the location parameter is the *median*, which is the quantile (1.17) relative to the specific cumulative probability $p \equiv 1/2$:

$$\operatorname{Med}\{X\} \equiv Q_X\left(\frac{1}{2}\right). \tag{1.26}$$

From (1.18), the median is defined equivalently by the following implicit equation:

$$\int_{-\infty}^{\operatorname{Med}\{X\}} f_X(x)\, dx = \frac{1}{2}. \tag{1.27}$$

As we prove in Appendix www.1.4, the median is affine equivariant, i.e. it satisfies (1.22). Therefore the median of a random variable is also a sensible parameter of location.

Consider a distribution that is *symmetrical* around some value $\widetilde{x}$, i.e. a distribution such that the probability density function f_X satisfies:

$$(\operatorname{Refl} \circ \operatorname{Shift}_{\widetilde{x}})[f_X] = \operatorname{Shift}_{\widetilde{x}}[f_X], \tag{1.28}$$

where the reflection and shift operators are defined in (B.32) and (B.33) respectively. In this case it is intuitive to assume that the symmetry point is a good parameter of location. Indeed, we prove in Appendix www.1.5 that the symmetry point coincides with both the median and the expected value:

$$\text{Med}\{X\} = \text{E}\{X\} = \widetilde{x}. \tag{1.29}$$

A third parameter of location is the *mode*, which refers to the shape of the probability density function f_X. Indeed, the mode is defined as the point that corresponds to the highest peak of the density function:

$$\text{Mod}\{X\} \equiv \underset{x\in\mathbb{R}}{\text{argmax}}\{f_X(x)\}. \tag{1.30}$$

By construction, the mode is peaked around the most likely outcomes. In Appendix www.1.4 we show that the mode is affine equivariant, i.e. it satisfies (1.22): therefore the mode of a random variable is also a sensible parameter of location. Nevertheless, there might exist two or more equally high global maxima, in which case the mode is not defined.

In the example (1.6) it is easy to see that the above three parameters of location, namely expected value, median and mode, coincide:

$$\text{E}\{X\} = \text{Med}\{X\} = \text{Mod}\{X\} = \widetilde{x}, \tag{1.31}$$

where $\widetilde{x} = 100$ is today's price (1.1) of the stock. This is a specific instance of a more general result, see Section 1.3.2.

We remark that the expected value summarizes "global" features of the distribution, in that the whole density f_X contributes to the result, see (1.25); the median only involves "half" of the distribution, see (1.27); the mode provides a "local" picture, in that only a specific value matters, see (1.30).

1.2.2 Dispersion

In this section we summarize in one number the degree of dispersion of the random variable X. In other words, we are looking for a *dispersion* parameter $\text{Dis}\{X\}$ that yields an indication of the extent to which the location parameter might be wrong in guessing the outcome of X.

As in the case of the location parameter, we require that the dispersion parameter display an intuitive property:

$$\text{Dis}\{a + bX\} = |b| \,\text{Dis}\{X\}, \tag{1.32}$$

where a and b are constants. Property (1.32) is called the *affine equivariance* of the dispersion parameter.

To understand the affine equivariance property of the dispersion parameter, imagine that the variable X is tomorrow's price of a stock in cents and that we assess a dispersion of, say 10 cents. Then the dispersion in dollars of the stock price should be 0.1 dollars:

$$\operatorname{Dis}\left\{\frac{X}{100}\right\} = 0.10\$ = \frac{\operatorname{Dis}\{X\}}{100}. \tag{1.33}$$

Furthermore, the dispersion of a portfolio made of that stock and a given amount m of cents in cash should be the same as the dispersion of the stock alone:

$$\operatorname{Dis}\{X\} = 10c = \operatorname{Dis}\{X+m\}. \tag{1.34}$$

In view of multivariate generalizations it is useful to reformulate (1.32) the affine equivariance property in a different way. First we define the *z-score* of the random variable X, which is a normalized version of X located in zero and with unitary dispersion:

$$Z_X \equiv \frac{X - \operatorname{Loc}\{X\}}{\operatorname{Dis}\{X\}}. \tag{1.35}$$

The affine equivariance property of the location parameter (1.22) and of the dispersion parameter (1.32) are equivalent to the condition that the squared z-score remain unaffected by affine transformations:

$$Z^2_{a+bX} = Z^2_X. \tag{1.36}$$

A popular dispersion parameter is the *interquantile range*, defined as the difference of two arbitrary quantiles

$$\operatorname{Ran}\{X\} \equiv Q_X(\overline{p}) - Q_X(\underline{p}), \tag{1.37}$$

where $\overline{p} > \underline{p}$. The standard choice is $\overline{p} \equiv 3/4$, which corresponds to the *upper quartile,* and $\underline{p} \equiv 1/4$, which corresponds to the *lower quartile.* We prove in Appendix www.1.4 that the range is affine equivariant, i.e. it satisfies (1.32).

To introduce another dispersion parameter, consider the *modal dispersion*:

$$\operatorname{MDis}\{X\} \equiv -\left.\frac{1}{\frac{d^2 \ln f_X}{dx^2}}\right|_{x=\operatorname{Mod}\{X\}}, \tag{1.38}$$

see O'Hagan (1994). As we prove in a more general multivariate setting in Appendix www.2.5, the square root of the modal dispersion is affine equivariant and thus it is a suitable dispersion parameter. To see the rationale of this definition, consider a second-order Taylor approximation of the probability density function of X in a neighborhood of the mode:

$$f_X(x) \approx f_X(\mathrm{Mod}\{X\}) + \left.\frac{d^2 f_X}{dx^2}\right|_{x=\mathrm{Mod}\{X\}} (x - \mathrm{Mod}\{X\})^2 . \tag{1.39}$$

The larger in absolute value the second derivative, which is negative around a maximum, the thinner the the probability density function around the mode, and thus the less the dispersion of X. Considering the logarithm of the pdf in the definition (1.38) and taking the square root of the result makes the ensuing parameter affine equivariant.

To define more dispersion parameters we notice that intuitively the dispersion of X is a sort of distance between X and its location parameter. We recall that the space L^p_X of functions of X is a vector space with the norm $\|\cdot\|_{X;p}$, see (B.57) and (B.58). Therefore we can define a dispersion parameter in a natural way as the distance between the random variable and its location parameter:

$$\mathrm{Dis}\{X\} \equiv \|X - \mathrm{Loc}\{X\}\|_{X;p} . \tag{1.40}$$

The general properties (A.7) of a norm imply that this definition of dispersion is affine equivariant, i.e. it satisfies (1.32).

In particular, if we set $p \equiv 1$ in (1.40) and we define the location parameter as the expected value (1.25), we obtain the *mean absolute deviation (MAD)*:

$$\begin{aligned} \mathrm{MAD}\{X\} &\equiv \mathrm{E}\{|X - \mathrm{E}\{X\}|\} \\ &= \int_{\mathbb{R}} |x - \mathrm{E}\{X\}| \, f_X(x) \, dx. \end{aligned} \tag{1.41}$$

On the other hand, if we set $p \equiv 2$ in (1.40) and again we define the location parameter as the expected value (1.25) we obtain the *standard deviation*:

$$\begin{aligned} \mathrm{Sd}\{X\} &\equiv \left(\mathrm{E}\left\{(X - \mathrm{E}\{X\})^2\right\}\right)^{\frac{1}{2}} \\ &= \sqrt{\int_{\mathbb{R}} (x - \mathrm{E}\{X\})^2 f_X(x) \, dx}. \end{aligned} \tag{1.42}$$

When the integral in (1.42) converges, the standard deviation is the benchmark dispersion parameter. The square of the standard deviation, which is very important in applications, is called the *variance*:

$$\mathrm{Var}\{X\} \equiv (\mathrm{Sd}\{X\})^2 = \int_{\mathbb{R}} (x - \mathrm{E}\{X\})^2 f_X(x) \, dx. \tag{1.43}$$

Whenever the characteristic function (1.12) of X is known and it is analytical, i.e. it can be recovered entirely from its Taylor series expansion, computing the variance is straightforward, see Appendix www.1.6.

In our example (1.6) the range reads:

$$\mathrm{Ran}\{X\} = \mathrm{erf}^{-1}\left(\frac{1}{2}\right) - \mathrm{erf}^{-1}\left(-\frac{1}{2}\right) \approx 0.95; \tag{1.44}$$

the mean absolute deviation reads:

$$\mathrm{MAD}\{X\} = \frac{1}{\sqrt{\pi}} \approx 0.56; \tag{1.45}$$

and the standard deviation reads:

$$\mathrm{Sd}\{X\} = \frac{1}{\sqrt{2}} \approx 0.71. \tag{1.46}$$

These are specific instances of more general results, see Section 1.3.2.

We remark that, similarly to the expected value, the standard deviation and the mean absolute deviation summarize global features of the distribution, in that the whole density f_X contributes to the result. On the other hand, similarly to the median, the range involves parts of the distribution. Finally, similarly to the mode, the modal dispersion provides a local picture, in that only a small neighborhood of a specific value matters.

1.2.3 Higher-order statistics

By means of the expectation operator (B.56) we can introduce the moments, summary statistics that provide more insight into the features of a distribution.

The *k-th raw moment* of a random variable X is the expectation of the k-th power of the random variable:

$$\mathrm{RM}_k^X \equiv \mathrm{E}\{X^k\}. \tag{1.47}$$

The *k-th central moment* of a random variable is a location-independent version of the respective raw moment:

$$\mathrm{CM}_k^X \equiv \mathrm{E}\left\{(X - \mathrm{E}\{X\})^k\right\}. \tag{1.48}$$

We already discussed the first raw moment of a random variable X, which is the expected value (1.25); we also discussed the second central moment, which is the variance (1.43).

The third central moment provides a measure of the degree of symmetry of the distribution of X. The standard measure of symmetry of a distribution is the *skewness*, which is the third central moment normalized by the standard deviation, in such a way to make it scale-independent:

$$\mathrm{Sk}\{X\} \equiv \frac{\mathrm{CM}_3^X}{(\mathrm{Sd}\{X\})^3}. \tag{1.49}$$

In particular, a distribution whose probability density function is symmetric around its expected value has null skewness. If the skewness is positive (negative), the distribution is bulkier on the left (right), see Figure 1.4.

In our example (1.6) we have:

$$\mathrm{Sk}\{X\} = 0. \tag{1.50}$$

This is a specific instance of a more general result, see Section 1.3.2. The result (1.50) is consistent with the symmetry of the probability density function (1.6).

The fourth moment provides a measure of the relative weight of the tails with respect to the central body of a distribution. The standard quantity to evaluate this balance is the *kurtosis*, defined as the normalized fourth central moment:

$$\mathrm{Ku}\{X\} \equiv \frac{\mathrm{CM}_4^X}{(\mathrm{Sd}\{X\})^4}. \tag{1.51}$$

The kurtosis gives an indication of how likely it is to observe a measurement far in the tails of the distribution: a large kurtosis implies that the distribution displays "*fat tails*".

In our example (1.6) we have:

$$\mathrm{Ku}\{X\} = 3. \tag{1.52}$$

This is a specific instance of a more general result, see Section 1.3.2.

We remark that all the above moments and summary statistics involve in general integrations. If the integral that defines the expectation operator (B.56) does not converge, the respective moment is not defined. Nevertheless, whenever the characteristic function of the distribution is known and analytical, i.e. it can be recovered entirely from its Taylor series expansion, we can compute these quantities by means of simple differentiation and some algebra, as we show in Appendix www.1.6.

1.2.4 Graphical representations

To obtain an immediate idea of the properties of location and dispersion of a random variable X it is useful to represent them graphically.

One way to do this is by means of a *box plot*, which is the plot of the first, second and third quartile: the box plot summarizes the location of the given distribution, in this case the median, and its dispersion, in this case the interquartile range. More in general, the plot of a few key quantiles gives an idea of the main features of the probability density function f_X, and thus of the distribution of X, see Figure 1.4. Furthermore, the box plot gives an idea of the degree of symmetry of the distribution: if the distance between lower quartile and median exceeds the distance between median and upper quartile the distribution is more spread below the median than it is above the median.

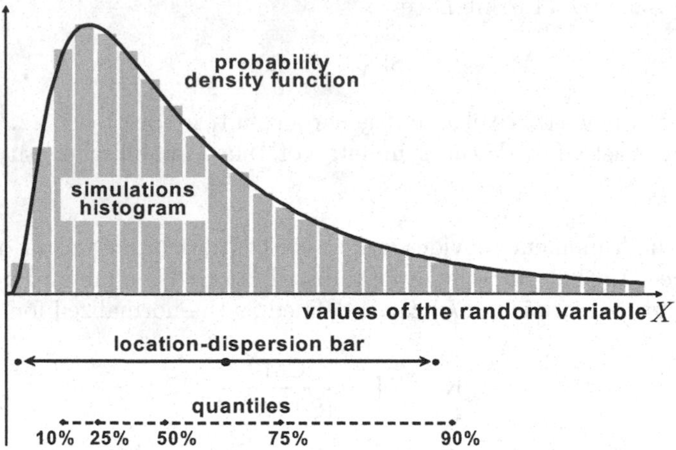

Fig. 1.4. Summary statistics of univariate distributions

Another way to summarize the main features of a distribution is by means of the location-dispersion bar, namely the set of points x which are not any farther from the location parameter of X than one dispersion:

$$\operatorname{Loc}\{X\} - \operatorname{Dis}\{X\} \leq x \leq \operatorname{Loc}\{X\} + \operatorname{Dis}\{X\}. \tag{1.53}$$

The location-dispersion bar is an interval centered on the location parameter and wide twice the dispersion parameter, see Figure 1.4. The dispersion bar becomes particularly useful in its generalization to a multivariate setting, see Section 2.4.3.

1.3 Taxonomy of distributions

In this section we discuss a few distributions that are useful in asset allocation applications. All the distribution introduced are special univariate cases of the more general distributions introduced in Section 2.6.

1.3.1 Uniform distribution

The *uniform distribution* models the situation where the realization of the random variable X is bound to take place on an interval $[a, b]$ and all the values within that interval are equally likely outcomes of the measurement of X.

We use the following notation to indicate that X is uniformly distributed on the interval $[a, b]$:

$$X \sim \mathrm{U}([a, b]). \tag{1.54}$$

Equivalent representations

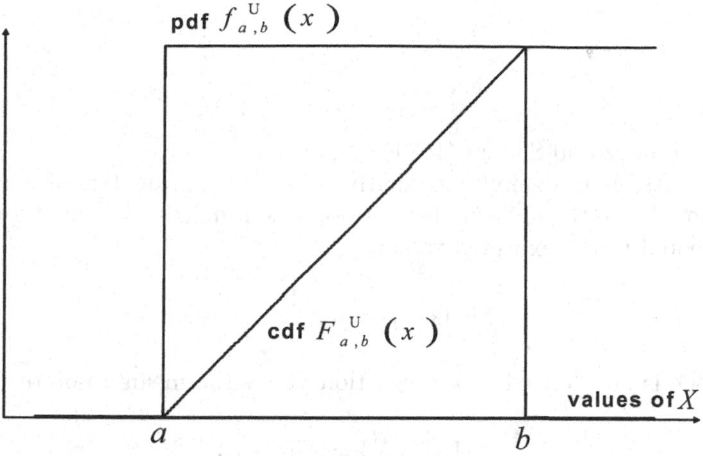

Fig. 1.5. Uniform distribution: pdf and cdf

The probability density function of the uniform distribution reads:

$$f^{U}_{a,b}(x) = \frac{1}{b-a}\mathbb{I}_{[a,b]}(x), \tag{1.55}$$

where $\mathbb{I}$ is the indicator function (B.72), see Figure 1.5.

The cumulative distribution function of the uniform distribution reads:

$$F^{U}_{a,b}(x) = \frac{x-a}{b-a}\mathbb{I}_{[a,b]}(x) + H^{(b)}(x), \tag{1.56}$$

where H is the Heaviside step function (B.73), see Figure 1.5.

The characteristic function of the uniform distribution reads:

$$\phi^{U}_{a,b}(\omega) = \frac{1}{\omega}\frac{2}{b-a}\sin\left(\frac{b-a}{2}\omega\right)e^{i\frac{a+b}{2}\omega}, \tag{1.57}$$

see Abramowitz and Stegun (1974).

Inverting (1.56) we obtain the quantile of the uniform distribution:

$$Q^{U}_{a,b}(p) = a + (b-a)\,p. \tag{1.58}$$

Summary statistics

The standard parameters that summarize the properties of the uniform distribution, namely expected value, standard deviation, skewness and kurtosis, read respectively:

$$\mathrm{E}\left\{X\right\} = a + \frac{1}{2}\left(b - a\right) \tag{1.59}$$

$$\mathrm{Sd}\left\{X\right\} = \frac{1}{\sqrt{12}}\left(b - a\right) \tag{1.60}$$

$$\mathrm{Sk}\left\{X\right\} = 0 \tag{1.61}$$

$$\mathrm{Ku}\left\{X\right\} = \frac{9}{5}, \tag{1.62}$$

see Abramowitz and Stegun (1974).

It is possible to compute explicitly also other parameters of location and dispersion. Since the uniform distribution is symmetrical, from (1.29) the median is equal to the expected value:

$$\mathrm{Med}\left\{X\right\} = a + \frac{1}{2}\left(b - a\right). \tag{1.63}$$

The mode is not defined. An integration yields the mean absolute deviation:

$$\mathrm{MAD}\left\{X\right\} = \frac{1}{8}\left(b - a\right). \tag{1.64}$$

The interquartile range is easily obtained from (1.58) and reads:

$$\mathrm{Ran}\left\{X\right\} = \frac{1}{2}\left(b - a\right). \tag{1.65}$$

1.3.2 Normal distribution

The *normal distribution* is by far the most used and studied distribution. Its bell-shaped profile and its analytical tractability make it the benchmark choice to describe random variables that are peaked around a given value but can take on values on the whole real axis. The normal distribution depends on two parameters μ and σ^2. The parameter μ is a location parameter that turns out to be the expected value and the parameter $|\sigma|$ is a dispersion parameter that turns out to be the standard deviation.

We use the following notation to indicate that X is normally distributed according to those parameters:

$$X \sim \mathrm{N}\left(\mu, \sigma^2\right). \tag{1.66}$$

The case $\mu \equiv 0$ and $\sigma^2 \equiv 1$ defines the *standard normal distribution.*

Equivalent representations

The probability density function of the normal distribution is defined as follows:

$$f^{\mathrm{N}}_{\mu,\sigma^2}\left(x\right) \equiv \frac{1}{\sqrt{2\pi\sigma^2}} e^{-\frac{(x-\mu)^2}{2\sigma^2}}, \tag{1.67}$$

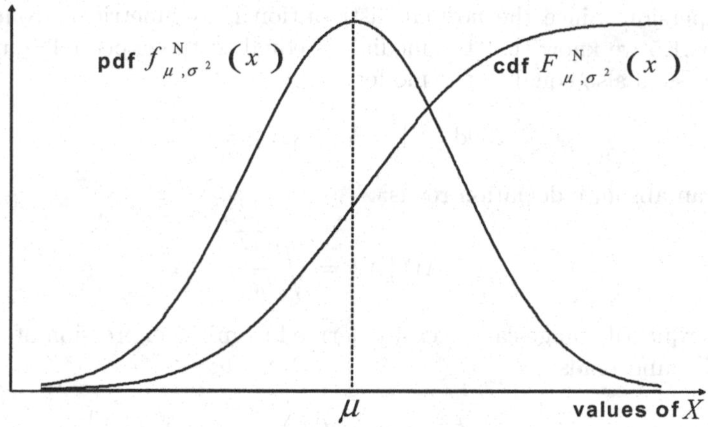

Fig. 1.6. Normal distribution: pdf and cdf

see Figure 1.6.

The cumulative distribution function of the normal distribution can be expressed in terms of the error function (B.75) as follows:

$$F^{\mathrm{N}}_{\mu,\sigma^2}(x) = \frac{1}{2}\left[1 + \operatorname{erf}\left(\frac{x-\mu}{\sqrt{2\sigma^2}}\right)\right], \tag{1.68}$$

see Figure 1.6.

The characteristic function of the normal distribution reads:

$$\phi^{\mathrm{N}}_{\mu,\sigma}(\omega) = e^{i\mu\omega - \frac{\sigma^2}{2}\omega^2}, \tag{1.69}$$

see Abramowitz and Stegun (1974).

Inverting (1.68) we obtain the quantile of the normal distribution:

$$Q^{\mathrm{N}}_{\mu,\sigma^2}(p) = \mu + \sqrt{2\sigma^2}\operatorname{erf}^{-1}(2p-1). \tag{1.70}$$

Summary statistics

The standard parameters that summarize the properties of the normal distribution, namely expected value, standard deviation, skewness and kurtosis, can be computed from the characteristic function (1.69) with the technique described in Appendix www.1.6, and read respectively:

$$\mathrm{E}\{X\} = \mu \tag{1.71}$$

$$\mathrm{Sd}\{X\} = \sqrt{\sigma^2} \tag{1.72}$$

$$\mathrm{Sk}\{X\} = 0 \tag{1.73}$$

$$\mathrm{Ku}\{X\} = 3. \tag{1.74}$$

It is possible to compute explicitly also other parameters of location and dispersion. Since the normal distribution is symmetrical, from Appendix www.1.5 we know that the median is equal to the expected value, which in this case is also equal to the mode:

$$\text{Med}\{X\} = \text{Mod}\{X\} = \mu. \tag{1.75}$$

The mean absolute deviation reads:

$$\text{MAD}\{X\} = \sqrt{\frac{2\sigma^2}{\pi}}. \tag{1.76}$$

The interquartile range can be easily derived from the expression of the quantile (1.70) and reads:

$$\text{Ran}\{X\} = \sqrt{2\sigma^2}\left[\text{erf}^{-1}\left(\frac{1}{2}\right) - \text{erf}^{-1}\left(-\frac{1}{2}\right)\right]. \tag{1.77}$$

1.3.3 Cauchy distribution

Like the normal distribution, the *Cauchy distribution* is bell-shaped and depends on two parameters μ and σ^2. The parameter μ is a location parameter that can take on any value and the parameter σ^2 is the square of a dispersion parameter $|\sigma|$.

We use the following notation to indicate that X is Cauchy distributed with the above parameters:

$$X \sim \text{Ca}\left(\mu, \sigma^2\right). \tag{1.78}$$

The case $\mu \equiv 0$ and $\sigma^2 \equiv 1$ is called the *standard Cauchy distribution.*

The Cauchy distribution is used instead of the normal distribution when extreme events are comparatively speaking more likely to occur than in the case of a normal distribution. This phenomenon is also known as *fat tails* behavior.

Equivalent representations

The probability density function of the Cauchy distribution, which we plot in Figure 1.7, is defined as follows:

$$f^{\text{Ca}}_{\mu,\sigma^2}(x) \equiv \frac{1}{\pi\sqrt{\sigma^2}}\left(1 + \frac{(x-\mu)^2}{\sigma^2}\right)^{-1}, \tag{1.79}$$

see Abramowitz and Stegun (1974) and `mathworld.com`.

The cumulative distribution function of the Cauchy distribution, which we plot in Figure 1.7, reads:

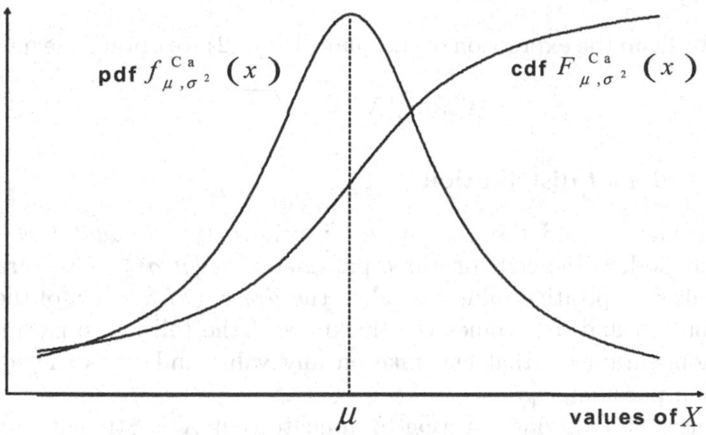

Fig. 1.7. Cauchy distribution: pdf and cdf

$$F^{\mathrm{Ca}}_{\mu,\sigma^2}(x) = \frac{1}{2} + \frac{1}{\pi} \arctan\left(\frac{x-\mu}{\sqrt{\sigma^2}}\right), \tag{1.80}$$

see e.g. `mathworld.com`.

The characteristic function of the Cauchy distribution reads:

$$\phi^{\mathrm{Ca}}_{\mu,\sigma^2}(\omega) = e^{i\mu\omega - \sqrt{\sigma^2}|\omega|}, \tag{1.81}$$

see e.g. Abramowitz and Stegun (1974) and `mathworld.com`.

The quantile of the Cauchy distribution is obtained inverting (1.80) and reads:

$$Q^{\mathrm{Ca}}_{\mu,\sigma^2}(p) = \mu + \sqrt{\sigma^2} \tan\left(\pi p - \frac{\pi}{2}\right). \tag{1.82}$$

Summary statistics

The moments of the Cauchy distribution are not defined. This happens because the probability density function (1.79) decays proportionally to x^{-2} in the tails. Therefore the computation of the generic moment of order k involves integrating a function of the order of x^{k-2} as $|x| \to \infty$, which does not converge for any positive integer k.

The fact that the moments are not defined is reflected also in the expression of the characteristic function (1.81), which is not differentiable in zero. Therefore in particular it cannot be expressed as a Taylor series in terms of the moments as in Appendix www.1.6.

Nevertheless, from the expression of the quantile (1.82) we obtain the median, which is also equal to the mode:

$$\text{Med}\{X\} = \text{Mod}\{X\} = \mu. \tag{1.83}$$

Similarly, from the expression of the quantile (1.82) we obtain the interquartile range:

$$\text{Ran}\{X\} = 2\sqrt{\sigma^2}. \tag{1.84}$$

1.3.4 Student t distribution

Like the normal and the Cauchy distributions, the *Student t distribution*, is bell-shaped. It depends on three parameters (ν, μ, σ^2). The parameter ν, which takes on positive values, is called the *degrees of freedom* of the Student t distribution and determines the thickness of the tails. The parameter μ is a location parameter that can take on any value and σ^2 is the square of a dispersion parameter $|\sigma|$.

We use the following notation to indicate that X is Student t distributed with the above parameters:

$$X \sim \text{St}\left(\nu, \mu, \sigma^2\right). \tag{1.85}$$

The case $\mu = 0$ and $\sigma^2 = 1$ is called the *standard Student t distribution.*

Equivalent representations

On `mathworld.com` we find the standard Student t probability density function. By applying formula (T.14) in Appendix www.1.2 we obtain the probability density function of the general Student t distribution, which reads:

$$f^{\text{St}}_{\nu,\mu,\sigma^2}(x) = \frac{\Gamma\left(\frac{\nu+1}{2}\right)}{\Gamma\left(\frac{\nu}{2}\right)} \frac{1}{\sqrt{\nu\pi\sigma^2}} \left(1 + \frac{1}{\nu}\frac{(x-\mu)^2}{\sigma^2}\right)^{-\frac{\nu+1}{2}} \tag{1.86}$$

where Γ is the gamma function (B.80). See in Figure 1.8 the bell-shaped profile of this function.

Similarly, we find on `mathworld.com` the standard Student t cumulative distribution function. By applying formula (T.15) in Appendix www.1.2 we obtain the cumulative distribution function of the general Student t distribution. In Figure 1.8 we plot this function, which reads explicitly:

$$\begin{aligned} F^{\text{St}}_{\nu,\mu,\sigma^2}(x) = \frac{1}{2} + \frac{1}{2}\Bigg[& I\left(1; \frac{\nu}{2}, \frac{1}{2}\right) \\ & - I\left(\frac{\nu\sigma^2}{\nu\sigma^2 + (x-\mu)^2}; \frac{\nu}{2}, \frac{1}{2}\right)\Bigg] \text{sign}\left(\frac{x-\mu}{\sigma}\right), \end{aligned} \tag{1.87}$$

where I is the regularized beta function (B.91).

The quantile of the Student t distribution cannot be expressed analytically.

On p. 948 of Abramowitz and Stegun (1974) we find the characteristic function of the standard Student t distribution. By applying formula (T.18)

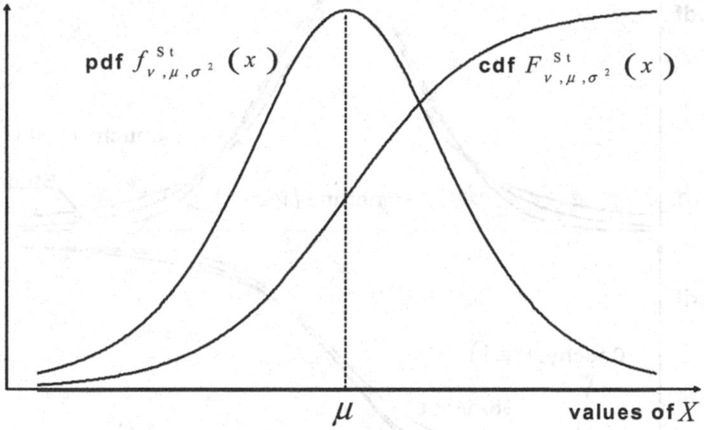

Fig. 1.8. Student t distribution: pdf and cdf

in Appendix www.1.2 we obtain the characteristic function of the general Student t distribution:

$$\phi^{St}_{\nu,\mu,\sigma^2} = \frac{e^{i\omega\mu}}{\pi\Gamma(\nu/2)} \left(\frac{\sigma^2\omega^2}{4\nu}\right)^{\frac{\nu}{4}} Y_{\frac{\nu}{2}}\left(\sqrt{\frac{\sigma^2\omega^2}{\nu}}\right), \tag{1.88}$$

where Γ denotes the gamma function (B.80) and Y_ν is the Bessel function of the second kind (B.93).

Summary statistics

The standard parameters that summarize the properties of the Student t distribution, namely expected value, standard deviation, skewness and kurtosis, are computed in Abramowitz and Stegun (1974) and read:

$$\mathrm{E}\{X\} = \mu \tag{1.89}$$

$$\mathrm{Sd}\{X\} = \sqrt{\frac{\nu}{\nu-2}\sigma^2} \tag{1.90}$$

$$\mathrm{Sk}\{X\} = 0 \tag{1.91}$$

$$\mathrm{Ku}\{X\} = 3 + \frac{6}{\nu-4}. \tag{1.92}$$

These parameters are defined for $\nu > 1$, 2, 3 and 4 respectively.

The Student t distribution includes the normal distribution and the Cauchy distribution as special cases. Indeed we show in Appendix www.2.14 in a more general context that the limit $\nu \to \infty$ of the Student t probability

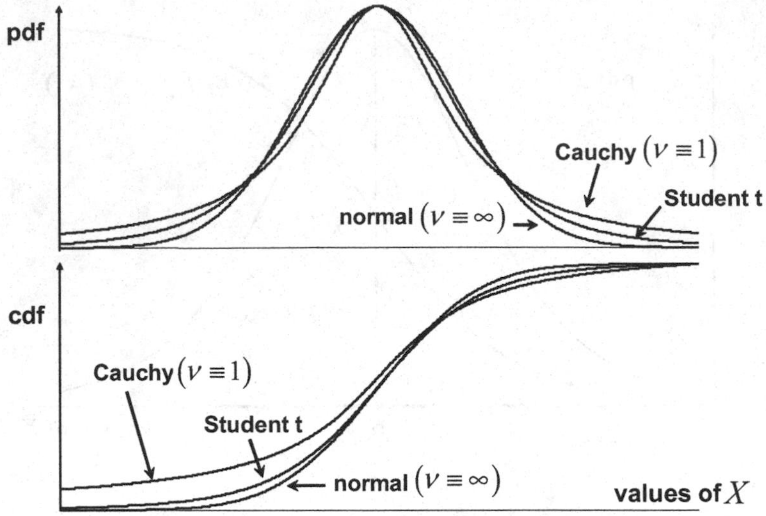

Fig. 1.9. Relations among Cauchy, normal, and Student t distributions

density function (1.86) yields the normal probability density function (1.67). On the other hand, if we set $\nu \equiv 1$ in (1.86) and recall (B.81) and (B.82), we obtain the Cauchy probability density function (1.79).

As we see in Figure 1.9, the lower the degrees of freedom, the "fatter" the tails of the probability density function and the flatter the cumulative distribution function. This is consistent with the above discussion of the Cauchy distribution and with the expression (1.92) of the kurtosis.

1.3.5 Lognormal distribution

The price of a security is a positive random variable. Furthermore, the random changes from the current price are better stated in percentage terms than in absolute terms. In other words, if the price now is, say, 1\$, the chance that the price will double, which corresponds to an absolute change of 1\$ is approximately equal to the chance that the price will become half, which corresponds to an absolute change of 0.5\$.

To model this feature, consider a random variable (the "percentage change") that is normally distributed:

$$Y \sim \mathrm{N}\left(\mu, \sigma^2\right). \tag{1.93}$$

The *lognormal distribution* is defined as the distribution of the variable $X \equiv e^Y$. The rationale behind this name is obviously the fact that by definition X is lognormally distributed if and only if its logarithm is normally distributed.

We use the following notation to indicate that X is lognormally distributed with the above parameters:

$$X \sim \text{LogN}\left(\mu, \sigma^2\right). \tag{1.94}$$

Equivalent representations

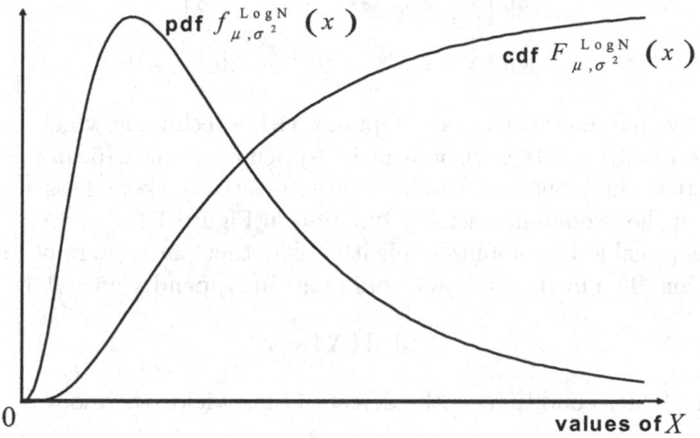

Fig. 1.10. Lognormal distribution: pdf and cdf

The probability density function of the lognormal distribution reads from (T.21) in Appendix www.1.1 as follows:

$$f_{\mu,\sigma^2}^{\text{LogN}}(x) = \frac{1}{x\sqrt{2\pi\sigma^2}} e^{-\frac{1}{2}\frac{(\ln(x)-\mu)^2}{\sigma^2}}. \tag{1.95}$$

We notice in Figure 1.10 that the lognormal pdf is not symmetrical.

Applying formula (T.22) in Appendix www.1.3 to the normal cumulative distribution function (1.68), we obtain the cumulative distribution function of the lognormal distribution, which we plot in Figure 1.10:

$$F_{\mu,\sigma^2}^{\text{LogN}}(x) = \frac{1}{2}\left(1 + \text{erf}\left(\frac{\ln(x) - \mu}{\sqrt{2\sigma^2}}\right)\right). \tag{1.96}$$

The characteristic function is not known in analytic form.

Applying formula (T.23) in Appendix www.1.3 to the normal quantile (1.70), we obtain the quantile of the lognormal distribution:

$$Q_{\mu,\sigma^2}^{\text{LogN}}(p) = e^{\mu + \sqrt{2\sigma^2}\,\text{erf}^{-1}(2p-1)}. \tag{1.97}$$

Summary statistics

The standard parameters that summarize the properties of the lognormal distribution, namely expected value, standard deviation, skewness and kurtosis read respectively:

$$\mathrm{E}\{X\} = e^{\mu+\frac{\sigma^2}{2}} \tag{1.98}$$

$$\mathrm{Sd}\{X\} = e^{\mu+\frac{\sigma^2}{2}}\sqrt{e^{\sigma^2}-1} \tag{1.99}$$

$$\mathrm{Sk}\{X\} = \sqrt{e^{\sigma^2}-1}\left(e^{\sigma^2}+2\right) \tag{1.100}$$

$$\mathrm{Ku}\{X\} = e^{4\sigma^2}+2e^{3\sigma^2}+3e^{2\sigma^2}-3. \tag{1.101}$$

The above parameters can be computed with a technique which we discuss in a general multivariate environment in Appendix www.2.16. In particular, we notice that the lognormal distribution is positively skewed, as we see in the profile of the probability density function in Figure 1.10.

It is possible to compute explicitly also other parameters of location and dispersion. The median follows from (T.9) in Appendix www.1.1:

$$\mathrm{Med}\{X\} = e^{\mu}. \tag{1.102}$$

The first-order condition on the density (1.95) yields the mode:

$$\mathrm{Mod}\{X\} = e^{\mu-\sigma^2}. \tag{1.103}$$

Notice that the three location parameters (1.98), (1.102) and (1.102) yield different results.

The expression of the interquartile range follows from the quantile (1.97) and reads:

$$\mathrm{Ran}\{X\} = e^{\mu}\left(e^{\sqrt{2\sigma^2}\,\mathrm{erf}^{-1}\left(\frac{1}{2}\right)} - e^{\sqrt{2\sigma^2}\,\mathrm{erf}^{-1}\left(-\frac{1}{2}\right)}\right). \tag{1.104}$$

1.3.6 Gamma distribution

We introduce here a distribution that is useful in Bayesian analysis, where the parameters of a distribution are considered as random variables. In particular, we will need a distribution to describe the variance, which is always non-negative. The gamma distribution proves particularly suitable in this respect.

Consider a set of ν random variables $(Y_1, \ldots, Y_\nu)$ that are normally identically distributed:

$$Y_t \sim \mathrm{N}\left(\mu, \sigma^2\right), \tag{1.105}$$

for all $t = 1, \ldots, \nu$. Furthermore, assume that these random variables are independent[1].

[1] Refer to Section 2.3 for a formal definition of dependence.

The *non-central gamma distribution* with ν degrees of freedom is defined as the distribution of the following variable:

$$X \equiv Y_1^2 + \cdots + Y_\nu^2. \tag{1.106}$$

As such, the non-central gamma distribution depends on three parameters (ν, μ, σ^2). The parameter ν is an integer and is called the *degrees of freedom* of the gamma distribution; the parameter μ can assume any value and is called the *non-centrality parameter*; the parameter σ^2 is a positive scalar and is called the *scale parameter.*

We use the following notation to indicate that X is distributed as a non-central gamma with the above parameters:

$$X \sim \text{Ga}\left(\nu, \mu, \sigma^2\right). \tag{1.107}$$

The special case where the non-centrality parameter is $\mu \equiv 0$ gives rise to the *central gamma distribution with ν degrees of freedom.* We use the following notation to indicate that X is central-gamma distributed with the above parameters:

$$X \sim \text{Ga}\left(\nu, \sigma^2\right). \tag{1.108}$$

The special case where the scale parameter is $\sigma^2 \equiv 1$ gives rise to the *(non-central) chi-square distribution with ν degrees of freedom.*

In particular, when $\mu \equiv 0$ and $\sigma^2 \equiv 1$ we obtain the *chi-square distribution with ν degrees of freedom*, which is denoted as follows:

$$X \sim \chi^2_\nu. \tag{1.109}$$

In view of generalizations to a multivariate setting and applications later on in the book, we focus below on the central gamma distribution, which includes the chi-square distribution as a special case.

Equivalent representations

The results and expressions that follow can be found on `mathworld.com`.

The probability density function of the central gamma distribution reads:

$$f^{\text{Ga}}_{\nu,\sigma^2}(x) = \frac{1}{\left(2\sigma^2\right)^{\frac{\nu}{2}} \Gamma\left(\frac{\nu}{2}\right)} x^{\frac{\nu}{2}-1} e^{-\frac{1}{2}\frac{x}{\sigma^2}}, \tag{1.110}$$

where Γ is the gamma function (B.80). We plot in Figure 1.11 the profile of this density.

The cumulative distribution function of the central gamma distribution reads:

$$F^{\text{Ga}}_{\nu,\sigma^2}(x) = P\left(\frac{x}{2\sigma^2}; \frac{\nu}{2}\right), \tag{1.111}$$

where P is the lower regularized gamma function (B.85), see Figure 1.11 for a plot.

The characteristic function of the central gamma distribution reads:

$$\phi^{\text{Ga}}_{\nu,\sigma^2}(\omega) = \left(1 - 2i\sigma^2\omega\right)^{-\frac{\nu}{2}}. \tag{1.112}$$

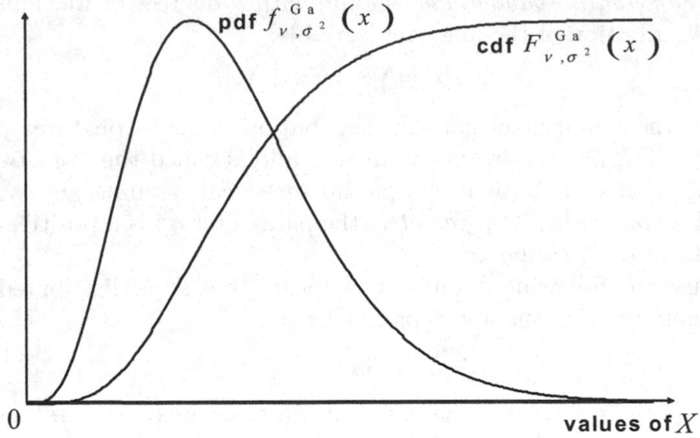

Fig. 1.11. Gamma distribution: pdf and cdf

Summary statistics

The standard parameters that summarize the properties of the gamma distribution, namely expected value, standard deviation, skewness and kurtosis read respectively:

$$\mathrm{E}\{X\} = \nu\sigma^2 \tag{1.113}$$

$$\mathrm{Sd}\{X\} = \sqrt{2\nu}\sigma^2 \tag{1.114}$$

$$\mathrm{Sk}\{X\} = \sqrt{\frac{8}{\nu}} \tag{1.115}$$

$$\mathrm{Ku}\{X\} = 3 + \frac{12}{\nu}. \tag{1.116}$$

The first-order condition on the probability density function yields the mode:

$$\mathrm{Mod}\{X\} = (\nu - 2)\,\sigma^2. \tag{1.117}$$

1.3.7 Empirical distribution

Suppose that our information i_T regarding the random variable X consists of T past measurements of this variable:

$$i_T \equiv \{x_1, \ldots, x_T\}. \tag{1.118}$$

Notice the lower-case notation in (1.118), since the measurements have already taken place, and therefore the outcomes are no longer random variables.

The *empirical distribution* provides a straightforward model for the basic assumption of statistics that we can learn about the future from the past: under the empirical distribution any of the past outcomes is assumed equally likely to occur again in future measurements of X, whereas any other value cannot occur.

We use the following notation to indicate that X is distributed according to an empirical distribution with the above observations:

$$X \sim \mathrm{Em}\left(i_T\right). \tag{1.119}$$

Equivalent representations

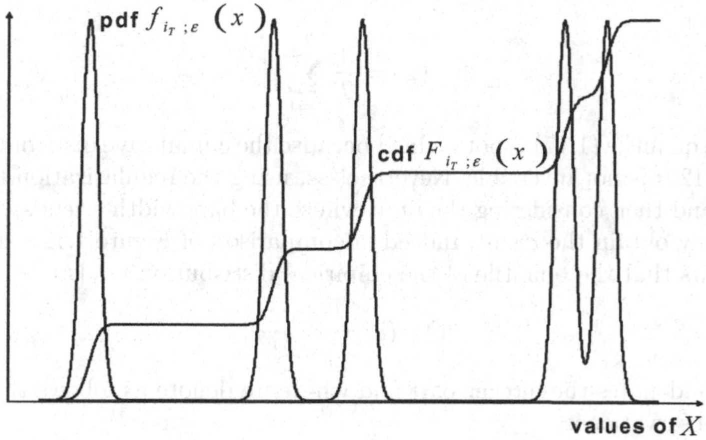

Fig. 1.12. Empirical distribution (regularized): pdf and cdf

The empirical distribution is discrete. Therefore its probability density function is a generalized function. As in (B.22), we can express the empirical pdf as follows:

$$f_{i_T}(x) = \frac{1}{T}\sum_{t=1}^{T} \delta^{(x_t)}(x), \tag{1.120}$$

where δ is the Dirac delta (B.16).

It is impossible to represent graphically this probability density function, unless we regularize it by means of the convolution as in (B.54). The regularized probability density function of the empirical distribution reads in terms of the smooth approximation (B.18) of the Dirac delta as follows:

$$f_{i_T;\epsilon} \equiv f_{i_T} * \delta_\epsilon^{(0)} = \frac{1}{T}\sum_{t=1}^{T} \delta_\epsilon^{(x_t)}, \tag{1.121}$$

where ϵ is a small bandwidth. We plot in Figure 1.12 the regularized version of the empirical probability density function.

From $(B.53)$ the empirical cumulative distribution function reads:

$$F_{i_T}(x) = \frac{1}{T}\sum_{t=1}^{T} H^{(x_t)}(x), \tag{1.122}$$

where H is the Heaviside step function $(B.73)$. In Figure 1.12 we plot the regularized cumulative distribution function ensuing from (1.121).

From the definition of the characteristic function (1.12) in terms of the expectation operator $(B.56)$, and from the property $(B.17)$ of the Dirac delta we obtain:

$$\phi_{i_T}(\omega) = \frac{1}{T}\sum_{t=1}^{T} e^{i\omega x_t}. \tag{1.123}$$

The quantile (1.17) is not defined because the cumulative distribution function (1.122) is not invertible. Nevertheless, using the regularization technique (1.20) and then considering the limit where the bandwidth ϵ tends to zero we can easily obtain the result. Indeed, a comparison of Figure 1.12 with Figure 1.2 shows that the quantile of the empirical distribution reads:

$$Q_{i_T}(p) = x_{[pT]:T}, \tag{1.124}$$

where $[\cdot]$ denotes the integer part and where we denote as follows the ordered set of observations:

$$\begin{aligned} x_{1:T} &\equiv \min\{x_1, \ldots, x_T\} \\ &\vdots \\ x_{T:T} &\equiv \max\{x_1, \ldots, x_T\}. \end{aligned} \tag{1.125}$$

Summary statistics

The standard parameters that summarize the properties of the empirical distribution, namely expected value, standard deviation, skewness and kurtosis, follow from the definition of the expectation operator $(B.56)$, and the property $(B.17)$ of the Dirac delta. We denote these parameters respectively as follows:

$$\widehat{\mathrm{E}}_{i_T} = \frac{1}{T}\sum_{t=1}^{T} x_t \tag{1.126}$$

$$\widehat{\mathrm{Sd}}_{i_T} = \sqrt{\frac{1}{T}\sum_{t=1}^{T} \left(x_t - \widehat{\mathrm{E}}_{i_T}\right)^2} \tag{1.127}$$

$$\widehat{\mathrm{Sk}}_{i_T} = \frac{1}{T}\sum_{t=1}^{T} \left(\frac{x_t - \widehat{\mathrm{E}}_{i_T}}{\widehat{\mathrm{Sd}}_{i_T}}\right)^3 \tag{1.128}$$

$$\widehat{\mathrm{Ku}}_{i_T} = \frac{1}{T}\sum_{t=1}^{T} \left(\frac{x_t - \widehat{\mathrm{E}}_{i_T}}{\widehat{\mathrm{Sd}}_{i_T}}\right)^4 . \tag{1.129}$$

These parameters are also called *sample mean*, *sample standard deviation*, *sample skewness* and *sample kurtosis* respectively.

The mode is not defined. From the expression for the quantile (1.124) we obtain the *sample median*:

$$\mathrm{Med}\{X\} = x_{[\frac{T}{2}]:T}. \tag{1.130}$$

Similarly, from the expression for the quantile we obtain the *sample interquartile range*:

$$\mathrm{Med}\{X\} = x_{[\frac{3}{4}T]:T} - x_{[\frac{1}{4}T]:T}. \tag{1.131}$$

2

Multivariate statistics

The financial markets contain many sources of risk. When dealing with several sources of risk at a time we cannot treat them separately: the joint structure of multi-dimensional randomness contains a wealth of information that goes beyond the juxtaposition of the information contained in each single variable.

In this chapter we discuss multivariate statistics. The structure of this chapter reflects that of Chapter 1: to ease the comprehension of the multivariate case refer to the respective section in that chapter. For more on this subject see also references such as Mardia, Kent, and Bibby (1979), Press (1982) and Morrison (2002).

In Section 2.1 we introduce the building blocks of multivariate distributions which are direct generalizations of the one-dimensional case. These include the three equivalent representations of a distribution in terms of the probability density function, the characteristic function and the cumulative distribution function.

In Section 2.2 we discuss the factorization of a distribution into its purely univariate components, namely the marginal distributions, and its purely joint component, namely the copula. To present copulas we use the leading example of vanilla options.

In Section 2.3 we introduce the concept of independence among random variables and the related concept of conditional distribution.

In Section 2.4 we discuss the location summary statistics of a distribution such as its expected value and its mode, and the dispersion summary statistics such as the covariance matrix and the modal dispersion. We detail the geometrical representations of these statistics in terms of the location-dispersion ellipsoid, and their probabilistic interpretations in terms of a multivariate version of Chebyshev's inequality. We conclude introducing more summary statistics such as the multivariate moments, which provide a deeper insight into the shape of a multivariate distribution.

In Section 2.5 we discuss summary statistics for the level of interdependence among the marginal components of a multivariate distribution. We introduce copula-driven measures of dependence such as the Schweizer-Wolff

A. Meucci, *Risk and Asset Allocation,* Springer Finance,

measure and copula-driven measures of concordance, such as Spearman's rho and Kendall's tau. We also analyze the advantages and potential pitfalls of using the correlation as a measure of interdependence.

In Section 2.6 we present a taxonomy of parametric distributions that represent the multivariate generalization of those introduced in Chapter 1. In particular, in view of their applications to estimation theory, we introduce matrix-variate distributions, such as the Wishart distribution, the matrix-variate normal, Cauchy and Student t distributions. In view of their applications to modeling prices, we introduce generic log-distributions, of which the lognormal is an example, along with a general technique to compute all the moments of these distributions.

In Section 2.7 we discuss a few broad classes of distributions that are very useful in applications, namely elliptical and symmetric stable distributions, which are symmetric and analytically tractable, and infinitely divisible distribution, that allow to model the financial markets at any investment horizon.

2.1 Building blocks

In this section we introduce the multivariate extension of the building blocks of univariate statistics discussed in Section 1.1, namely the concept of multivariate distribution and its equivalent representations in terms of the joint probability density function, the joint cumulative distribution function and the joint characteristic function.

A *random variable* $\mathbf{X}$ of dimension N is a vector that corresponds to a joint measurement of N variables that has yet to take place:

$$\mathbf{X} \equiv (X_1, \ldots, X_N)' . \tag{2.1}$$

A joint measurement corresponds to one point in the space $\mathbb{R}^N$. Therefore the joint measurements of $\mathbf{X}$ can assume a range of values in various regions of $\mathbb{R}^N$, and each of these values has a specific probability to occur.

For example, consider two stocks that trade today on the exchange at the following prices (e.g. in dollars):

$$\widetilde{x}_1 \equiv 100, \quad \widetilde{x}_2 \equiv 50. \tag{2.2}$$

Tomorrow's prices $\mathbf{X} \equiv (X_1, X_2)'$ for these stocks are a bivariate random variable. A joint measurement is a point in the plane $\mathbb{R}^2$ and with each point on the plane is associated a different probability.

The stochastic features of the different possible measurements of a random variable $\mathbf{X}$ can be described in terms of a *multivariate distribution*. A distribution is characterized by a *space of events* $\mathfrak{E}$ and a *probability* $\mathbb{P}$.

The unknown outcome $\mathbf{x}$ of the joint measurement of the entries of $\mathbf{X}$ corresponds to one specific event $\mathfrak{e}$ among many that can take place in a space of events $\mathfrak{E}$. Therefore, a multivariate random variable is a function from the space of events to the range of measurements in $\mathbb{R}^N$: if a specific event $\mathfrak{e}$ takes place, the measurement will take on the value $\mathbf{x} \equiv \mathbf{X}(\mathfrak{e})$. In a different universe a different event $\mathfrak{e}'$ might have taken place and thus the measurement would have assumed a different value $\mathbf{x}' \equiv \mathbf{X}(\mathfrak{e}')$.

The likelihood of different possible events is described by a probability $\mathbb{P}$, which is a measure on the space of events. The following notation stands for the probability of all the events $\mathfrak{e}$ in the space of events $\mathfrak{E}$ that give rise to a joint measurement of $\mathbf{X}$ in the region $\mathcal{R}$ of the space $\mathbb{R}^N$:

$$\mathbb{P}\{\mathbf{X} \in \mathcal{R}\} \equiv \mathbb{P}\left\{\mathfrak{e} \in \mathfrak{E} \text{ such that } \mathbf{X}(\mathfrak{e}) \in \mathcal{R} \subset \mathbb{R}^N\right\}. \tag{2.3}$$

This expression generalizes (1.2).

As in the one-dimensional case, a distribution can be represented in three equivalent ways.

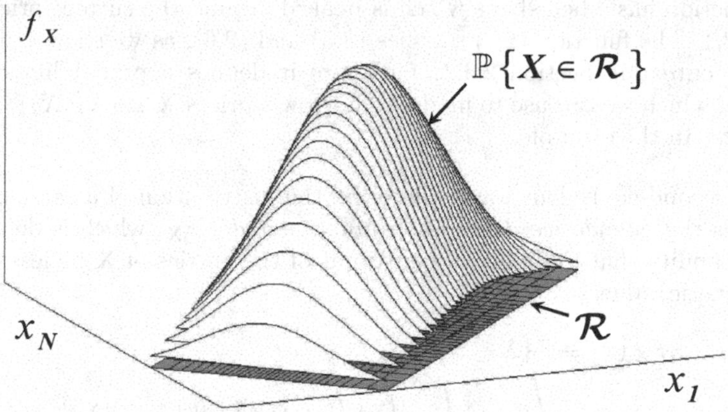

Fig. 2.1. Multivariate probability density function

The most intuitive way to represent the distribution of the random variable $\mathbf{X}$ is through the *probability density function (pdf)* $f_{\mathbf{X}}$. Intuitively, the pdf shows a peak where the outcome of the measurement of $\mathbf{X}$ is more likely to occur. More formally, the probability density function is defined in such a way that the probability that a measurement takes place in a generic region $\mathcal{R}$ is the volume comprised between the region and the density, see Figure 2.1:

$$\mathbb{P}\{\mathbf{X} \in \mathcal{R}\} \equiv \int_{\mathcal{R}} f_{\mathbf{X}}(\mathbf{x})\, d\mathbf{x}. \tag{2.4}$$

In particular, since a probability is non-negative, the probability density function is non-negative:

$$f_{\mathbf{X}}(\mathbf{x}) \geq 0. \tag{2.5}$$

Furthermore, since the joint measurement of $\mathbf{X}$ must assume a value in $\mathbb{R}^N$, the following normalization condition must hold:

$$\int_{\mathbb{R}^N} f_{\mathbf{X}}(\mathbf{x})\, d\mathbf{x} = 1. \tag{2.6}$$

For instance, consider the following function:

$$f_{\mathbf{X}}(x_1, x_2) \equiv \frac{\sqrt{5}}{\sqrt{8}\pi} e^{-\frac{1}{2}u(x_1,x_2)}, \tag{2.7}$$

where u is the following quadratic form:

$$u(x_1, x_2) \equiv \begin{pmatrix} x_1 - \widetilde{x}_1 \\ x_2 - \widetilde{x}_2 \end{pmatrix}' \begin{pmatrix} \frac{10}{3} & -\frac{2}{3}\sqrt{10} \\ -\frac{2}{3}\sqrt{10} & \frac{10}{3} \end{pmatrix} \begin{pmatrix} x_1 - \widetilde{x}_1 \\ x_2 - \widetilde{x}_2 \end{pmatrix}; \tag{2.8}$$

and where $(\widetilde{x}_1, \widetilde{x}_2)$ are the current prices (2.2) of the two stocks in our example. This function has a bell shape which is peaked around the current prices, see Figure 2.1. The function (2.7) satisfies (2.5) and (2.6), as we show in a more general context in Section 2.6.2. Therefore it defines a probability density function, which we can use to model tomorrow's prices $\mathbf{X} \equiv (X_1, X_2)'$ for the two stocks in the example.

The second equivalent way to describe the distribution of a random variable $\mathbf{X}$ is the *cumulative distribution function (cdf)* $F_{\mathbf{X}}$, which is defined as the probability that the joint measurement of the entries of $\mathbf{X}$ be less than a given generic value:

$$\begin{aligned} F_{\mathbf{X}}(\mathbf{x}) &\equiv \mathbb{P}\{\mathbf{X} \leq \mathbf{x}\} \qquad (2.9) \\ &= \int_{-\infty}^{x_1} \cdots \int_{-\infty}^{x_N} f_{\mathbf{X}}(u_1, \ldots, u_N)\, du_1 \cdots du_N. \end{aligned}$$

The cumulative distribution function is obtained from the probability density function by applying the combined integration operators $(B.27)$ as follows:

$$F_{\mathbf{X}} = (\mathcal{I}_1 \circ \cdots \circ \mathcal{I}_N)[f_{\mathbf{X}}]. \tag{2.10}$$

In turn, the probability density function can be recovered from the cumulative distribution function by applying the combined differentiation operators $(B.25)$ as follows:

$$f_{\mathbf{X}} = (\mathcal{D}_1 \circ \cdots \circ \mathcal{D}_N)[F_{\mathbf{X}}]. \tag{2.11}$$

Therefore the two representations in terms of pdf and cdf are equivalent. The positivity condition (2.5) and the normalization condition (2.6) on the pdf transfer to the cdf in a way similar to the one-dimensional case (1.10). Indeed $F_{\mathbf{X}}$ is an increasing function of each coordinate and satisfies the following normalization conditions:

$$F_{\mathbf{X}}(x_1, \ldots, -\infty, \ldots, x_N) = 0, \quad F_{\mathbf{X}}(+\infty, \ldots, +\infty) = 1. \tag{2.12}$$

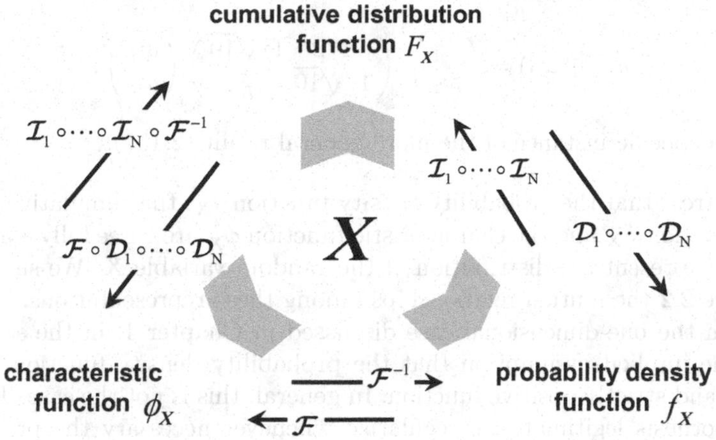

Fig. 2.2. Equivalent representations of a multivariate distribution

The third way to describe the properties of a distribution is by means of the *characteristic function (cf)* $\phi_{\mathbf{X}}$, defined in terms of the expectation operator (B.56) as follows:

$$\phi_{\mathbf{X}}(\boldsymbol{\omega}) \equiv \mathrm{E}\left\{e^{i\boldsymbol{\omega}'\mathbf{X}}\right\}, \tag{2.13}$$

where $i \equiv \sqrt{-1}$ is the imaginary unit. The characteristic function assumes values in the complex plane.

A comparison of (2.13) with (B.34) and (B.56) shows that the characteristic function is the Fourier transform of the probability density function:

$$\phi_{\mathbf{X}} = \mathcal{F}[f_{\mathbf{X}}]. \tag{2.14}$$

Therefore the probability density function can be recovered by means of the inverse Fourier transform (B.40) from the characteristic function:

$$f_{\mathbf{X}} = \mathcal{F}^{-1}[\phi_{\mathbf{X}}]. \tag{2.15}$$

At times the characteristic function proves to be the easiest way to represent a distribution.

The characteristic function of the distribution of the example (2.7) reads:

$$\phi_{\mathbf{X}}(\omega_1, \omega_2) = e^{i(\omega_1 \widetilde{x}_1 + \omega_2 \widetilde{x}_2)} e^{-\frac{1}{2} w(\omega_1, \omega_2)}, \tag{2.16}$$

where $(\widetilde{x}_1, \widetilde{x}_2)$ are the current prices (2.2) of the stocks and where w is the following quadratic form:

$$w(\omega_1, \omega_2) = \begin{pmatrix} \omega_1 \\ \omega_2 \end{pmatrix}' \begin{pmatrix} 1/2 & 1/\sqrt{10} \\ 1/\sqrt{10} & 1/2 \end{pmatrix} \begin{pmatrix} \omega_1 \\ \omega_2 \end{pmatrix}. \tag{2.17}$$

This is a specific instance of the more general result (2.157).

We stress that the probability density function $f_{\mathbf{X}}$, the cumulative distribution function $F_{\mathbf{X}}$ and the characteristic function $\phi_{\mathbf{X}}$ are three fully equivalent ways to represent the distribution of the random variable $\mathbf{X}$. We summarize in Figure 2.2 the mutual relationships among these representations.

As in the one-dimensional case discussed in Chapter 1, in the sequel we make the implicit assumption that the probability density function $f_{\mathbf{X}}$ is a smooth and strictly positive function. In general, this is not the case. To make our hypothesis legitimate we regularize whenever necessary the probability density function as discussed in Appendix B.4:

$$f_{\mathbf{X}} \mapsto f_{\mathbf{X};\epsilon} \equiv \frac{1}{(2\pi)^{\frac{N}{2}} \epsilon^N} \int_{\mathbb{R}^N} e^{-\frac{(\mathbf{y}-\mathbf{x})'(\mathbf{y}-\mathbf{x})}{2\epsilon^2}} f(\mathbf{y}) \, d\mathbf{y}. \tag{2.18}$$

For the practical as well as "philosophical" motivations behind the regularization, see $(B.54)$ and comments thereafter.

2.2 Factorization of a distribution

The distribution of a multivariate random variable $\mathbf{X}$ can be factored into two separate components. On the one hand the marginal distributions of each entry of the vector $\mathbf{X}$, which represent the purely univariate features of $\mathbf{X}$. On the other hand the copula, a standardized distribution which summarizes the purely "joint" component of the distribution of $\mathbf{X}$. We summarize this schematically as follows:

$$\boxed{\textbf{multivariate} = \textbf{"1-dim" (marginals)} + \textbf{"joint"(copula)}} \tag{2.19}$$

2.2.1 Marginal distribution

Consider an N-dimensional random variable $\mathbf{X}$. We split $\mathbf{X}$ in two sub-sets: the K-dimensional random variable $\mathbf{X}_A$ made of the first K entries and the

$(N-K)$-dimensional random variable $\mathbf{X}_B$ made of the remaining entries:

$$\mathbf{X} \equiv \begin{pmatrix} \mathbf{X}_A \\ \mathbf{X}_B \end{pmatrix}. \tag{2.20}$$

The *marginal distribution* of the variable $\mathbf{X}_B$ is the distribution of $\mathbf{X}_B$ obtained disregarding the existence of $\mathbf{X}_A$. In particular, we obtain the marginal distribution of the generic entry X_n by disregarding the remaining $N-1$ entries.

Consider the bivariate example (2.7), which describes the joint stochastic behavior of two stock prices. The marginal distribution of the first stock must be the univariate example (1.6) of Chapter 1, which describes the stochastic behavior of the first stock only. Otherwise, the two models are in contradiction with each other and one of them must be wrong.

We can represent the marginal distribution of $\mathbf{X}_B$ by means of its cumulative distribution function:

$$\begin{aligned} F_{\mathbf{X}_B}(\mathbf{x}_B) &\equiv \mathbb{P}\{\mathbf{X}_B \leq \mathbf{x}_B\} = \mathbb{P}\{\mathbf{X}_A \leq +\infty, \mathbf{X}_B \leq \mathbf{x}_B\} \\ &\equiv F_{\mathbf{X}}(+\infty, \mathbf{x}_B). \end{aligned} \tag{2.21}$$

In words, the marginal cumulative distribution function is the joint cumulative distribution function, where the variables we intend to disregard are set to infinity.

Equivalently, we can represent the marginal distribution of $\mathbf{X}_B$ by means of its probability density function. Applying the differentiation operator to the cumulative distribution function (2.21) as in (2.11) we obtain:

$$f_{\mathbf{X}_B}(\mathbf{x}_B) \equiv \int_{\mathbb{R}^K} f_{\mathbf{X}}(\mathbf{x}_A, \mathbf{x}_B)\, d\mathbf{x}_A. \tag{2.22}$$

In words, the marginal pdf averages out of the joint pdf the variables that we intend to disregard.

In our example, the integration of the joint pdf (2.7) yields:

$$f_{X_1}(x_1) = \int_{-\infty}^{+\infty} f_{\mathbf{X}}(x_1, x_2)\, dx_2 = \frac{1}{\sqrt{\pi}} e^{-(x-\widetilde{x}_1)^2}. \tag{2.23}$$

This computation is a specific instance of the more general result (2.162). Not surprisingly (2.23) is the one-dimensional pdf (1.6) of the first stock price.

Finally, we can represent the marginal distribution of $\mathbf{X}_B$ by means of its characteristic function:

$$\begin{aligned} \phi_{\mathbf{X}_B}(\boldsymbol{\omega}) &\equiv \mathrm{E}\left\{e^{i\boldsymbol{\omega}'\mathbf{X}_B}\right\} = \mathrm{E}\left\{e^{i\boldsymbol{\psi}'\mathbf{X}_A + \boldsymbol{\omega}'\mathbf{X}_B}\right\}\Big|_{\boldsymbol{\psi}=\mathbf{0}} \\ &\equiv \phi_{\mathbf{X}}(\mathbf{0}, \boldsymbol{\omega}). \end{aligned} \tag{2.24}$$

In words, the marginal characteristic function is the joint characteristic function, where the variables we intend to disregard are set to zero.

2.2.2 Copulas

In this section we introduce copulas. For more on this subject consult references such as Nelsen (1999).

Definition

The copula represents the true interdependence structure of a random variable, which in our applications is the market. Intuitively, the copula is a standardized version of the purely joint features of a multivariate distribution, which is obtained by filtering out all the purely one-dimensional features, namely the marginal distribution of each entry X_n.

In order to factor out the marginal components, we simply transform deterministically each entry X_n in a new random variable U_n, whose distribution is the same for each entry. Since the distribution of each U_n is normalized this way, we lose track of the specific marginal distribution of X_n.

In order to map a generic one-dimensional random variable X into a random variable U which has a distribution of our choice, consider the cumulative distribution function F_X defined in (1.7). By means of the function F_X we can define a new random variable, called the *grade* of X:

$$U \equiv F_X(X) . \tag{2.25}$$

The grade of X is a deterministic transformation of the random variable X that assumes values in the interval $[0,1]$. We prove in we Appendix www.2.1 that the grade is uniformly distributed on this interval:

$$U \sim \mathrm{U}([0,1]) . \tag{2.26}$$

To obtain a random variable Z with a distribution of our choice, we prove in Appendix www.2.1 that it suffices to compute the quantile function Q_Z of that distribution as in (1.17), and then to define Z as the quantile applied to the grade U:

$$Z \equiv Q_Z(U) . \tag{2.27}$$

In Figure 2.3 we display the graphical interpretation of the above operations.[1]

In particular, we can standardize each marginal component X_n of the original random variable $\mathbf{X}$ by means of the uniform distribution. Therefore, we consider the vector of the grades:

[1] This technique also allows us to simulate univariate distributions of any kind starting with a uniform random number generator.

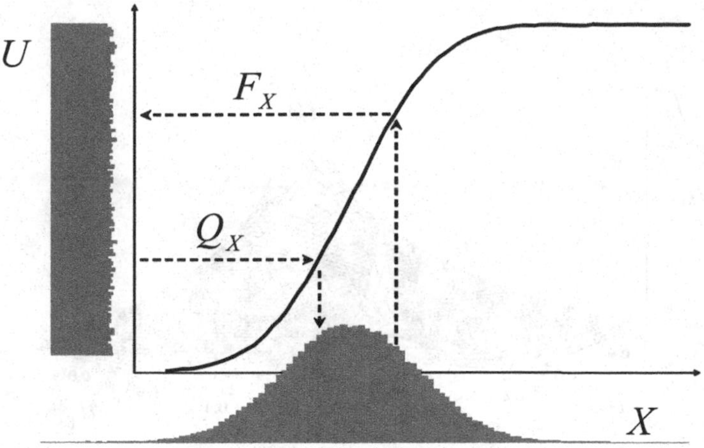

Fig. 2.3. Distribution of the grades: relation with cdf and quantile

$$\mathbf{U} \equiv \begin{pmatrix} U_1 \\ \vdots \\ U_N \end{pmatrix} \equiv \begin{pmatrix} F_{X_1}(X_1) \\ \vdots \\ F_{X_N}(X_N) \end{pmatrix}. \tag{2.28}$$

This random variable assumes values on the unit hypercube:

$$[0,1]^N \equiv [0,1] \times \cdots \times [0,1]. \tag{2.29}$$

The *copula* of the multivariate random variable $\mathbf{X}$ is the joint distribution of its grades (2.28).

Representations

Since the copula is a distribution, namely the distribution of the grades $\mathbf{U}$, we can represent it in terms of the probability density function or the cumulative distribution function, or the characteristic function.

In Appendix www.2.3 we prove that the pdf of the copula reads:

$$f_{\mathbf{U}}(u_1, \ldots, u_N) = \frac{f_{\mathbf{X}}(Q_{X_1}(u_1), \ldots, Q_{X_N}(u_N))}{f_{X_1}(Q_{X_1}(u_1)) \cdots f_{X_N}(Q_{X_N}(u_N))}, \tag{2.30}$$

where Q_{X_n} is the quantile (1.17) of the generic n-th marginal entry of $\mathbf{X}$.

In Figure 2.4 we plot the probability density function of the copula of the leading example (2.7), which we compute explicitly in a more general setting in (2.176).

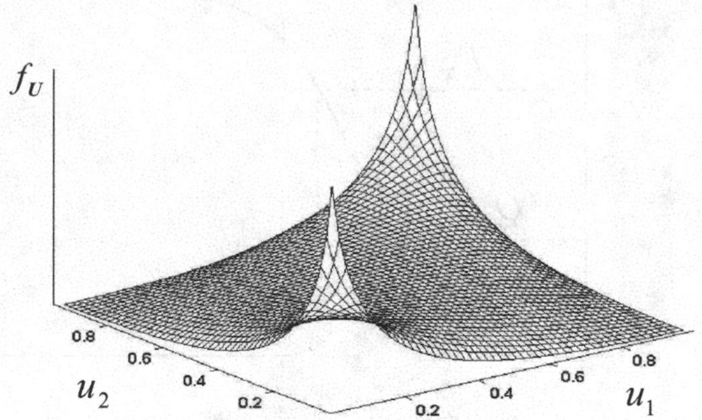

Fig. 2.4. Copula: probability density function

We can also represent the copula of the random variable $\mathbf{X}$ equivalently in terms of its cumulative distribution function. We prove in Appendix www.2.3 that the cdf of the copula of $\mathbf{X}$ reads:

$$F_{\mathbf{U}}(u_1, \ldots, u_N) = F_{\mathbf{X}}\left(Q_{X_1}(u_1), \ldots, Q_{X_N}(u_N)\right). \tag{2.31}$$

In particular, since the marginal distribution of the generic n-th entry is uniform, from (2.21) and (1.56) we obtain:

$$F_{\mathbf{U}}(1, \ldots, u_n, \ldots, 1) = u_n, \tag{2.32}$$

see Figure 2.10 for a few examples.

Properties

We can write (2.30) as follows:

$$f_{\mathbf{X}}(x_1, \ldots, x_N) = f_{\mathbf{U}}\left(F_{X_1}(x_1), \ldots, F_{X_N}(x_N)\right) \prod_{n=1}^{N} f_{X_n}(x_n). \tag{2.33}$$

This expression formalizes the loose expression (2.19): the joint pdf of a generic variable $\mathbf{X}$ is the product of the pdf of its copula and the pdf of the marginal densities of its entries. In other words, the copula factors out the purely marginal features of a distribution.

The copula contains all the information about the joint features of a distribution in a standardized form. Indeed, given the copula of $\mathbf{X}$, i.e. the distribution of the grades $\mathbf{U}$, from (2.28) we can reconstruct the distribution of $\mathbf{X}$ with a *deterministic* transformation of each grade separately:

$$\mathbf{X} \stackrel{d}{=} \begin{pmatrix} Q_{X_1}(U_1) \\ \vdots \\ Q_{X_N}(U_N) \end{pmatrix}. \tag{2.34}$$

Therefore, the copula is a standardized distribution that summarizes the purely joint features behind a multivariate random variable.

The purely joint features of a distribution characterize the true structure of randomness of a multivariate random variable. In other words, the copula allows to detect the true interdependence structure behind a generic multivariate random variable **X**. In practical terms, the copula provides an effective tool to monitor and *hedge* the risks in the markets.

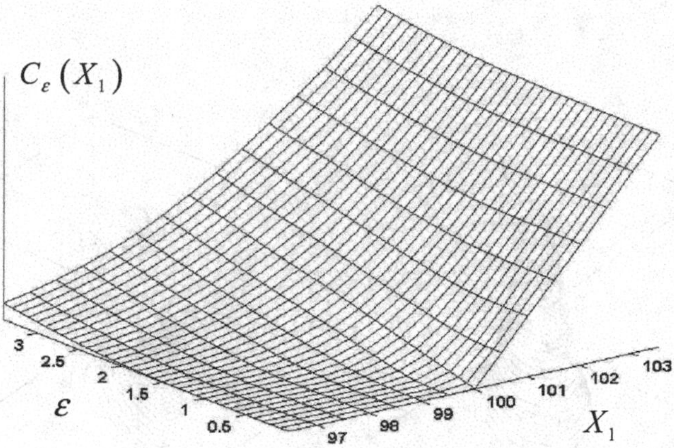

Fig. 2.5. Regularization of call option payoff

To see this, consider two *co-monotonic* random variables **X** and **Y**, namely random variables such that:

$$\begin{pmatrix} Y_1 \\ \vdots \\ Y_N \end{pmatrix} = \begin{pmatrix} g_1(X_1) \\ \vdots \\ g_N(X_N) \end{pmatrix}, \tag{2.35}$$

where each g_n is an increasing invertible function of its argument.

For instance, in our example (2.7) of two stock prices $\mathbf{X} \equiv (X_1, X_2)'$, consider the payoff of a call option on the first stock with strike K, i.e. the following random variable:

$$C_1 \equiv \max(X_1 - K, 0), \tag{2.36}$$

where the strike price is, say, $K \equiv 100$.

The function C_1 is not strictly increasing in its argument X_1, but it becomes so if we replace it with a regularized version by means of (B.49). In Appendix www.2.7 we show that the regularized call option payoff reads:

$$C_{1;\epsilon} \equiv \frac{(X_1 - K)}{2}\left(1 + \operatorname{erf}\left(\frac{X_1 - K}{\sqrt{2\epsilon^2}}\right)\right) + \frac{\epsilon}{\sqrt{2\pi}} e^{-\frac{(X_1-K)^2}{2\epsilon^2}}. \tag{2.37}$$

This profile is smooth, strictly increasing in X_1, and tends to the exact profile (2.36) as the bandwidth ϵ tends to zero, see Figure 2.5. Therefore the stock price X_1 and the regularized call option payoff C_1 are co-monotonic and so are the pairs (X_1, X_2) and (C_1, X_2).

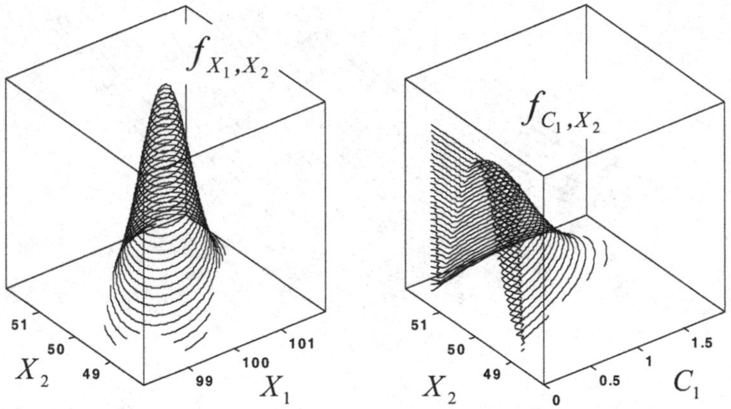

Fig. 2.6. Co-monotonic transformations: effects on the joint distribution

The joint distributions of co-monotonic variables are not equal, see Figure 2.6. Yet, the sources of randomness behind two co-monotonic random variables are the same. The common feature of these variables is their copula, as we show in Appendix www.2.3:

$$(\mathbf{X}, \mathbf{Y}) \text{ co-monotonic} \Leftrightarrow \text{copula of } \mathbf{X} = \text{copula of } \mathbf{Y}. \tag{2.38}$$

In our example the joint distribution of the first stock price and the second stock price (X_1, X_2) is different than the joint distribution of the call option on the first stock and the second stock price (C_1, X_2). We see this in Figure 2.6, where we plot the two different probability density functions. Nevertheless, the copula of (X_1, X_2) is the same as the copula of (C_1, X_2) and is represented by the probability density function in Figure 2.4.

2.3 Dependence

Loosely speaking, two random variables are independent if any information on either variable does not affect the distribution of the other random variable. To introduce formally the concept of dependence, it is more intuitive to first define conditional distributions.

Consider an N-dimensional random variable $\mathbf{X}$. We split $\mathbf{X}$ in two subsets: the K-dimensional random variable $\mathbf{X}_A$ of the first K entries and the $(N-K)$-dimensional random variable $\mathbf{X}_B$ of the remaining entries:

$$\mathbf{X} \equiv \begin{pmatrix} \mathbf{X}_A \\ \mathbf{X}_B \end{pmatrix}. \tag{2.39}$$

The *conditional distribution* of the variable $\mathbf{X}_B$ given $\mathbf{x}_A$ is the distribution of $\mathbf{X}_B$ knowing that the realization of $\mathbf{X}_A$ is the specific value $\mathbf{x}_A$. We denote the conditioned random variable equivalently as $\mathbf{X}_B|\mathbf{x}_A$ or $\mathbf{X}_B|\mathbf{X}_A = \mathbf{x}_A$.

Suppose that in our example (2.7) the two stock prices $\mathbf{X} \equiv (X_1, X_2)'$ appear almost, but not quite, simultaneously on the screen. Before we look at the screen, the probability distribution of second stock price X_2 is represented by its marginal distribution. After we see the price of the first stock we have more information available. The distribution that describes the second stock price X_2, knowing that the price of the first stock is $X_1 \equiv x_1$, is the conditional distribution $X_2|x_1$.

The most intuitive way to represent the conditional distribution is the probability density function:

$$f_{\mathbf{X}_B|\mathbf{x}_A}(\mathbf{x}_B) = \frac{f_{\mathbf{X}}(\mathbf{x}_A, \mathbf{x}_B)}{\int f_{\mathbf{X}}(\mathbf{x}_A, \mathbf{x}_B)\, d\mathbf{x}_B} = \frac{f_{\mathbf{X}}(\mathbf{x}_A, \mathbf{x}_B)}{f_{\mathbf{X}_A}(\mathbf{x}_A)}. \tag{2.40}$$

In words, the conditional pdf of $\mathbf{X}_B$ given knowledge of $\mathbf{X}_A$ is the joint pdf of $\mathbf{X}_A$ and $\mathbf{X}_B$ divided by the marginal pdf of $\mathbf{X}_A$ evaluated at the known point $\mathbf{x}_A$. Geometrically, the conditional pdf of $\mathbf{X}_B$ is a (rescaled) section of the joint pdf, which passes through the known point $\mathbf{x}_A$, see Figure 2.7. Equivalently, we could represent the conditional distribution with the respective cumulative density function or characteristic function, but the representation would be less intuitive.

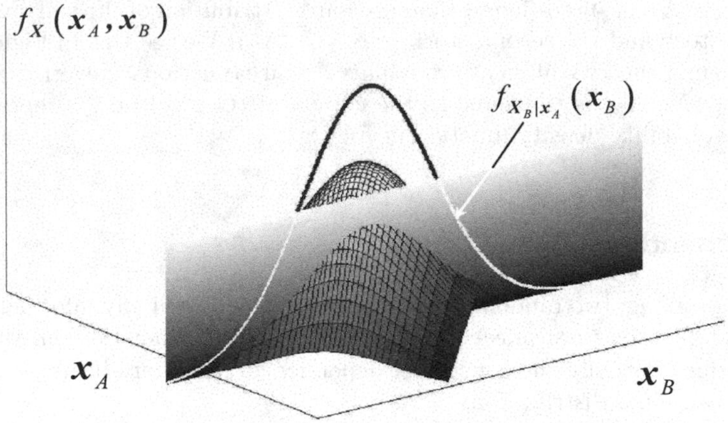

Fig. 2.7. Conditional probability density function

In our example, dividing the joint pdf of the two stock prices (2.7) by the marginal pdf of the first stock price (2.23) and simplifying, we obtain:

$$f_{X_2|x_1}(x_2) = \frac{1}{\sqrt{2\pi\sigma_C^2}} e^{-\frac{1}{2\sigma_C^2}(x_2-\mu_C)^2}, \tag{2.41}$$

where

$$\mu_C \equiv \widetilde{x}_2 + \sqrt{\frac{2}{5}}(x_1 - \widetilde{x}_1), \quad \sigma_C^2 \equiv \frac{3}{10}, \tag{2.42}$$

and where $(\widetilde{x}_1, \widetilde{x}_2)$ are the current prices (2.2). This computation is a specific instance of the more general result (2.173). The conditional pdf of the second stock price depends explicitly on the value x_1 of the first stock price, which is known by assumption.

From (2.40) we derive *Bayes' rule*, which is of the utmost importance in many financial applications:

$$\begin{aligned} f_{\mathbf{X}_A|\mathbf{x}_B}(\mathbf{x}_A) &= \frac{f_{\mathbf{X}}(\mathbf{x}_A, \mathbf{x}_B)}{\int f_{\mathbf{X}}(\mathbf{x}_A, \mathbf{x}_B)\, d\mathbf{x}_A} \\ &= \frac{f_{\mathbf{X}_B|\mathbf{x}_A}(\mathbf{x}_B) f_{\mathbf{X}_A}(\mathbf{x}_A)}{\int f_{\mathbf{X}_B|\mathbf{x}_A}(\mathbf{x}_B) f_{\mathbf{X}_A}(\mathbf{x}_A)\, d\mathbf{x}_A}. \end{aligned} \tag{2.43}$$

Bayes' rule expresses the conditional distribution of $\mathbf{X}_A$ given $\mathbf{x}_B$ in terms of the conditional distribution of $\mathbf{X}_B$ given $\mathbf{x}_A$ and the marginal distribution of $\mathbf{X}_A$.

At this point we have the tools to introduce the concept of (in)dependence among random variables. Splitting the multivariate random variable $\mathbf{X}$ into two sub-sets $(\mathbf{X}_A, \mathbf{X}_B)$ as in (2.39), we say that $\mathbf{X}_B$ is *independent* of $\mathbf{X}_A$ if the conditional distribution of $\mathbf{X}_B$ given $\mathbf{x}_A$ does not contain any more information than the marginal distribution of $\mathbf{X}_B$. More precisely, the variable $\mathbf{X}_B$ is independent of the variable $\mathbf{X}_A$ if for arbitrary functions g and h the marginal distribution of $g(\mathbf{X}_B)$ and the conditional distribution of $g(\mathbf{X}_B)$ given $h(\mathbf{x}_A)$ are the same.

The two stock prices in our example are independent if knowing the price (or the return, or any other function) of one stock does not add information regarding the distribution of the other stock and viceversa.

We can check for independence among variables in terms of their probability density function. Indeed, it can be proved that the mutual independence of $\mathbf{X}_A$ and $\mathbf{X}_B$ is equivalent to the joint pdf of $\mathbf{X}_A$ and $\mathbf{X}_B$ being the product of the marginal pdf of $\mathbf{X}_A$ and the marginal pdf of $\mathbf{X}_B$:

$$(\mathbf{X}_A, \mathbf{X}_B) \text{ independent} \Leftrightarrow f_{\mathbf{X}}(\mathbf{x}_A, \mathbf{x}_B) = f_{\mathbf{X}_A}(\mathbf{x}_A) f_{\mathbf{X}_B}(\mathbf{x}_B), \tag{2.44}$$

see Shirayaev (1989).

In particular, (2.40) and (2.44) imply the intuitive result that the marginal distribution of a variable and its conditional distribution given the realization of an independent variable are the same:

$$(\mathbf{X}_A, \mathbf{X}_B) \text{ independent} \Rightarrow f_{\mathbf{X}_B|\mathbf{x}_A}(\mathbf{x}_B) = f_{\mathbf{X}_B}(\mathbf{x}_B). \tag{2.45}$$

In our example the two stock prices are not independent, since the conditional distribution of one stock price (2.41) depends on the other stock price.

Similarly, we can check for independence among variables in terms of their cumulative distribution function. Indeed, substituting (2.44) in the definition of the cdf (2.9) and integrating, the mutual independence of $\mathbf{X}_A$ and $\mathbf{X}_B$ is equivalent to the joint cdf of $\mathbf{X}_A$ and $\mathbf{X}_B$ being the product of the marginal cdf of $\mathbf{X}_A$ and the marginal cdf of $\mathbf{X}_B$:

$$(\mathbf{X}_A, \mathbf{X}_B) \text{ independent} \Leftrightarrow F_{\mathbf{X}}(\mathbf{x}_A, \mathbf{x}_B) = F_{\mathbf{X}_A}(\mathbf{x}_A) F_{\mathbf{X}_B}(\mathbf{x}_B). \tag{2.46}$$

Finally, we can check for independence among variables in terms of their characteristic function. Indeed, from (2.44) for any functions g and h the expectation operator $(B.56)$ can be factored as follows:

$$\begin{aligned}(\mathbf{X}_A, \mathbf{X}_B) &\text{ independent} \Rightarrow \\ \mathrm{E}\{g(\mathbf{X}_A) h(\mathbf{X}_B)\} &= \mathrm{E}\{g(\mathbf{X}_A)\} \mathrm{E}\{h(\mathbf{X}_B)\}.\end{aligned} \tag{2.47}$$

Therefore, from the definition of the characteristic function (2.13) we obtain that the mutual independence of $\mathbf{X}_A$ and $\mathbf{X}_B$ is equivalent to the joint characteristic function of $\mathbf{X}_A$ and $\mathbf{X}_B$ being the product of the marginal cf of $\mathbf{X}_A$ and the marginal cf of $\mathbf{X}_B$:

$$(\mathbf{X}_A, \mathbf{X}_B) \text{ independent} \Leftrightarrow \phi_{\mathbf{X}}(\boldsymbol{\omega}_A, \boldsymbol{\omega}_B) = \phi_{\mathbf{X}_A}(\boldsymbol{\omega}_A)\, \phi_{\mathbf{X}_B}(\boldsymbol{\omega}_B). \tag{2.48}$$

2.4 Shape summary statistics

In this section we discuss multivariate parameters of location and dispersion that summarize the main properties of a multivariate distribution. As in the one-dimensional case, these parameters provide an easy-to-interpret picture of the main properties of a multivariate distribution. After discussing their definition and properties we present a geometrical interpretation that recurs throughout the book. We conclude with a brief introduction to higher-order summary statistics.

2.4.1 Location

Consider an N-dimensional random variable $\mathbf{X}$. Our purpose is to summarize the whole distribution of $\mathbf{X}$ into one *location parameter* $\text{Loc}\{\mathbf{X}\}$, similarly to what we did in Section 1.2.1 for the univariate case.

Theory

As in the one-dimensional case, we require that the location parameter display some intuitive features. For instance, if the distribution is peaked around a specific value, the location parameter should be close to that peak. In particular, a constant $\mathbf{m}$ can be seen as an infinitely peaked random variable, see (B.22) and comments thereafter. Thus the location of a constant should be the constant itself:

$$\text{Loc}\{\mathbf{m}\} = \mathbf{m}. \tag{2.49}$$

This implies that the location parameter must be an N-dimensional vector.

Furthermore, consider a generic *affine transformation*:

$$\mathbf{X} \mapsto \mathbf{Y} \equiv \mathbf{a} + \mathbf{B}\mathbf{X}, \tag{2.50}$$

where $\mathbf{a}$ is a vector and $\mathbf{B}$ is a conformable matrix. A sensible parameter of location should track any invertible affine transformation of the original variable, i.e. a transformation such as (2.50), where $\mathbf{B}$ is an invertible matrix. In other words, if $\mathbf{B}$ is invertible, the location parameters should satisfy the following property:

$$\text{Loc}\{\mathbf{a} + \mathbf{B}\mathbf{X}\} = \mathbf{a} + \mathbf{B}\,\text{Loc}\{\mathbf{X}\}. \tag{2.51}$$

Property (2.51) is called the *affine equivariance* of the location parameter. For the rationale behind this requirement refer to the one dimensional case (1.24).

Examples

An example of location parameter is the multivariate *mode*, defined as the multivariate generalization of (1.30), namely as the highest peak of the joint probability density function:

$$\text{Mod}\{\mathbf{X}\} \equiv \underset{\mathbf{x}\in\mathbb{R}^N}{\text{argmax}}\{f_{\mathbf{X}}(\mathbf{x})\}. \tag{2.52}$$

We prove in Appendix www.2.5 that the mode is affine equivariant, i.e. it satisfies (2.51).

Consider our leading example (2.7) of two stock prices. From the first-order conditions on the joint pdf we obtain:

$$\text{Mod}\{\mathbf{X}\} = (\widetilde{x}_1, \widetilde{x}_2)', \tag{2.53}$$

where $(\widetilde{x}_1, \widetilde{x}_2)$ are the current prices (2.2). This is a specific instance of the more general result (2.158).

Another multivariate location parameter is the multivariate *expected value*, defined as the juxtaposition of the expected value (1.25) of the marginal distribution of each entry:

$$\text{E}\{\mathbf{X}\} \equiv (\text{E}\{X_1\}, \ldots, \text{E}\{X_N\})'. \tag{2.54}$$

Indeed, we prove in Appendix www.2.6 that the expected value is affine equivariant, i.e. it satisfies (2.51).

In our example (2.7) we have:

$$\text{E}\{\mathbf{X}\} = (\widetilde{x}_1, \widetilde{x}_2)', \tag{2.55}$$

where $(\widetilde{x}_1, \widetilde{x}_2)$ are the current prices (2.2). This is a specific instance of the more general result (2.158).

On the other hand, the juxtaposition of the median, or any other quantile, of each entry of a random variable does not satisfy (2.51) and therefore it does not define a suitable location parameter.

Mode and expected value might not be defined: the expectation integral might not converge in the case of the expected value, and the maximum of the probability density function might not be unique in the case of the mode. If they are defined, they both represent suitable location parameters.

Nevertheless, the expected value (2.54) is the benchmark multivariate location parameter.

In the first place, as in the one-dimensional case the expected value is a global parameter that includes information from the whole distribution, whereas the mode is a local parameter that depends on the value of the probability density function at one single point.

Secondly, the expected value enjoys a purely multivariate feature: the affine equivariance property holds for generic, i.e. not necessarily invertible, affine transformations. In other words the following equality holds for *any* conformable matrix $\widetilde{\mathbf{B}}$ and vector $\widetilde{\mathbf{a}}$:

$$\mathrm{E}\left\{\widetilde{\mathbf{a}}+\widetilde{\mathbf{B}}\mathbf{X}\right\}=\widetilde{\mathbf{a}}+\widetilde{\mathbf{B}}\,\mathrm{E}\left\{\mathbf{X}\right\}, \tag{2.56}$$

see Appendix www.2.6. This is not true for other parameters of location.

For example the mode of the sum of two variables in general is not the sum of the modes:

$$\operatorname{Mod}\left\{X+Y\right\}\neq\operatorname{Mod}\left\{X\right\}+\operatorname{Mod}\left\{Y\right\}. \tag{2.57}$$

This implies that the affine equivariance for generic affine transformations (2.56) does not hold for the mode even in the simple case $\mathbf{a}\equiv 0$ and $\mathbf{B}\equiv(1,1)$.

Finally, whenever the characteristic function of $\mathbf{X}$ is known and analytical, i.e. it can be recovered entirely from its Taylor series expansion, computing the expected value is straightforward, as we show in Appendix www.2.10.

2.4.2 Dispersion

Consider an N-dimensional random variable $\mathbf{X}$. Here we extend to a multivariate environment the concept of *dispersion parameter* discussed in Section 1.2.2 for the univariate case.

Theory

As in the univariate case discussed in Chapter 1, we require that the dispersion parameter behaves suitably under invertible affine transformations:

$$\mathbf{X}\mapsto\mathbf{Y}\equiv\mathbf{a}+\mathbf{B}\mathbf{X}, \tag{2.58}$$

where $\mathbf{a}$ is a vector and $\mathbf{B}$ is a conformable invertible matrix.

To determine the nature of the required behavior, we recall the definition (1.35) of the absolute value of the z-score of the variable X in the univariate case:

$$|Z_X|\equiv\sqrt{(X-\operatorname{Loc}\{X\})\frac{1}{\operatorname{Dis}\{X\}^2}(X-\operatorname{Loc}\{X\})}. \tag{2.59}$$

In that context, the dispersion parameter $\operatorname{Dis}\{X\}$ is properly defined if the absolute value of the z-score is unaffected by affine transformations:

$$|Z_{a+bX}|=|Z_X|, \tag{2.60}$$

see (1.36).

To generalize the absolute value of the z-score to a multivariate environment, we introduce the *Mahalanobis distance* of the point $\mathbf{x}$ from the point $\boldsymbol{\mu}$ through the metric $\boldsymbol{\Sigma}$, denoted and defined as follows:

$$\mathrm{Ma}(\mathbf{x}, \boldsymbol{\mu}, \boldsymbol{\Sigma}) \equiv \sqrt{(\mathbf{x} - \boldsymbol{\mu})' \boldsymbol{\Sigma}^{-1} (\mathbf{x} - \boldsymbol{\mu})}, \tag{2.61}$$

where the metric $\boldsymbol{\Sigma}$ is a symmetric and positive matrix. The points $\mathbf{x}$ which share the same Mahalanobis distance from $\boldsymbol{\mu}$ lie on the surface of an ellipsoid centered in $\boldsymbol{\mu}$, see (A.73). The larger (the eigenvalues of) $\boldsymbol{\Sigma}$, the smaller the Mahalanobis distance of the generic point $\mathbf{x}$ from the center $\boldsymbol{\mu}$. Therefore the matrix $\boldsymbol{\Sigma}$ indeed provides a metric to measure distances.

Comparing (2.59) with (2.61) we see that in a multivariate environment the absolute value of the z-score is replaced by the Mahalanobis distance from the location parameter through the metric provided by the yet to be defined "squared" dispersion parameter:

$$\mathrm{Ma}_{\mathbf{X}} \equiv \mathrm{Ma}(\mathbf{X}, \mathrm{Loc}\{\mathbf{X}\}, \mathrm{DisSq}\{\mathbf{X}\}). \tag{2.62}$$

We remark that considering (2.62) is intuitive, since a natural formulation of the dispersion of the variable $\mathbf{X}$ requires the dispersion parameter to represent a metric, i.e. a distance, between the variable and its location parameter.

In this context the dispersion parameter $\mathrm{DisSq}\{\mathbf{X}\}$ is properly defined if it satisfies two properties.

In the first place $\mathrm{DisSq}\{\mathbf{X}\}$ must be a symmetric and positive matrix, in such a way to define a metric in (2.62).

Secondly, $\mathrm{DisSq}\{\mathbf{X}\}$ must be such that the Mahalanobis distance (2.62) is invariant under invertible affine transformations:

$$\mathrm{Ma}_{\mathbf{a}+\mathbf{BX}} = \mathrm{Ma}_{\mathbf{X}}. \tag{2.63}$$

Given the affine equivariant properties of the location parameter (2.51), this is true if and only if for all invertible affine transformations (2.58) the dispersion parameter satisfies:

$$\mathrm{DisSq}\{\mathbf{a} + \mathbf{BX}\} = \mathbf{B}\,\mathrm{DisSq}\{\mathbf{X}\}\,\mathbf{B}'. \tag{2.64}$$

We call this property the *affine equivariance* of a multivariate dispersion parameter.

To summarize, a *dispersion matrix*, or *dispersion parameter*, or *scatter matrix* or *scatter parameter* is a symmetric and positive matrix $\mathrm{DisSq}\{\mathbf{X}\}$ that is affine equivariant, i.e. it satisfies (2.64).

Examples

An example of scatter matrix is the *modal dispersion*:

$$\mathrm{MDis}\left\{\mathbf{X}\right\} \equiv -\left(\left. \frac{\partial^2 \ln f_{\mathbf{X}}}{\partial \mathbf{x} \partial \mathbf{x}'} \right|_{\mathbf{x}=\mathrm{Mod}\{\mathbf{X}\}} \right)^{-1}, \tag{2.65}$$

see e.g. O'Hagan (1994). In Appendix www.2.5 we prove that the modal dispersion is indeed a scatter matrix, i.e. it is a symmetric and positive matrix that is affine equivariant. The rationale behind the modal dispersion follows from a second-order Taylor expansion of the pdf $f_{\mathbf{X}}$ around its mode, see (1.39) for the univariate case: the larger in absolute value the (always negative) second derivative in (2.65), the thinner the probability density function of $\mathbf{X}$ around its mode and thus the less disperse the distribution.

Consider our leading example (2.7). From a direct computation of the second derivatives of the log-pdf at the mode (2.53) we obtain:

$$\mathrm{MDis}\left\{\mathbf{X}\right\} = \begin{pmatrix} 1/2 & 1/\sqrt{10} \\ 1/\sqrt{10} & 1/2 \end{pmatrix}. \tag{2.66}$$

Another example of scatter parameter is the *covariance* matrix, defined as follows:

$$\mathrm{Cov}\left\{\mathbf{X}\right\} \equiv \mathrm{E}\left\{ \left(\mathbf{X} - \mathrm{E}\left\{\mathbf{X}\right\}\right) \left(\mathbf{X} - \mathrm{E}\left\{\mathbf{X}\right\}\right)' \right\}, \tag{2.67}$$

or component-wise:

$$\begin{aligned} \mathrm{Cov}\left\{X_m, X_n\right\} &\equiv \left[\mathrm{Cov}\left\{\mathbf{X}\right\}\right]_{mn} \\ &\equiv \mathrm{E}\left\{ \left(X_m - \mathrm{E}\left\{X_m\right\}\right) \left(X_n - \mathrm{E}\left\{X_n\right\}\right) \right\}. \end{aligned} \tag{2.68}$$

In Appendix www.2.6 we prove that the covariance is a scatter matrix, i.e. it is symmetric, positive and affine equivariant.

In our leading example (2.7) we obtain:

$$\mathrm{Cov}\left\{\mathbf{X}\right\} = \begin{pmatrix} 1/2 & 1/\sqrt{10} \\ 1/\sqrt{10} & 1/2 \end{pmatrix}. \tag{2.69}$$

This is a specific instance of a more general result, see Section 2.6.2.

Modal dispersion and covariance matrix might not be defined: the expectation integral might not converge in the case of the covariance, and the mode might not be unique in the case of the modal dispersion. When they are defined, they both represent suitable dispersion parameters.

Nevertheless, the covariance is the benchmark multivariate scatter parameter.

In the first place, like the variance in the one-dimensional case, the covariance is a global parameter that includes information from the whole distribution, whereas the modal dispersion is a local parameter that depends on the shape of the probability density function around one single point, i.e. the mode.

Secondly, from the factorization of the expectation operator in the presence of independent variables (2.47) and the component-wise definition of the covariance matrix (2.68) we obtain that the covariance of independent variables is null:

$$(X_m, X_n) \text{ independent} \Rightarrow \operatorname{Cov}\{X_m, X_n\} = 0. \tag{2.70}$$

This result motivates the name "covariance", as independent variables do not "co-vary".

In the third place, whenever the characteristic function of $\mathbf{X}$ is known and analytical, i.e. it can be recovered entirely from its Taylor series expansion, computing the covariance matrix is straightforward, as we show in Appendix www.2.10.

Finally, the affine equivariance property (2.64) holds in the case of the covariance even for generic, i.e. not necessarily invertible, affine transformations. In other words, the following identity holds for *any* conformable matrix $\widetilde{\mathbf{B}}$ and vector $\widetilde{\mathbf{a}}$:

$$\operatorname{Cov}\left\{\widetilde{\mathbf{a}} + \widetilde{\mathbf{B}}\mathbf{X}\right\} = \widetilde{\mathbf{B}} \operatorname{Cov}\{\mathbf{X}\} \widetilde{\mathbf{B}}', \tag{2.71}$$

see Appendix www.2.6. This is not true for other dispersion parameters.

For example, since from (2.57) the mode is not affine equivariant for non-invertible transformations, neither can be the modal dispersion.

The generic affine equivariance (2.56) and (2.71) of the expected value and covariance matrix respectively also allows us to build a dispersion parameter with a more intuitive "bottom up" approach.

Indeed, consider a specific type of non-invertible affine transformations, i.e. a linear combinations $\boldsymbol{\alpha}'\mathbf{X}$, where $\boldsymbol{\alpha}$ is an N-dimensional vector of constants. A linear combination of random variables is a univariate random variable. Therefore we can compute the dispersion parameter (1.40) defined in terms of the expectation operator:

$$\operatorname{Dis}\{\boldsymbol{\alpha}'\mathbf{X}\} \equiv \left(\mathrm{E}\left\{|\boldsymbol{\alpha}'(\mathbf{X} - \mathrm{E}\{\mathbf{X}\})|^p\right\}\right)^{\frac{1}{p}}. \tag{2.72}$$

For a general value of p, there exists no result concerning linear combinations that involve equalities. Nevertheless, in the case $p \equiv 2$ the dispersion in (2.72) becomes the standard deviation and a few algebraic manipulations show that there exists a matrix $\mathbf{S}$ such that

$$\operatorname{Sd}\{\boldsymbol{\alpha}'\mathbf{X}\} = \sqrt{\boldsymbol{\alpha}'\mathbf{S}\boldsymbol{\alpha}}. \tag{2.73}$$

From (B.65) and (B.68) the matrix $\mathbf{S}$ coincides with the covariance (2.67). In particular, from (2.73) we obtain that the diagonal elements of the covariance matrix are the variances of the marginal distributions of each entry:

$$\operatorname{Cov}\{X_n, X_n\} = (\operatorname{Sd}\{X_n\})^2 = \operatorname{Var}\{X_n\}. \tag{2.74}$$

2.4.3 Location-dispersion ellipsoid

Consider an N-dimensional random variable $\mathbf{X}$. In this section we propose a graphical interpretation of the parameters of location and dispersion of $\mathbf{X}$. In particular, we will develop our discussion around the benchmark parameters, i.e. the expected value $\mathrm{E}\{\mathbf{X}\}$ defined in (2.54) and the covariance matrix $\mathrm{Cov}\{\mathbf{X}\}$ defined in (2.67), which we denote here as E and Cov respectively to ease the notation.

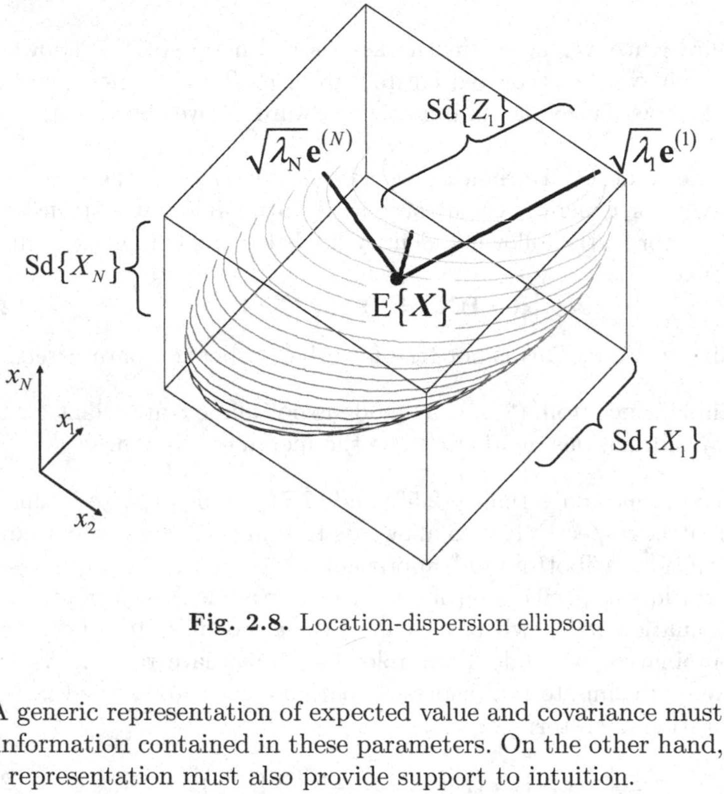

Fig. 2.8. Location-dispersion ellipsoid

A generic representation of expected value and covariance must convey all the information contained in these parameters. On the other hand, a geometrical representation must also provide support to intuition.

We state here and motivate in the sequel that we can effectively represent geometrically E and Cov by means of the *location-dispersion ellipsoid*, defined as follows:

$$\mathcal{E}_{\mathrm{E},\mathrm{Cov}} \equiv \left\{ \mathbf{x} \text{ such that } (\mathbf{x}-\mathrm{E})' \, \mathrm{Cov}^{-1} (\mathbf{x}-\mathrm{E}) \leq 1 \right\}, \tag{2.75}$$

see Figure 2.8.

First of all, we remark that this is indeed the implicit equation of an ellipsoid. The expected value is a vector and the covariance matrix is symmetric and positive definite. Therefore, from (A.73) the locus $\mathcal{E}_{\mathrm{E},\mathrm{Cov}}$ is an ellipsoid centered in the location parameter E.

The fact that the location-dispersion ellipsoid $\mathcal{E}_{\mathrm{E},\mathrm{Cov}}$ is centered in the expected value E shows that on the one hand the ellipsoid conveys all the information about E, and on the other hand the ellipsoid supports intuition regarding the meaning of E, which is the average location of the random variable $\mathbf{X}$.

As far as the dispersion parameter Cov is concerned, we already know from the discussion in Appendix A.5 that the ellipsoid $\mathcal{E}_{\mathrm{E},\mathrm{Cov}}$ conveys all the information contained in the covariance matrix Cov.

To show that it also support intuition regarding the dispersion properties of the random variable $\mathbf{X}$ we rephrase in this context the analysis of Appendix A.5. Consider the spectral decomposition (A.70) of the covariance matrix:

$$\mathrm{Cov}\{\mathbf{X}\} = \mathbf{E}\boldsymbol{\Lambda}\mathbf{E}'. \tag{2.76}$$

In this expression $\boldsymbol{\Lambda}$ is the diagonal matrix of the eigenvalues of the covariance sorted in decreasing order:

$$\boldsymbol{\Lambda} \equiv \mathrm{diag}\,(\lambda_1, \ldots, \lambda_N)\,; \tag{2.77}$$

and $\mathbf{E}$ is the juxtaposition of the respective eigenvectors:

$$\mathbf{E} \equiv \left(\mathbf{e}^{(1)}, \ldots, \mathbf{e}^{(N)}\right), \tag{2.78}$$

which satisfies $\mathbf{E}\mathbf{E}' = \mathbf{I}_N$, the identity matrix.

We know from Appendix A.5 that the principal axes of the location-dispersion ellipsoid $\mathcal{E}_{\mathrm{E},\mathrm{Cov}}$ are parallel to the the eigenvectors $\left\{\mathbf{e}^{(1)}, \ldots, \mathbf{e}^{(N)}\right\}$ of the covariance. On the other hand, in this context the eigenvectors define very special directions, namely the directions along which the randomness in $\mathbf{X}$ displays zero covariance. In other words, consider the following random variable:

$$\mathbf{Z} \equiv \mathbf{E}'\mathbf{X} = \begin{pmatrix} \left[\mathbf{e}^{(1)}\right]'\mathbf{X} \\ \vdots \\ \left[\mathbf{e}^{(N)}\right]'\mathbf{X} \end{pmatrix}. \tag{2.79}$$

Each entry of the vector $\mathbf{Z}$ is the projection of the random variable $\mathbf{X}$ on one eigenvector. From $\mathbf{E}\mathbf{E}' = \mathbf{I}_N$, for any $n \neq m$ we have:

$$\mathrm{Cov}\{Z_m, Z_n\} = \left[\mathbf{e}^{(m)}\right]' \mathbf{E}\boldsymbol{\Lambda}\mathbf{E}' \left[\mathbf{e}^{(n)}\right] = [\boldsymbol{\Lambda}]_{mn} = 0. \tag{2.80}$$

Thus the principal axes of the location-dispersion ellipsoid $\mathcal{E}_{\mathrm{E},\mathrm{Cov}}$ define the directions along which the randomness in $\mathbf{X}$ displays zero covariance.

Furthermore from Appendix A.5 the length of the principal axes of the location-dispersion ellipsoid $\mathcal{E}_{\mathrm{E},\mathrm{Cov}}$ are the square root of the eigenvalues of the covariance. On the other hand, in this context the eigenvalues have a very special meaning, namely they represent the variance of $\mathbf{X}$ along the direction of the eigenvectors:

$$\operatorname{Var}\{Z_n\} = \left[\mathbf{e}^{(n)}\right]' \mathbf{E}\boldsymbol{\Lambda}\mathbf{E}' \left[\mathbf{e}^{(n)}\right] = \lambda_n. \tag{2.81}$$

Thus from (2.74) the length of the principal axes of the location-dispersion ellipsoid $\mathcal{E}_{\mathrm{E,Cov}}$ represent the standard deviation of $\mathbf{X}$ along the direction of the principal axes.

In particular, from (A.68) the first eigenvalue corresponds the maximum variance achievable with a projection:

$$\lambda_1 = \max_{\|\mathbf{e}\|=1} \{\operatorname{Var}\{\mathbf{e}'\mathbf{X}\}\}; \tag{2.82}$$

and the first eigenvector $\mathbf{e}^{(1)}$ is the direction of maximal variation, i.e. it satisfies:

$$\mathbf{e}^{(1)} = \underset{\|\mathbf{e}\|=1}{\operatorname{argmax}} \{\operatorname{Var}\{\mathbf{e}'\mathbf{X}\}\}. \tag{2.83}$$

Similarly, from (A.69) the last eigenvalue corresponds the minimum variance achievable with a projection:

$$\lambda_N = \min_{\|\mathbf{e}\|=1} \{\operatorname{Var}\{\mathbf{e}'\mathbf{X}\}\}; \tag{2.84}$$

and the last eigenvector $\mathbf{e}^{(N)}$ is the direction of minimal variation, i.e. it satisfies:

$$\mathbf{e}^{(N)} = \underset{\|\mathbf{e}\|=1}{\operatorname{argmin}} \{\operatorname{Var}\{\mathbf{e}'\mathbf{X}\}\}. \tag{2.85}$$

Moreover, the location-dispersion ellipsoid $\mathcal{E}_{\mathrm{E,Cov}}$ is a suitable generalization of the one-dimensional location-dispersion bar defined in (1.53). Indeed, consider the rectangle with sides parallel to the reference axes of $\mathbb{R}^N$ which enshrouds the ellipsoid, see Figure 2.8. We prove in Appendix www.2.8 that the generic n-th side of this rectangle is centered on the expected value $\mathrm{E}\{X_n\}$ of the n-th marginal component and is long twice the standard deviation $\mathrm{Sd}\{X_n\}$ of the n-th marginal component. In other words, the enshrouding rectangle is defined by the following set of N equations:

$$\mathrm{E}\{X_n\} - \mathrm{Sd}\{X_n\} \leq x_n \leq \mathrm{E}\{X_n\} + \mathrm{Sd}\{X_n\}. \tag{2.86}$$

Each of these equations represents the location-dispersion bar (1.53) of the respective marginal distribution.

Finally, the location-dispersion ellipsoid $\mathcal{E}_{\mathrm{E,Cov}}$ is, among all the ellipsoids of equal volume, the one that contains the highest probability of occurrence of the random variable $\mathbf{X}$ within its boundaries. To make this statement precise, we consider the locus:

$$\mathcal{E}^q_{\mathrm{E,Cov}} \equiv \left\{\mathbf{x} \text{ such that } (\mathbf{x} - \mathrm{E})' \operatorname{Cov}^{-1} (\mathbf{x} - \mathrm{E}) \leq q^2\right\}. \tag{2.87}$$

This locus represents a rescaled version of the location-dispersion ellipsoid (2.75), where all the principal axis are multiplied by a factor q, see Figure 2.9. In Appendix www.2.9 we prove the following results.

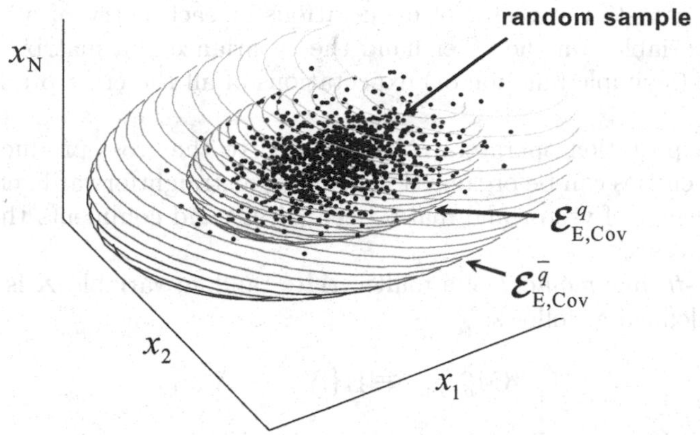

Fig. 2.9. Multivariate Chebyshev inequality

By the *Chebyshev inequality* for *any* vector $\mathbf{v}$ and *any* symmetric and positive matrix $\mathbf{U}$ the probability that observations occur outside the ellipsoid $\mathcal{E}^{q}_{\mathbf{v},\mathbf{U}}$ with principal axes proportional to q decays as the square of q:

$$\mathbb{P}\left\{\mathbf{X} \notin \mathcal{E}^{q}_{\mathbf{v},\mathbf{U}}\right\} \leq \frac{a_{\mathbf{v},\mathbf{U}}}{q^2}, \tag{2.88}$$

where the constant a is the expected squared Mahalanobis distance (2.61) of the random variable $\mathbf{X}$ from the point $\mathbf{v}$ through the metric $\mathbf{U}$:

$$a_{\mathbf{v},\mathbf{U}} \equiv \mathrm{E}\left\{\mathrm{Ma}^2\left(\mathbf{X}, \mathbf{v}, \mathbf{U}\right)\right\}. \tag{2.89}$$

Nevertheless, if we set $\mathbf{v}$ equal to the expected value and $\mathbf{U}$ equal to the covariance matrix in (2.88), the function a reaches a minimum, and is equal to the dimension the random variable $\mathbf{X}$. Therefore the probability of $\mathbf{X}$ not occurring in the ellipsoid is uniformly the minimum possible and reads:

$$\mathbb{P}\left\{\mathbf{X} \notin \mathcal{E}^{q}_{\mathrm{E,Cov}}\right\} \leq \frac{N}{q^2}. \tag{2.90}$$

In other words, the location-dispersion ellipsoid $\mathcal{E}_{\mathrm{E,Cov}}$ is the one ellipsoid among those of equal volume that enshrouds the most probability.

2.4.4 Higher-order statistics

Similarly to the one-dimensional case discussed in Section 1.2.3, we can gain more insight into the statistical features of a multivariate distribution from the moments of that distribution of order higher than the expected value and the covariance matrix.

To introduce the higher moments, we recall that the expected value is a vector, namely the vector of expectations of each entry of a multivariate random variable. On the other hand the covariance is a matrix, namely the matrix of (a simple function of) expectations of all the cross products of two entries.

The expectation operator $(B.56)$ applied to the cross products of three, four, etc. entries can be organized in tensors, a straightforward generalization of the concept of vector and matrix, see $(A.92)$ and comments thereafter for a quick review.

The *k-th raw moment* of a multivariate random variable $\mathbf{X}$ is a tensor of order k, defined as follows:

$$\mathrm{RM}^{\mathbf{X}}_{n_1 \cdots n_k} \equiv \mathrm{E}\left\{X_{n_1} \cdots X_{n_k}\right\}. \tag{2.91}$$

This definition generalizes the one-dimensional raw-moment (1.47). In particular, the expected value (2.54) is the first raw moment.

The *k-th central moment* of a random variable is a location-independent version of the respective raw moment:

$$\mathrm{CM}^{\mathbf{X}}_{n_1 \cdots n_k} \equiv \mathrm{E}\left\{(X_{n_1} - \mathrm{E}\left\{X_{n_1}\right\}) \cdots (X_{n_k} - \mathrm{E}\left\{X_{n_k}\right\})\right\}. \tag{2.92}$$

This definition generalizes the one-dimensional central-moment (1.48). In particular, the covariance matrix (2.68) is the second central moment.

The central moments of a distribution are tensors that enjoy special transformation properties. For instance, from (2.71) the covariance matrix is equivariant under any, not necessarily invertible, affine transformation. From the linearity of the expectation operator $(B.56)$ and the definition of the central moments (2.92), it follows that all the central moments are affine equivariant, in that for any M-dimensional vector $\mathbf{a}$ and any $M \times N$ matrix $\mathbf{B}$ the following relation holds:

$$\mathrm{CM}^{\mathbf{a}+\mathbf{BX}}_{m_1 \cdots m_k} = \sum_{n_1, \ldots, n_k = 1}^{N} B_{m_1, n_1} \cdots B_{m_k, n_k} \, \mathrm{CM}^{\mathbf{X}}_{n_1 \cdots n_k}. \tag{2.93}$$

For example, consider $\mathbf{a} \equiv \mathbf{0}$ and $\mathbf{B}' \equiv \mathbf{b}$, an N-dimensional vector. In this case the affine-equivariance property (2.93) yields the expression for the central moments (1.48) of the one-dimensional variable $\mathbf{b}'\mathbf{X}$. For instance, the third central moment reads:

$$\mathrm{CM}^{\mathbf{b}'\mathbf{X}}_3 = \sum_{l,m,n=1}^{N} b_l b_m b_n \, \mathrm{CM}^{\mathbf{X}}_{lmn}. \tag{2.94}$$

Similarly to the univariate case, it is possible to define normalized version of the higher central moments.

The *co-skewness* is the following three-dimensional tensor:

$$\begin{aligned}\mathrm{Sk}\{X_l, X_m, X_n\} &\equiv [\mathrm{Sk}\{\mathbf{X}\}]_{lmn} \\ &\equiv \frac{\mathrm{CM}^{\mathbf{X}}_{lmn}}{\mathrm{Sd}\{X_l\}\,\mathrm{Sd}\{X_m\}\,\mathrm{Sd}\{X_n\}},\end{aligned} \tag{2.95}$$

which generalizes the univariate skewness (1.49). The co-skewness provides information on the symmetry of the distribution of $\mathbf{X}$. It is also possible to summarize the information provided by the co-skewness in one overall index of symmetry, see Mardia (1970).

The *co-kurtosis* is the following four-dimensional tensor:

$$\begin{aligned}\mathrm{Ku}\{X_l, X_m, X_n, X_p\} &\equiv [\mathrm{Ku}\{\mathbf{X}\}]_{lmnp} \\ &\equiv \frac{\mathrm{CM}^{\mathbf{X}}_{lmnp}}{\mathrm{Sd}\{X_l\}\,\mathrm{Sd}\{X_m\}\,\mathrm{Sd}\{X_n\}\,\mathrm{Sd}\{X_p\}},\end{aligned} \tag{2.96}$$

which generalizes the univariate kurtosis (1.51). The co-kurtosis provides information on the thickness of the tails of the distribution of $\mathbf{X}$. It is also possible to summarize the information provided by the co-kurtosis in one overall index of tail thickness, see Mardia (1970).

Computing the above summary statistics involves in general integrations. Nevertheless, whenever the characteristic function of $\mathbf{X}$ is known and analytical, i.e. it can be recovered entirely from its Taylor series expansion, we can compute these quantities by means of simple differentiation and some algebra, as we show Appendix www.2.10.

Nevertheless, the number of parameters in the higher moments grows as N^k, where N is the dimension of the multivariate distribution of $\mathbf{X}$ and k is the order of the moment. This number becomes intractable for $k > 2$ in any practical application.

2.5 Dependence summary statistics

The N entries of a random variable $\mathbf{X}$ display in general a complex dependence structure that it is important to monitor in view of hedging and managing risk. In this section we describe how to summarize in one number the dependence between two generic entries X_m and X_n. We refer the reader to references such as Nelsen (1999) for more results on this subject.

2.5.1 Measures of dependence

A *measure of dependence* $\mathrm{Dep}\{X_m, X_n\}$ between two random variables X_m and X_n should be a function of the distribution of the variables, normalized in such a way to make it easy to interpret, for instance as follows:

$$0 \leq \mathrm{Dep}\{X_m, X_n\} \leq 1. \tag{2.97}$$

Furthermore, it should display a minimal set of intuitive features, such as the following:

1. Total independence represents one extreme of the spectrum of possible values:
$$(X_m, X_n) \text{ independent} \Leftrightarrow \mathrm{Dep}\{X_m, X_n\} \equiv 0. \tag{2.98}$$
2. Total dependence represents the other extreme of the spectrum of possible values:
$$(X_m, X_n) \text{ co-monotonic} \Leftrightarrow \mathrm{Dep}\{X_m, X_n\} \equiv 1, \tag{2.99}$$
where co-monotonicity is defined in (2.35).
3. The measure of dependence spots the core interdependence structure. In other words, assume that the random variable X_m is a deterministic invertible function of a random variable Y_m, i.e. they are in one-to-one correspondence, and that an analogous relation holds between X_n and another random variable Y_n. The dependence between the first set of variables should be the same as the dependence between the second set of variables:
$$\left.\begin{array}{l}(X_m, Y_m) \text{ one-to-one} \\ (X_n, Y_n) \text{ one-to-one}\end{array}\right\} \Rightarrow \mathrm{Dep}\{X_m, X_n\} = \mathrm{Dep}\{Y_m, Y_n\}. \tag{2.100}$$

In Section 2.2.2 we determined that the core interdependence structure between two generic variables X_m and X_n is driven by their copula. We recall that the copula is the joint distribution of the grades:

$$\begin{pmatrix} U_m \\ U_n \end{pmatrix} \equiv \begin{pmatrix} F_{X_m}(X_m) \\ F_{X_n}(X_n) \end{pmatrix}, \tag{2.101}$$

where F_{X_m} is the cumulative distribution function of X_m, see (2.28). Therefore in order to define a measure of dependence between $(X_m, X_n)'$ it is natural to turn to their copula, which we represent in terms of the cumulative distribution function F_{U_m,U_n}.

As far as the property on independence (2.98) is concerned, since the marginal distribution of each of the grades (2.101) is uniform, from (1.56) and (2.46) we see that X_m and X_n are independent if and only if their copula is uniformly distributed on the unit square, in which case the cumulative distribution function of the copula reads:

$$\Pi(u_m, u_n) \equiv u_m u_n, \tag{2.102}$$

see Figure 2.10.

Intuitively, the measure of dependence between X_m and X_n should be a distance between their copula, as represented by F_{U_m,U_n}, and the copula of two independent variables, as represented by (2.102): the larger the distance,

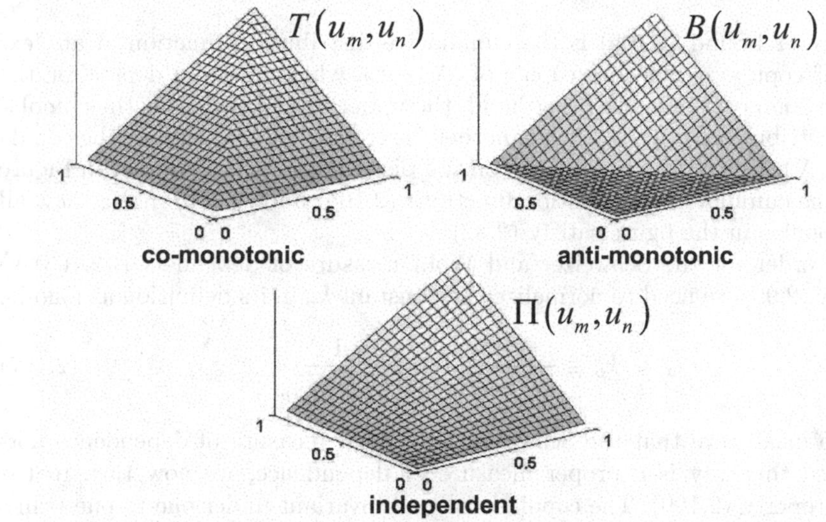

Fig. 2.10. Cumulative distribution function of special bivariate copulas

the higher the level of dependence. We can introduce a distance between these two functions by means of the L_p-norm (B.12), defined in this case on the unit square $\mathbb{Q} \equiv [0,1] \times [0,1]$. This way we obtain the following family of measures of dependence, called the *Schweizer-Wolff measures of dependence*:

$$\begin{aligned} \mathrm{SW}\{X_m, X_n\} &\equiv k_p \left\| F_{U_m,U_n} - \Pi \right\|_p \qquad (2.103) \\ &\equiv k_p \left(\int_{\mathbb{Q}} \left| F_{U_m,U_n}(u_m, u_n) - \Pi(u_m, u_n) \right|^p du_m du_n \right)^{\frac{1}{p}}, \end{aligned}$$

where $p \geq 1$ and k_p is a constant yet to be defined. By construction, this measure satisfies (2.98), which is the first property required of a measure of dependence.

To determine the constant in (2.103) we turn to (2.99), the property of a generic measure of dependence which regards total dependence. It can be proved that the *Frechet-Hoeffding bounds* hold on the cumulative distribution function of a generic copula:

$$B(u_m, u_n) \leq F_{U_m,U_n}(u_m, u_n) \leq T(u_m, u_n), \qquad (2.104)$$

where the "bottom" bound is defined as follows:

$$B(u_m, u_n) \equiv \max(u_m + u_n - 1, 0); \qquad (2.105)$$

and the "top" bound is defined as follows:

$$T(u_m, u_n) \equiv \min(u_m, u_n). \tag{2.106}$$

The lower bound (2.105) is the cumulative distribution function of an "extreme" copula, namely the copula of $(X, -X)$, which does not depend on the distribution of X. On the other hand, the upper bound (2.106) is the cumulative distribution function of the another "extreme" copula, namely the copula of (X, X), which does not depend on the distribution of X. We plot in Figure 2.10 the cumulative distribution functions (2.105) and (2.106): notice how all the copulas in the figure satisfy (2.32).

In order for the Schweizer and Wolff measure of dependence (2.103) to satisfy (2.99) we need to normalize the constant k_p in its definition as follows:

$$k_p \equiv \frac{1}{\|B - \Pi\|_p} = \frac{1}{\|T - \Pi\|_p}. \tag{2.107}$$

To make sure that the Schweizer and Wolff measure of dependence normalized this way is a proper measure of dependence, we now turn to the last property (2.100). The copula is *almost* invariant under one-to-one transformations such as those that appear in (2.100). Indeed, consider two new variables:

$$\begin{pmatrix} Y_m \\ Y_n \end{pmatrix} \equiv \begin{pmatrix} g(X_m) \\ h(X_n) \end{pmatrix}, \tag{2.108}$$

where g and h are *increasing* invertible functions. In other, words (X_m, X_n) and (Y_m, Y_n) are co-monotonic, see (2.35). Now consider the copula of (Y_m, Y_n), which is the distribution of the grades:

$$\begin{pmatrix} V_m \\ V_n \end{pmatrix} \equiv \begin{pmatrix} F_{Y_m}(Y_m) \\ F_{Y_n}(Y_n) \end{pmatrix}. \tag{2.109}$$

Since from (2.38) the copula of (X_m, X_n) is the same as the copula of (Y_m, Y_n), the following relation holds:

$$F_{V_m,V_n} = F_{U_m,U_n}. \tag{2.110}$$

Therefore the Schweizer and Wolff measure of dependence automatically satisfies (2.100) for increasing one-to-one correspondences among the variables. If on the other hand one of the two variables, say Y_n, is an invertible *decreasing* function of X_n then (2.110) must be replaced by the following expression:

$$F_{V_m,V_n} = u_m - F_{U_m,U_n}(u_m, 1 - u_n). \tag{2.111}$$

Nevertheless, the integral in (2.103) is not affected by this change. Therefore the Schweizer and Wolff measure of dependence also satisfies (2.100).

Consider our leading example (2.7) of two stock prices (X_1, X_2). As in (2.36) consider a call option on the first stock price with strike K, i.e. the following random variable:

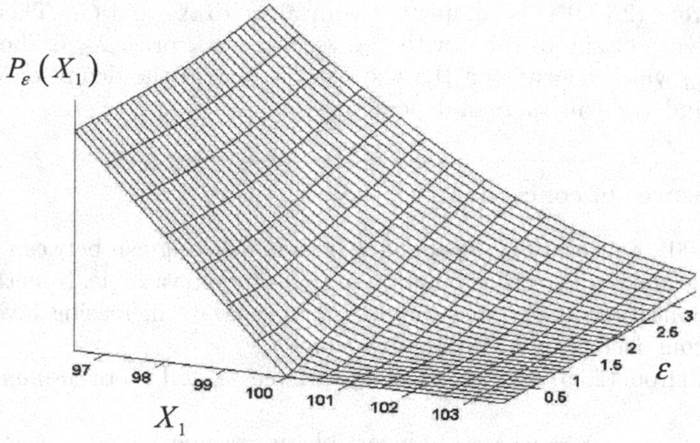

Fig. 2.11. Regularization of put option payoff

$$C_1 \equiv \max(X_1 - K, 0) . \tag{2.112}$$

The call option payoff is a strictly increasing function of the underlying X_1, once we replace it by its regularized version as in (2.37), see also Figure 2.5. Consider the payoff of a put option on the first stock with strike K, i.e. the new random variable:

$$P_1 \equiv -\min(X_1 - K, 0) . \tag{2.113}$$

The put option payoff is a strictly decreasing function of the underlying X_1, once we replace it by its regularized version by means of $(B.49)$:

$$P_\epsilon \equiv -\frac{(X_1 - K)}{2}\left(1 - \operatorname{erf}\left(\frac{X_1 - K}{\sqrt{2\epsilon^2}}\right)\right) + \frac{\epsilon}{\sqrt{2\pi}} e^{-\frac{1}{2\epsilon^2}(X_1 - K)^2}, \tag{2.114}$$

see Figure 2.11 for the plot and Appendix www.2.7 for the proof.

We summarize the Schweizer and Wolff measure of dependence (2.103) between any two of the above securities in the following table:

SW	X_1	C_1	P_1	X_2
X_1	1	1	1	γ_p
C_1		1	1	γ_p
P_1			1	γ_p
X_2				1

(2.115)

The first stock price X_1 and the (regularized) call option payoff C_1 are co-monotonic functions: from (2.99) their measure of dependence is one. The (regularized) put option payoff P_1 is an invertible function of the first stock price X_1: from (2.100) it is completely equivalent to X_1 and C_1. Therefore, the dependence of any of them with the second stock price X_2 is the same constant γ_p, which depends on the choice of $p \geq 1$ in the definition of the Schweizer and Wolff measure of dependence.

2.5.2 Measures of concordance

Due to (2.100), a measure of dependence does not distinguish between a random variable and any invertible function of that random variable. Nonetheless, in many applications it becomes important to separate increasing invertible functions from decreasing invertible functions.

We recall from (2.35) that two random variables X and Y are *co-monotonic* if

$$Y = g(X), \quad g \text{ invertible, increasing.} \tag{2.116}$$

Similarly, we define two random variables X and Y as *anti-monotonic* if

$$Y = g(X), \quad g \text{ invertible, decreasing.} \tag{2.117}$$

In our example, the (regularized) call option payoff (2.112) and the price of the underlying stock are co-monotonic; the (regularized) put option payoff (2.113) and the price of the underlying stock are anti-monotonic; the (regularized) call option payoff and the (regularized) put option payoff are anti-monotonic. The interests of an investor who owns the call option are very different than the interests of an investor who owns the put option. A dependence parameter such as the Schweizer and Wolff measure does not distinguish between calls and puts, as we see in (2.115).

Therefore, we are led to consider *measures of concordance*, that convey more information than measures of dependence. Ideally, a measure of concordance $\text{Con}\{X_m, X_n\}$ between two random variables X_m and X_n should be a function of their distribution, normalized in such a way to make it easy to interpret, for instance as follows:

$$-1 \leq \text{Con}\{X_m, X_n\} \leq 1. \tag{2.118}$$

Furthermore, it should displays a set of intuitive features:

1'. Independence represents the middle of the spectrum of possible values:

$$(X_m, X_n) \text{ independent} \Leftrightarrow \text{Con}\{X_m, X_n\} = 0. \tag{2.119}$$

2a'. Total concordance represents one extreme of the spectrum of possible values:

$$(X_m, X_n) \text{ co-monotonic} \Leftrightarrow \text{Con}\{X_m, X_n\} = 1. \tag{2.120}$$

2b'. Total discordance represents the other extreme of the spectrum of possible values:

$$(X_m, X_n) \text{ anti-monotonic} \Leftrightarrow \text{Con}\{X_m, X_n\} = -1. \tag{2.121}$$

3. The measure of concordance spots the core interdependence structure:

$$\left.\begin{array}{l}(X_m, Y_m) \text{ co-monotonic} \\ (X_n, Y_n) \text{ co-monotonic}\end{array}\right\} \Rightarrow \text{Con}\{X_m, X_n\} = \text{Con}\{Y_m, Y_n\}. \tag{2.122}$$

4. Concordance and discordance play a symmetric role:

$$\text{Con}\{X_m, -X_n\} = -\text{Con}\{X_m, X_n\}. \tag{2.123}$$

A comparison with the properties of the measures of dependence shows that a measure of concordance satisfying Properties 1'-4 would indeed convey all the information contained in a measure of dependence, and more.

Unfortunately, (2.119) cannot be satisfied together with the other properties. Intuitively, if we want to measure the "direction" of the dependence between two variables with a single number, variables that in some scenarios are concordant and in some other scenarios are discordant display the same amount of concordance as independent variables, although they are not. Therefore, (2.119) must be weakened as follows:

1. Independence implies the middle of the spectrum of possible values:

$$(X_m, X_n) \text{ independent} \Rightarrow \text{Con}\{X_m, X_n\} = 0. \tag{2.124}$$

For similar reasons (2.120) and (2.121) must be weakened as follows:

2a. Total concordance implies one extreme of the spectrum of possible values:

$$(X_m, X_n) \text{ co-monotonic} \Rightarrow \text{Con}\{X_m, X_n\} = 1. \tag{2.125}$$

2b. Total discordance implies the other extreme of the spectrum of possible values:

$$(X_m, X_n) \text{ anti-monotonic} \Rightarrow \text{Con}\{X_m, X_n\} = -1. \tag{2.126}$$

Just like in the case of measures of dependence, to define measures of concordance between X_m and X_n we turn to their copula, i.e. the joint distribution of the grades:

$$\begin{pmatrix} U_m \\ U_n \end{pmatrix} \equiv \begin{pmatrix} F_{X_m}(X_m) \\ F_{X_n}(X_n) \end{pmatrix}, \tag{2.127}$$

which we represent in terms of its cumulative distribution function F_{U_m,U_n}, or its probability density function f_{U_m,U_n}.

A popular measure of concordance is *Kendall's tau*. Kendall's tau is a normalized weighed average of the distance with sign of the cdf of the copula of (X_m, X_n) from the cdf (2.102) of the copula of two independent variables:

$$\tau\{X_m, X_n\} \equiv 4 \int_{\mathbb{Q}} \left(F_{U_m,U_n}(u_m, u_n) - \frac{1}{4} \right) f_{U_m,U_n}(u_m, u_n)\, du_m du_n, \tag{2.128}$$

where $\mathbb{Q} \equiv [0,1] \times [0,1]$ is the unit square.

This definition reminds us of the Schweizer and Wolff measure of dependence (2.103), but it is different in two respects. One difference is minor: Kendall's tau is a *weighted* average of the difference of the two functions, the weights being provided by the pdf of the copula. The second difference is conceptually more relevant: Kendall's tau evaluates the difference *with sign*, not in absolute value. It can be checked that due to this last feature, Kendall's tau satisfies Properties 1-4 above and thus it defines a suitable measure of concordance.

Consider our leading example (2.7) of two stock prices (X_1, X_2), together with a call option (2.112) and a put option (2.113) on the first stock. We summarize Kendall's τ between any two of the above securities in the following table:

τ	X_1	C_1	P_1	X_2
X_1	1	1	-1	$\frac{2}{\pi}\arcsin\left(\sqrt{\frac{2}{5}}\right)$
C_1		1	-1	$\frac{2}{\pi}\arcsin\left(\sqrt{\frac{2}{5}}\right)$
P_1			1	$-\frac{2}{\pi}\arcsin\left(\sqrt{\frac{2}{5}}\right)$
X_2				1

(2.129)

The first stock price X_1, and the (regularized) call option payoff C_1 are co-monotonic, and therefore due to (2.125) their concordance is 1. The (regularized) put option payoff P_1 and the first stock price X_1 are anti-monotonic, and therefore due to (2.126) their concordance is -1. Similarly, the (regularized) put option payoff P_1 and the (regularized) call option payoff C_1 are anti-monotonic, and therefore their concordance is -1. The value of Kendall's τ between the first stock price X_1 and the second stock price X_2 is a specific instance of the more general result (2.178). Since the (regularized) call option payoff C_1 and the first stock price X_1 are co-monotonic, due to (2.122) the concordance of the second stock price X_2 with C_1 is the same as the concordance of X_2 with X_1. On the other hand, since the (regularized) put option payoff P_1 and the first stock price X_1 are anti-monotonic, due to (2.123) the concordance of the second stock price X_2 with P_1 is the opposite of the concordance of X_2 with X_1.

We mention another popular measure of concordance, *Spearman's rho*, which is the correlation of the grades:

$$\rho\{X_m, X_n\} \equiv \frac{\text{Cov}\{U_m, U_n\}}{\text{Sd}\{U_m\}\,\text{Sd}\{U_n\}}, \tag{2.130}$$

see Section 2.5.3 below for a discussion of the correlation. It is possible to check that Spearman's rho satisfies Properties 1-4 above and therefore it defines a suitable measure of concordance.

Spearman's rho and Kendall's tau evaluation of the concordance between two variables is in general different, although this difference is bounded as follows:

$$\frac{3\tau - 1}{2} \leq \rho \leq \frac{1 + 2\tau - \tau^2}{2}, \tag{2.131}$$

whenever $\tau \geq 0$; and

$$\frac{\tau^2 + 2\tau - 1}{2} \leq \rho \leq \frac{1 + 3\tau}{2}, \tag{2.132}$$

whenever $\tau \leq 0$.

2.5.3 Correlation

In this section we draw a bridge between the concordance summary statistics of a generic multivariate random variable $\mathbf{X}$ discussed above and the location-dispersion summary statistics of $\mathbf{X}$ introduced in Section 2.4, in particular the expected value (2.54) and the covariance matrix (2.67).

In defining the concordance summary statistics we relied on copulas, because copulas capture the core interdependence among variables: indeed the copula of one random variable with any of a set of co-monotonic variables is the same, although the co-monotonic variables might have very different marginal distributions.

The expected value is a purely "marginal" parameter, since it is the juxtaposition of the expected values of the single marginal entries X_n. Therefore, we cannot find any relation between expected value and parameters of concordance.

On the other hand, the covariance matrix displays both "marginal" and "joint" features.

From (2.74) the diagonal entries of the covariance matrix are the square of the standard deviation of the marginal entries. We can get rid of these purely "marginal" features by normalizing the covariance matrix into what is called the *correlation matrix*:

$$\text{Cor}\{X_m, X_n\} \equiv [\text{Cor}\{\mathbf{X}\}]_{mn} \equiv \frac{\text{Cov}\{X_m, X_n\}}{\text{Sd}\{X_m\}\,\text{Sd}\{X_n\}}. \tag{2.133}$$

The correlation is an extremely popular parameter among finance practitioners.

In our leading example of two stock prices (X_1, X_2) we derive from (2.69) their correlation:

$$\operatorname{Cor}\{X_1, X_2\} = \sqrt{\frac{2}{5}}. \tag{2.134}$$

The correlation displays some features that remind us of the properties of the measures of concordance.

Indeed the correlation is a normalized parameter:

$$-1 \leq \operatorname{Cor}\{X_m, X_n\} \leq 1. \tag{2.135}$$

This follows from the Cauchy-Schwartz inequality (B.69). Furthermore, the following holds.

1. Independence implies the middle of the spectrum of possible correlation values:

$$(X_m, X_n) \text{ independent} \Rightarrow \operatorname{Cor}\{X_m, X_n\} = 0. \tag{2.136}$$

   This is true since the covariance of independent variables is zero, see (2.70).

2a. Positive affine concordance represents one extreme of the spectrum of possible correlation values:

$$X_m = a + bX_n \Leftrightarrow \operatorname{Cor}\{X_m, X_n\} = 1, \tag{2.137}$$

   where a is a scalar and b is a positive scalar. This follows from (B.70).

2b. Negative affine concordance represents the other extreme of the spectrum of possible correlation values:

$$X_m = a - bX_n \Leftrightarrow \operatorname{Cor}\{X_m, X_n\} = -1, \tag{2.138}$$

   where a is a scalar and b is a positive scalar. This follows from (B.71).

3. Correlations are unaffected by positive affine transformations:

$$\left.\begin{array}{l} Y_m \equiv a + bX_m \\ Y_n \equiv c + dX_n \end{array}\right\} \Rightarrow \operatorname{Cor}\{X_m, X_n\} = \operatorname{Cor}\{Y_m, Y_n\}, \tag{2.139}$$

   where (a, c) are scalars and (b, d) are positive scalars. This follows from the affine equivariance property (2.71) of the covariance matrix.

4. Correlation and anti-correlation play a symmetric role:

$$\operatorname{Cor}\{X_m, -X_n\} = -\operatorname{Cor}\{X_m, X_n\}. \tag{2.140}$$

   This follows from the affine equivariance property (2.71) of the covariance matrix.

A comparison of these properties with the respective properties of the measures of concordance shows that the correlation fails to be a suitable measure of concordance because it only covers affine transformations, whereas the measures of concordance cover more general invertible transformations.

Consider again our leading example (2.7) of two stock prices (X_1, X_2), along with the call option (2.112) and the put option (2.113) on the first stock. We summarize in the following table the correlation between any two of the above securities, which we computed by means of simulations:

Cor	X_1	C_1	P_1	X_2
X_1	1	.86	−.86	.63
C_1		1	−.47	.54
P_1			1	−.54
X_2				1

(2.141)

Although the first stock price X_1, and the (regularized) call option payoff C_1 are co-monotonic, their correlation is not 1. Although the (regularized) put option payoff P_1 and the first stock price X_1 are anti-monotonic, their correlation is not -1. Similarly, although the (regularized) put option payoff P_1 and the (regularized) call option payoff C_1 are anti-monotonic, their correlation is not -1. As far as the second stock is concerned, although the (regularized) call option payoff C_1 and the first stock price X_1 are co-monotonic, the correlation of the second stock price X_2 with C_1 is not the same as the correlation of X_2 with X_1. Similarly, although the (regularized) put option payoff P_1 and the first stock price X_1 are anti-monotonic, the correlation of the second stock price X_2 with P_1 is not the opposite of the correlation of X_2 with X_1.

Furthermore, a measure of concordance is defined in terms of the copula, and as such is not influenced by the marginal distribution of the variables involved. On the other hand, the set of possible values of the correlation does depend on the marginal distributions between the variables involved.

For example, consider two normally distributed random variables:

$$X_1 \sim \mathrm{N}\left(\mu_1, \sigma_1^2\right), \quad X_2 \equiv \mathrm{N}\left(\mu_2, \sigma_2^2\right). \tag{2.142}$$

It is possible to show that the correlation between these variables can take on any value in the interval $[-1, 1]$, see (2.169).

On the other hand, consider two lognormal variables:

$$Y_1 \equiv e^{X_1}, \quad Y_2 \equiv e^{X_2}. \tag{2.143}$$

The correlation between these variables is bounded within an interval smaller than $[-1, 1]$. For instance, the fact that both variables are positive implies that the correlation between the variables (2.143) cannot equal -1.

After the above critiques, one might wonder why correlation is such a popular tool. In the first place, the correlation indeed draws a bridge between the location-dispersion parameters and the dependence-concordance summary statistics. Secondly, for an important class of distribution the correlation completely defines the dependence structure, see Section 2.7.1.

2.6 Taxonomy of distributions

In this section we provide a taxonomy of multivariate distributions, stressing only the features that are needed in the sequel to tackle financial applications. Except for the order statistics, all the distribution introduced are generalizations of the one-dimensional distributions introduced in Section 1.3, to which the reader is referred to support intuition.

2.6.1 Uniform distribution

The simplest multivariate distribution is the uniform distribution. The *uniform distribution* models the situation where the only information available about the N-dimensional random variable $\mathbf{X}$ is that its realization is bound to take place on a given range in $\mathbb{R}^N$, and that all points in that range are equally likely outcomes for the realization.

In particular, consider an ellipsoid $\mathcal{E}_{\boldsymbol{\mu},\boldsymbol{\Sigma}}$ centered in $\boldsymbol{\mu}$ with shape defined by the symmetric and positive matrix $\boldsymbol{\Sigma}$ as in $(A.73)$. We use the following notation to indicate that $\mathbf{X}$ is uniformly distributed on the ellipsoid $\mathcal{E}_{\boldsymbol{\mu},\boldsymbol{\Sigma}}$:

$$\mathbf{X} \sim \mathrm{U}\left(\mathcal{E}_{\boldsymbol{\mu},\boldsymbol{\Sigma}}\right). \tag{2.144}$$

In Appendix www.2.11 we follow Fang, Kotz, and Ng (1990) to prove the results in the sequel.

The probability density function of the uniform distribution on the ellipsoid reads:

$$f^{\mathrm{U}}_{\boldsymbol{\mu},\Sigma}(\mathbf{x}) = \frac{\Gamma\left(\frac{N}{2}+1\right)}{\pi^{\frac{N}{2}}\left|\boldsymbol{\Sigma}\right|^{\frac{1}{2}}} \mathbb{I}_{\mathcal{E}_{\boldsymbol{\mu},\boldsymbol{\Sigma}}}(\mathbf{x}), \tag{2.145}$$

where Γ is the gamma function $(B.80)$ and $\mathbb{I}$ is the indicator function $(B.72)$.

The characteristic function of the uniform distribution on the ellipsoid reads:

$$\phi^{\mathrm{U}}_{\boldsymbol{\mu},\Sigma}(\boldsymbol{\omega}) = e^{i\boldsymbol{\omega}'\boldsymbol{\mu}} \psi\left(\boldsymbol{\omega}'\boldsymbol{\Sigma}\boldsymbol{\omega}\right), \tag{2.146}$$

where the function ψ is defined in terms of the beta function $(B.88)$ as follows[2]:

$$\psi(\gamma) \equiv \frac{2}{B\left(\frac{1}{2},\frac{N+1}{2}\right)} \int_0^1 \cos\left(\sqrt{\gamma} z\right)\left(1-z^2\right)^{\frac{N-1}{2}} dz. \tag{2.147}$$

[2] There are two minor typos in Fang, Kotz, and Ng (1990)

The mode is not defined, but the standard location parameter, i.e. the expected value, is defined and reads:

$$\mathrm{E}\{\mathbf{X}\} = \boldsymbol{\mu}. \tag{2.148}$$

The modal dispersion is not defined, but the standard scatter parameter, i.e. the covariance matrix, is defined and reads:

$$\mathrm{Cov}\{\mathbf{X}\} = \frac{1}{N+2}\boldsymbol{\Sigma}. \tag{2.149}$$

Now we split $\mathbf{X}$ in two sub-sets: the K-dimensional random variable $\mathbf{X}_A$ made of the first K entries and the $(N-K)$-dimensional random variable $\mathbf{X}_B$ made of the remaining entries. The marginal distribution of $\mathbf{X}_A$ is not uniform. The conditional distribution of $\mathbf{X}_B$ given $\mathbf{X}_A$ is uniform on an ellipsoid of lower dimension.

Bivariate standard uniform distribution

To gain more insight in the properties of the multivariate uniform distribution, we consider more in detail the bivariate uniform distribution on the unit circle.

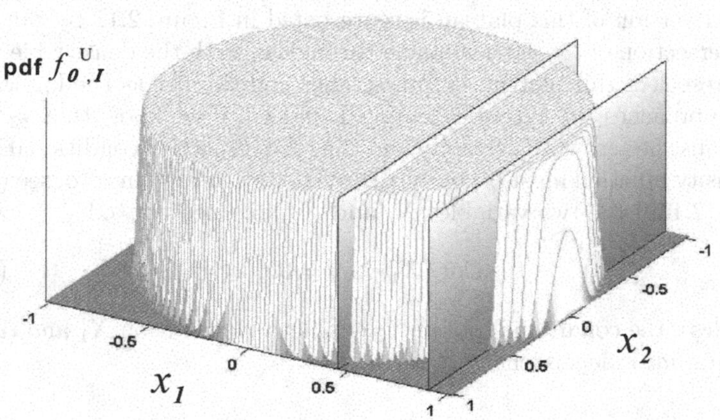

Fig. 2.12. Uniform distribution on the unit circle

The probability density function (2.145) is zero outside the unit circle and constant on the circle:

$$f_{X_1,X_2}(x_1,x_2) = \frac{1}{\pi}\mathbb{I}_{\{x_1^2+x_2^2\leq 1\}}(x_1,x_2), \tag{2.150}$$

see Figure 2.12.

We compute explicitly the marginal density of X_1. From Figure 2.12 we see that if $|x_1| > 1$ the marginal pdf is zero. When $|x_1| \leq 1$ the marginal density in x_1 is proportional to the area of the intersection of the vertical plane through x_1 with the density pie in Figure 2.12. This area is zero in $x_1 \equiv \pm 1$ and it reaches its maximum in $x_1 \equiv 0$. Indeed:

$$f_{X_1}(x_1) \equiv \int_{-\sqrt{1-x_1^2}}^{+\sqrt{1-x_1^2}} \frac{1}{\pi} dx_2 = \frac{2}{\pi}\sqrt{1-x_1^2}. \tag{2.151}$$

This formula shows that the marginal distribution of a uniform distribution is not uniform.

As for the conditional density of X_2 given x_1, we see in Figure 2.12 that the conditional pdf is non-zero only in the following domain:

$$-\sqrt{1-x_1^2} \leq x_2 \leq \sqrt{1-x_1^2}. \tag{2.152}$$

In this region the conditional pdf of X_2 given x_1 reads:

$$f_{X_2|x_1}(x_2) = \frac{f_{X_1,X_2}(x_1,x_2)}{f_{X_1}(x_1)} = \frac{1}{2\sqrt{1-x_1^2}}. \tag{2.153}$$

Since it does not depend on its argument x_2, this function describes a plateau. A rescaled version of this plateau is represented in Figure 2.12 by the profile of the intersection of the vertical plane through x_1 with the density pie. When suitably rescaled, this plateau becomes taller and thinner as the known variable x_1 approaches the extremes $x_1 \equiv \pm 1$: indeed, if we know that $x_1 \equiv \pm 1$, then X_2 must be zero with certainty and thus the respective conditional probability density function must becomes a Dirac delta centered in zero, see (B.22).

From (2.149) the two variables X_1 and X_2 are uncorrelated:

$$\text{Cor}\{X_1, X_2\} = 0. \tag{2.154}$$

Nevertheless, the conditional pdf of X_2 explicitly depends on X_1 and thus X_1 and X_2 are *not* independent.

2.6.2 Normal distribution

The *normal distribution* is the most widely used model to describe the statistical properties of a random variable $\mathbf{X}$ that can take on values in the whole space $\mathbb{R}^N$ in a symmetrical way around a peak.

The normal distribution depends on two parameters: an N-dimensional location vector $\boldsymbol{\mu}$ that determines the peak of the distribution, and an $N \times N$ symmetric and positive scatter matrix $\boldsymbol{\Sigma}$ that determines the shape of the distribution around its peak.

We use the following notation to indicate that $\mathbf{X}$ is normally distributed with the above parameters:

$$\mathbf{X} \sim \mathrm{N}(\boldsymbol{\mu}, \boldsymbol{\Sigma}). \tag{2.155}$$

The *standard normal distribution* corresponds to the specific case $\boldsymbol{\mu} \equiv \mathbf{0}$ and $\boldsymbol{\Sigma} \equiv \mathbf{I}$, the identity matrix.

The following results and more on the normal distribution can be found e.g. in Mardia, Kent, and Bibby (1979), Press (1982) and Morrison (2002).

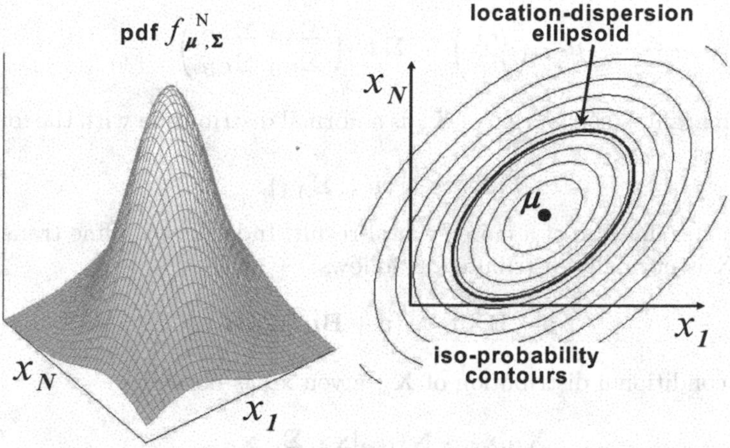

Fig. 2.13. Normal distribution

The multivariate normal probability density function reads:

$$f^{\mathrm{N}}_{\boldsymbol{\mu},\boldsymbol{\Sigma}}(\mathbf{x}) = (2\pi)^{-\frac{N}{2}} |\boldsymbol{\Sigma}|^{-\frac{1}{2}} e^{-\frac{1}{2}(\mathbf{x}-\boldsymbol{\mu})'\boldsymbol{\Sigma}^{-1}(\mathbf{x}-\boldsymbol{\mu})}, \tag{2.156}$$

see the left portion of Figure 2.13 for a plot in the bivariate case, and the right portion of that figure for the projection on the plane of the points that share the same values of the pdf.

The characteristic function of the normal distribution reads:

$$\phi^{\mathrm{N}}_{\boldsymbol{\mu},\boldsymbol{\Sigma}}(\boldsymbol{\omega}) = e^{i\boldsymbol{\mu}'\boldsymbol{\omega}-\frac{1}{2}\boldsymbol{\omega}'\boldsymbol{\Sigma}\boldsymbol{\omega}}, \tag{2.157}$$

see also Appendix www.2.12. The expected value and the mode coincide and read:

$$\mathrm{E}\{\mathbf{X}\} = \mathrm{Mod}\{\mathbf{X}\} = \boldsymbol{\mu}. \tag{2.158}$$

The covariance matrix and the modal dispersion are both defined. They coincide and read:

$$\operatorname{Cov}\{\mathbf{X}\} = \operatorname{MDis}\{\mathbf{X}\} = \boldsymbol{\Sigma}. \tag{2.159}$$

In the right portion of Figure 2.13 we plot for the bivariate case the location-dispersion ellipsoid $\mathcal{E}_{\mathrm{E,Cov}}$ defined in (2.75), see the discussion in Section 2.4.3.

Now we split $\mathbf{X}$ in two sub-sets: the K-dimensional random variable $\mathbf{X}_A$ made of the first K entries and the $(N-K)$-dimensional random variable $\mathbf{X}_B$ made of the remaining entries:

$$\mathbf{X} \equiv \begin{pmatrix} \mathbf{X}_A \\ \mathbf{X}_B \end{pmatrix}. \tag{2.160}$$

We split accordingly the location and the scatter parameters:

$$\boldsymbol{\mu} \equiv \begin{pmatrix} \boldsymbol{\mu}_A \\ \boldsymbol{\mu}_B \end{pmatrix}, \quad \boldsymbol{\Sigma} \equiv \begin{pmatrix} \boldsymbol{\Sigma}_{AA} & \boldsymbol{\Sigma}_{AB} \\ \boldsymbol{\Sigma}_{BA} & \boldsymbol{\Sigma}_{BB} \end{pmatrix}. \tag{2.161}$$

The marginal distribution of $\mathbf{X}_A$ is a normal distribution with the following parameters:

$$\mathbf{X}_A \sim \mathrm{N}(\boldsymbol{\mu}_A, \boldsymbol{\Sigma}_{AA}). \tag{2.162}$$

This is a specific case of a more general result. Indeed, any affine transformation of $\mathbf{X}$ is normally distributed as follows:

$$\mathbf{a} + \mathbf{B}\mathbf{X} \sim \mathrm{N}(\mathbf{a} + \mathbf{B}\boldsymbol{\mu}, \mathbf{B}\boldsymbol{\Sigma}\mathbf{B}'). \tag{2.163}$$

The conditional distribution of $\mathbf{X}_B$ given $\mathbf{x}_A$ is normal:

$$\mathbf{X}_B|\mathbf{x}_A \sim \mathrm{N}(\boldsymbol{\mu}_B|\mathbf{x}_A, \boldsymbol{\Sigma}_B|\mathbf{x}_A), \tag{2.164}$$

where

$$\boldsymbol{\mu}_B|\mathbf{x}_A \equiv \boldsymbol{\mu}_B + \boldsymbol{\Sigma}_{BA}\boldsymbol{\Sigma}_{AA}^{-1}(\mathbf{x}_A - \boldsymbol{\mu}_A) \tag{2.165}$$

$$\boldsymbol{\Sigma}_B|\mathbf{x}_A \equiv \boldsymbol{\Sigma}_{BB} - \boldsymbol{\Sigma}_{BA}\boldsymbol{\Sigma}_{AA}^{-1}\boldsymbol{\Sigma}_{AB}. \tag{2.166}$$

Notice that the expression of the conditional covariance does not depend on the known variable $\mathbf{x}_A$.

As for independence, two jointly normal random variables are independent if and only if their covariance, or equivalently their correlation, is null:

$$(X_m, X_n) \text{ independent} \Leftrightarrow \operatorname{Cov}\{X_m, X_n\} = 0. \tag{2.167}$$

This is another very special feature of the normal distribution. In general the much weaker relation (2.70) holds.

Bivariate normal distribution

To better understand the properties of the multivariate normal distribution, we consider the bivariate case.

We write the scatter parameter component-wise as follows:

$$\boldsymbol{\Sigma} \equiv \begin{pmatrix} \sigma_1^2 & \rho\sigma_1\sigma_2 \\ \rho\sigma_1\sigma_2 & \sigma_2^2 \end{pmatrix}, \tag{2.168}$$

where $|\rho| \leq 1$. This is the most general parametrization of a symmetric and positive 2×2 matrix. Also, it is convenient for notational purposes to assume $\sigma_1, \sigma_2 \geq 0$.

Since from (2.159) the matrix $\boldsymbol{\Sigma}$ is the covariance, it follows immediately that

$$\mathrm{Cor}\{X_1, X_2\} = \rho, \tag{2.169}$$

which shows that the correlation of two jointly normal variables can assume any value in the interval $[-1, 1]$.

In this notation, the expression of the normal probability density function (2.156) reads

$$f^{\mathrm{N}}_{\mu,\Sigma}(x_1, x_2) = \frac{1}{2\pi\sqrt{\sigma_1^2\sigma_2^2(1-\rho^2)}} e^{-\frac{1}{2}\frac{z_1^2 - 2\rho z_1 z_2 + z_2^2}{(1-\rho^2)}} \tag{2.170}$$

where (z_1, z_2) are the z-scores, i.e. the standardized variables:

$$z_i \equiv \frac{x_i - \mu_i}{\sigma_i}, \quad i = 1, 2. \tag{2.171}$$

From (2.162) the marginal distribution of X_1 is normal:

$$X_1 \sim \mathrm{N}\left(\mu_1, \sigma_1^2\right). \tag{2.172}$$

From Figure 2.13, this result is intuitive. Indeed, the marginal density in x_1 is proportional to the area underneath the joint probability density function cut by the vertical plane through x_1: this area decreases at infinity and has a peak at the point $x_1 \equiv \mu_1$.

From (2.164) the conditional distribution of X_2 given x_1 is also normal:

$$X_2|x_1 \sim \mathrm{N}\left(\mu_2|x_1, (\sigma_2|x_1)^2\right). \tag{2.173}$$

The above parameters read explicitly:

$$\mu_2|x_1 \equiv \mu_2 + \rho\frac{\sigma_2}{\sigma_1}(x_1 - \mu_1) \tag{2.174}$$

$$\sigma_2|x_1 \equiv \sigma_2\sqrt{1-\rho^2}. \tag{2.175}$$

From Figure 2.13, this result is intuitive. Indeed, the (rescaled) profile of the conditional density of X_2 given x_1 is given by the intersection of the vertical plane through x_1 with the joint probability density function: this intersection has a bell shape peaked in $\mu_2|x_1$.

We now consider independence. If in (2.170) we set $\rho \equiv 0$, the pdf can be factored into the product of the pdf of the two marginal distributions of X_1 and X_2. In other words, from (2.169) we see that the two variables are independent if and only if their correlation is zero, which is stated more generally in (2.167).

To gain further insight in the dependence structure of the bivariate normal distribution, we consider the copula. In Appendix www.2.12 we prove that the probability density function of the copula reads in terms of the inverse of the error function (B.75) as follows:

$$f^{\mathrm{N}}_{U_1,U_2}(u_1,u_2) = \frac{1}{\sqrt{1-\rho^2}} e^{g_\rho\left(\mathrm{erf}^{-1}(2u_1-1),\mathrm{erf}^{-1}(2u_2-1)\right)}, \tag{2.176}$$

where g is defined as follows:

$$g_\rho(v_1,v_2) \equiv -\frac{\rho}{1-\rho^2}\begin{pmatrix} v_1 \\ v_2 \end{pmatrix}' \begin{pmatrix} \rho & -1 \\ -1 & \rho \end{pmatrix} \begin{pmatrix} v_1 \\ v_2 \end{pmatrix}. \tag{2.177}$$

From this expression we see that the copula of two jointly normal variables is completely determined by their correlation.

Therefore it is not surprising that Kendall's tau, the measure of concordance defined in (2.128), reads:

$$\tau\{X_1,X_2\} = \frac{2}{\pi}\arcsin(\rho). \tag{2.178}$$

In other words, the concordance of two jointly normal variables is completely determined by their correlation.

Matrix-variate normal distribution

Consider an $(N \times K)$-matrix-valued random variable:

$$\mathbf{X} \equiv \left(\mathbf{X}^{(1)},\ldots,\mathbf{X}^{(K)}\right) \equiv \begin{pmatrix} \mathbf{X}_{(1)} \\ \vdots \\ \mathbf{X}_{(N)} \end{pmatrix}, \tag{2.179}$$

where each column $\mathbf{X}^{(k)}$ is an N-dimensional random variable and each row $\mathbf{X}_{(n)}$ is a K-dimensional random variable. The random matrix $\mathbf{X}$ has a *matrix-variate normal distribution* if

$$\mathrm{vec}(\mathbf{X}) \sim \mathrm{N}(\mathrm{vec}(\mathbf{M}), \mathbf{S} \otimes \boldsymbol{\Sigma}), \tag{2.180}$$

where vec is the operator (A.104) that stacks the columns of a matrix into a vector; $\mathbf{S}$ is a $K \times K$ symmetric and positive definite matrix; $\boldsymbol{\Sigma}$ is an $N \times N$ symmetric and positive definite matrix; and $\otimes$ denotes the Kronecker product (A.96). We denote a matrix-variate normal distribution with the above parameters as follows:

$$\mathbf{X} \sim \mathrm{N}(\mathbf{M}, \mathbf{\Sigma}, \mathbf{S}). \tag{2.181}$$

The following results are proved in Appendix www.2.13.

The probability density function of the matrix-valued random variable (2.181) can be conveniently expressed as follows:

$$f^{\mathrm{N}}_{\boldsymbol{\mu},\mathbf{\Sigma},\mathbf{S}}(\mathbf{X}) \equiv (2\pi)^{-\frac{NK}{2}} |\mathbf{\Sigma}|^{-\frac{K}{2}} |\mathbf{S}|^{-\frac{N}{2}} \\ e^{-\frac{1}{2}\operatorname{tr}\left\{\mathbf{S}^{-1}(\mathbf{X}-\mathbf{M})'\mathbf{\Sigma}^{-1}(\mathbf{X}-\mathbf{M})\right\}}. \tag{2.182}$$

Notice that this density generalizes the vector-variate normal probability density function (2.156). Therefore the multivariate normal distribution (2.155) can be seen as the following special case of the matrix-variate normal distribution:

$$\mathrm{N}(\boldsymbol{\mu}, \mathbf{\Sigma}) = \mathrm{N}(\boldsymbol{\mu}, \mathbf{\Sigma}, 1). \tag{2.183}$$

From the definition (2.180) we see immediately that the matrix $\mathbf{M}$ is the expected value of $\mathbf{X}$:

$$\mathrm{E}\{\mathbf{X}\} = \mathbf{M}. \tag{2.184}$$

The matrix $\mathbf{\Sigma}$ defines the overall covariance structure between any two N-dimensional columns $\mathbf{X}^{(j)}, \mathbf{X}^{(k)}$ among the K that constitute the random matrix $\mathbf{X}$:

$$\mathrm{Cov}\left\{\mathbf{X}^{(j)}, \mathbf{X}^{(k)}\right\} = S_{jk}\mathbf{\Sigma}. \tag{2.185}$$

Similarly, the matrix $\mathbf{S}$ defines the overall covariance structure between any two K-dimensional rows $\mathbf{X}_{(m)}, \mathbf{X}_{(n)}$ among the N that constitute the random matrix $\mathbf{X}$:

$$\mathrm{Cov}\left\{\mathbf{X}_{(m)}, \mathbf{X}_{(n)}\right\} = \Sigma_{mn}\mathbf{S}. \tag{2.186}$$

2.6.3 Student t distribution

The *Student t distribution* is another model that describes the statistical properties of a random variable $\mathbf{X}$ which can assume values on the whole space $\mathbb{R}^N$ in a symmetrical way around a peak.

Similarly to the normal distribution, the Student t distribution depends on an N-dimensional location parameter $\boldsymbol{\mu}$ that determines the peak of the distribution, and an $N \times N$ symmetric and positive scatter matrix $\mathbf{\Sigma}$ that determines the shape of the distribution around its peak. It also depends on an additional parameter ν, the *degrees of freedom* of the Student t distribution, whose positive value determines the relative importance of the peak of the distribution with respect to its tails.

We use the following notation to indicate that $\mathbf{X}$ has a Student t-distribution with the above parameters:

$$\mathbf{X} \sim \mathrm{St}(\nu, \boldsymbol{\mu}, \mathbf{\Sigma}). \tag{2.187}$$

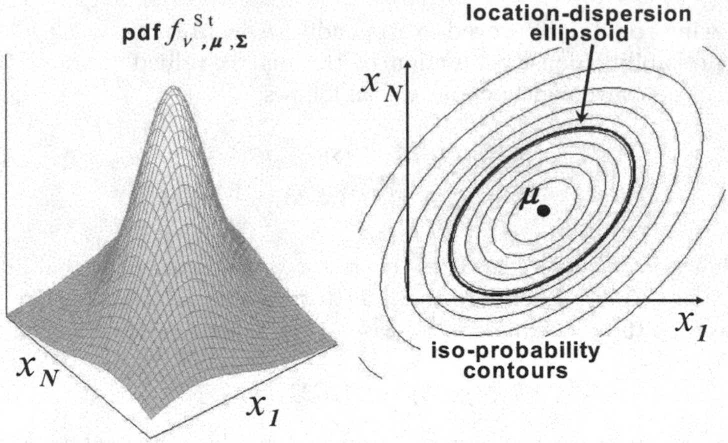

Fig. 2.14. Student t distribution

The *standard Student t distribution* corresponds to the specific case $\boldsymbol{\mu} \equiv \mathbf{0}$ and $\boldsymbol{\Sigma} \equiv \mathbf{I}$, the identity matrix.

The multivariate Student t probability density function reads:

$$f^{\mathrm{St}}_{\nu,\boldsymbol{\mu},\boldsymbol{\Sigma}}(\mathbf{x}) = (\nu\pi)^{-\frac{N}{2}} \frac{\Gamma\left(\frac{\nu+N}{2}\right)}{\Gamma\left(\frac{\nu}{2}\right)} |\boldsymbol{\Sigma}|^{-\frac{1}{2}} \left(1 + \frac{1}{\nu}(\mathbf{x}-\boldsymbol{\mu})' \boldsymbol{\Sigma}^{-1} (\mathbf{x}-\boldsymbol{\mu})\right)^{-\frac{\nu+N}{2}}, \tag{2.188}$$

where Γ denotes the gamma function (B.80), see Kotz and Nadarajah (2004) and Fang, Kotz, and Ng (1990). In the left portion of Figure 2.14 we plot the bivariate case, and in the right portion we plot the projection on the plane of the points that share the same values of the pdf.

The characteristic function of the Student t distribution is computed in Sutradhar (1986) and Sutradhar (1988). The characteristic function of the Student t distribution assumes a different form depending on whether the degrees of freedom ν are odd or even. We report here the expression for odd degrees of freedom:

$$\phi^{\mathrm{St}}_{\nu,\boldsymbol{\mu},\boldsymbol{\Sigma}}(\boldsymbol{\omega}) = \frac{\sqrt{\pi}\,\Gamma\left(\frac{\nu+1}{2}\right)}{2^{\nu-1}\Gamma\left(\frac{\nu}{2}\right)} e^{\left(i\boldsymbol{\mu}'\boldsymbol{\omega} - \sqrt{\nu\boldsymbol{\omega}'\boldsymbol{\Sigma}\boldsymbol{\omega}}\right)} \sum_{r=1}^{\frac{\nu+1}{2}} \binom{\nu - r}{\frac{\nu+1}{2} - r} \frac{\left(2\sqrt{\nu\boldsymbol{\omega}'\boldsymbol{\Sigma}\boldsymbol{\omega}}\right)^{r-1}}{(r-1)!}, \tag{2.189}$$

where Γ is the gamma function (B.80).

The expected value and the mode coincide and read:

$$\mathrm{E}\{\mathbf{X}\} = \mathrm{Mod}\{\mathbf{X}\} = \boldsymbol{\mu}. \tag{2.190}$$

The covariance matrix is defined if $\nu > 0$ and reads:

$$\mathrm{Cov}\{\mathbf{X}\} = \frac{\nu}{\nu - 2}\boldsymbol{\Sigma}. \tag{2.191}$$

In the right portion of Figure 2.14 we plot for the bivariate case the location-dispersion ellipsoid $\mathcal{E}_{\boldsymbol{\mu},\boldsymbol{\Sigma}}$ defined in (2.75), see the discussion in Section 2.4.3.

Now we split $\mathbf{X}$ in two sub-sets: the K-dimensional random variable $\mathbf{X}_A$ made of the first K entries and the $(N-K)$-dimensional random variable $\mathbf{X}_B$ made of the remaining entries:

$$\mathbf{X} \equiv \begin{pmatrix} \mathbf{X}_A \\ \mathbf{X}_B \end{pmatrix}. \tag{2.192}$$

We split accordingly the location and the scatter parameters:

$$\boldsymbol{\mu} \equiv \begin{pmatrix} \boldsymbol{\mu}_A \\ \boldsymbol{\mu}_B \end{pmatrix}, \quad \boldsymbol{\Sigma} \equiv \begin{pmatrix} \boldsymbol{\Sigma}_{AA} & \boldsymbol{\Sigma}_{AB} \\ \boldsymbol{\Sigma}_{BA} & \boldsymbol{\Sigma}_{BB} \end{pmatrix}. \tag{2.193}$$

The marginal distribution of $\mathbf{X}_A$ is a Student t distribution with the following parameters:

$$\mathbf{X}_A \sim \mathrm{St}(\nu, \boldsymbol{\mu}_A, \boldsymbol{\Sigma}_{AA}). \tag{2.194}$$

This is a specific case of a more general result. Indeed, any affine transformation of $\mathbf{X}$ is Student t distributed as follows:

$$\mathbf{a} + \mathbf{B}\mathbf{X} \sim \mathrm{St}\left(\nu, \mathbf{a} + \mathbf{B}\boldsymbol{\mu}, \mathbf{B}\boldsymbol{\Sigma}\mathbf{B}'\right). \tag{2.195}$$

On the other hand, unlike in the normal case, the conditional distribution of a Student t distribution is in general not a Student t distribution. Indeed, from the expression of the joint pdf (2.188) and the fact that from (2.194) the marginal pdf is in the form (2.188) it is immediate to compute the pdf of the conditional distribution as the ratio of the joint pdf and the marginal pdf. Nevertheless, the conditional pdf is not of the form (2.188).

As far as independence is concerned, since the generic conditional distribution is not a Student t distribution and the generic marginal distribution is a Student t distribution it follows that marginal and conditional distribution cannot coincide. Therefore random variables that are jointly Student t distributed are not independent.

Just like in the one-dimensional case, the Student t distribution encompasses the normal distribution as a special case. Indeed, as we show more in general in Appendix www.2.14, in the limit $\nu \to \infty$ the Student t probability density function (2.188) yields the normal probability density function (2.156) and thus:

$$\mathrm{St}\,(\infty, \boldsymbol{\mu}, \boldsymbol{\Sigma}) = \mathrm{N}\,(\boldsymbol{\mu}, \boldsymbol{\Sigma}) . \tag{2.196}$$

As the degrees of freedom ν decrease, the tails in the pdf (2.188) of the distribution become thicker and thicker. We can see this by comparing Figure 2.14 with Figure 2.13, see also Section 2.6.4 and refer to Figure 1.9 for the univariate case.

Matrix-variate Student t distribution

The matrix-variate Student t distribution was introduced by Dickey (1967), see Appendix www.2.14 for the relation with the notation in the original paper.

Consider an $(N \times K)$-matrix-valued random variable

$$\mathbf{X} \equiv \left(\mathbf{X}^{(1)}, \ldots, \mathbf{X}^{(K)}\right) \equiv \begin{pmatrix} \mathbf{X}_{(1)} \\ \vdots \\ \mathbf{X}_{(N)} \end{pmatrix}, \tag{2.197}$$

where each column $\mathbf{X}^{(k)}$ is an N-dimensional random variable and each row $\mathbf{X}_{(n)}$ is a K-dimensional random variable.

The random matrix $\mathbf{X}$ is distributed according to a *matrix-valued Student t distribution* with the following parameters

$$\mathbf{X} \sim \mathrm{St}\,(\nu, \mathbf{M}, \boldsymbol{\Sigma}, \mathbf{S}) , \tag{2.198}$$

if its probability density function reads:

$$\begin{aligned} f^{\mathrm{St}}_{\nu,\boldsymbol{\mu},\boldsymbol{\Sigma},\mathbf{S}}\,(\mathbf{X}) \equiv\ & \gamma\, |\boldsymbol{\Sigma}|^{-\frac{K}{2}}\, |\mathbf{S}|^{-\frac{N}{2}} \\ & \left| \mathbf{I}_K + \mathbf{S}^{-1} (\mathbf{X} - \mathbf{M})' \frac{\boldsymbol{\Sigma}^{-1}}{\nu} (\mathbf{X} - \mathbf{M}) \right|^{-\frac{\nu+N}{2}} . \end{aligned} \tag{2.199}$$

In this expression $\mathbf{M}$ is an $N \times K$ matrix; $\boldsymbol{\Sigma}$ is an $N \times N$ symmetric and positive definite matrix; $\mathbf{S}$ is a $K \times K$ symmetric and positive definite matrix; ν is a positive integer; and γ is a normalization constant defined in terms of the gamma function (B.80) as follows:

$$\gamma \equiv (\nu\pi)^{-\frac{NK}{2}} \frac{\Gamma\left(\frac{\nu+N}{2}\right)}{\Gamma\left(\frac{\nu}{2}\right)} \frac{\Gamma\left(\frac{\nu+N-1}{2}\right)}{\Gamma\left(\frac{\nu-1}{2}\right)} \cdots \frac{\Gamma\left(\frac{\nu+N-K+1}{2}\right)}{\Gamma\left(\frac{\nu-K+1}{2}\right)} . \tag{2.200}$$

Notice that the density (2.199) generalizes the vector-variate Student t probability density function (2.188). Therefore the multivariate Student t distribution (2.187) can be seen as the following special case of the matrix-variate Student t distribution:

$$\mathrm{St}\,(\nu, \boldsymbol{\mu}, \boldsymbol{\Sigma}) = \mathrm{St}\,(\nu, \boldsymbol{\mu}, \boldsymbol{\Sigma}, 1) . \tag{2.201}$$

Unlike in the normal case (2.180), by stacking the columns of the matrix $\mathbf{X}$ in (2.198) with the vec operator (A.104) we do not obtain a Student t-distributed variable:

$$\operatorname{vec}(\mathbf{X}) \nsim \operatorname{St}(\nu, \operatorname{vec}(\mathbf{M}), \mathbf{S} \otimes \boldsymbol{\Sigma}). \tag{2.202}$$

Nevertheless the following results hold:

$$\operatorname{E}\{\mathbf{X}\} = \mathbf{M}, \tag{2.203}$$

which generalizes (2.190); and

$$\operatorname{Cov}\{\operatorname{vec}(\mathbf{X})\} = \frac{\nu}{\nu - 2} \mathbf{S} \otimes \boldsymbol{\Sigma}, \tag{2.204}$$

which generalizes (2.191). Therefore the matrix $\boldsymbol{\Sigma}$ defines the overall covariance structure between any two N-dimensional columns $\mathbf{X}^{(j)}, \mathbf{X}^{(k)}$ among the K that constitute the random matrix $\mathbf{X}$:

$$\operatorname{Cov}\left\{\mathbf{X}^{(j)}, \mathbf{X}^{(k)}\right\} = \frac{\nu}{\nu - 2} S_{jk} \boldsymbol{\Sigma}. \tag{2.205}$$

Similarly, the matrix $\mathbf{S}$ defines the overall covariance structure between any two K-dimensional rows $\mathbf{X}_{(m)}, \mathbf{X}_{(n)}$ among the N that constitute the random matrix $\mathbf{X}$:

$$\operatorname{Cov}\left\{\mathbf{X}_{(m)}, \mathbf{X}_{(n)}\right\} = \frac{\nu}{\nu - 2} \Sigma_{mn} \mathbf{S}. \tag{2.206}$$

These result parallel (2.184)-(2.186) for the normal distribution. Indeed, in the limit $\nu \to \infty$ the matrix-variate Student t distribution (2.198) becomes the matrix-variate normal distribution (2.181):

$$\operatorname{St}(\infty, \mathbf{M}, \boldsymbol{\Sigma}, \mathbf{S}) = \operatorname{N}(\mathbf{M}, \boldsymbol{\Sigma}, \mathbf{S}), \tag{2.207}$$

see the proof in Appendix www.2.14.

2.6.4 Cauchy distribution

As in the univariate setting, the special case of the Student t distribution with $\nu \equiv 1$ degrees of freedom is called the *Cauchy distribution*, which we denote as follows:

$$\operatorname{Ca}(\boldsymbol{\mu}, \boldsymbol{\Sigma}) \equiv \operatorname{St}(1, \boldsymbol{\mu}, \boldsymbol{\Sigma}). \tag{2.208}$$

The *standard Cauchy distribution* corresponds to the specific case $\boldsymbol{\mu} \equiv \mathbf{0}$ and $\boldsymbol{\Sigma} \equiv \mathbf{I}$, the identity matrix.

From (2.188), the probability density function of the Cauchy distribution reads:

$$f^{\mathrm{Ca}}_{\boldsymbol{\mu},\boldsymbol{\Sigma}}(\mathbf{x}) = \frac{\Gamma\left(\frac{1+N}{2}\right)}{\pi^{\frac{N}{2}} \Gamma\left(\frac{1}{2}\right)} |\boldsymbol{\Sigma}|^{-\frac{1}{2}} \left(1 + (\mathbf{x} - \boldsymbol{\mu})' \boldsymbol{\Sigma}^{-1} (\mathbf{x} - \boldsymbol{\mu})\right)^{-\frac{N+1}{2}}, \tag{2.209}$$

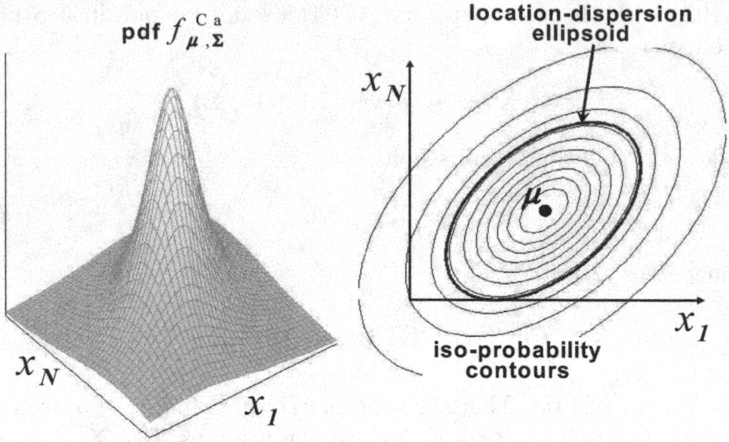

Fig. 2.15. Cauchy distribution

see the left portion of Figure 2.15 for a plot in the bivariate case, and the right portion of that figure for the projection on the plane of the points that share the same values of the pdf.

From (2.189), the characteristic function of the Cauchy distribution reads:

$$\phi^{\mathrm{Ca}}_{\boldsymbol{\mu},\boldsymbol{\Sigma}}(\boldsymbol{\omega}) = e^{i\boldsymbol{\mu}'\boldsymbol{\omega}-\sqrt{\boldsymbol{\omega}'\boldsymbol{\Sigma}\boldsymbol{\omega}}}. \tag{2.210}$$

The tails in the density (2.209) are so thick that the moments are not defined. Nevertheless, the mode is defined and reads:

$$\mathrm{Mod}\{\mathbf{X}\} = \boldsymbol{\mu}. \tag{2.211}$$

Similarly, the modal dispersion (2.65) is defined and reads:

$$\mathrm{MDis}\{\mathbf{X}\} = \frac{1}{N+1}\boldsymbol{\Sigma}, \tag{2.212}$$

see Appendix www.2.15. In the right portion of Figure 2.15 we plot for the bivariate case the location-dispersion ellipsoid $\mathcal{E}_{\boldsymbol{\mu},\boldsymbol{\Sigma}}$ defined in (2.75), see the discussion in Section 2.4.3.

2.6.5 Log-distributions

Log-distributions are defined as the exponential of other parametric distributions. As such, they are suitable to model positive quantities such as prices of limited-liability securities.

More precisely, consider a random variable $\mathbf{Y}$, whose distribution is represented by its pdf $f_{\mathbf{Y}}$, or its cdf $F_{\mathbf{Y}}$, or its characteristic function $\phi_{\mathbf{Y}}$. The variable $\mathbf{X} \equiv e^{\mathbf{Y}}$, where the exponential acts component-wise, is *log-*$\mathbf{Y}$ *distributed*, because by definition the logarithm of $\mathbf{X}$ has the same distribution as $\mathbf{Y}$.

The following results are discussed in Appendix www.2.16.

The probability density function of a log-$\mathbf{Y}$ distribution reads:

$$f_{\mathbf{X}}(\mathbf{x}) = \frac{f_{\mathbf{Y}}(\ln(\mathbf{x}))}{\prod_{n=1}^{N} x_n}. \tag{2.213}$$

The raw moments of a log-$\mathbf{Y}$ distribution read:

$$\mathrm{E}\{X_{n_1} \cdots X_{n_k}\} = \phi_{\mathbf{Y}}(\boldsymbol{\omega}_{n_1 \cdots n_k}), \tag{2.214}$$

where the vector $\boldsymbol{\omega}$ is defined in terms of the canonical basis $(A.15)$ as follows:

$$\boldsymbol{\omega}_{n_1 \cdots n_k} \equiv \frac{1}{i}\left(\boldsymbol{\delta}^{(n_1)} + \cdots + \boldsymbol{\delta}^{(n_k)}\right). \tag{2.215}$$

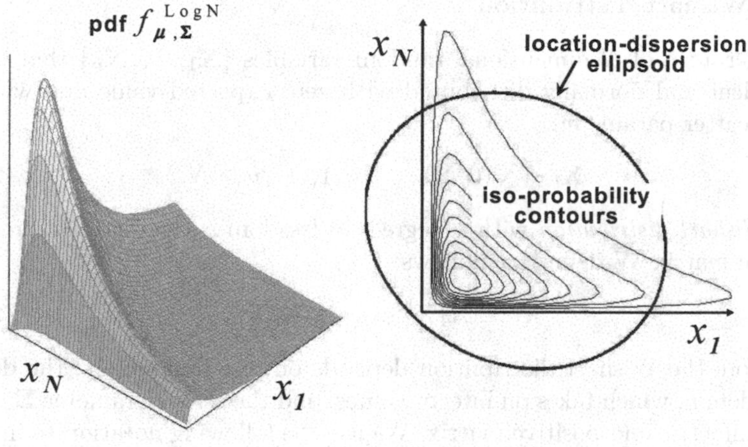

Fig. 2.16. Lognormal distribution

In particular, consider a random variable $\mathbf{Y}$ that is normally distributed with expected value $\boldsymbol{\mu}$ and covariance matrix $\boldsymbol{\Sigma}$:

$$\mathbf{Y} \sim \mathrm{N}(\boldsymbol{\mu}, \boldsymbol{\Sigma}). \tag{2.216}$$

We use the following notation to indicate that $\mathbf{X} \equiv e^{\mathbf{Y}}$ has a *lognormal distribution* with the above parameters:

$$\mathbf{X} \sim \operatorname{LogN}(\boldsymbol{\mu}, \boldsymbol{\Sigma}). \tag{2.217}$$

The probability density function of the lognormal distribution follows from (2.213) and the pdf (2.156) of the normal distribution:

$$f_{\boldsymbol{\mu},\Sigma}^{\mathrm{LogN}}(\mathbf{x}) = \frac{(2\pi)^{-\frac{N}{2}} |\boldsymbol{\Sigma}|^{-\frac{1}{2}}}{\prod_{n=1}^{N} x_n} e^{-\frac{1}{2}(\ln(\mathbf{x})-\boldsymbol{\mu})' \boldsymbol{\Sigma}^{-1}(\ln(\mathbf{x})-\boldsymbol{\mu})}, \tag{2.218}$$

see the left portion of Figure 2.16 for a plot in the bivariate case, and the right portion of that figure for the projection on the plane of the points that share the same values of the pdf.

Expected values and covariances of the lognormal distribution follow from (2.214) and the characteristic function (2.157) of the normal distribution:

$$\mathrm{E}\{X_n\} = e^{\mu_n + \frac{\Sigma_{nn}}{2}} \tag{2.219}$$

$$\operatorname{Cov}\{X_m, X_n\} = e^{\mu_m + \mu_n + \frac{\Sigma_{mm}}{2} + \frac{\Sigma_{nn}}{2}} \left(e^{\Sigma_{mn}} - 1\right). \tag{2.220}$$

In the right portion of Figure 2.16 we plot for the bivariate case the location-dispersion ellipsoid $\mathcal{E}_{\mathrm{E,Cov}}$ defined in (2.75), see the discussion in Section 2.4.3.

2.6.6 Wishart distribution

Consider a set of N-dimensional random variables $\{\mathbf{X}_1, \ldots, \mathbf{X}_\nu\}$ that are independent and normally distributed with zero expected value and with the same scatter parameter:

$$\mathbf{X}_t \sim \mathrm{N}(\mathbf{0}, \boldsymbol{\Sigma}), \quad t = 1, \ldots, \nu \geq N. \tag{2.221}$$

The *Wishart distribution* with ν degrees of freedom is the distribution of the random matrix $\mathbf{W}$ defined as follows:

$$\mathbf{W} \equiv \mathbf{X}_1 \mathbf{X}_1' + \cdots + \mathbf{X}_\nu \mathbf{X}_\nu'. \tag{2.222}$$

Therefore the Wishart distribution depends on two parameters: the degrees of freedom ν, which takes on integer values, and the scale parameter $\boldsymbol{\Sigma}$, which is a symmetric and positive matrix. We use the following notation to indicate that $\mathbf{W}$ is a Wishart-distributed matrix with the above parameters:

$$\mathbf{W} \sim \mathrm{W}(\nu, \boldsymbol{\Sigma}). \tag{2.223}$$

Notice that by construction $\mathbf{W}$ is a symmetric and positive matrix-valued random variable. This distribution plays a major role in the analysis of the estimation of covariance matrices.

The following results on the Wishart distribution can be found in Anderson (1984) and Mardia, Kent, and Bibby (1979).

The probability density function of the Wishart distribution reads:

$$f_{\nu,\boldsymbol{\Sigma}}^{\mathrm{W}}(\mathbf{W}) = \frac{1}{\kappa} |\boldsymbol{\Sigma}|^{-\frac{\nu}{2}} |\mathbf{W}|^{\frac{\nu-N-1}{2}} e^{-\frac{1}{2}\operatorname{tr}(\boldsymbol{\Sigma}^{-1}\mathbf{W})}, \tag{2.224}$$

where κ is a normalization constant defined in terms of the the gamma function (B.80) as follows:

$$\kappa \equiv 2^{\frac{\nu N}{2}} \pi^{\frac{N(N-1)}{4}} \Gamma\left(\frac{\nu}{2}\right) \Gamma\left(\frac{\nu-1}{2}\right) \cdots \Gamma\left(\frac{\nu-N+1}{2}\right). \tag{2.225}$$

The characteristic function of the Wishart distribution reads:

$$\phi_{\nu,\boldsymbol{\Sigma}}^{\mathrm{W}}(\boldsymbol{\Omega}) \equiv \mathrm{E}\left\{e^{i\operatorname{tr}(\mathbf{W}\boldsymbol{\Omega})}\right\} = |\mathbf{I} - 2i\boldsymbol{\Sigma}\boldsymbol{\Omega}|^{-\frac{\nu}{2}}. \tag{2.226}$$

The expected value, which is the standard parameter of location, reads component-wise as follows:

$$\mathrm{E}\{W_{mn}\} = \nu \Sigma_{mn}. \tag{2.227}$$

The cross-covariances, which determine the dispersion of $\mathbf{W}$, read:

$$\operatorname{Cov}\{W_{mn}, W_{pq}\} = \nu\left(\Sigma_{mp}\Sigma_{nq} + \Sigma_{mq}\Sigma_{np}\right). \tag{2.228}$$

As in Magnus and Neudecker (1979), we can express this in compact notation as follows:

$$\operatorname{Cov}\{\operatorname{vec}[\mathbf{W}]\} = \nu\left(\mathbf{I}_{N^2} + \mathbf{K}_{NN}\right)(\boldsymbol{\Sigma} \otimes \boldsymbol{\Sigma}), \tag{2.229}$$

where vec is the operator (A.104) that stacks the columns of $\mathbf{W}$ into a vector, $\mathbf{I}$ is the identity matrix, $\mathbf{K}$ is the commutation matrix (A.108) and $\otimes$ is the Kronecker product (A.96).

A comparison of (2.224) with (1.110) shows that the Wishart distribution is the multivariate generalization of the gamma distribution (1.108). Furthermore, for a generic vector $\mathbf{a}$ we obtain:

$$\mathbf{W} \sim \mathrm{W}(\nu, \boldsymbol{\Sigma}) \Rightarrow \mathbf{a}'\mathbf{W}\mathbf{a} \sim \mathrm{Ga}(\nu, \mathbf{a}'\boldsymbol{\Sigma}\mathbf{a}), \tag{2.230}$$

see Appendix www.2.17.

Since the inverse of a symmetric and positive matrix is a symmetric and positive matrix, the Wishart distribution can be used to model a symmetric and positive matrix also through its inverse. In other words, assume that the inverse of a random matrix $\mathbf{Z}$ is Wishart-distributed:

$$\mathbf{Z}^{-1} \sim \mathrm{W}\left(\nu, \boldsymbol{\Psi}^{-1}\right). \tag{2.231}$$

Then the distribution of $\mathbf{Z}$ is called *inverse-Wishart*, and is denoted as follows:

$$\mathbf{Z} \sim \mathrm{IW}(\nu, \boldsymbol{\Psi}). \tag{2.232}$$

We stress that $\mathbf{Z}$ is, like $\mathbf{Z}^{-1}$, a matrix-valued random variable that is symmetric and positive.

In Appendix www.2.17 we prove that the probability density function of the inverse-Wishart distribution reads:

$$f^{\mathrm{IW}}_{\nu,\mathbf{\Psi}}(\mathbf{Z}) = \frac{1}{\kappa} |\mathbf{\Psi}|^{\frac{\nu}{2}} |\mathbf{Z}|^{-\frac{\nu+N+1}{2}} e^{-\frac{1}{2}\operatorname{tr}(\mathbf{\Psi}\mathbf{Z}^{-1})}, \tag{2.233}$$

where κ is the normalization constant (2.225).

The 2 × 2 Wishart distribution

To better understand the Wishart distribution we consider the case of 2×2 matrices:

$$\mathbf{W} \equiv \begin{pmatrix} W_{11} & W_{12} \\ W_{21} & W_{22} \end{pmatrix} \sim \mathrm{W}\left(\nu; \begin{pmatrix} \sigma_1^2 & \rho\sigma_1\sigma_2 \\ \rho\sigma_1\sigma_2 & \sigma_2^2 \end{pmatrix}\right), \tag{2.234}$$

where $|\rho| \leq 1$.

The symmetry of $\mathbf{W}$ implies $W_{12} \equiv W_{21}$. Therefore this random matrix is completely determined by the three entries (W_{11}, W_{12}, W_{22}).

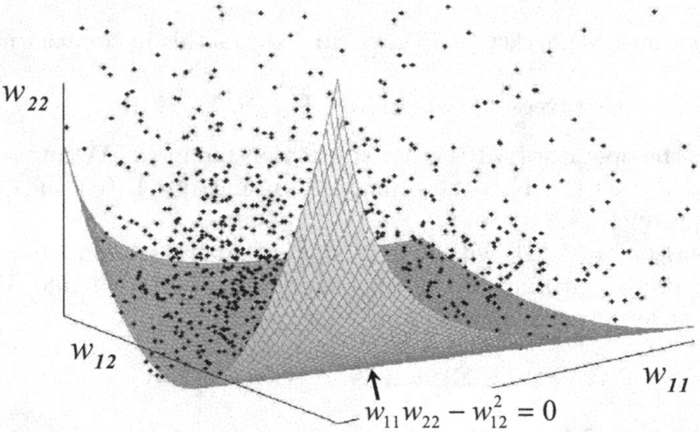

Fig. 2.17. Wishart distribution

Furthermore a symmetric matrix is positive if and only if its eigenvalues are positive. In the 2×2 case, denoting as λ_1 and λ_2 the two eigenvalues, these are positive if and only if the following inequalities are satisfied:

$$\lambda_1\lambda_2 > 0, \qquad \lambda_1 + \lambda_2 > 0. \tag{2.235}$$

On the other hand, the product of the eigenvalues is the determinant of $\mathbf{W}$ and the sum of the eigenvalues is the trace of $\mathbf{W}$, which are both invariants, see Appendix A.4. Therefore the positivity condition is equivalent to the two conditions below:

$$|\mathbf{W}| \equiv W_{11}W_{22} - W_{12}^2 \geq 0 \tag{2.236}$$

$$\mathrm{tr}\,(\mathbf{W}) \equiv W_{11} + W_{22} \geq 0, \tag{2.237}$$

where the first expression follows from $(A.41)$.

In Figure 2.17 we plot a few outcomes of a simulation of (2.234). Notice that all the outcomes lie above the surface $w_{11}w_{22} - w_{12}^2 = 0$: therefore (2.236) is satisfied. Furthermore, all the outcomes satisfy $w_{22} \geq -w_{11}$: therefore (2.237) is also satisfied. In other words, each outcome corresponds to a symmetric and positive 2×2 matrix.

2.6.7 Empirical distribution

The generalization to the multivariate case of the empirical distribution is immediate. Suppose that we can access T past measurements of the N-dimensional random variable $\mathbf{X}$:

$$i_T \equiv \{\mathbf{x}_1, \ldots, \mathbf{x}_T\}, \tag{2.238}$$

where we use the lower-case notation because these measurements have already taken place and thus they no longer represent random variables.

The *empirical distribution* models in the most simplistic way the basic assumption of statistics that we can learn from past experience. More precisely, under this distribution any of the past occurrences is an equally likely potential outcome of future measurements of $\mathbf{X}$, whereas different realizations cannot occur.

We use the following notation to indicate that $\mathbf{X}$ is distributed according to an empirical distribution stemming from the above observations:

$$\mathbf{X} \sim \mathrm{Em}\,(i_T)\,. \tag{2.239}$$

The empirical distribution is discrete. Therefore its probability density function is a generalized function. As in $(B.22)$, we can express the empirical pdf as follows:

$$f_{i_T}(\mathbf{x}) = \frac{1}{T}\sum_{t=1}^{T} \delta^{(\mathbf{x}_t)}(\mathbf{x}), \tag{2.240}$$

where δ is the Dirac delta $(B.16)$.

To visualize this probability density function we regularize it by means of the convolution as in $(B.54)$. The regularized probability density function of the empirical distribution reads in terms of the smooth approximation $(B.18)$ of the Dirac delta as follows:

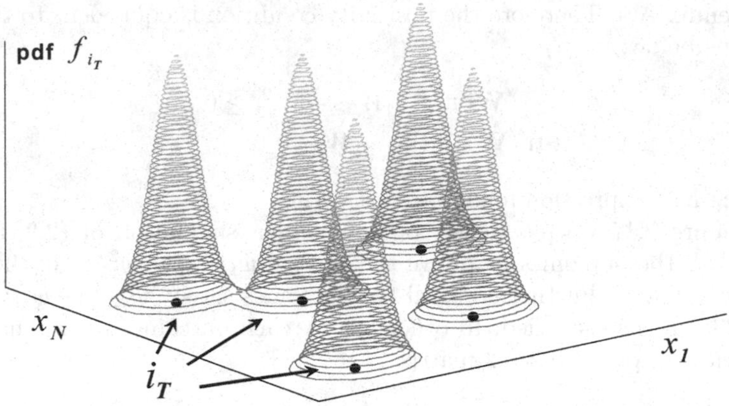

Fig. 2.18. Empirical distribution (regularized)

$$f_{i_T;\epsilon} \equiv f_{i_T} * \delta_{\epsilon}^{(\mathbf{0})} = \frac{1}{T}\sum_{t=1}^{T} \delta_{\epsilon}^{(\mathbf{x}_t)}, \tag{2.241}$$

where ϵ is a small bandwidth, see Figure 2.18.

From (B.53) the empirical cumulative distribution function reads:

$$F_{i_T} = \frac{1}{T}\sum_{t=1}^{T} H^{(\mathbf{x}_t)}, \tag{2.242}$$

where H is the Heaviside step function (B.73).

From the definition of the characteristic function (2.13) in terms of the expectation operator (B.56), and from the property (B.17) of the Dirac delta we obtain the characteristic function of the empirical distribution:

$$\phi_{i_T}(\boldsymbol{\omega}) = \frac{1}{T}\sum_{t=1}^{T} e^{i\boldsymbol{\omega}'\mathbf{x}_t}. \tag{2.243}$$

From the same rationale we also obtain the moments of any order of the empirical distribution. In particular, the expected value is called the *sample mean*, which we denote as follows:

$$\widehat{\mathrm{E}}_{i_T} \equiv \frac{1}{T}\sum_{t=1}^{T} \mathbf{x}_t. \tag{2.244}$$

Similarly, the covariance matrix of the empirical distribution is called the *sample covariance*, which we denote as follows:

$$\widehat{\mathrm{Cov}}_{iT} \equiv \frac{1}{T} \sum_{t=1}^{T} \left(\mathbf{x}_t - \widehat{\mathrm{E}}_{iT} \right) \left(\mathbf{x}_t - \widehat{\mathrm{E}}_{iT} \right)' . \tag{2.245}$$

2.6.8 Order statistics

The order statistics are useful in the context of nonparametric estimation. The following results and more can be found in David (1981).

Consider T independent and identically distributed univariate random variables and their respective realizations:

$$\{X_1, \ldots, X_T\}, \quad \{x_1, \ldots, x_T\}, \tag{2.246}$$

where as usual the upper-case notation indicates the random variable, and the lower-case notation indicates the respective realization.

Consider the smallest among the realized variables: this is, say, the realization of the second variable x_2. In a different scenario, the smallest realization might have been the realization of a different random variable, say x_4. In general, the value x_2 in the first scenario is different than the value x_4 in the second scenario. In other words, the minimum among the random variables (2.246) is a random variable.

Similarly, the maximum among the random variables (2.246) is a random variable. More in general, consider the whole set of ordered random variables:

$$\begin{aligned} X_{1:T} &\equiv \min \{X_1, \ldots, X_T\} \\ &\vdots \\ X_{T:T} &\equiv \max \{X_1, \ldots, X_T\} . \end{aligned} \tag{2.247}$$

The generic r-th element $X_{r:T}$, i.e. the r-th smallest random variable, is called the *r-th order statistic*.

The probability density function of the order statistics reads:

$$f_{X_{r:T}}(x) = \frac{T!}{(r-1)!\,(T-r)!} F_X^{r-1}(x) \left(1 - F_X(x)\right)^{T-r} f_X(x), \tag{2.248}$$

where f_X and F_X denote respectively the common probability density function and the common cumulative distribution function respectively of all the variables (2.246).

The cumulative distribution function of the order statistics reads:

$$F_{X_{r:T}}(x) = I\left(F_X(x), r, T-r+1\right), \tag{2.249}$$

where I is the regularized beta function (B.91).

When defined, the expected value of the generic r-th order statistic can be expressed in terms of the common quantile function Q_X of the variables (2.246) as follows:

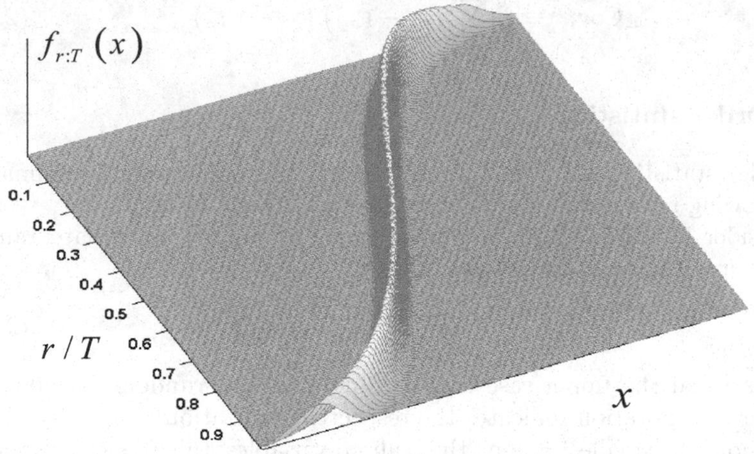

Fig. 2.19. Probability density function of order statistics

$$\mathrm{E}\{X_{r:T}\} = \int_{\mathbb{R}} Q_X(u)\,\widetilde{\delta}_{r,T}(u)\,du, \tag{2.250}$$

where the function δ is defined in terms of the indicator function (B.72) and reads:

$$\widetilde{\delta}_{r,T}(u) \equiv \frac{T!}{(r-1)!\,(T-r)!} u^{r-1} (1-u)^{T-r}\, \mathbb{I}_{[0,1]}(u). \tag{2.251}$$

In the limit of a large sample T this function is a smooth approximation to the Dirac delta (B.16):

$$\widetilde{\delta}_{r,T} \overset{T\to\infty}{\longrightarrow} \delta^{(r/T)}. \tag{2.252}$$

Therefore, when it is defined, the expected value of the r-th order statistic can be approximated by the quantile of any of the variables (2.246) as follows:

$$\mathrm{E}\{X_{r:T}\} \approx Q_X\left(\frac{r}{T}\right), \tag{2.253}$$

see Figure 2.19 and compare with Figure 1.2.

The concentration of the distribution of the order statistics around its expected value and the accuracy of the approximation (2.253) increases with the size T of the sample according to (2.252).

An important case of order statistics are those of the uniform distribution. Consider a set of T random variables that are independent and identically uniformly distributed on the unit interval:

$$U_t \sim \mathrm{U}([0,1]), \quad t = 1,\ldots,T. \tag{2.254}$$

The order statistics of the uniform distribution are important because they represent the grade of any order statistics. In other words, the order statistic from a generic distribution (2.247) has the same distribution as the quantile of the respective order statistics from the uniform distribution:

$$(X_{1:T}, \ldots, X_{T:T}) \stackrel{d}{=} (Q_X(U_{1:T}), \ldots, Q_X(U_{T:T})). \tag{2.255}$$

This result is a straightforward consequence of the definition of quantile, see (2.27) and Figure 2.3.

2.7 Special classes of distributions

In this section we put the distributions presented in Section 2.6 in a more general perspective in view of their applications. Refer to Figure 2.20 throughout the discussion.

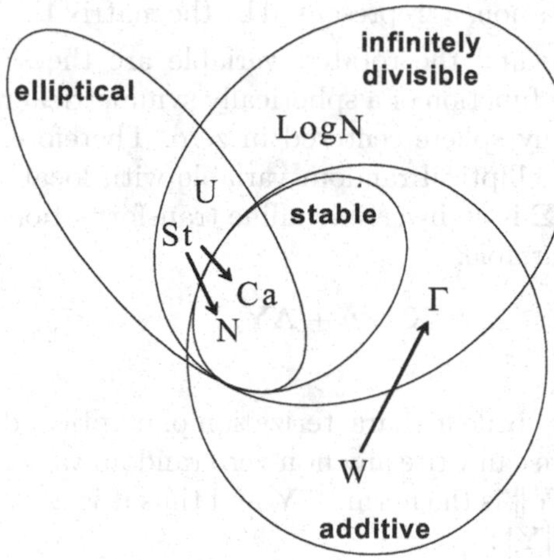

Fig. 2.20. Special classes of distributions

2.7.1 Elliptical distributions

Elliptical distributions are highly symmetrical distributions that are analytically tractable and yet flexible enough to model a wide range of situations. Refer to Fang, Kotz, and Ng (1990) and Fang and Zhang (1990) for more details.

Consider an N-dimensional random variable $\mathbf{X}$, whose distribution we represent by means of its probability density function $f_{\mathbf{X}}$. Consider the iso-probability contours:

$$C_L \equiv \{\mathbf{x} \text{ such that } f_{\mathbf{X}}(\mathbf{x}) = L\}. \tag{2.256}$$

The random variable $\mathbf{X}$ is *elliptically distributed* with location parameter $\boldsymbol{\mu}$ and scatter matrix $\boldsymbol{\Sigma}$ if for all levels $L \in (0, \infty)$ the iso-probability contour is the surface of the following ellipsoid:

$$\mathcal{E}_{\boldsymbol{\mu},\boldsymbol{\Sigma}}^{q(L)} \equiv \left\{\mathbf{x} \text{ such that } (\mathbf{x}-\boldsymbol{\mu})' \boldsymbol{\Sigma}^{-1} (\mathbf{x}-\boldsymbol{\mu}) \leq q(L)^2\right\}, \tag{2.257}$$

for a suitable function $q(L)$, see ($A.73$) for the details of the geometrical interpretation of this locus.

Examples of such distributions are the normal, Student t and Cauchy distributions respectively, as we see from the right portion of Figure 2.13, Figure 2.14 and Figure 2.15 respectively.

An equivalent characterization of an elliptical distribution is the following. Consider a random variable $\mathbf{Y}$ whose distribution is *spherically symmetrical*, i.e. such that for any rotation, as represented by the matrix $\boldsymbol{\Gamma}$, the distributions of the original variable and the rotated variable are the same: $\mathbf{Y} \stackrel{d}{=} \boldsymbol{\Gamma}\mathbf{Y}$. The probability density function of a spherically symmetrical random variable must be constant on any sphere centered in zero. Therefore, as we show in Appendix www.2.4, an elliptical random variable with location parameter $\boldsymbol{\mu}$ and scatter parameter $\boldsymbol{\Sigma}$ is an invertible affine transformation of a spherically symmetrical random variable:

$$\mathbf{X} \equiv \boldsymbol{\mu} + \mathbf{A}\mathbf{Y}, \tag{2.258}$$

where $\mathbf{A}\mathbf{A}' = \boldsymbol{\Sigma}$.

To obtain a final, equivalent characterization of elliptical distributions, we notice that in general we can write any non-zero random variable $\mathbf{Y}$ as follows: $\mathbf{Y} = R\mathbf{U}$, where $R \equiv \|\mathbf{Y}\|$ is the norm of $\mathbf{Y}$ and thus it is a univariate random variable, and $\mathbf{U} \equiv \mathbf{Y}/\|\mathbf{Y}\|$.

It can be proved that if $\mathbf{Y}$ is spherically symmetrical, then R and $\mathbf{U}$ are independent and $\mathbf{U}$ is uniformly distributed on the surface of the unit ball $\mathcal{E}_{\mathbf{0},\mathbf{I}}$ in N dimensions. Therefore a final equivalent definition of an elliptical distribution with location parameter $\boldsymbol{\mu}$ and scatter matrix $\boldsymbol{\Sigma}$ is the following:

$$\mathbf{X} \equiv \boldsymbol{\mu} + R\mathbf{A}\mathbf{U}. \tag{2.259}$$

In this expression

$$\mathbf{A}\mathbf{A}' \equiv \boldsymbol{\Sigma}, \quad R \equiv \left\|\mathbf{A}^{-1}(\mathbf{X}-\boldsymbol{\mu})\right\|, \quad \mathbf{U} \equiv \frac{\mathbf{A}^{-1}(\mathbf{X}-\boldsymbol{\mu})}{\left\|\mathbf{A}^{-1}(\mathbf{X}-\boldsymbol{\mu})\right\|}, \tag{2.260}$$

and $\mathbf{U}$ is uniformly distributed on the surface of the unit ball and is independent of R.

We show in Appendix www.2.18 that the generic elliptical probability density function must be of the form:

$$f_{\boldsymbol{\mu},\boldsymbol{\Sigma}}(\mathbf{x}) = |\boldsymbol{\Sigma}|^{-\frac{1}{2}} g_N\left(\mathrm{Ma}^2(\mathbf{x},\boldsymbol{\mu},\boldsymbol{\Sigma})\right), \tag{2.261}$$

where g_N is a non-negative univariate function that satisfies

$$\int_0^{\infty} v^{\frac{N}{2}-1} g_N(v)\, dv < \infty; \tag{2.262}$$

the parameter $\boldsymbol{\mu}$ is the center of the ellipsoid $\mathcal{E}_{\boldsymbol{\mu},\boldsymbol{\Sigma}}$; the parameter $\boldsymbol{\Sigma}$ is a symmetric and positive matrix that determines the shape of the ellipsoid $\mathcal{E}_{\boldsymbol{\mu},\boldsymbol{\Sigma}}$; and Ma is the Mahalanobis distance of the point $\mathbf{x}$ from $\boldsymbol{\mu}$ through the metric $\boldsymbol{\Sigma}$, as defined in (2.61).

For example, for the uniform distribution from (2.145) we obtain:

$$g_N^{\mathrm{U}}\left(\mathrm{Ma}^2\right) \equiv \frac{\Gamma\left(\frac{N}{2}+1\right)}{\pi^{\frac{N}{2}}} \mathbb{I}_{[0,1]}\left(\mathrm{Ma}^2\right). \tag{2.263}$$

For the normal distribution from (2.156) we obtain:

$$g_N^{\mathrm{N}}\left(\mathrm{Ma}^2\right) \equiv \frac{e^{-\frac{\mathrm{Ma}^2}{2}}}{(2\pi)^{\frac{N}{2}}}. \tag{2.264}$$

For the Student t distribution from (2.188) we obtain:

$$g_N^{\mathrm{St}}\left(\mathrm{Ma}^2\right) \equiv \frac{\Gamma\left(\frac{\nu+N}{2}\right)}{\Gamma\left(\frac{\nu}{2}\right)(\nu\pi)^{\frac{N}{2}}} \left(1+\frac{\mathrm{Ma}^2}{\nu}\right)^{-\frac{\nu+N}{2}}, \tag{2.265}$$

which also covers the Cauchy distribution as the special case $\nu \equiv 1$, see (2.209).
Therefore, all the above are elliptical distributions.

Equivalently, elliptical distributions can be represented in terms of their characteristic function. The generic elliptical characteristic function has the following form:

$$\phi_{\boldsymbol{\mu},\boldsymbol{\Sigma}}(\boldsymbol{\omega}) \equiv e^{i\boldsymbol{\omega}'\boldsymbol{\mu}} \psi\left(\boldsymbol{\omega}'\boldsymbol{\Sigma}\boldsymbol{\omega}\right), \tag{2.266}$$

where ψ is a suitable real-valued function.

For example, we see from (2.157) that for the normal distribution we have:

$$\psi^{\mathrm{N}}(\gamma) \equiv e^{-\frac{\gamma}{2}}. \tag{2.267}$$

The expression of ψ for the uniform distribution is given in (2.147). It is immediate to derive the expression of ψ for the Cauchy distribution from (2.210) and for the Student t distribution from (2.189).

Since an elliptical distribution is fully determined by the location parameter $\boldsymbol{\mu}$, the dispersion parameter $\boldsymbol{\Sigma}$ and the generator g of the probability density function (or equivalently the generator ψ of the characteristic function), we use the following notation to denote that a variable $\mathbf{X}$ is elliptically distributed with the above parameters:

$$\mathbf{X} \sim \mathrm{El}\left(\boldsymbol{\mu}, \boldsymbol{\Sigma}, g_N\right), \tag{2.268}$$

where we emphasized that the generator g depends on the dimension N of the random variable $\mathbf{X}$.

For example, the normal distribution is elliptical and thus from (2.264) the following notations are equivalent:

$$\mathrm{N}\left(\boldsymbol{\mu}, \boldsymbol{\Sigma}\right) \equiv \mathrm{El}\left(\boldsymbol{\mu}, \boldsymbol{\Sigma}, \frac{e^{-\frac{1}{2}\cdot}}{(2\pi)^{\frac{N}{2}}}\right). \tag{2.269}$$

Among the most remarkable properties of elliptical distributions we mention their behavior under affine transformations. Indeed, affine transformations of elliptically distributed random variables are elliptical distributed and the new location-dispersion parameters are easily computed in terms of the original ones.

More precisely, if $\mathbf{X}$ is an N-dimensional elliptical variable as in (2.268), then for any K-dimensional vector $\mathbf{a}$ and any $K \times N$ matrix $\mathbf{B}$ the following relation holds:

$$\mathbf{a} + \mathbf{B}\mathbf{X} \sim \mathrm{El}\left(\mathbf{a} + \mathbf{B}\boldsymbol{\mu}, \mathbf{B}\boldsymbol{\Sigma}\mathbf{B}', g_K\right). \tag{2.270}$$

Notice nonetheless that the new generator g_K has in general a very different functional form than the original generator g_N.

For example, consider a the bivariate uniform distribution on the unit circle. In the above notation, its distribution reads:

$$\left(X_1, X_2\right)' \sim \mathrm{El}\left(\mathbf{0}, \mathbf{I}_2, g_2\right), \tag{2.271}$$

where $\mathbf{I}$ is the identity matrix and from (2.150) the two-dimensional generator is defined in terms of the indicator function (B.72) as follows:

$$g_2\left(r^2\right) \equiv \frac{1}{\pi}\mathbb{I}_{[0,1]}\left(r^2\right). \tag{2.272}$$

Now consider the affine transformation determined by the following choice:

$$\mathbf{a} \equiv 0, \quad \mathbf{B} \equiv (1, 0). \tag{2.273}$$

The outcome of the transformation is the marginal distribution of the first variable X_1. From (2.270) we obtain:

$$X_1 \sim \mathrm{El}\left(0, 1, g_1\right), \tag{2.274}$$

where from (2.151) the one-dimensional generator reads:

$$g_1\left(r^2\right) \equiv \frac{2}{\pi}\sqrt{1-r^2}. \tag{2.275}$$

Therefore the marginal distribution of a uniform random variable is elliptical, but it is not uniform.

Further remarkable properties of the elliptical distributions regard the moments, when these are defined. As we show in Appendix www.2.18 the following relation holds:

$$\mathrm{E}\left\{\mathbf{X}\right\} = \boldsymbol{\mu}, \quad \mathrm{Cov}\left\{\mathbf{X}\right\} = \frac{\mathrm{E}\left\{R^2\right\}}{N}\boldsymbol{\Sigma}, \tag{2.276}$$

where R is defined in (2.260).

More in general, for the central moments of any order (2.92) we obtain:

$$\begin{aligned}\mathrm{CM}^{\mathbf{X}}_{m_1 \cdots m_k} &= \mathrm{E}\left\{R^k\right\} \\ &\sum_{n_1,\ldots,n_k=1}^{N} A_{m_1 n_1} \cdots A_{m_k n_k}\, \mathrm{E}\left\{U_{n_1} \cdots U_{n_k}\right\}.\end{aligned} \tag{2.277}$$

In this expression $\mathbf{A}$ and $\mathbf{U}$ are defined in (2.260). The moments of the uniform distribution on the surface of the unit ball are null if any variable appears an odd number of times; otherwise they read:

$$\mathrm{E}\left\{U_1^{2s_1} \cdots U_N^{2s_N}\right\} = \frac{\prod_{n=1}^{N} \frac{(2s_n)!}{4^{s_n} s_n!}}{\frac{N}{2}\left(\frac{N}{2}+1\right)\cdots\left(\frac{N}{2}+\left(\sum_{n=1}^{N} s_n\right)-1\right)}. \tag{2.278}$$

Since the copula of a distribution does not depend on purely marginal parameters such as the expected value and the standard deviation, for elliptical random variable the copula is fully determined by the correlations, see for instance the normal case (2.176). As a consequence, since the measures of concordance are defined in terms of the copula of a distribution, the measures of concordance between the entries of an elliptical random variable $\mathbf{X}$ are fully determined by the correlation matrix. For instance, Lindskog, McNeil, and Schmock (2003) prove that Kendall's tau (2.128) is the following function of correlation:

$$\tau\left\{X_m, X_n\right\} = \frac{2}{\pi}\arcsin\left(\mathrm{Cor}\left\{X_m, X_n\right\}\right), \tag{2.279}$$

which extends the result for the normal case (2.178) to generic elliptical variables.

2.7.2 Stable distributions

In view of our applications, *stable distributions* are analytically tractable distributions that can be projected to specific horizons in the future, see Figure 3.11. For more results on stable distributions see e.g. Embrechts, Klueppelberg, and Mikosch (1997) and references therein.

Consider three independent random variables $(\mathbf{X}, \mathbf{Y}, \mathbf{Z})$ with the same multivariate distribution. That distribution is stable if for any positive constants α and β there exist constants γ and δ such that the following holds:

$$\alpha\mathbf{X} + \beta\mathbf{Y} \stackrel{d}{=} \gamma + \delta\mathbf{Z}, \tag{2.280}$$

where "$\stackrel{d}{=}$" denotes "equal in distribution". In other words, the distribution is closed under linear combinations.

For example, assume that the three variables are independently normally distributed:

$$(\mathbf{X}, \mathbf{Y}, \mathbf{Z}) \sim \mathrm{N}(\boldsymbol{\mu}, \boldsymbol{\Sigma}). \tag{2.281}$$

Then from (2.163) we obtain:

$$\alpha\mathbf{X} + \beta\mathbf{Y} \sim \mathrm{N}\left((\alpha + \beta)\boldsymbol{\mu}, (\alpha^2 + \beta^2)\boldsymbol{\Sigma}\right).$$

Using (2.163) again we can verify that setting $\gamma \equiv \left(\alpha + \beta - \sqrt{\alpha^2 + \beta^2}\right)\boldsymbol{\mu}$ and $\delta \equiv \sqrt{\alpha^2 + \beta^2}$ the relation (2.280) is satisfied and thus the normal distribution is stable.

The Cauchy distribution (2.208) is stable. The lognormal distribution (2.217) is not stable, as the sum of lognormal variables is not lognormal. Similarly, for a generic number ν of degrees of freedom the Student t distribution (2.187) is not stable.

In view of our applications we are particularly interested in symmetric stable distributions, such as the normal distribution and the Cauchy distribution. Symmetric stable distributions are best represented in terms of their characteristic function. Indeed, a random variable $\mathbf{X}$ has a *symmetric stable distribution* if and only if its characteristic function has the following form:

$$\begin{aligned} \phi_{\mathbf{X}}(\boldsymbol{\omega}) &\equiv \mathrm{E}\left\{e^{i\boldsymbol{\omega}'\mathbf{X}}\right\} \\ &= e^{i\boldsymbol{\omega}'\boldsymbol{\mu}} \exp\left(-\int_{\mathbb{R}^N} |\boldsymbol{\omega}'\mathbf{s}|^{\alpha} m_{\boldsymbol{\Sigma}}(\mathbf{s})\, d\mathbf{s}\right). \end{aligned} \tag{2.282}$$

In this expression the parameter $\boldsymbol{\mu}$ is a location vector and the parameter α is a scalar that determines such features as the thickness of the tails of the distribution. The (generalized) function $m_{\boldsymbol{\Sigma}}$ defines a symmetric measure that is non-zero on the surface of the ellipsoid $\mathcal{E}_{\mathbf{0},\boldsymbol{\Sigma}}$ with shape parameter $\boldsymbol{\Sigma}$ centered in zero, see (A.73). In formulas:

$$m_{\boldsymbol{\Sigma}}(\mathbf{s}) = m_{\boldsymbol{\Sigma}}(-\mathbf{s}), \text{ for all } \mathbf{s} \in \mathbb{R}^N, \tag{2.283}$$

and

$$m_{\mathbf{\Sigma}}(\mathbf{s}) \equiv 0 \text{ for all } \mathbf{s} \text{ such that } \mathbf{s}'\mathbf{\Sigma}^{-1}\mathbf{s} \neq 1. \tag{2.284}$$

We use the following notation to indicate that $\mathbf{X}$ has a symmetric stable distribution with the above parameters:

$$\mathbf{X} \sim \mathrm{SS}(\alpha, \boldsymbol{\mu}, m_{\mathbf{\Sigma}}). \tag{2.285}$$

Symmetric stable distributions are also called *symmetric-alpha-stable* ($s\alpha s$) *distributions*.

For example, consider a normally distributed random variable:

$$\mathbf{X} \sim \mathrm{N}(\boldsymbol{\mu}, \mathbf{\Sigma}). \tag{2.286}$$

Consider the spectral decomposition ($A.70$) of the covariance matrix:

$$\mathbf{\Sigma} \equiv \mathbf{E}\mathbf{\Lambda}^{\frac{1}{2}}\mathbf{\Lambda}^{\frac{1}{2}}\mathbf{E}', \tag{2.287}$$

where $\mathbf{\Lambda}$ is the diagonal matrix of the eigenvalues of $\mathbf{S}$:

$$\mathbf{\Lambda} \equiv \mathrm{diag}(\lambda_1, \ldots, \lambda_N); \tag{2.288}$$

and $\mathbf{E}$ is the juxtaposition of the respective eigenvectors:

$$\mathbf{E} \equiv \left(\mathbf{e}^{(1)}, \ldots, \mathbf{e}^{(N)}\right). \tag{2.289}$$

Define N vectors $\{\mathbf{v}^{(1)}, \ldots \mathbf{v}^{(N)}\}$ as follows:

$$\left(\mathbf{v}^{(1)}, \ldots, \mathbf{v}^{(N)}\right) \equiv \mathbf{V} \equiv \mathbf{E}\mathbf{\Lambda}^{\frac{1}{2}}. \tag{2.290}$$

Define the following measure:

$$m_{\mathbf{\Sigma}} \equiv \frac{1}{4}\sum_{n=1}^{N}\left(\delta^{(\mathbf{v}_n)} + \delta^{(-\mathbf{v}_n)}\right), \tag{2.291}$$

where $\delta^{(\mathbf{x})}$ is the Dirac delta centered in $\mathbf{x}$ as defined in ($B.16$).

We prove in Appendix www.2.19 the following results. The measure $m_{\mathbf{\Sigma}}$ satisfies (2.283) and (2.284). In turn, the characteristic function (2.157) of the normal distribution can be written as follows:

$$\phi^{\mathrm{N}}_{\boldsymbol{\mu},\Sigma}(\boldsymbol{\omega}) = e^{i\boldsymbol{\mu}'\boldsymbol{\omega}} \exp\left(-\int_{\mathbb{R}^N} |\boldsymbol{\omega}'\mathbf{s}|^2 \, m_{\mathbf{\Sigma}}(\mathbf{s})\, d\mathbf{s}\right). \tag{2.292}$$

Therefore the following notations are equivalent:

$$\mathrm{N}(\boldsymbol{\mu}, \mathbf{\Sigma}) \equiv \mathrm{SS}\left(2, \boldsymbol{\mu}, \frac{1}{4}\sum_{n=1}^{N}\left(\delta^{(\mathbf{v}_n)} + \delta^{(-\mathbf{v}_n)}\right)\right). \tag{2.293}$$

We remark that one should not confuse stability with additivity: a distribution is *additive* if the sum of two variables with that distribution belongs to the same class of distributions. Indeed, stable distributions are additive, but the reverse implication is not true.

For example, consider three independent random matrices that are Wishart-distributed with the same scale factor:

$$(\mathbf{W}, \mathbf{S}, \boldsymbol{\Omega}) \sim \mathrm{W}(\nu, \boldsymbol{\Sigma}). \tag{2.294}$$

Then:

$$\mathbf{W} + \mathbf{S} \sim \mathrm{W}(2\nu, \boldsymbol{\Sigma}). \tag{2.295}$$

This follows easily from the definition (2.222) of the Wishart distribution. Therefore the Wishart distribution for a given scale parameter is additive. Nevertheless

$$\mathbf{W} + \mathbf{S} \overset{d}{\neq} \gamma + \delta\boldsymbol{\Omega}. \tag{2.296}$$

Therefore the Wishart distribution for a given scale parameter is not stable.

2.7.3 Infinitely divisible distributions

In view of our applications, *infinitely divisible* distributions can be projected to a generic investment horizon, see Figure 3.11, although the computation might not be straightforward.

More formally, the distribution of a random variable $\mathbf{X}$ is infinitely divisible if, for any integer T, the distribution of $\mathbf{X}$ is the same as the distribution of the sum of T suitably chosen independent and identically distributed random variables:

$$\mathbf{X} \overset{d}{=} \mathbf{Y}_1 + \cdots + \mathbf{Y}_T. \tag{2.297}$$

For example, assume that $\mathbf{X}$ is normally distributed:

$$\mathbf{X} \sim \mathrm{N}(\boldsymbol{\mu}, \boldsymbol{\Sigma}). \tag{2.298}$$

For an arbitrary integer T consider the following set of independent and identically distributed normal random variables:

$$\mathbf{Y}_t \sim \mathrm{N}\left(\frac{\boldsymbol{\mu}}{T}, \frac{\boldsymbol{\Sigma}}{T}\right). \tag{2.299}$$

It is immediate to check that these variables satisfy (2.297). Therefore the normal distribution is stable.

Unlike in the normal case, although for a given T the distribution of all the terms $\mathbf{Y}_\nu$ in (2.297) is the same, this distribution need not be the same for all values of T.

For instance, the lognormal distribution is infinitely divisible, see Thorin (1977). Nevertheless, unlike in the normal case (2.299), the distribution of the divisors depends on T.

Many common distributional models are infinitely divisible. For instance, the elliptical distributions discussed in this book are infinitely divisible.

On the other hand, not all distributions are infinitely divisible. For example, the Wishart distribution is not infinitely divisible, except in the univariate case. Indeed the gamma distribution, which is the one-dimensional Wishart distribution, is infinitely divisible, see Cuppens (1975).

3

Modeling the market

In this chapter we model the market. The definition of a market depends on the investor, who focuses on a specific pool of assets.

For a trader of Eurodollar futures the market are the "reds", "greens", "blues" and "golds" (nicknames of the contracts that expire after one, two, three and four years, respectively). For a retiree, the market is a set of mutual funds.

Furthermore, in general the investor has a specific *investment horizon*.

A day-trader aims at cashing profits within a few hours from the investment decision. A retiree has an investment horizon of the order of a few years.

Therefore the market for an investor is represented by a set of N securities and an investment horizon τ. These securities can be any tradable asset: bonds, commodities, mutual funds, currencies, etc. We denote the value, or the price, at the generic time t of the securities in the market by the N-dimensional vector $\mathbf{P}_t$. We denote as T the time when the allocation decision is made.

In view of making the best possible asset allocation decision, the investor is interested in modeling the value of the securities in his market at his investment horizon. The prices at the investment horizon $\mathbf{P}_{T+\tau}$ are a multivariate random variable: therefore modeling the market means determining the distribution of $\mathbf{P}_{T+\tau}$.

In a stochastic environment apparently almost any distribution seems suitable to describe the market. If something unexpected happens, one might always blame the non-zero probability of that specific event. Nevertheless, a rational approach should link the market model, i.e. the distribution of the prices at the investment horizon, with the observations, i.e. the past realizations of some market observables.

The bridge between past and future consists of four conceptual building blocks.

1. Detecting the invariants

The market displays some phenomena that repeat themselves identically throughout history: we call these phenomena *invariants*. The first step consists in detecting the invariants, i.e. the market variables that can be modeled as the realization of a set of *independent and identically distributed* random variables.

For example the weekly returns are invariants for the stock market.

We tackle in Section 3.1 the search for the invariants in different markets: equities, commodities, foreign exchange, fixed-income securities and derivatives.

2. Determining the distribution of the invariants

Due to the repetitive behavior of the market invariants, it is possible by means of statistical procedures to infer their distribution.

For example in the stock market the stochastic behavior of the weekly returns can be modeled by a multivariate Student t distribution.

Fitting a distribution to the empirical observations of the invariants is a very broad subject. We devote Chapter 4 to this problem.

3. Projecting the invariants into the future

The estimated distribution of the invariants refers to a specific estimation interval. This distribution needs to be projected to the generic investment horizon τ that is relevant to the investor.

For example from the distribution of weekly returns we need to compute the distribution of monthly returns.

We discuss in Section 3.2 how to determine the distribution of the invariants that refer to the generic investment horizon: it turns out that the projection formula is easily handled in terms of the characteristic function of the invariants.

4. Mapping the invariants into the market prices

Since the invariants are not the market prices, we need to translate the distribution of the invariants into the distribution of the prices of the securities in the market at the investment horizon.

For example from the distribution of monthly returns we need to compute the distribution of the stock prices one month in the future.

We discuss this point in Section 3.3. We also present a shortcut to compute all the moments of the distribution of the market prices directly in terms of the characteristic function of the invariants. This shortcut is particularly convenient in the context of mean-variance optimization.

The above steps allow us to model the market when the total number of securities is limited. In practical applications, the number of securities involved in asset allocation problems is typically large. In these cases the actual dimension of randomness in the market is much lower than the number of securities. In Section 3.4 we discuss the main dimension-reduction techniques: explicit-factor approaches, such as regression analysis, and hidden-factor approaches, such as principal component analysis and idiosyncratic factors. To support intuition we stress the geometric interpretation of these approaches in terms of the location-dispersion ellipsoid. Finally we present a useful routine to perform dimension reduction in practice in a variety of contexts, including portfolio replication.

To conclude, in Section 3.5 we present a non-trivial implementation of all the above steps in the swap market. By setting the problem in the continuum we provide a frequency-based interpretation of the classical "level-slope-hump" principal component factorization. From this we compute the distribution of the swap prices exactly and by means of the duration-convexity approximation.

To summarize, in this chapter we detect the market invariants, we project their distribution to a generic horizon in the future and we translate this projection into the distribution of the market prices at the investment horizon, possibly after reducing the dimension of the market.

In the above analysis we take for granted the distribution of the invariants at a fixed estimation interval. In reality, this distribution can only be estimated with some approximation, as discussed in Chapter 4. We tackle the many dangers of estimation risk in the third part of the book.

3.1 The quest for invariance

In this section we show how to process the information available in the market to determine the market invariants.

In order to do so, we need a more precise definition of the concept of invariant. Consider a starting point $\widetilde{t}$ and a time interval $\widetilde{\tau}$, which we call the *estimation interval*. Consider the set of equally-spaced dates:

$$\mathcal{D}_{\widetilde{t},\widetilde{\tau}} \equiv \{\widetilde{t}, \widetilde{t}+\widetilde{\tau}, \widetilde{t}+2\widetilde{\tau}, \ldots\}. \tag{3.1}$$

Consider a set of random variables:

$$X_t, \quad t \in \mathcal{D}_{\widetilde{t},\widetilde{\tau}}. \tag{3.2}$$

The random variables X_t are *market invariants* for the starting point $\widetilde{t}$ and the estimation interval $\widetilde{\tau}$ if they are *independent and identically distributed* and if the realization x_t of X_t becomes available at time t.

For example, assume that the estimation interval $\widetilde{\tau}$ is one week and the starting point $\widetilde{t}$ is the first Wednesday after January 1st 2000. In this case $\mathcal{D}_{\widetilde{t},\widetilde{\tau}}$ is the set of all Wednesdays since January 1st 2000. Consider flipping a fair coin once every Wednesday since January 1st 2000. One outcome is independent of the other, they are identically distributed (50% head, 50% tail), and the result of each outcome becomes available immediately. Therefore, the outcomes of our coin-flipping game are invariants for the starting point "first Wednesday after January 1st 2000", and a weekly estimation interval.

A *time homogenous invariant* is an invariant whose distribution does not depend on the reference time $\widetilde{t}$. In our quest for invariance, we will always look for time-homogeneous invariants.

In the previous example, it does not matter whether the coins are flipped each Wednesday or each Thursday. Thus the outcomes of the coin-flipping game are time-homogeneous invariants.

To detect invariance, we look into the time series of the financial data available. The *time series* of a generic set of random variables is the set of past realizations of those random variables. Denoting as T the current time, the time series is the set

$$x_t, \quad t = \widetilde{t}, \widetilde{t} + \widetilde{\tau}, \ldots, T, \tag{3.3}$$

where the lower case notation indicates that x_t is the specific realization of the random variable X_t occurred at time t in the past.

For example the time series in the coin-flipping game is the record of heads and tails flipped since the first Wednesday after January 1st 2000 until last Wednesday.

In order to detect invariance, we perform two simple graphical tests.

The first test consists in splitting the time series (3.3) into two series:

$$x_t, \quad t = \widetilde{t}, \ldots, \widetilde{t} + [\frac{T - \widetilde{t}}{2\widetilde{\tau}}]\widetilde{\tau} \tag{3.4}$$

$$x_t, \quad t = \left([\frac{T - \widetilde{t}}{2\widetilde{\tau}}] + 1 \right) \widetilde{\tau}, \ldots, T, \tag{3.5}$$

where $[\cdot]$ denotes the integer part. Then we compare the respective histograms. If X_t is an invariant, in particular all the terms in the series are identically distributed: therefore the two histograms should look very similar to each other.

The second test consists of the scatter-plot of the time series (3.3) on one axis against its lagged values on the other axis. In other words, we compare the following two series:

$$x_t \text{ versus } x_{t-\widetilde{\tau}}, \qquad t = \widetilde{t} + \widetilde{\tau}, \ldots, T. \tag{3.6}$$

If X_t is an invariant, in particular all the terms in the series are independent of each other: therefore the scatter plot must be symmetrical with respect to the reference axes. Furthermore, since all the terms are identically distributed, the scatter plot must resemble a circular cloud.

These tests are sufficient to support our arguments. For more on this subject, see e.g. Hamilton (1994), Campbell, Lo, and MacKinlay (1997), Lo and MacKinlay (2002).

3.1.1 Equities, commodities, exchange rates

In this section we pursue the quest for invariance in the stock market. Nevertheless the present discussion applies to other tradable assets, such as commodities and currency exchange rates.

We make the standard assumption that the securities do not yield any cash-flow. This does not affect the generality of the discussion: it is always possible to assume that cash-flows such as dividends are immediately re-invested in the same security.

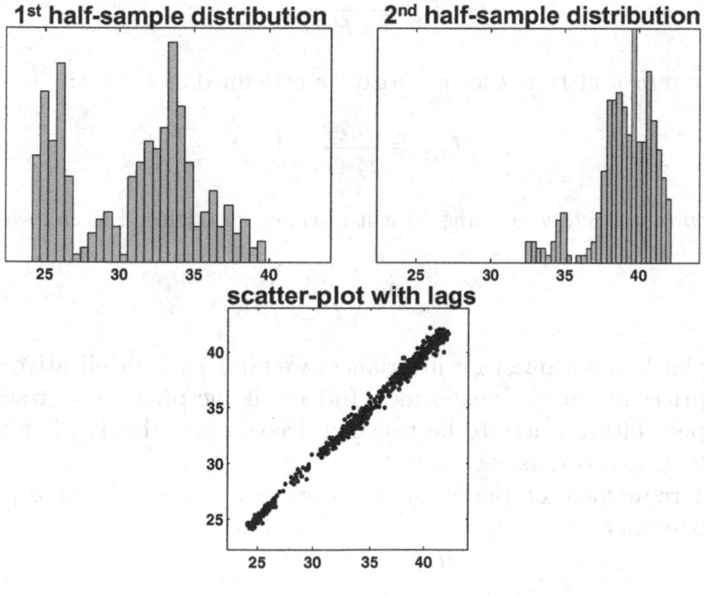

Fig. 3.1. Stock prices are not market invariants

Consider one stock. We assume that we know the stock price at all past times. The first question is whether the price can be considered a market invariant. To ascertain this, we fix an estimation interval $\widetilde{\tau}$ (e.g. one week) and a starting point $\widetilde{t}$ (e.g. five years ago) and we consider the set of stock prices at the equally spaced estimation times (3.1):

$$P_t, \quad t \in \mathcal{D}_{\widetilde{t},\widetilde{\tau}}. \tag{3.7}$$

Each of these random variables becomes available at the respective time t. To see if they are independent and identically distributed we analyze the time series of their realization up to the investment decision time:

$$p_t, \quad t = \widetilde{t}, \widetilde{t} + \widetilde{\tau}, \ldots, T. \tag{3.8}$$

If the stock price were an invariant, the histogram of the first half of the time series would be similar to the histogram of the second half of the time series. Furthermore, the scatter-plot of the price series with its lagged values would resemble a circular cloud. In Figure 3.1 we see that this is not the case: stock prices are not market invariants.

Before we continue, we need to introduce some terminology. The *total return* at time t for a horizon τ on any asset (equity, fixed income, etc.) that trades at the price P_t at the generic time t is defined as the following multiplicative factor between two subsequent prices:

$$H_{t,\tau} \equiv \frac{P_t}{P_{t-\tau}}. \tag{3.9}$$

The *linear return* at time t for a horizon τ is defined as follows:

$$L_{t,\tau} \equiv \frac{P_t}{P_{t-\tau}} - 1. \tag{3.10}$$

The *compounded return* at time t for a horizon τ is defined as follows:

$$C_{t,\tau} \equiv \ln\left(\frac{P_t}{P_{t-\tau}}\right). \tag{3.11}$$

Going back to our quest for invariance, we notice a multiplicative relation between prices at two different times. Indeed, if the prices were rescaled we would expect future prices to be rescaled accordingly: this is what happens when a stock split occurs.

Therefore we focus on the set of non-overlapping total returns as potential market invariants:

$$H_{t,\widetilde{\tau}}, \quad t \in \mathcal{D}_{\widetilde{t},\widetilde{\tau}}. \tag{3.12}$$

Each of these random variables becomes available at the respective time t. To see if they are independent and identically distributed we perform the tests described in the introduction to Section 3.1 on the time series of the past observations of the non-overlapping total returns:

$$h_{t,\widetilde{\tau}}, \quad t = \widetilde{t}, \widetilde{t} + \widetilde{\tau}, \ldots, T. \tag{3.13}$$

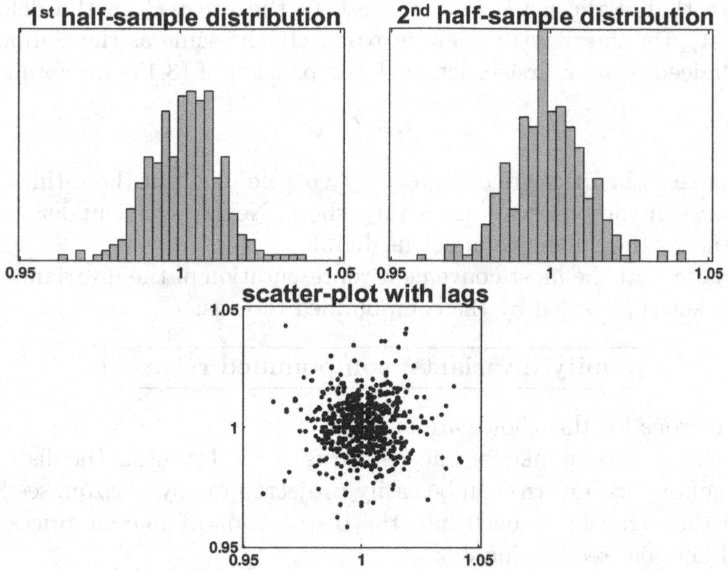

Fig. 3.2. Stock returns are market invariants

First we split the series (3.13) in two halves and plot the histogram of each half. If all the $H_{t,\widetilde{\tau}}$ are identically distributed, the histogram from the first sample of the series must resemble the histogram from the second sample. In Figure 3.2 we see that this is the case.

Then we move on to the second test: we scatter-plot the time series of the total returns against the same time series lagged by one estimation interval. If $H_{t,\widetilde{\tau}}$ is independent of $H_{t+\widetilde{\tau},\widetilde{\tau}}$ and they are identically distributed, the scatter plot must resemble a circular cloud. In Figure 3.2 we see that this is indeed the case.

Therefore we accept the set of non-overlapping total returns as invariants for the equity market. More in general, any function g of the total returns defines new invariants for the equity market:

$$g\left(H_{t,\widetilde{\tau}}\right), \quad t \in \mathcal{D}_{\widetilde{t},\widetilde{\tau}}. \tag{3.14}$$

Indeed, if the set of $H_{t,\widetilde{\tau}}$ are independent and identically distributed random variables that become known at time t, so are the variables (3.14).

In particular, the linear returns (3.10) and the compounded returns (3.11) are functions of the total returns, as well as of one another:

$$L = e^{C} - 1 = H - 1, \quad C = \ln\left(1 + L\right) = \ln\left(H\right). \tag{3.15}$$

Therefore, both linear returns and compounded returns are invariants for the stock market.

Notice that if the price $P_{t-\tau}$ is close to the price P_t in the definitions (3.9)-(3.11), the linear return is approximately the same as the compounded return. Indeed, from a first-order Taylor expansion of (3.15) we obtain:

$$L \approx C. \tag{3.16}$$

This happens when the price is not very volatile or when the estimation interval between the observations is very short. Nevertheless, under standard circumstances the difference is not negligible.

We claim that the most convenient representation of the invariants for the stock market is provided by the compounded returns:

$$\boxed{\textbf{equity invariants: compounded returns}} \tag{3.17}$$

The reasons for this choice are twofold.

In the first place, unlike for linear returns or total returns, the distribution of the compounded returns can be easily projected to any horizon, see Section 3.2, and then translated back into the distribution of market prices at the specified horizon, see Section 3.3.

Secondly, the distribution of either linear returns or total returns is not symmetrical: for example we see from (3.9) that total returns cannot be negative, whereas their range is unbounded from above. Instead, compounded returns have an approximately symmetrical distribution. This makes it easier to model the distribution of the compounded returns.

For example, from the time series analysis of the stock prices over a weekly estimation interval $\widetilde{\tau}$ we derive that the distribution of the compounded returns (3.11) on a given stock can be fitted to a normal distribution:

$$C_{t,\widetilde{\tau}} \equiv \ln\left(\frac{P_t}{P_{t-\widetilde{\tau}}}\right) \sim \mathrm{N}\left(\mu, \sigma^2\right). \tag{3.18}$$

Notice that (3.18) is the benchmark assumption in continuous-time finance and economics, see Black and Scholes (1973) and Merton (1992). Measuring time in years we obtain

$$\widetilde{\tau} \equiv \frac{1}{52} \tag{3.19}$$

and, say,

$$\mu \equiv 9.6 \times 10^{-2}, \quad \sigma^2 \equiv 7.7 \times 10^{-4}. \tag{3.20}$$

The distribution of the original invariants, i.e. the total returns (3.12), is lognormal with the same parameters:

$$H_{t,\widetilde{\tau}} \equiv \frac{P_t}{P_{t-\widetilde{\tau}}} \sim \mathrm{LogN}\left(\mu, \sigma^2\right). \tag{3.21}$$

This distribution is not as analytically tractable as (3.18).

The symmetry of the compounded returns becomes especially important in a multivariate setting, where we can model the joint distribution of these invariants with flexible, yet parsimonious, parametric models that are analytically tractable. For instance, we can model the compounded returns of a set of stocks as members of the class of elliptical distributions:

$$\mathbf{X}_{t,\widetilde{\tau}} \equiv \mathbf{C}_{t,\widetilde{\tau}} \sim \mathrm{El}\left(\boldsymbol{\mu}, \boldsymbol{\Sigma}, g\right), \tag{3.22}$$

for suitable choices of the location parameter $\boldsymbol{\mu}$, the scatter parameter $\boldsymbol{\Sigma}$ and the probability density generator g, see (2.268). Alternatively, we can model the compounded returns of a set of stocks as members of the class of symmetric stable distributions:

$$\mathbf{X}_{t,\widetilde{\tau}} \equiv \mathbf{C}_{t,\widetilde{\tau}} \sim \mathrm{SS}\left(\alpha, \boldsymbol{\mu}, m_{\boldsymbol{\Sigma}}\right), \tag{3.23}$$

for suitable choices of the tail parameter α, the location parameter $\boldsymbol{\mu}$, the scatter parameter $\boldsymbol{\Sigma}$ and the measure m, see (2.285).

We mention that in a multivariate context it is not unusual to detect certain functions of the returns, such as linear combinations, which are not independent across time. This gives rise to the phenomenon of *cointegration*, which has been exploited by practitioners to try to predict the market movements of certain portfolios. For instance, trading strategies such as *equity pairs* are based on cointegration, see e.g. Alexander and Dimitriu (2002). A discussion of this subject is beyond the scope of the book and the interested reader should consult references such as Hamilton (1994).

3.1.2 Fixed-income market

In this section we pursue the quest for invariance in the fixed-income market. Without loss of generality, we focus on zero-coupon bonds, which are the building blocks of the whole fixed-income market.

A zero-coupon bond is a fixed-term loan: a certain amount of money $Z_t^{(E)}$ is turned in at the generic time t and a (larger) determined amount is received back at a later, specified *maturity date* E. Since the amount to be received is determined, we can normalize it as follows without loss of generality:

$$Z_E^{(E)} \equiv 1. \tag{3.24}$$

As in the equity market, the first question is whether bond prices can be market invariants. In other words, we fix an estimation interval $\widetilde{\tau}$ (e.g. one week) and a starting point $\widetilde{t}$ (e.g. five years ago) and we consider the set of bond prices:

$$Z_t^{(E)}, \quad t \in \mathcal{D}_{\widetilde{t},\widetilde{\tau}}, \tag{3.25}$$

where the set of equally spaced estimation intervals is defined in (3.1). Each of these random variables becomes available at the respective time t. Nevertheless, the constraint (3.24) affects the evolution of the price: as we see in Figure

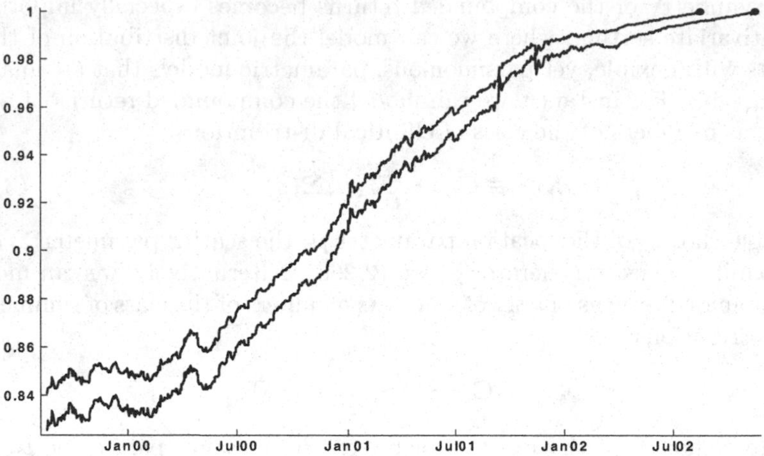

Fig. 3.3. Lack of time-homogeneity of bond prices

3.3 the time series of a bond price $Z_t^{(E)}$ converges to the redemption value, as the maturity approaches. Therefore bond prices cannot be market invariants, because the convergence to the redemption value at maturity breaks the time homogeneity of the set of variables (3.25).

As a second attempt, we notice that, like in the equity market, there exists a multiplicative relation between the prices at two different times. Therefore, we are led to consider the set of non-overlapping total returns on the generic bond whose time *of* maturity is E:

$$H_{t,\widetilde{\tau}}^{(E)} \equiv \frac{Z_t^{(E)}}{Z_{t-\widetilde{\tau}}^{(E)}}, \quad t \in \mathcal{D}_{\widetilde{t},\widetilde{\tau}}. \tag{3.26}$$

Each of these random variables becomes available at the respective time t. Nevertheless, the total returns cannot be invariants, because the convergence to the redemption value of the prices also breaks the time homogeneity of the set of variables (3.26).

To find an invariant, we must formulate the problem in a time-homogenous framework by eliminating the redemption date. Suppose that there exists a zero-coupon bond for all possible maturities. We can compare the price $Z_t^{(E)}$ of the bond we are interested in with the price $Z_{t-\widetilde{\tau}}^{(E-\widetilde{\tau})}$ of another bond that expires at a date which is equally far in the future, i.e. with the same time *to* maturity. This series is time-homogeneous, as we see in Figure 3.4, where we plot the price of the bond that at each point of the time series expires five years in the future.

Therefore, we consider the set of non-overlapping "total returns" on bond prices with the same time v *to* maturity:

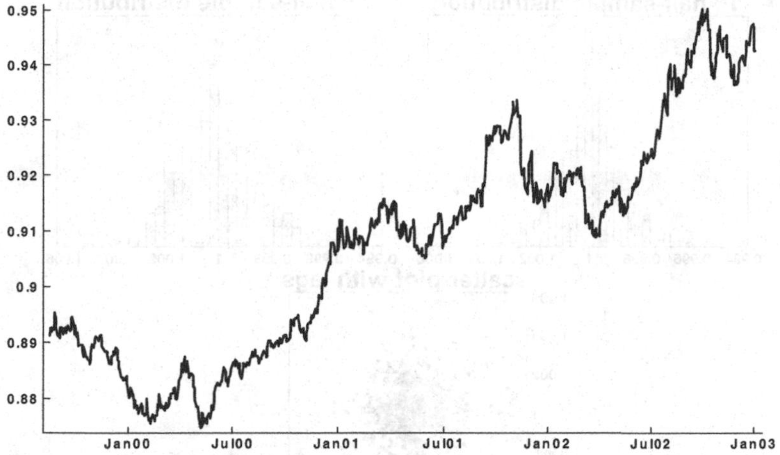

Fig. 3.4. Time-homogeneity of bond prices with fixed time to maturtity

$$R_{t,\widetilde{\tau}}^{(v)} \equiv \frac{Z_t^{(t+v)}}{Z_{t-\widetilde{\tau}}^{(t+v-\widetilde{\tau})}}, \quad t \in \mathcal{D}_{\widetilde{t},\widetilde{\tau}}. \tag{3.27}$$

Notice that these variables do not depend on the fixed expiry E and thus they are time-homogeneous. We stress that these "total returns to maturity" do not represent real returns on a security, since they are the ratio of the prices of two different securities.

Each of the random variables in (3.27) becomes available at the respective time t. To see if they qualify as invariants for the fixed-income market, we perform the two simple tests discussed in the introduction to Section 3.1 on the time series of the past realizations of these random variables:

$$r_{t,\widetilde{\tau}}^{(v)}, \quad t = \widetilde{t}, \widetilde{t} + \widetilde{\tau}, \ldots, T. \tag{3.28}$$

First we split the series (3.28) in two halves and plot the histogram of each half. If all the $R_{t,\widetilde{\tau}}^{(v)}$ are identically distributed, the histogram from the first sample of the series must resemble the histogram from the second sample. In Figure 3.5 we see that this is the case.

Then we move on to the second test: we scatter-plot the time series (3.28) against the same time series lagged by one estimation interval. If each $R_{t,\widetilde{\tau}}^{(v)}$ is independent of $R_{t+\widetilde{\tau},\widetilde{\tau}}^{(v)}$ and they are identically distributed, the scatter plot must resemble a circular cloud. In Figure 3.5 we see that this is indeed the case.

Therefore we accept (3.27) as invariants for the fixed-income market. More in general, any function g of R defines new invariants for the fixed-income market:

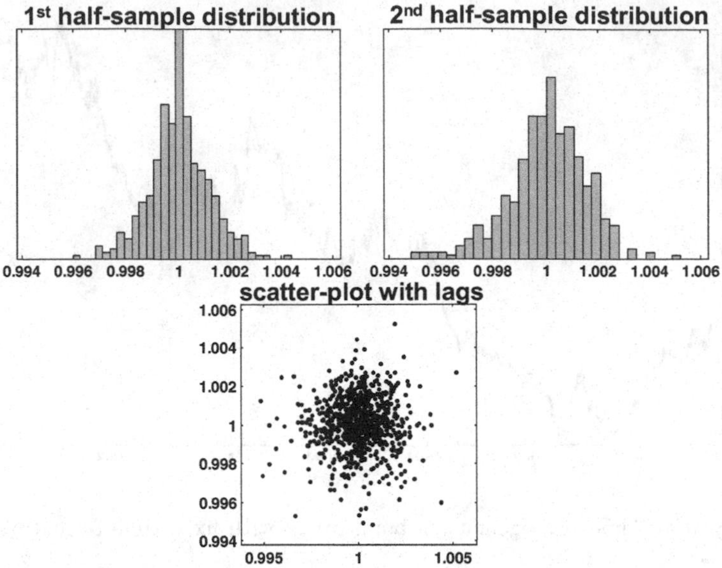

Fig. 3.5. Fixed-income market invariants

$$g\left(R_{t,\tilde{\tau}}^{(v)}\right), \quad t \in \mathcal{D}_{\tilde{t},\tilde{\tau}}. \tag{3.29}$$

Indeed, also (3.29) are independent and identically distributed random variables that become known at time t.

To determine the most convenient representation of the market invariants, i.e. the best function g in (3.29), we need some terminology. Consider a generic time t and a zero-coupon bond that expires at time $t+v$ and thus trades at the price $Z_t^{(t+v)}$. The *yield to maturity* v of this bond is defined as follows:

$$Y_t^{(v)} \equiv -\frac{1}{v}\ln\left(Z_t^{(t+v)}\right). \tag{3.30}$$

The graph of the yield to maturity as a function of the maturity is called the *yield curve*. A comparison of (3.30) with (3.11) shows that the yield to maturity times the time to maturity is the compounded return of a zero-coupon bond over a horizon equal to its entire life. In particular if, as it is customary in the fixed-income world, time is measured in years, then the yield to maturity can be interpreted as the annualized return of the bond.

It is easy to relate the fixed-income invariant (3.27) to the yield to maturity (3.30). Consider the changes in yield to maturity:

$$X_{t,\tilde{\tau}}^{(v)} \equiv Y_t^{(v)} - Y_{t-\tilde{\tau}}^{(v)} = -\frac{1}{v}\ln\left(R_{t,\tilde{\tau}}^{(v)}\right). \tag{3.31}$$

Since R is an invariant, so is X.

Notice that the changes in yield to maturity do not refer to a specific bond, as each invariant (3.31) is defined in terms of two bonds with different maturities. Instead, each invariant is specific to a given sector v of the yield curve.

We claim that the most convenient representation of the invariants for the fixed-income market is provided by the changes in yield to maturity:

$$\boxed{\textbf{fixed-income invariants: changes in yield to maturity}} \tag{3.32}$$

The reasons for this choice are two-fold.

In the first place, unlike the original invariants (3.27), the distribution of changes in yield to maturity can be easily projected to any horizon, see Section 3.2, and then translated back into the distribution of bond prices at the specified horizon, see Section 3.3.

Secondly, the distribution of the original invariants (3.27) is not symmetrical: for example those invariants cannot be negative. Instead, the distribution of the changes in yield to maturity is symmetrical.[1]. This makes it easier to model the distribution of the changes in yield to maturity.

For example from weekly time series analysis we derive that the distribution of the changes in yield to maturity (3.31) for the three-year sector of the bond market can be fitted to a normal distribution:

$$X_{t,\widetilde{\tau}}^{(v)} \equiv Y_t^{(v)} - Y_{t-\widetilde{\tau}}^{(v)} \sim \mathrm{N}\left(\mu, \sigma^2\right). \tag{3.33}$$

Measuring time in years we have

$$\widetilde{\tau} \equiv \frac{1}{52}, \quad v \equiv 3 \tag{3.34}$$

and, say,

$$\mu \equiv 0, \quad \sigma^2 \equiv 2 \times 10^{-5}. \tag{3.35}$$

The distribution of the original invariants (3.27) is lognormal with the following parameters:

$$R_{t,\widetilde{\tau}}^{(v)} = e^{-vX_{t,\widetilde{\tau}}^{(v)}} \sim \mathrm{LogN}\left(-v\mu, v^2\sigma^2\right). \tag{3.36}$$

This distribution is not as analytically tractable as (3.33).

---

[1] Apparently, this is not correct. The bond is a loan: as such the money lent cannot exceed the money returned when the loan expires, which prevents the yield to maturity from being negative. Therefore the change in yield to maturity must satisfy the constraint $X_t \geq -Y_{t-\widetilde{\tau}}$. We can bypass this problem by considering as invariant the changes in the "shadow yield" S, a variable that can take any value and such that $Y_t^{(v)} \equiv \max\left(S_t^{(v)}, 0\right)$, see Black (1995).

The symmetry of the changes in yield to maturity becomes especially important in a multivariate setting, where we can model the joint distribution of the changes in yield to maturity, together with other symmetric invariants such as the compounded returns for the stock market, by means of flexible, yet parsimonious, parametric models that are analytically tractable. For instance, we can model these invariants as members of the class of elliptical distributions:

$$\mathbf{X}_{t,\widetilde{\tau}} \sim \mathrm{El}\left(\boldsymbol{\mu}, \boldsymbol{\Sigma}, g\right), \tag{3.37}$$

for suitable choices of the location parameter $\boldsymbol{\mu}$, the scatter parameter $\boldsymbol{\Sigma}$ and the probability density generator g, see (2.268). Alternatively, we can model the changes in yield to maturity of a set of bonds, together with other symmetrical invariants, as members of the class of symmetric stable distributions:

$$\mathbf{X}_{t,\widetilde{\tau}} \sim \mathrm{SS}\left(\alpha, \boldsymbol{\mu}, m_{\boldsymbol{\Sigma}}\right), \tag{3.38}$$

for suitable choices of the tail parameter α, the location parameter $\boldsymbol{\mu}$, the scatter parameter $\boldsymbol{\Sigma}$ and the measure m, see (2.285).

We mention that in a multivariate context it is not unusual to detect certain functions of the changes in yield to maturity, such as linear combinations, which are not independent across time. This gives rise to the phenomenon of *cointegration*, see e.g. Anderson, Granger, and Hall (1990) and Stock and Watson (1988). This phenomenon has been exploited by practitioners. For instance, cointegration is the foundation of a trading strategy known as *PCA trading*. A discussion of this subject is beyond the scope of the book.

3.1.3 Derivatives

In this section we pursue the quest for invariance in the derivatives market, see Wilmott (1998) and Hull (2002) for more on this subject. Although our approach is as general as possible, this market is very heterogeneous, and therefore each case must be analyzed independently.

Although "raw" securities such as stocks and zero-coupon bonds constitute the building blocks of the market, there exist financial products that cannot be analyzed in terms of the building blocks only: the *derivatives* of the raw securities.

There exist several kinds of derivatives, but the most liquid derivatives are the *vanilla European options*, tradable products defined and priced as functions of the price of one or more underlying raw securities and/or some extra market variables. In other words, a vanilla European derivative is a security whose price D_t at the generic time t can be expressed as follows:

$$D_t = h\left(\mathbf{V}_t\right), \tag{3.39}$$

where h is a specific pricing function that might depend on a set of parameters and $\mathbf{V}_t$ is the price at time t of a set of market variables.

The most liquid vanilla European options are the call option and put option.

A *European call option* with *strike* K and *expiry date* E on an *underlying* whose price at the generic time t we denote as U_t is a security whose price at time $t \leq E$ reads[2]:

$$C_t^{(K,E)} \equiv C^{BS}\left(E-t, K, U_t, Z_t^{(E)}, \sigma_t^{(K,E)}\right). \tag{3.40}$$

In this expression $Z_t^{(E)}$ is the price at time t of a zero-coupon bond that matures at time E; and $\sigma_t^{(K,E)}$ is called the *implied percentage volatility* at time t of the underlying U relative to the strike K and to the expiry E. The implied volatility is a new market variable which we discuss further below.

The function C^{BS} in (3.40) is the pricing formula of Black and Scholes (1973). The Black-Scholes formula can be expressed in terms of the error function (B.75) as follows:

$$\begin{aligned} C^{BS}(\tau, K, U, Z, \sigma) \equiv & \frac{1}{2} U\left(1+\operatorname{erf}\left(\frac{d_1}{\sqrt{2}}\right)\right) \\ & -\frac{1}{2} Z K\left(1+\operatorname{erf}\left(\frac{d_2}{\sqrt{2}}\right)\right), \end{aligned} \tag{3.41}$$

where the two ancillary variables (d_1, d_2) are defined as follows:

$$d_1 \equiv \frac{1}{\sigma\sqrt{\tau}}\left\{\ln\left(\frac{U}{ZK}\right)+\frac{\sigma^2 \tau}{2}\right\} \tag{3.42}$$

$$d_2 \equiv d_1 - \sigma\sqrt{\tau}. \tag{3.43}$$

The call option price (3.40) is of the form (3.39), where the market variables are the price of the underlying, the zero-coupon bond price and the implied percentage volatility:

$$\mathbf{V}_t \equiv \left(U_t, Z_t^{(E)}, \sigma_t^{(K,E)}\right)'. \tag{3.44}$$

The *payoff* of an option is its value at expiry. The payoff of the call option only depends on the underlying, as (3.40) reduces at expiry to the following simpler function:

$$C_E^{(K,E)} = \max\left(U_E - K, 0\right). \tag{3.45}$$

[2] We introduce the value of the call option (3.40) from a trader's perspective, according to which the implied volatility is an exogenous market variable. The standard textbook approach first models the "right" process for the underlying U and then derives the "right" pricing formula from non-arbitrage arguments. Formula (3.40) is a specific instance of the textbook approach first developed in Black and Scholes (1973), where the process for the underlying is assumed lognormal. In this approach σ is the constant percentage volatility of the underlying.

A *European put option* with strike K and expiry E on an underlying whose price at the generic time t we denote as U_t is a security whose price at time $t \leq E$ reads:

$$P_t^{(K,E)} = C^{BS}\left(E - t, K, U_t, Z_t^{(E)}, \sigma_t^{(K,E)}\right) - U_t + Z_t^{(E)} K, \tag{3.46}$$

where C^{BS} is the Black-Scholes pricing function (3.40) of the call option with the same strike and expiry. The pricing relation (3.46) is called *put-call parity*. Since the call price is of the form (3.39), so is the put price (3.46), for the same market variables (3.44).

Similarly to the call option, the payoff of the put option only depends on the underlying, as (3.46) reduces at expiry to the following simpler function:

$$P_E^{(K,E)} = -\min\left(U_E - K, 0\right). \tag{3.47}$$

We can now proceed in our quest for invariance in the derivatives market. We have already detected in Sections 3.1.1 and 3.1.2 the invariants behind two among the three market variables (3.44) involved in pricing derivatives, namely the bond Z and the underlying U, whether this is a commodity, a foreign exchange rate, a stock, or a fixed-income security.

Therefore, in order to complete the study of the invariance in the derivatives market, we have to analyze the invariance behind the implied percentage volatility σ of the underlying. There exist several studies in the financial literature regarding the evolution of the implied volatility in the so-called risk neutral measure, a synthetic environment that allows to compute no-arbitrage prices for securities, see e.g. Schoenbucher (1999), Amerio, Fusai, and Vulcano (2002), Brace, Goldys, Van der Hoek, and Womersley (2002). In our case we are interested in the econometric study of the patterns of the implied volatility, see also Fengler, Haerdle, and Schmidt (2003).

In particular, we consider the *at-the-money-forward* (ATMF) implied percentage volatility of the underlying, which is the implied percentage volatility of an option whose strike is equal to the *forward price* of the underlying at expiry:

$$K_t \equiv \frac{U_t}{Z_t^{(E)}}. \tag{3.48}$$

We focus on the ATMF volatility because ATMF options are the most liquid.

As in the other markets, we first consider whether the ATMF volatility is itself a market invariant. In other words, we fix an estimation interval $\widetilde{\tau}$ (e.g. one week) and a starting point $\widetilde{t}$ (e.g. five years ago) and we consider the set of ATMF implied percentage volatility:

$$\sigma_t^{(K_t,E)}, \quad t \in \mathcal{D}_{\widetilde{t},\widetilde{\tau}}, \tag{3.49}$$

where the observation dates are equally spaced as in (3.1). Each of these random variables becomes available at the respective time t. Nevertheless, implied volatilities cannot be market invariants, because the convergence to the payoff at expiry breaks the time-homogeneity of the set of variables (3.49).

As in the case of bonds, we must formulate the problem in a time-homogenous framework by eliminating the expiration date. Therefore we consider the set of implied percentage volatilities with the same time v *to* expiry:

$$\sigma_t^{(K_t,t+v)}, \quad t \in \mathcal{D}_{\widetilde{t},\widetilde{\tau}}. \tag{3.50}$$

As we show in Appendix www.3.1 the following approximation holds:

$$\sigma_t^{(K_t,t+v)} \approx \sqrt{\frac{2\pi}{v}} \frac{C_t^{(K_t,t+v)}}{U_t}. \tag{3.51}$$

In other words, the variables (3.50) represent the prices of time-homogeneous contracts divided by the underlying. If the underlying displays an unstable, say explosive, pattern, the price of the respective time-homogeneous contract also displays an unstable pattern. Once we normalize the contract by the value of the underlying as in (3.51), the result displays a time-homogenous and stable pattern.

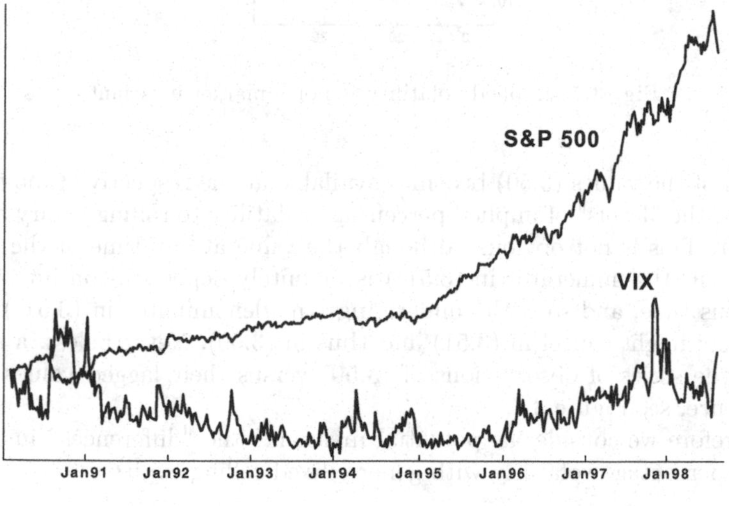

Fig. 3.6. Implied volatility versus price of underlying

For example, consider options in the stock market. The *VIX index* is the rolling ATMF implied percentage volatility of the S&P 500, i.e. the left-hand side in (3.51) and the S&P 500 index is the underlying, i.e. the denominator in the right-hand side of (3.51). In Figure 3.6 we plot the VIX index and the

S&P 500. Although the underlying displays an explosive pattern, the VIX index is stable.

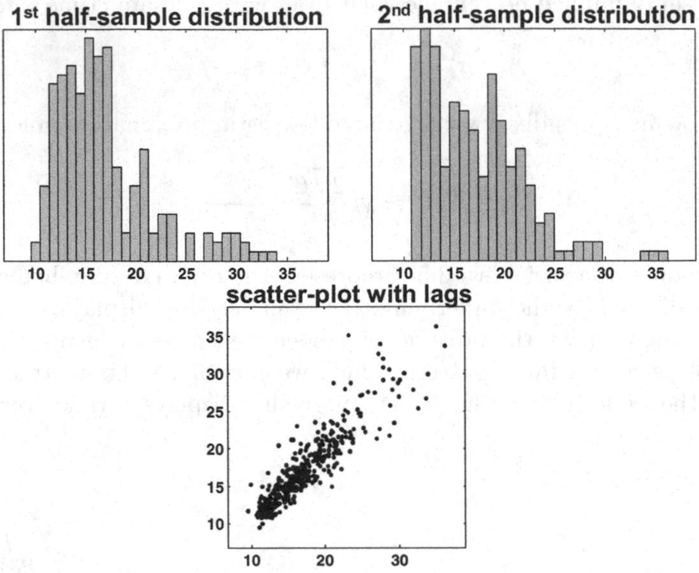

Fig. 3.7. Implied volatility is not a market invariant

Each of the values (3.50) becomes available at the respective time t. Nevertheless, the "levels" of implied percentage volatility to rolling expiry are not invariant. This is not obvious: although the value at any time of the rolling ATMF call (the numerator in (3.51)) is definitely dependent on its value at a previous time, and so is the underlying (the denominator in (3.51)), these two effects might cancel in (3.51) and thus in (3.50). Nevertheless, a scatter plot of the series of observations of (3.50) versus their lagged values shows dependence, see Figure 3.7.

Therefore we consider as potential invariants the "differences" in ATMF implied percentage volatility with generic fixed rolling expiry v:

$$X_{t,\widetilde{\tau}}^{(v)} \equiv \sigma_{t}^{(K_t, t+v)} - \sigma_{t-\widetilde{\tau}}^{(K_{t-\widetilde{\tau}}, t-\widetilde{\tau}+v)}, \quad t \in \mathcal{D}_{\widetilde{t},\widetilde{\tau}}. \tag{3.52}$$

Each of these random variables becomes available at the respective time t. To check whether they qualify as invariants for the derivatives market, we perform the two simple tests discussed in the introduction to Section 3.1 on the past realizations of the random variables (3.52):

$$x_{t,\widetilde{\tau}}^{(v)}, \quad t = \widetilde{t}, \widetilde{t} + \widetilde{\tau}, \ldots, T. \tag{3.53}$$

First we split the series (3.53) in two halves and plot the histogram of each half. If all the $X_{t,\tilde{\tau}}^{(v)}$ are identically distributed, the histogram from the first sample of the series must resemble the histogram from the second sample. In Figure 3.8 we see that this is the case.

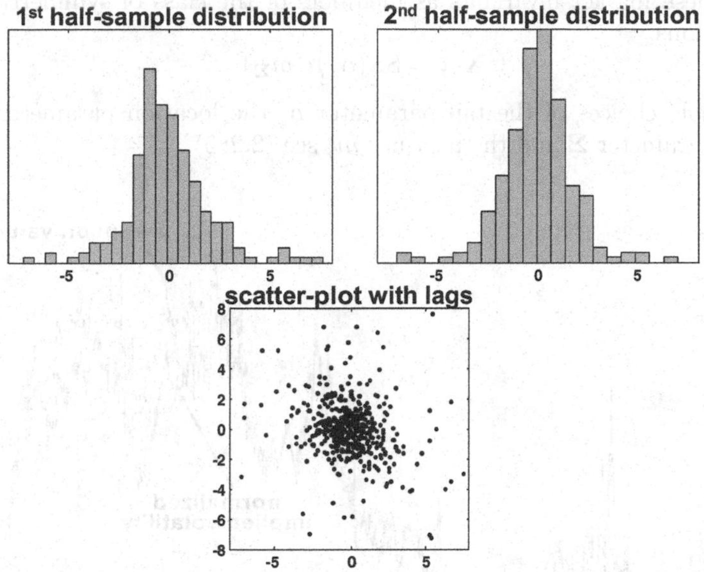

Fig. 3.8. Changes in implied volatility are market invariants

Then we move on to the second test: we scatter-plot the time series (3.53) against the same series lagged by one estimation interval. If each $X_{t+\tilde{\tau},\tilde{\tau}}^{(v)}$ is independent of $X_{t,\tilde{\tau}}^{(v)}$ and they are identically distributed, the scatter plot must resemble a circular cloud. In Figure 3.8 we see that this is indeed the case.

Therefore we accept the set of changes in the rolling at–the-money forward implied volatility (3.52) as invariants for the derivatives market:

$$\boxed{\textbf{derivatives invariants: changes in roll. ATMF impl. vol.}} \qquad (3.54)$$

As for the market invariants in the equity and in the fixed-income world, the distribution of changes in ATMF implied percentage volatility to rolling expiry can be easily projected to any horizon, see Section 3.2, and then translated back into option prices at the specified horizon, see Section 3.3.

Furthermore, the distribution of the changes in ATMF implied percentage volatility to rolling expiry is symmetrical. This feature becomes especially important in a multivariate setting, where we can model the joint distribution of these and possibly other symmetrical invariants by means of flexible, yet parsimonious, parametric models that are analytically tractable. For instance, we can model these market invariants as members of the class of elliptical distributions:

$$\mathbf{X}_{t,\widetilde{\tau}} \sim \mathrm{El}\left(\boldsymbol{\mu}, \boldsymbol{\Sigma}, g\right), \tag{3.55}$$

for suitable choices of the location parameter $\boldsymbol{\mu}$, the scatter parameter $\boldsymbol{\Sigma}$ and the probability density generator g, see (2.268). Alternatively, we can model these market invariants as members of the class of symmetric stable distributions:

$$\mathbf{X}_{t,\widetilde{\tau}} \sim \mathrm{SS}\left(\alpha, \boldsymbol{\mu}, m_{\boldsymbol{\Sigma}}\right), \tag{3.56}$$

for suitable choices of the tail parameter α, the location parameter $\boldsymbol{\mu}$, the scatter parameter $\boldsymbol{\Sigma}$ and the measure m, see (2.285).

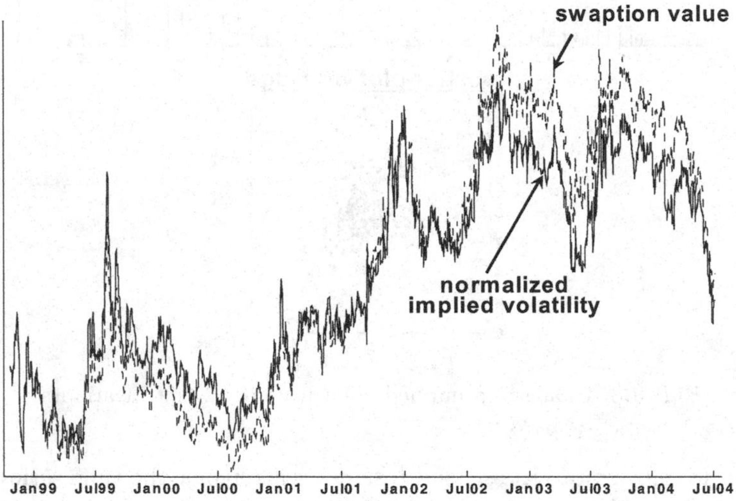

Fig. 3.9. Normalized volatility as proxy of swaption value

Before concluding we mention a variation of the invariants (3.52) that is popular among swaption traders. First we need some terminology. The *v_a-into-v_b forward par swap rate* $S_t^{(v_a,v_b)}$ is defined as follows in terms of the zero-coupon bond prices Z and an additional fixed parameter ρ, which in the US swap market is three months:

$$S_t^{(v_a,v_b)} \equiv \frac{Z_t^{(t+v_a)} - Z_t^{(t+v_a+v_b)}}{\rho \sum_{k=1}^{v_b/\rho} Z_t^{(t+v_a+k\rho)}}. \tag{3.57}$$

The parameter v_a is called *term*. The parameter v_b is called *tenor*. The forward par swap rate (3.57) is the fixed rate that makes the respective forward swap contract worthless at inception, see (3.203) and comments thereafter.

A vanilla υ_a*-into-*υ_b *payer swaption* is a call option like (3.40), where the underlying is a maturing forward par swap rate $S_t^{(E-t;\upsilon_b)}$, and the option expires one term ahead of the time T when the contract is signed:

$$E \equiv T + \upsilon_a. \tag{3.58}$$

Similarly, a vanilla υ_a*-into-*υ_b *receiver swaption* is a put option like (3.46), with underlying and expiration date as in the payer swaption. See Rebonato (1998) or Brigo and Mercurio (2001) for more on the swaption market.

Swaption traders focus on the *normalized implied volatility*, also known as *basis point implied volatility*, or "b.p. vol", which is the ATMF implied percentage volatility multiplied by the underlying, i.e. the forward par rate:

$$\sigma_t^{BP} \equiv S_t^{(\upsilon_a,\upsilon_b)} \sigma_t^{(K_t,t+\upsilon_a;\upsilon_b)}. \tag{3.59}$$

Notice that the implied volatility depends on the extra-parameter υ_b, i.e. the tenor.

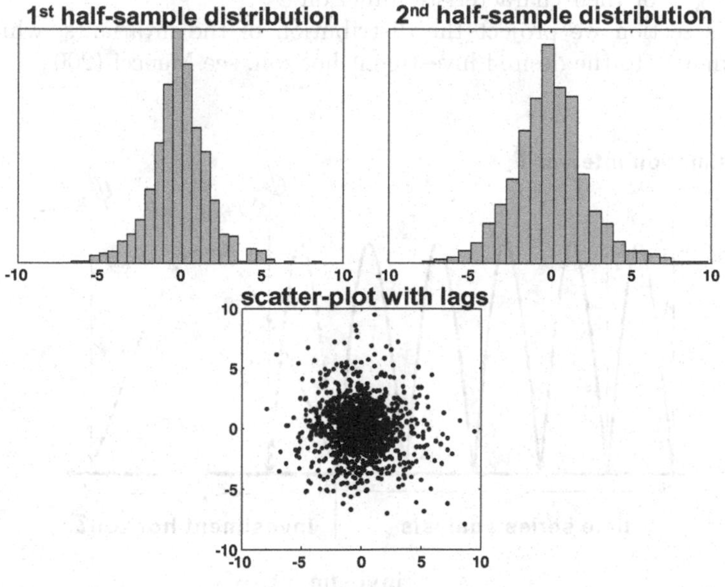

Fig. 3.10. Changes in normalized volatility are market invariants

From (3.51) the basis point volatility closely tracks the price of the ATMF swaption value.

For example, in Figure 3.9 we consider the case of the one-into-five year ATMF receiver swaption in the US market. We plot the daily values of both the ATMF implied basis point volatility (3.59) and the ATMF swaption price.

In the swaption world the underlying rate (3.57) has a bounded range and thus it does not display the explosive pattern typical of a stock price. Therefore the swaption prices are also stable, see Figure 3.9, and compare with Figure 3.6. This implies that in (3.51) we do not need to normalize the swaption price with the underlying in order to obtain stable patterns. Therefore in the swaption world the changes in ATMF implied basis point volatility are market invariants, as the two simple tests discussed in the introduction to Section 3.1 show, see Figure 3.10.

3.2 Projection of the invariants to the investment horizon

In Section 3.1 we detected the invariants $\mathbf{X}_{t,\widetilde{\tau}}$ for our market relative to the estimation interval $\widetilde{\tau}$. In Chapter 4 we show how to estimate the distribution of these invariants. The estimation process yields the representation of the distribution of the invariants, in the form of either their probability density function $f_{\mathbf{X}_{t,\widetilde{\tau}}}$ or their characteristic function $\phi_{\mathbf{X}_{t,\widetilde{\tau}}}$.

In this section we project the distribution of the invariants, which we assume known, to the desired investment horizon, see Meucci (2004).

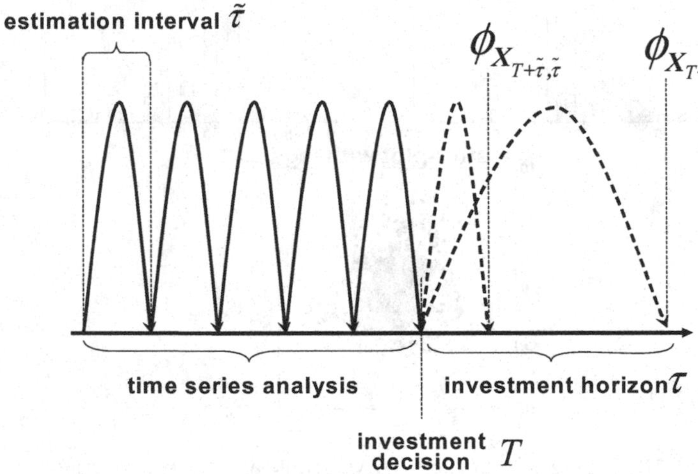

Fig. 3.11. Projection of the market invariants to the investment horizon

The distribution of the invariants as estimated in Chapter 4 is the same for all the generic times t. Denoting as T the time the investment decision is made, the estimation process yields the distribution of the "next step" invariants $\mathbf{X}_{T+\widetilde{\tau},\widetilde{\tau}}$, which become known with certainty at time $T+\widetilde{\tau}$, see Figure 3.11. This distribution contains all the information on the market for the specific horizon $\widetilde{\tau}$ that we can possibly obtain from historical analysis.

Nevertheless, the investment horizon τ is in general different, typically larger, than the estimation interval $\widetilde{\tau}$. In order to proceed with an allocation decision, we need to determine the distribution of $\mathbf{X}_{T+\tau,\tau}$, where τ is the generic desired investment horizon. This random variable, which only becomes known with certainty at the investment horizon, contains all the information on the market for that horizon that we can possibly obtain from historical analysis. Therefore our aim is determining either the probability density function $f_{\mathbf{X}_{T+\tau,\tau}}$ or the characteristic function $\phi_{\mathbf{X}_{T+\tau,\tau}}$ of the investment-horizon invariants, see Figure 3.11.

Due to the specification of the market invariants it is easy to derive this distribution. Indeed, consider first an investment horizon τ that is a multiple of the estimation horizon $\widetilde{\tau}$. The invariants are *additive*, i.e. they satisfy the following relation:

$$\mathbf{X}_{T+\tau,\tau} = \mathbf{X}_{T+\tau,\widetilde{\tau}} + \mathbf{X}_{T+\tau-\widetilde{\tau},\widetilde{\tau}} + \cdots + \mathbf{X}_{T+\widetilde{\tau},\widetilde{\tau}}. \tag{3.60}$$

This follows easily from the fact that all the invariants are in the form of differences: in the equity market (or the commodity market, or the foreign exchange market) the compounded returns (3.11) satisfy:

$$\mathbf{X}_{t,\tau} \equiv \ln(\mathbf{P}_t) - \ln(\mathbf{P}_{t-\tau}); \tag{3.61}$$

in the fixed-income market the changes in yield to maturity (3.31) satisfy:

$$\mathbf{X}_{t,\tau} \equiv \mathbf{Y}_t - \mathbf{Y}_{t-\tau}, \tag{3.62}$$

where each entry correspond to a different time to maturity; in the derivatives market the changes in implied volatilities (3.52) satisfy:

$$\mathbf{X}_{t,\tau} \equiv \boldsymbol{\sigma}_t - \boldsymbol{\sigma}_{t-\tau}, \tag{3.63}$$

where each entry refers to a specific ATMF time to expiry. Therefore we can factor the investment-horizon difference into the sum of the estimation-interval differences, which is (3.60).

Since the terms in the sum (3.60) are invariants relative to non-overlapping time intervals, they are independent and identically distributed random variables. This makes it straightforward to compute the distribution of the investment horizon invariants. Indeed, as we show in Appendix www.3.2, the investment-horizon characteristic function is simply a power of the estimated characteristic function:

$$\phi_{\mathbf{X}_{T+\tau,\tau}} = \left(\phi_{\mathbf{X}_{t,\widetilde{\tau}}}\right)^{\frac{\tau}{\widetilde{\tau}}}, \tag{3.64}$$

where the characteristic function on the right hand side does not depend on the specific time t. Representations involving either the investment-horizon pdf $f_{\mathbf{X}_{T+\tau,\tau}}$ or the estimation-interval pdf $f_{\mathbf{X}_{t,\widetilde{\tau}}}$ can be easily derived from this expression by means of the generic relations (2.14) and (2.15) between the probability density function and the characteristic function, which we report here:

$$\phi_{\mathbf{X}} = \mathcal{F}[f_{\mathbf{X}}], \quad f_{\mathbf{X}} = \mathcal{F}^{-1}[\phi_{\mathbf{X}}], \tag{3.65}$$

where $\mathcal{F}$ denotes the Fourier transform (B.34) and $\mathcal{F}^{-1}$ denotes the inverse Fourier transform (B.40).

Expression (3.64) and its equivalent formulations represent the projection of the invariants from the estimation interval $\widetilde{\tau}$ to the investment horizon τ.

We remark that we formulated the projection to the horizon assuming that the investment horizon τ was a multiple of the estimation interval $\widetilde{\tau}$. This assumption does not seem to play any role in the projection formula (3.64). Indeed, we can drop that hypothesis, and freely use the projection formula for any horizon, as long as the distribution of the estimated invariant is infinitely divisible, see Section 2.7.3. If this is not the case, the expression on the right-hand side of (3.64) might not be a viable characteristic function: in such circumstances formula (3.64) only holds for investment horizons that are multiple of the estimation interval.

Consider the normally distributed weekly compounded returns on a stock (3.18) and the three-year sector of the curve with normally distributed weekly yield changes (3.33). In other words, consider the following two market invariants:

$$\mathbf{X}_{t,\widetilde{\tau}} \equiv \begin{pmatrix} C_{t,\widetilde{\tau}} \\ X^{\upsilon}_{t,\widetilde{\tau}} \end{pmatrix} \equiv \begin{pmatrix} \ln P_t - \ln P_{t-\widetilde{\tau}} \\ Y^{(\upsilon)}_t - Y^{(\upsilon)}_{t-\widetilde{\tau}} \end{pmatrix}, \tag{3.66}$$

where υ denotes the three-year sector of the curve as in (3.34). Assume that their distribution is jointly normal:

$$\mathbf{X}_{t,\widetilde{\tau}} \sim \mathrm{N}(\boldsymbol{\mu}, \boldsymbol{\Sigma}), \tag{3.67}$$

where

$$\boldsymbol{\mu} \equiv \begin{pmatrix} \mu_C \\ \mu_X \end{pmatrix}, \quad \boldsymbol{\Sigma} \equiv \begin{pmatrix} \sigma^2_C & \rho\sigma_C\sigma_X \\ \rho\sigma_C\sigma_X & \sigma^2_X \end{pmatrix}; \tag{3.68}$$

and where (μ_C, σ^2_C) are estimated in (3.20), (μ_X, σ^2_X) are estimated in (3.35) and the correlation is estimated as, say,

$$\rho \equiv 35\%. \tag{3.69}$$

From (2.157) we obtain the characteristic function of the weekly invariants:

$$\phi_{\mathbf{X}_{t,\widetilde{\tau}}}(\boldsymbol{\omega}) = e^{i\boldsymbol{\omega}'\boldsymbol{\mu} - \frac{1}{2}\boldsymbol{\omega}'\boldsymbol{\Sigma}\boldsymbol{\omega}}. \tag{3.70}$$

Assume that the investment horizon, measured in years, is four and a half weeks:

$$\widetilde{\tau} \equiv \frac{1}{52}, \quad \tau \equiv \frac{4.5}{52}. \tag{3.71}$$

Notice that $\tau/\widetilde{\tau}$ is not an integer, but from (2.298) the normal distribution is infinitely divisible and therefore we do not need to worry about this issue.

We are interested in the distribution of the invariants relative to the investment horizon:

$$\mathbf{X}_{T+\tau,\tau} \equiv \begin{pmatrix} C_{T+\tau,\tau} \\ X^{v}_{T+\tau,\tau} \end{pmatrix} \equiv \begin{pmatrix} \ln P_{T+\tau} - \ln P_{T} \\ Y^{(v)}_{T+\tau} - Y^{(v)}_{T} \end{pmatrix}. \tag{3.72}$$

To obtain their distribution we use (3.64) to project the characteristic function (3.70) to the investment horizon:

$$\phi_{\mathbf{X}_{T+\tau,\tau}}(\boldsymbol{\omega}) = e^{i\boldsymbol{\omega}'\frac{\tau}{\widetilde{\tau}}\boldsymbol{\mu} - \frac{1}{2}\boldsymbol{\omega}'\frac{\tau}{\widetilde{\tau}}\boldsymbol{\Sigma}\boldsymbol{\omega}}. \tag{3.73}$$

This formula shows that the compounded return on the stock and the change in yield to maturity of the three-year sector at the investment horizon have a joint normal distribution with the following parameters:

$$\mathbf{X}_{T+\tau,\tau} \sim \mathrm{N}\left(\frac{\tau}{\widetilde{\tau}}\boldsymbol{\mu}, \frac{\tau}{\widetilde{\tau}}\boldsymbol{\Sigma}\right). \tag{3.74}$$

The projection formula (3.64) implies a special relation between the projected moments and the estimated moments of the invariants. As we prove in Appendix www.3.3, when the expected value is defined the following result holds:

$$\mathrm{E}\left\{\mathbf{X}_{T+\tau,\tau}\right\} = \frac{\tau}{\widetilde{\tau}}\mathrm{E}\left\{\mathbf{X}_{t,\widetilde{\tau}}\right\}, \tag{3.75}$$

where the right hand side does not depend on the specific date t. Also, when the covariance is defined the following result holds:

$$\mathrm{Cov}\left\{\mathbf{X}_{T+\tau,\tau}\right\} = \frac{\tau}{\widetilde{\tau}}\mathrm{Cov}\left\{\mathbf{X}_{t,\widetilde{\tau}}\right\}, \tag{3.76}$$

where again the right hand side does not depend on the specific date t.

In particular, we recall from (2.74) that the diagonal elements of the covariance matrix are the square of the standard deviation of the respective entries. Therefore (3.76) implies:

$$\mathrm{Sd}\left\{\mathbf{X}_{T+\tau,\tau}\right\} = \sqrt{\tau}\,\mathrm{Sd}\left\{\mathbf{X}\right\}, \tag{3.77}$$

where in the right hand side we dropped the specific date t, which does not play a role, and we set the reference horizon $\widetilde{\tau} \equiv 1$, measuring time in years and dropping it from the notation. This identity is known among practitioners as the *square-root rule*. Specifically, in the case of equities it reads "the standard deviation of the compounded return of a stock at a given horizon is the square root of the horizon times the annualized standard deviation of the compounded return". In the case of fixed-income securities it reads: "the standard deviation of the change in yield to maturity in a given time span is the square root of the time span times the annualized standard deviation of the change in yield to maturity".

We remark that the simplicity of the projection formula (3.64) is due to the particular formulation for the market invariants that we chose in Section 3.1. For instance, if we had chosen as invariants for the stock market the linear returns (3.10) instead of the compounded returns, we would have obtained instead of (3.60) the following projection formula:

$$\mathbf{L}_{T+\tau,\tau} = \operatorname{diag}\left(\mathbf{1} + \mathbf{L}_{T+\tau,\widetilde{\tau}}\right) \cdots \operatorname{diag}\left(\mathbf{1} + \mathbf{L}_{T+\widetilde{\tau},\widetilde{\tau}}\right) - \mathbf{1}. \tag{3.78}$$

The distribution of $\mathbf{L}_{T,\tau}$ in terms of the distribution of $\mathbf{L}_{t,\widetilde{\tau}}$ cannot be represented in closed form as in (3.64). Similarly, the projection formula must be adapted in an ad-hoc way for more complex market dynamics than those discussed in Section 3.1.

We conclude pointing out that the simplicity of the projection formula (3.64) hides the dangers of *estimation risk*. In other words, the distribution at the investment horizon is given precisely by (3.64) *if* the estimation-horizon distribution is known exactly. Since by definition an estimate is only an approximation to reality, the distribution at the investment horizon cannot be precise. In fact, the farther in the future the investment horizon, the larger the effect of the estimation error. We discuss estimation risk and how to cope with it extensively in the third part of the book.

3.3 From invariants to market prices

In general the market, i.e. the prices at the investment horizon of the securities that we are considering, is a function of the investment-horizon invariants:

$$\mathbf{P} = \mathbf{g}\left(\mathbf{X}\right), \tag{3.79}$$

where in this section we use the short-hand notation $\mathbf{P}$ for $\mathbf{P}_{T+\tau}$ and $\mathbf{X}$ for $\mathbf{X}_{T+\tau,\tau}$.

In this section we discuss how to recover the distribution of the market from the distribution of the investment-horizon invariants, as obtained in (3.64). We analyze separately raw securities and derivatives.

3.3.1 Raw securities

Obtaining the distribution of the prices of the raw securities is particularly simple.

In the case of equities, foreign exchange rates and commodities, discussed in Section 3.1.1, the invariants are the compounded returns (3.11) and therefore the pricing formula (3.79) takes the following form:

$$P_{T+\tau} = P_T e^X. \tag{3.80}$$

Consider now the fixed-income securities discussed in Section 3.1.2. From (3.27) and (3.31) we obtain the pricing function of the generic zero-coupon bond with maturity E:

$$Z_{T+\tau}^{(E)} = Z_T^{(E-\tau)} e^{-X^{(E-T-\tau)}(E-T-\tau)}. \tag{3.81}$$

We see that in the case of raw securities, the pricing function (3.79) has the following simple form:

$$\mathbf{P} = e^{\mathbf{Y}}, \tag{3.82}$$

where the ancillary variable $\mathbf{Y}$ is an affine transformation of the market invariants:

$$\mathbf{Y} \equiv \boldsymbol{\gamma} + \operatorname{diag}(\boldsymbol{\varepsilon})\mathbf{X}. \tag{3.83}$$

The constant vectors $\boldsymbol{\gamma}$ and $\boldsymbol{\varepsilon}$ in this expression read respectively component-wise:

$$\gamma_n \equiv \begin{cases} \ln(P_T), \text{ if the } n\text{-th security is a stock} \\ \ln\left(Z_T^{(E-\tau)}\right), \text{ if the } n\text{-th security is bond} \end{cases} \tag{3.84}$$

and

$$\varepsilon_n \equiv \begin{cases} 1, \text{ if the } n\text{-th security is a stock} \\ -(E-T-\tau), \text{ if the } n\text{-th security is bond.} \end{cases} \tag{3.85}$$

For example, consider the two-security market relative to the invariants (3.72). In other words, one security is a stock and the other one is a zero-coupon bond with maturity:

$$E \equiv T + \tau + \upsilon, \tag{3.86}$$

where υ is the three-years sector of the curve. In this case (3.82)-(3.83) read:

$$\mathbf{P} \equiv \begin{pmatrix} P_{T+\tau} \\ Z_{T+\tau}^{(E)} \end{pmatrix} = e^{\boldsymbol{\gamma} + \operatorname{diag}(\boldsymbol{\varepsilon})\mathbf{X}}, \tag{3.87}$$

where $\mathbf{X}$ is (3.72) and from (3.84) and (3.85) we obtain:

$$\boldsymbol{\gamma} \equiv \begin{pmatrix} \ln(P_T) \\ \ln\left(Z_T^{(T+\upsilon)}\right) \end{pmatrix}, \quad \boldsymbol{\varepsilon} \equiv \begin{pmatrix} 1 \\ -\upsilon \end{pmatrix}. \tag{3.88}$$

Since the ancillary variable (3.83) is a simple affine transformation of the market invariants, computing its distribution from that of the market invariants $\mathbf{X}$ is straightforward, see Appendix 2.4. For example, in terms of the characteristic function we obtain:

$$\phi_{\mathbf{Y}}(\boldsymbol{\omega}) = e^{i\boldsymbol{\omega}'\boldsymbol{\gamma}} \phi_{\mathbf{X}}(\operatorname{diag}(\boldsymbol{\varepsilon})\boldsymbol{\omega}). \tag{3.89}$$

In our example the characteristic function of the horizon invariants is (3.73). Therefore from (3.89) the characteristic function of the ancillary variable $\mathbf{Y}$ reads:

$$\phi_{\mathbf{Y}}(\boldsymbol{\omega}) = e^{i\boldsymbol{\omega}'\left[\boldsymbol{\gamma}+\frac{\tau}{\widetilde{\tau}}\operatorname{diag}(\boldsymbol{\varepsilon})\boldsymbol{\mu}\right]-\frac{1}{2}\frac{\tau}{\widetilde{\tau}}\boldsymbol{\omega}'\operatorname{diag}(\boldsymbol{\varepsilon})\boldsymbol{\Sigma}\operatorname{diag}(\boldsymbol{\varepsilon})\boldsymbol{\omega}}, \tag{3.90}$$

where $\boldsymbol{\mu}$ and $\boldsymbol{\Sigma}$ are given in (3.68) and $\boldsymbol{\gamma}$ and $\boldsymbol{\varepsilon}$ are given in (3.88).

In other words, the ancillary variable $\mathbf{Y}$ is normally distributed with the following parameters:

$$\mathbf{Y} \sim \mathrm{N}\left(\boldsymbol{\gamma} + \frac{\tau}{\widetilde{\tau}}\operatorname{diag}(\boldsymbol{\varepsilon})\boldsymbol{\mu}, \frac{\tau}{\widetilde{\tau}}\operatorname{diag}(\boldsymbol{\varepsilon})\boldsymbol{\Sigma}\operatorname{diag}(\boldsymbol{\varepsilon})\right). \tag{3.91}$$

Notice that we could have obtained this result also from (3.74) and the affine property (2.163) of the normal distribution.

To compute the distribution of the prices, we notice from (3.82) that the prices $\mathbf{P} \equiv e^{\mathbf{Y}}$ have a log-$\mathbf{Y}$ distribution, see Section 2.6.5. In some cases this distribution can be computed explicitly.

In our example, since the ancillary variable $\mathbf{Y}$ in (3.91) is normal, the variable $\mathbf{P}$ is by definition lognormal with the same parameters:

$$\mathbf{P} \sim \mathrm{LogN}\left(\boldsymbol{\gamma} + \frac{\tau}{\widetilde{\tau}}\operatorname{diag}(\boldsymbol{\varepsilon})\boldsymbol{\mu}, \frac{\tau}{\widetilde{\tau}}\operatorname{diag}(\boldsymbol{\varepsilon})\boldsymbol{\Sigma}\operatorname{diag}(\boldsymbol{\varepsilon})\right), \tag{3.92}$$

where $\boldsymbol{\mu}$ and $\boldsymbol{\Sigma}$ are given in (3.68) and $\boldsymbol{\gamma}$ and $\boldsymbol{\varepsilon}$ are given in (3.88).

In most cases it is not possible to compute the distribution of the prices in closed form. Nevertheless, in practical allocation problems only the first few moments of the distribution of the prices are required. We can easily compute all the moments of the distribution of $\mathbf{P}$ directly from the characteristic function of the market invariants.

Indeed, dropping the horizon to ease the notation, from (2.214) and (3.89) the generic raw moment of the prices of the securities reads:

$$\mathrm{E}\{P_{n_1}\cdots P_{n_k}\} = e^{i\boldsymbol{\gamma}'\boldsymbol{\omega}_{n_1\cdots n_k}}\phi_{\mathbf{X}}\left(\operatorname{diag}(\boldsymbol{\varepsilon})\boldsymbol{\omega}_{n_1\cdots n_k}\right), \tag{3.93}$$

where the vector $\boldsymbol{\omega}$ is defined in terms of the canonical basis (A.15) as follows:

$$\boldsymbol{\omega}_{n_1\cdots n_k} \equiv \frac{1}{i}\left(\boldsymbol{\delta}^{(n_1)} + \cdots + \boldsymbol{\delta}^{(n_k)}\right). \tag{3.94}$$

In particular we can compute the expected value of the prices of the generic n-th security:

$$\mathrm{E}\{P_n\} = e^{\gamma_n}\phi_{\mathbf{X}}\left(-i\varepsilon_n\boldsymbol{\delta}^{(n)}\right). \tag{3.95}$$

Similarly, we can compute the covariance of the prices of the generic m-th and n-th securities:

$$\operatorname{Cov}\{P_m, P_n\} = \operatorname{E}\{P_m P_n\} - \operatorname{E}\{P_m\}\operatorname{E}\{P_n\}, \tag{3.96}$$

where

$$\operatorname{E}\{P_m P_n\} = e^{\gamma_m+\gamma_n}\phi_{\mathbf{X}}\left(-i\varepsilon_m\boldsymbol{\delta}^{(m)} - i\varepsilon_n\boldsymbol{\delta}^{(n)}\right). \tag{3.97}$$

Formulas (3.95) and (3.96) are particularly useful in the mean-variance allocation framework, which we discuss in Chapter 6.

For example, the stock price $P_{T+\tau}$ at the investment horizon is the first entry of the vector $\mathbf{P}$ in our example (3.87). Substituting (3.88) in (3.95) we obtain:

$$\operatorname{E}\{P_{T+\tau}\} = P_T\phi_{\mathbf{X}}\begin{pmatrix}-i\\0\end{pmatrix}. \tag{3.98}$$

From the expression of the characteristic function (3.73) of the investment-horizon invariants this means:

$$\operatorname{E}\{P_{T+\tau}\} = P_T e^{\frac{\tau}{\tilde{\tau}}\mu_C + \frac{\tau}{\tilde{\tau}}\frac{\sigma_C^2}{2}}, \tag{3.99}$$

where (μ_C, σ_C^2) are estimated in (3.20). This formula is in accordance with the expected value of the first entry of the joint lognormal variable (3.92), as computed in (2.219).

We remark that this technique is very general, because it allows to compute *all* the moments of the prices from a *generic* distribution of investment-horizon invariants, as represented by the characteristic function.

Furthermore, we can replace the simple expression (3.64) of the characteristic function at the investment horizon $\phi_{\mathbf{X}}$ in (3.93) and directly compute all the moments of the distribution of the market prices from the estimated characteristic function:

$$\operatorname{E}\{P_{n_1}\cdots P_{n_k}\} = e^{i\boldsymbol{\gamma}'\boldsymbol{\omega}_{n_1\cdots n_k}}\left[\phi_{\mathbf{X}_{t,\tilde{\tau}}}\left(\operatorname{diag}(\boldsymbol{\varepsilon})\,\boldsymbol{\omega}_{n_1\cdots n_k}\right)\right]^{\frac{\tau}{\tilde{\tau}}}, \tag{3.100}$$

where the right hand side does not depend on the specific time t and $\boldsymbol{\omega}$ is given in (3.94).

For example, we could have derived (3.99) by means of (3.100) directly from the expression for the estimation-interval characteristic function (3.70). The check is left to the reader.

We stress again that the simplicity of expressions such as (3.93) and (3.100) hides the dangers of *estimation risk*, which we discuss in the third part of the book.

3.3.2 Derivatives

In the case of derivatives, the prices at the investment horizon $\mathbf{P}$ do not have a simple log-distribution. If the generic entry of the price vector $\mathbf{P}$ corresponds to a derivative, the investment-horizon pricing function (3.79) reads:

$$P = g(\mathbf{X}), \tag{3.101}$$

where g is in general a complicated function of several investment-horizon invariants.

For example, consider a call option with strike K that expires at time E on a stock that trades at price U_t. From (3.40) we obtain:

$$C_{T+\tau}^{(K,E)} \equiv C^{BS}\left(v, K, U_{T+\tau}, Z_{T+\tau}^{(E)}, \sigma_{T+\tau}^{(K,E)}\right), \tag{3.102}$$

where C^{BS} is the Black-Scholes formula (3.41) and

$$v \equiv (E - T - \tau). \tag{3.103}$$

The three market variables (U, Z, σ) all admit invariants and thus can be expressed as functions of the respective horizon-invariant. For the stock from (3.80) we have:

$$U_{T+\tau} = U_T e^{X_1}, \tag{3.104}$$

where X_1 is the compounded return to the investment horizon.

For the zero-coupon bond, from (3.81) we have:

$$Z_{T+\tau}^{(E)} = Z_T^{(E-\tau)} e^{-X_2 v}, \tag{3.105}$$

where X_2 is the change until the investment horizon in yield for the v-sector of the yield curve.

For the implied volatility from (3.52) we have[3]:

$$\sigma_{T+\tau}^{(K,E)} = \sigma_T^{(K_T, E-\tau)} + X_3, \tag{3.106}$$

where K_T is the ATMF strike (3.48) and X_3 is the change over the investment horizon in ATMF implied percentage volatility with fixed rolling expiry (3.103).

Therefore the investment-horizon pricing function (3.101) reads:

$$C_{T+\tau}^{(K,E)}(\mathbf{X}) = C^{BS}\left(v, K, U_T e^{X_1}, Z_T^{(E-\tau)} e^{-X_2 v}, \sigma_T^{(K_T, E-\tau)} + X_3\right). \tag{3.107}$$

In the general case, given the complexity of the pricing formula at the investment horizon (3.101), it is close to impossible to compute the exact distribution of the prices from the market invariants. Nevertheless, the pricing formula may be approximated by its Taylor expansion:

[3] More accurately, the right-hand side in (3.106) is $\sigma_{T+\tau}^{(K_{T+\tau}, E)}$. The difference between the two sides is the *smile* of the implied voltility, see e.g. Hull (2002)

$$P = g(\mathbf{m}) + (\mathbf{X} - \mathbf{m})' \, \partial_{\mathbf{x}} g|_{\mathbf{x}=\mathbf{m}} + \frac{1}{2} (\mathbf{X} - \mathbf{m})' \, \partial^2_{\mathbf{xx}} g\big|_{\mathbf{x}=\mathbf{m}} (\mathbf{X} - \mathbf{m}) + \cdots, \tag{3.108}$$

where $\mathbf{m}$ is a significative value of the invariants. One standard choice is zero:

$$\mathbf{m} \equiv \mathbf{0}. \tag{3.109}$$

Another standard choice is the expected value:

$$\mathbf{m} \equiv \mathrm{E}\{\mathbf{X}\}. \tag{3.110}$$

If the approximation in (3.108) is performed up to the first order, the market prices at the horizon are a linear function of the invariants. If the approximation is carried on up to the second order, the market prices are quadratic functions of the invariants. In either case, the distribution of the market prices becomes a tractable expression of the distribution of the invariants.

Depending on its end users, the approximation (3.108) is known under different names.

In the derivatives world the expansion up to order zero is called the *theta* approximation. The expansion up to order one is called the *delta-vega approximation*. The *delta* is the first derivative (mathematical operation) of the investment-horizon pricing function of the derivative (financial contract) with respect to the underlying, whereas the *vega* is the first derivative (mathematical operation) of the investment-horizon pricing function of the derivative (financial contract) with respect to the implied volatility. The expansion up to order two is called the *gamma approximation*. The *gamma* is the second derivative (mathematical operation) of the investment-horizon pricing function of the derivative (financial contract) with respect to the underlying.

In the fixed-income world the expansion up to order zero in (3.108) is known as the *roll-down* or *slide approximation*. The expansion up to order one is known as the *PVBP* or *duration approximation*. The expansion up to order two is known as the *convexity approximation*, see Section 3.5 for a thorough case-study.

We stress again that the accuracy of (3.108) is jeopardized by the hidden threat of *estimation risk*, which we discuss in the third part of the book.

3.4 Dimension reduction

According to (3.79), the prices at the investment horizon of the securities in our market are a function of the randomness in the market:

$$\mathbf{P}_{T+\tau} = \mathbf{g}(\mathbf{X}_{T+\tau,\tau}), \tag{3.111}$$

where $\mathbf{X}_{t,\tau}$ denotes the generic set of market invariants relative to the interval τ that becomes known at time t.

In a generic market of a large number of securities, the following two phenomena typically occur.

In the first place the actual dimension of the market is less than the number of securities. This is due to the joint presence in the market of derivatives and underlying securities. Such phenomena can be analyzed in terms of the copula of the market and the related dependence summary statistics, as discussed in Section 2.5.

For example, consider a market of two products: a stock which trades at the generic time t at the price S_t and a call option on that stock with strike $\widetilde{S}$ that trades at the price C_t. If the investment horizon coincides with the expiry of the option, the market is one-dimensional:

$$\mathbf{P}_{T+\tau} = \begin{pmatrix} S_{T+\tau} \\ \max\left(S_{T+\tau} - \widetilde{S}, 0\right) \end{pmatrix}, \tag{3.112}$$

see Figure 2.5. From Table 2.115, the Schweizer and Wolff measure of dependence between these two securities is one, i.e. the maximum possible value.

In the second place, the actual dimension of the randomness in the market, i.e. the actual dimension of the N-dimensional vector of investment-horizon invariants $\mathbf{X}$, is less than N. This is the subject of the remainder of this section.

We aim at expressing the vector of invariants $\mathbf{X}$ as a function of two sets of variables: a vector $\mathbf{F}$ of a few *common factors* that are responsible for most of the randomness in the market; and a residual vector $\mathbf{U}$ of *perturbations* that have a marginal effect:

$$\mathbf{X}_{t,\tau} \equiv \mathbf{h}\left(\mathbf{F}_{t,\tau}\right) + \mathbf{U}_{t,\tau}. \tag{3.113}$$

In this expression the vector of factors $\mathbf{F}$ should have a much lower dimension than the market invariants:

$$K \equiv \dim\left(\mathbf{F}_{t,\tau}\right) \ll N \equiv \dim\left(\mathbf{X}_{t,\tau}\right). \tag{3.114}$$

We remark that, since $\mathbf{X}_{t,\tau}$ represents the market invariants, i.e. it is a vector of independent and identically distributed random variables that become known at time t, both factors $\mathbf{F}_{t,\tau}$ and perturbations $\mathbf{U}_{t,\tau}$ must also be market invariants. In the sequel we drop the generic time t and the generic interval τ from the notation.

Intuitively, the factors should affect all the invariants and be responsible for most of the randomness in the market. In other words the invariants recovered through the factors should be very close to the original market invariants:

$$\widetilde{\mathbf{X}} \equiv \mathbf{h}\left(\mathbf{F}\right) \approx \mathbf{X}. \tag{3.115}$$

To measure the goodness of this approximation we use the *generalized r-square*, which we define as follows:

$$R^2\left\{\mathbf{X},\widetilde{\mathbf{X}}\right\} \equiv 1 - \frac{\mathrm{E}\left\{\left(\mathbf{X}-\widetilde{\mathbf{X}}\right)'\left(\mathbf{X}-\widetilde{\mathbf{X}}\right)\right\}}{\mathrm{tr}\left\{\mathrm{Cov}\left\{\mathbf{X}\right\}\right\}}. \tag{3.116}$$

The term in the numerator is a measure of the amount of randomness in the residual, which is zero if and only if the approximation (3.115) is exact. The term in the denominator is a measure of the amount of randomness in the original invariants, as it is proportional to the average of the variances of all the invariants. The factor model (3.113) is viable if the generalized r-square approaches one. An r-square close to zero or even negative indicates that the factor model performs poorly.

The generic factor model (3.113) is too broad. In the sequel we will restrict our models to linear functions. In other words, we express the invariants in the following form:

$$\mathbf{X} \equiv \mathbf{BF} + \mathbf{U}. \tag{3.117}$$

The K columns of the $N \times K$ matrix $\mathbf{B}$ are called the *factor loadings*: they transfer the effect of each of the K factors in $\mathbf{F}$ to the N invariants in $\mathbf{X}$. Notice that (3.117) represents a first-order Taylor approximations of the general formula (3.113), if we include a constant among the factors.

Ideally, common factors and perturbations should be independent variables. For practical purposes this requirement is too restrictive, therefore we only impose that common factors and perturbation be uncorrelated:

$$\mathrm{Cor}\left\{\mathbf{F},\mathbf{U}\right\} = \mathbf{0}_{K\times N}, \tag{3.118}$$

which is a weaker assumption, see (2.136). The two assumptions (3.117) and (3.118) encompass the vast majority of the factor models considered in the financial literature.

Factor models for the market invariants can be obtained in two ways: either the factors are measurable market invariants, in which case we obtain an explicit factor model, or they are synthetic variables defined in terms of the original market invariants, in which case we obtain a hidden factor model. In either case, the perturbations are defined as the residual term.

3.4.1 Explicit factors

Here we assume that the factors $\mathbf{F}$ in the linear factor model (3.117) are explicit market variables. In other words, for any choice of the $N \times K$ matrix $\mathbf{B}$ of the factor loadings we obtain a linear model that defines the residuals as follows:

$$\mathbf{X} \equiv \mathbf{BF} + \mathbf{U}. \tag{3.119}$$

The *regression factor loadings* correspond to the best choice of the coefficients $\mathbf{B}$ in terms of the generalized r-square criterion (3.116). By definition the regression factor loadings solve:

$$\mathbf{B}_r \equiv \underset{\mathbf{B}}{\operatorname{argmax}}\, R^2\left\{\mathbf{X}, \mathbf{BF}\right\}, \tag{3.120}$$

where "r" stands for "regression".

As we show in Appendix www.3.4, the regression factor loadings read:

$$\mathbf{B}_r \equiv \mathrm{E}\left\{\mathbf{XF}'\right\} \mathrm{E}\left\{\mathbf{FF}'\right\}^{-1}. \tag{3.121}$$

The regression factor loadings in turn yield the recovered invariants $\widetilde{\mathbf{X}}_r \equiv \mathbf{B}_r\mathbf{F}$ and the perturbations, i.e. the residuals $\mathbf{U}_r \equiv \mathbf{X} - \widetilde{\mathbf{X}}_r$. Unfortunately, the perturbations do not display zero correlation with the explicit factors unless the factors have zero expected value:

$$\mathrm{E}\left\{\mathbf{F}\right\} = \mathbf{0} \Rightarrow \mathrm{Cor}\left\{\mathbf{F}, \mathbf{U}\right\} = \mathbf{0}_{K\times N}. \tag{3.122}$$

For example, consider an invariant and a factor that are jointly normally distributed:

$$\begin{pmatrix} X \\ F \end{pmatrix} \sim \mathrm{N}\left(\begin{pmatrix} \mu_X \\ \mu_F \end{pmatrix}, \begin{pmatrix} \sigma_X^2 & \rho\sigma_X\sigma_F \\ \rho\sigma_X\sigma_F & \sigma_F^2 \end{pmatrix} \right). \tag{3.123}$$

In this case the regression factor loading reads:

$$b_r = \frac{\mu_X\mu_F + \rho\sigma_X\sigma_F}{\mu_F^2 + \sigma_F^2}. \tag{3.124}$$

From the more general formulas of Appendix www.3.4 we obtain:

$$\mathrm{Cov}\left\{U, F\right\} = \rho\sigma_X\sigma_F \left(1 - \frac{1}{1 + \mu_F^2/\sigma_F^2}\right) - \frac{\mu_F\mu_X}{1 + \mu_F^2/\sigma_F^2}, \tag{3.125}$$

which is null if $\mu_F \equiv 0$.

Nevertheless, we can always include a constant among the factors:

$$\mathbf{F} \mapsto \begin{pmatrix} 1 \\ \mathbf{F} \end{pmatrix}. \tag{3.126}$$

We show in Appendix www.3.4 that in this case the regression coefficients (3.121) yield the following recovered invariants:

$$\widetilde{\mathbf{X}}_r \equiv \mathrm{E}\left\{\mathbf{X}\right\} + \mathrm{Cov}\left\{\mathbf{X}, \mathbf{F}\right\} \mathrm{Cov}\left\{\mathbf{F}\right\}^{-1} \left(\mathbf{F} - \mathrm{E}\left\{\mathbf{F}\right\}\right), \tag{3.127}$$

see Figure 3.12. The perturbations, which are defined as the residuals $\mathbf{U}_r \equiv \mathbf{X} - \widetilde{\mathbf{X}}_r$, have zero expected value and display zero correlation with the factors:

$$\mathrm{E}\left\{\mathbf{U}_r\right\} = \mathbf{0}, \quad \mathrm{Cor}\left\{\mathbf{F}, \mathbf{U}\right\} = \mathbf{0}_{K\times N}. \tag{3.128}$$

Furthermore, the covariance of the residual reads:

$$\mathrm{Cov}\left\{\mathbf{U}_r\right\} = \mathrm{Cov}\left\{\mathbf{X}\right\} - \mathrm{Cov}\left\{\mathbf{X}, \mathbf{F}\right\} \mathrm{Cov}\left\{\mathbf{F}\right\}^{-1} \mathrm{Cov}\left\{\mathbf{F}, \mathbf{X}\right\}. \tag{3.129}$$

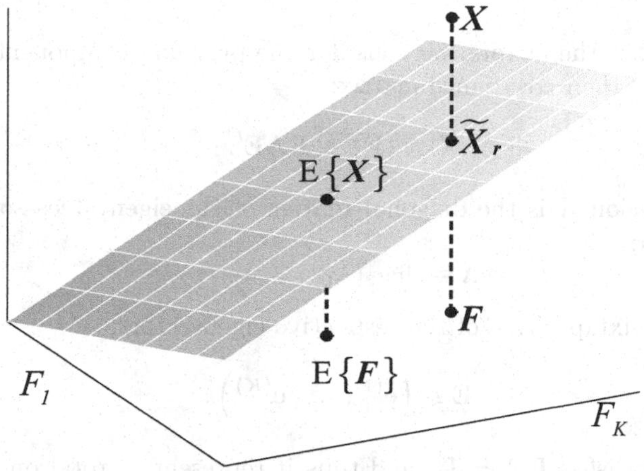

Fig. 3.12. Explicit factor dimension reduction: regression

Notice the similarities between the recovered invariants (3.127) and the expected value of the conditional normal distribution (2.165) on the one hand, and the covariance of the residuals (3.129) and the covariance of the conditional normal distribution (2.166) on the other hand.

If in our example (3.123) we add a constant we obtain from (3.127) the following recovered invariant:

$$\widetilde{X}_r \equiv \mu_X + \rho \frac{\sigma_X}{\sigma_F} (F - \mu_F) . \tag{3.130}$$

This is the expected value of the conditional distribution of the invariant given the factor (2.174). Similarly, the variance of the residual reads:

$$\operatorname{Var}\left\{\widetilde{U}_r\right\} = \sigma_X^2 \left(1 - \rho^2\right), \tag{3.131}$$

which is the variance of the conditional distribution of the invariant given the factor (2.175).

In order to evaluate the quality of an explicit factor model, it is better to reformulate our model in a scale-independent fashion.

First of all we normalize the market invariants by means of their z-scores $\mathbf{Z}_X$, which from (1.35) read component-wise:

$$Z_X^{(n)} \equiv \frac{X_n - \mathrm{E}\{X_n\}}{\sqrt{\operatorname{Cov}\{X_n, X_n\}}}. \tag{3.132}$$

The z-scores have zero expected value and unit standard-deviation: therefore they represent a scale- and location-independent version of the market invariants.

To normalize the factors, we consider the principal component decomposition (2.76) of their covariance matrix:

$$\mathrm{Cov}\{\mathbf{F}\} \equiv \mathbf{E}\boldsymbol{\Lambda}\mathbf{E}'. \tag{3.133}$$

In this expression $\boldsymbol{\Lambda}$ is the diagonal matrix of the eigenvalues sorted in decreasing order:

$$\boldsymbol{\Lambda} \equiv \mathrm{diag}(\lambda_1, \ldots, \lambda_K); \tag{3.134}$$

and $\mathbf{E}$ is the juxtaposition of the respective eigenvectors:

$$\mathbf{E} \equiv \left(\mathbf{e}^{(1)}, \ldots, \mathbf{e}^{(K)}\right). \tag{3.135}$$

This matrix satisfies $\mathbf{E}\mathbf{E}' = \mathbf{I}_K$ and thus it represents a rotation, see Figure A.4. In terms of the principal component decomposition we can normalize the factors as follows:

$$\mathbf{Z}_F \equiv \boldsymbol{\Lambda}^{-\frac{1}{2}}\mathbf{E}'(\mathbf{F} - \mathrm{E}\{\mathbf{F}\}). \tag{3.136}$$

These are the z-scores of the factors, rotated in a way that decorrelates them:

$$\mathrm{Cov}\{\mathbf{Z}_F\} = \mathbf{I}_K, \tag{3.137}$$

see the proof in Appendix www.3.4.

In terms of the normalized variables (3.132) and (3.136), the recovered invariants (3.127) read:

$$\widetilde{\mathbf{Z}}_X = \mathbf{C}_{XF}\mathbf{Z}_F, \tag{3.138}$$

where the matrix $\mathbf{C}_{XF}$ is the correlation between the market invariants and the (rotated) explicit factors:

$$\mathbf{C}_{XF} \equiv \mathrm{Cor}\{\mathbf{X}, \mathbf{E}'\mathbf{F}\}. \tag{3.139}$$

The correlation $\mathbf{C}_{XF}$ in (3.138) is responsible for transferring the randomness of the factors into the recovered invariants. Indeed, we show in Appendix www.3.4 that the generalized r-square (3.116) of the explicit factor model can be expressed as an average correlation:

$$R^2\left\{\mathbf{X}, \widetilde{\mathbf{X}}_r\right\} = \frac{\mathrm{tr}(\mathbf{C}_{XF}\mathbf{C}'_{XF})}{N}. \tag{3.140}$$

Therefore the factors should be chosen as correlated as possible to the market invariants, in order to increase their explanatory power.

In our example (3.123), where there exists only one factor, (3.139) reads:

$$\mathbf{C}_{XF} \equiv \operatorname{Cor}\{X, F\} = \rho. \tag{3.141}$$

Therefore in our simple one-factor model the generalized r-square (3.140) is the square of the correlation between the factor and the invariant:

$$R^2 = \rho^2. \tag{3.142}$$

Indeed, from (3.131) when the factor and the invariant are highly correlated, the residual is minimal and thus the explanatory power of the factor model is maximal.

Adding factors trivially improves the quality of the result. Nevertheless, the number of factors should be kept at a minimum, in order not to defeat the purpose of dimension reduction.

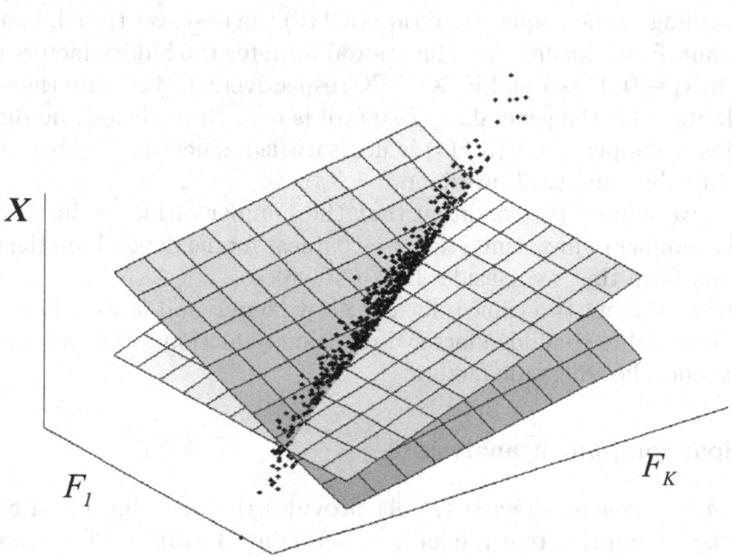

Fig. 3.13. Collinearity: the regression plane is not defined

Furthermore, the factors should be chosen as diversified as possible, in order to avoid the problem of *collinearity*. Indeed, when the K factors are not diversified they span a hyperplane of dimension less than K. This makes it impossible to identify the regression hyperplane, see Figure 3.13.

Several criteria have been developed in the statistical and financial literature to select the most suitable among a pool of potential explicit factors, such as the *Akaike information criterion* and the *Bayesian information criterion*. We refer the reader to references such as Parzen, Tanabe, and Kitagawa

(1998), see also Connor and Korajczyk (1993) for financial applications. To implement the selection in practice once a suitable criterion has been determined see Section 3.4.5.

3.4.2 Hidden factors

In a linear model with hidden factors we assume that the factors are not explicit market variables. Instead, they are functions of the original invariants that summarize as much information about the invariants as possible. Including a constant among the hidden factors, (3.117) reads:

$$\mathbf{X} \equiv \mathbf{q} + \mathbf{B}\mathbf{F}(\mathbf{X}) + \mathbf{U}. \tag{3.143}$$

For any choice of the constant $\mathbf{q}$ and of the factor loading matrix $\mathbf{B}$ and for any choice $\mathbf{F}(\cdot)$ of the functional form that summarizes the invariants $\mathbf{X}$ into the synthetic factors, we obtain a model that defines the residuals $\mathbf{U}$.

According to the r-square criterion (3.116) the best, yet trivial, joint choice of constant, factor loadings and functional form for the hidden factors is represented by $\mathbf{q} \equiv \mathbf{0}$, $\mathbf{B} \equiv \mathbf{I}$ and $\mathbf{F}(\mathbf{X}) \equiv \mathbf{X}$ respectively. In this case the residuals are null and thus the generalized r-square is one. Nevertheless, no dimension reduction takes place, i.e. (3.114) is not satisfied, since the number of factors is equal to the number of invariants.

Once we impose the condition that the number of hidden factors be less than the number of invariants, the "best" linear model depends on the possible functional form that we consider for the factors.

Here we present two choices for the above functional form, which give rise to two approaches to hidden factor dimension reduction: principal component analysis and idiosyncractic factors.

Principal component analysis

Principal component analysis (*PCA*) provides the best dimension reduction under the assumption that the hidden factors in (3.143) be affine transformations of the invariants:

$$\mathbf{F}_p \equiv \mathbf{d}_p + \mathbf{A}_p'\mathbf{X}, \tag{3.144}$$

where $\mathbf{d}$ is a K-dimensional vector, $\mathbf{A}$ is an $N \times K$ matrix and "p" stands for "PCA". Notice that this is a first-order Taylor expansion of the more general functional form $\mathbf{F}(\mathbf{X})$ that appears in (3.143).

Under the above assumption, from (3.143) the optimally recovered invariants must be an affine transformation of the original invariants:

$$\widetilde{\mathbf{X}}_p \equiv \mathbf{m}_p + \mathbf{B}_p\mathbf{A}_p'\mathbf{X}, \tag{3.145}$$

where

$$\mathbf{m}_p \equiv \mathbf{q} + \mathbf{B}_p\mathbf{d}_p. \tag{3.146}$$

Therefore, the PCA solution is represented by the following set of factor loadings and coefficients:

$$(\mathbf{B}_p, \mathbf{A}_p, \mathbf{m}_p) \equiv \underset{\mathbf{B},\mathbf{A},\mathbf{m}}{\operatorname{argmax}} R^2 \left\{ \mathbf{X}, \mathbf{m} + \mathbf{B}\mathbf{A}'\mathbf{X} \right\}. \tag{3.147}$$

From this solution we can identify the coefficients $\mathbf{q}$ and $\mathbf{d}$ by imposing for instance the following condition:

$$\mathrm{E}\{\mathbf{F}\} \equiv \mathbf{0}. \tag{3.148}$$

To present the solution to this problem, we consider the spectral decomposition of the covariance matrix (2.76), which we report here:

$$\operatorname{Cov}\{\mathbf{X}\} \equiv \mathbf{E}\boldsymbol{\Lambda}\mathbf{E}'. \tag{3.149}$$

In this expression $\boldsymbol{\Lambda}$ is the diagonal matrix of the decreasing, positive eigenvalues of the covariance:

$$\boldsymbol{\Lambda} \equiv \operatorname{diag}(\lambda_1, \ldots, \lambda_N); \tag{3.150}$$

and $\mathbf{E}$ is the juxtaposition of the respective eigenvectors:

$$\mathbf{E} \equiv \left(\mathbf{e}^{(1)}, \ldots, \mathbf{e}^{(N)} \right), \tag{3.151}$$

which satisfies $\mathbf{E}\mathbf{E}' = \mathbf{I}_N$. Also, we consider the location-dispersion ellipsoid (2.75) associated with the expected value and the covariance matrix, see Figure 3.14.

First, we present a heuristic argument under the assumption that we only require one factor, i.e. $K \equiv 1$. We guess that this factor reads:

$$F \equiv \left[\mathbf{e}^{(1)} \right]' \mathbf{X}. \tag{3.152}$$

Indeed, from (2.82)-(2.83) the one-dimensional variable F captures the most randomness contained in the invariants that is possible by means of a linear transformation. The variable F represents the orthogonal projection of the variable $\mathbf{X}$ onto the direction defined by the first eigenvector, i.e. the longest principal axis in the location-dispersion ellipsoid.

To recover the N-dimensional invariant $\mathbf{X}$ with an affine transformation of F we must proceed as follows: we choose a fixed vector, i.e. a direction in $\mathbb{R}^N$; we multiply this vector by F; and we add a constant vector $\mathbf{m}$, i.e. we "center" the newly defined recovered variable.

Since the direction that contains most of the randomness in $\mathbf{X}$ is the longest principal axis, we let the random variable F vary along that direction by multiplying it by the first eigenvector $\mathbf{e}^{(1)}$. From (3.152) this means that the recovered invariants become the following affine function of the original invariants:

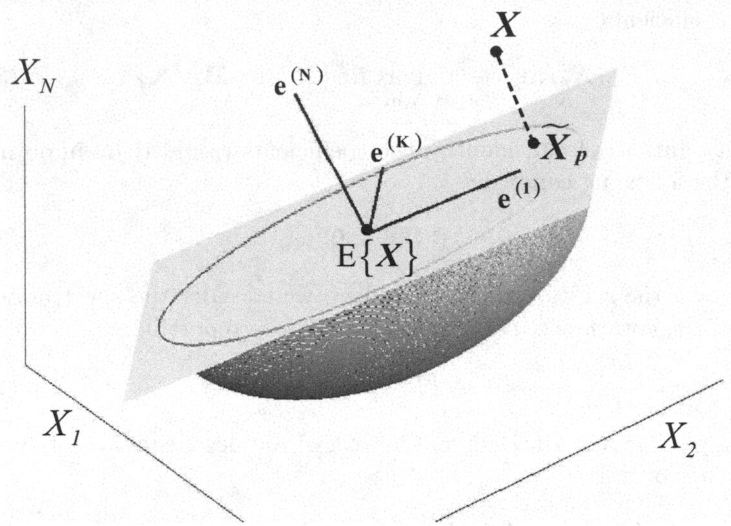

Fig. 3.14. Hidden factor dimension reduction: PCA

$$\widetilde{\mathbf{X}} \equiv \mathbf{m} + \mathbf{e}^{(1)} \left[\mathbf{e}^{(1)}\right]' \mathbf{X}. \tag{3.153}$$

To properly choose $\mathbf{m}$, i.e. to properly center the above recovered invariants, we impose that the expected values of both the original and the recovered invariants be the same. From this condition we immediately obtain:

$$\mathbf{m} \equiv \left(\mathbf{I}_N - \left[\mathbf{e}^{(1)}\right] \left[\mathbf{e}^{(1)}\right]'\right) \mathrm{E}\{\mathbf{X}\}, \tag{3.154}$$

where $\mathbf{I}_N$ is the identity matrix. Notice that with this choice of $\mathbf{m}$ the optimally recovered invariants (3.153) become the orthogonal projection of the original invariants along the direction of the longest principal axis of the location-dispersion ellipsoid. This is the line that contains the maximum possible randomness of the original invariants, i.e. the line that contains the maximum information about the original invariants.

Since (3.153)-(3.154) are in the form (3.145) we would argue that they provide the PCA dimension reduction (3.147) by means of one factor:

$$\{\mathbf{B}_p, \mathbf{A}_p, \mathbf{m}_p\} \equiv \left\{\mathbf{e}^{(1)}, \mathbf{e}^{(1)}, \left(\mathbf{I}_N - \left[\mathbf{e}^{(1)}\right] \left[\mathbf{e}^{(1)}\right]'\right) \mathrm{E}\{\mathbf{X}\}\right\}. \tag{3.155}$$

As far as the factor (3.152) is concerned, in order to satisfy (3.148), we shift it by a scalar as follows:

$$F \equiv \left[\mathbf{e}^{(1)}\right]' \mathbf{X} - \left[\mathbf{e}^{(1)}\right]' \mathrm{E}\{\mathbf{X}\}. \tag{3.156}$$

Since this factor is in the form (3.144), we would argue that it represents the PCA factor, when only one factor is required.

It turns out that the above heuristic arguments and conjectures are correct. Furthermore, they can be generalized to any number K of factors. Indeed, the following statements and results hold, see Brillinger (2001).

Consider the $N \times K$ matrix defined as the juxtaposition of the first K eigenvectors:

$$\mathbf{E}_K \equiv \left(\mathbf{e}^{(1)}, \ldots, \mathbf{e}^{(K)}\right). \tag{3.157}$$

The solution to the PCA dimension reduction problem (3.147) reads:

$$\{\mathbf{B}_p, \mathbf{A}_p, \mathbf{m}_p\} \equiv \{\mathbf{E}_K, \mathbf{E}_K, (\mathbf{I}_N - \mathbf{E}_K\mathbf{E}'_K)\,\mathrm{E}\{\mathbf{X}\}\}, \tag{3.158}$$

which generalizes (3.155).

The hidden factors that optimally summarize the most information in the invariants by means of affine transformations read:

$$\mathbf{F}_p \equiv \mathbf{E}'_K (\mathbf{X} - \mathrm{E}\{\mathbf{X}\}). \tag{3.159}$$

This expression generalizes (3.156).

From the solution (3.158) we also obtain the expression of the PCA-recovered invariants:

$$\widetilde{\mathbf{X}}_p \equiv \mathrm{E}\{\mathbf{X}\} + \mathbf{E}_K\mathbf{E}'_K (\mathbf{X} - \mathrm{E}\{\mathbf{X}\}). \tag{3.160}$$

This expression generalizes (3.153)-(3.154). As we show in Appendix www.3.5, this expression represents the orthogonal projection of the original invariants onto the hyperplane spanned by the K longest principal axes, i.e. the K-dimensional hyperplane that contains the maximum information about the original invariants, see Figure 3.14.

Furthermore, the perturbations in the PCA dimension reduction model, defined as the residuals $\mathbf{U}_p \equiv \mathbf{X} - \widetilde{\mathbf{X}}_p$, have zero expected value and display zero correlation with the factors:

$$\mathrm{E}\{\mathbf{U}_p\} = \mathbf{0}, \quad \mathrm{Cor}\{\mathbf{F}_p, \mathbf{U}_p\} = \mathbf{0}_{K \times N}, \tag{3.161}$$

see Appendix www.3.5.

Quite obviously, the quality of the approximation provided by the recovered invariants (3.160) depends on the number K of factors. Indeed, we prove in Appendix www.3.5 that the generalized r-square (3.116) can be expressed in terms of the eigenvalues (3.150) of the covariance matrix as follows:

$$R^2\left\{\mathbf{X}, \widetilde{\mathbf{X}}_p\right\} = \frac{\sum_{n=1}^{K} \lambda_n}{\sum_{n=1}^{N} \lambda_n}. \tag{3.162}$$

This expression is intuitive. Adding the generic K-th factor to a $(K-1)$-factor PCA analysis corresponds to adding one dimension to the hyperplane on which the invariants are projected, namely the direction of the K-th largest principal axis of the location-dispersion ellipsoid. On the other hand, the K-th eigenvalue is the variance of the K-th factor:

$$\operatorname{Var}\{F_n\} = \left[\mathbf{e}^{(n)}\right]' \mathbf{E}\boldsymbol{\Lambda}\mathbf{E}' \left[\mathbf{e}^{(n)}\right] = \lambda_n. \tag{3.163}$$

We can thus interpret the K-th eigenvalue as the contribution to the total recovered randomness obtained by adding the K-th dimension of randomness.

In this respect, the numerator in (3.162) is the cumulative contribution to total randomness from the K main dimensions of randomness. Similarly, the denominator is the cumulative contribution to total risk from all the factors, i.e. the denominator represents the total randomness in the invariants.

To summarize, the generalized r-square is the percentage cumulative contribution to total randomness from the K main dimensions of randomness. Notice that the eigenvalues are sorted in decreasing order. Therefore, the marginal contribution of adding one factor decreases with the number of factors.

Idiosyncratic perturbations

Principal component analysis is not the only way to specify the linear hidden-factor model (3.143), which we report here:

$$\mathbf{X} \equiv \mathbf{q} + \mathbf{B}\mathbf{F}(\mathbf{X}) + \mathbf{U}. \tag{3.164}$$

Among other options, one can impose that each of the residual perturbations refer to one and only one invariant, i.e. that the entries of $\mathbf{U}$ be independent of one another.

Imposing this constraint corresponds to factoring the randomness in the market into K contributions common to all the market invariants and N *idiosyncratic perturbations* each of which affects only one invariant.

Nevertheless, the assumption that the perturbations be independent of one another is too strong in general markets. Even the much weaker assumption that the perturbations be uncorrelated is too strong. Indeed this hypothesis, together with the standard assumption (3.118) that factors and perturbations be uncorrelated, is equivalent to the following condition:

$$\operatorname{Cov}\{X_m, X_n\} = \left[\mathbf{B}\operatorname{Cov}\{\mathbf{F}(\mathbf{X})\}\mathbf{B}'\right]_{mn}, \quad \text{for all } m \neq n. \tag{3.165}$$

This condition can be satisfied in general only in approximation.

Furthermore, the common factors and factor loadings can be identified only modulo an invertible transformation, which we can, but do not have to, assume linear. In other words, if a pair $(\mathbf{F}, \mathbf{B})$ yields a viable model (3.164), so does the pair $\left(\mathbf{A}\mathbf{F}, \mathbf{B}\mathbf{A}^{-1}\right)$ for any conformable invertible matrix $\mathbf{A}$.

3.4.3 Explicit vs. hidden factors

At this point the legitimate question might arise, whether in order to summarize the randomness in the market is it better to use explicit factors, as discussed in Section 3.4.1, or hidden factors, as discussed in Section 3.4.2.

In general, explicit factor models are easier to interpret, whereas hidden factor models tend to provide a better explanatory power. The first statement is straightforward, therefore we focus on the comparison of the explanatory power of the two methods. Nevertheless, each situation should be evaluated independently.

Consider a generic PCA dimension reduction on the first K factors of an N-dimensional set of invariants $\mathbf{X}$. From (3.160) this process recovers the following invariants:

$$\widetilde{\mathbf{X}}_p \equiv \mathrm{E}\{\mathbf{X}\} + \mathbf{E}_K \mathbf{E}_K' (\mathbf{X} - \mathrm{E}\{\mathbf{X}\}). \tag{3.166}$$

We recall that the recovered invariants represent the projection of the original invariants onto the K-dimensional hyperplane of maximum randomness spanned by the first K principal axes of the location-dispersion ellipsoid, see Figure 3.15 and compare with Figure 3.14.

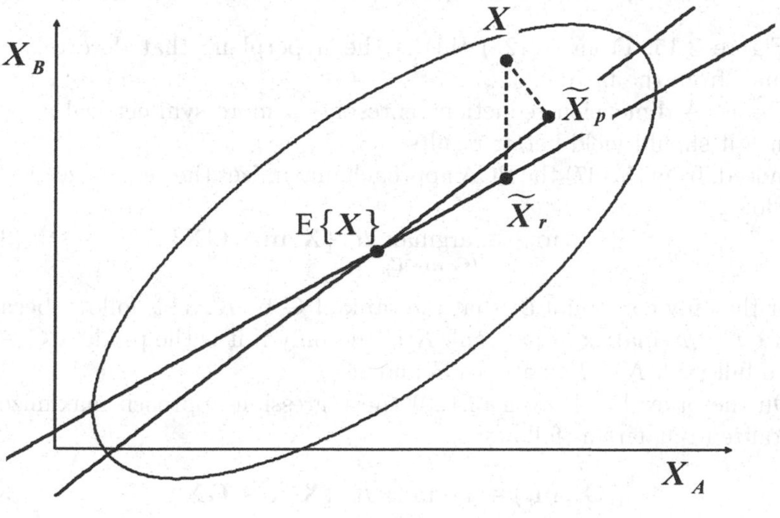

Fig. 3.15. Regression vs. PCA dimension reduction

In order to compare the PCA results with an explicit-factor model we need to restrict our analysis to endogenous explicit-factor models. In other words, first we split the invariants into two subsets:

$$\mathbf{X} \equiv \begin{pmatrix} \mathbf{X}_A \\ \mathbf{X}_B \end{pmatrix}, \tag{3.167}$$

where $\mathbf{X}_A$ is a set of K among the N entries of $\mathbf{X}$, and $\mathbf{X}_B$ is the set of the remaining entries. As factors, we consider the variables $\mathbf{X}_A$ and a constant:

$$\mathbf{F} \equiv \begin{pmatrix} 1 \\ \mathbf{X}_A \end{pmatrix}. \tag{3.168}$$

This regression model is completely endogenous, in that the factors are a function of the original invariants. From (3.127) the recovered invariants read:

$$\widetilde{\mathbf{X}}_r \equiv \mathrm{E}\{\mathbf{X}\} + \mathbf{V}_{A,B}(\mathbf{X} - \mathrm{E}\{\mathbf{X}\}), \tag{3.169}$$

where

$$\mathbf{V}_{A,B} \equiv \left(\begin{array}{c|c} \mathbf{I}_K & \mathbf{0}_{K,N-K} \\ \hline \mathrm{Cov}\{\mathbf{X}_B, \mathbf{X}_A\} \mathrm{Cov}\{\mathbf{X}_A\}^{-1} & \mathbf{0}_{N-K,N-K} \end{array} \right). \tag{3.170}$$

Geometrically, the recovered invariants represent the projection of the original invariants along the direction defined by the reference axes of $\mathbf{X}_B$ onto the K-dimensional hyperplane that passes through the expected value and satisfies the following parametric equation:

$$\mathbf{x}_B = \mathrm{E}\{\mathbf{X}_B\} + \mathrm{Cov}\{\mathbf{X}_B, \mathbf{X}_A\} \mathrm{Cov}\{\mathbf{X}_A\}^{-1}(\mathbf{x}_A - \mathrm{E}\{\mathbf{X}_A\}), \tag{3.171}$$

see Figure 3.15. From (3.128) this is the hyperplane that decorrelates the residuals from the factors.

The PCA dimension reduction represents a more symmetrical approach. As such it should yield better results.

Indeed, from (3.147) the PCA approach maximizes the generalized r-square as follows:

$$(\mathbf{G}_p, \mathbf{m}_p) \equiv \underset{(\mathbf{G},\mathbf{m}) \in \mathcal{C}_p}{\mathrm{argmax}}\, R^2\{\mathbf{X}, \mathbf{m} + \mathbf{G}\mathbf{X}\}, \tag{3.172}$$

under the only constraint $\mathcal{C}_p$ that the rank of $\mathbf{G}$ be K. This follows because a generic $N \times N$ matrix $\mathbf{G}$ has rank K if and only if it is the product $\mathbf{G} \equiv \mathbf{B}\mathbf{E}'$ of two full-rank $N \times K$ matrices $\mathbf{B}$ and $\mathbf{E}$.

On the other hand, from (3.120) the regression approach maximizes the generalized r-square as follows:

$$(\mathbf{G}_r, \mathbf{m}_r) \equiv \underset{(\mathbf{G},\mathbf{m}) \in \mathcal{C}_r}{\mathrm{argmax}}\, R^2\{\mathbf{X}, \mathbf{m} + \mathbf{G}\mathbf{X}\}, \tag{3.173}$$

under a much stronger set of constraints:

$$\mathcal{C}_r : \begin{cases} \mathbf{m}_A \equiv \mathbf{0}_K \\ \mathbf{G}_{AA} \equiv \mathbf{I}_K \\ \mathbf{G}_{AB} \equiv \mathbf{0}_{K,N-K} \\ \mathbf{G}_{BB} \equiv \mathbf{0}_{N-K,N-K}. \end{cases} \tag{3.174}$$

Therefore the PCA approach yields better results:

$$R^2\left\{\mathbf{X},\widetilde{\mathbf{X}}_p\right\} \geq R^2\left\{\mathbf{X},\widetilde{\mathbf{X}}_r\right\}. \tag{3.175}$$

Nevertheless, the regression approach displays other advantages. For instance, the regression dimension reduction is invariant under any rescaling of the factors (3.168), whereas the PCA approach is only invariant under a global rescaling of all the invariants. In practice, one has to be careful to measure the variables in homogenous units when implementing PCA dimension reduction, whereas this is not necessary when implementing regression dimension reduction.

Furthermore, the explanatory power as summarized by the generalized r-square is a statistical identity, whereas the word "explaining" is closely related to the word "understanding": in other words the interpretation of the K invariants $\mathbf{X}_A$ is clear, whereas the interpretation of the K PCA factors might be more obscure.

Finally, if the explanatory variables $\mathbf{X}_A$ are chosen appropriately among the invariants $\mathbf{X}$, regression and principal component analysis yield similar reductions, i.e. (3.175) approaches an equality.

3.4.4 Notable examples

We present here a few notable examples of dimension reduction in the financial markets by means of the techniques discussed in this section: a model for equities, based on one explicit factor and related to the Capital Asset Pricing Model; another model for equity, namely the Fama-French regression, based on three explicit factors; a model for the fixed income market, based on three hidden factors, namely the level-slope-hump PCA decomposition of the yield curve; and a hidden-factor model with idiosyncratic perturbations, related to the Arbitrage Pricing Theory.

Explicit factors and the Capital Asset Pricing Model

Consider a broad stock index like the S&P 500, whose value at the generic time t we denote as M_t. Consider as invariants for a market of N stocks the linear returns (3.10):

$$L_{t,\tau}^{(n)} \equiv \frac{P_t^{(n)}}{P_{t-\tau}^{(n)}} - 1, \quad n = 1,\ldots,N. \tag{3.176}$$

Consider an explicit factors linear model (3.119) based on a constant and one explicit factor, defined as the linear return on the market index:

$$F_{t,\tau}^{M} \equiv \frac{M_t}{M_{t-\tau}} - 1. \tag{3.177}$$

In this case the regression (3.127) recovers the following portion of the stock returns:

$$\widetilde{L}_{t,\tau}^{(n)} \equiv \mathrm{E}\left\{L_{t,\tau}^{(n)}\right\} + \beta_{\tau}^{(n)} \left(F_{t,\tau}^{M} - \mathrm{E}\left\{F_{t,\tau}^{M}\right\}\right), \tag{3.178}$$

where the regression coefficient β is called the *beta* of the stock. From (3.127), the beta is defined as follows:

$$\beta_{\tau}^{(n)} \equiv \frac{\mathrm{Cov}\left\{L_{t,\tau}^{(n)}, F_{t,\tau}^{M}\right\}}{\mathrm{Var}\left\{F_{t,\tau}^{M}\right\}}. \tag{3.179}$$

Notice that the beta depends on the interval τ. Had we used compounded returns as invariants instead, the "square-root rule" (3.76) would have made the beta independent of the interval.

Suppose that the distribution of the linear returns of each stock satisfies the following additional constraint:

$$\mathrm{E}\left\{L_{t,\tau}^{(n)}\right\} = \beta_{\tau}^{(n)}\,\mathrm{E}\left\{F_{t,\tau}^{M}\right\} + \left(1 - \beta_{\tau}^{(n)}\right) R_{t,\tau}^{f}. \tag{3.180}$$

In this expression the *risk-free rate* R is the return on a zero-coupon bond from a period τ before maturity until maturity, which in the notation of Section 3.1.2 reads:

$$R_{t,\tau}^{f} \equiv \left(\frac{1}{Z_{t-\tau}^{(t)}} - 1\right). \tag{3.181}$$

Then the explicit factor model (3.178) becomes the *Capital Asset Pricing Model (CAPM)* of Sharpe (1964), and Lintner (1965), a general equilibrium model for the markets which recovers the following portion of the stock returns:

$$\widetilde{L}_{t,\tau}^{(n)} \equiv R_{t,\tau}^{f} + \beta_{\tau}^{(n)} \left(F_{t,\tau}^{M} - R_{t,\tau}^{f}\right). \tag{3.182}$$

See Ingersoll (1987) for an introduction to the CAPM.

Market-size-type explicit factors

A notable three-factor model for linear returns on stocks is discussed in Fama and French (1993). We consider a set of N stocks, where the generic n-th stock trades at time t at the price $P_t^{(n)}$ and we specify the invariants as the compounded returns (3.11) on these stocks:

$$C_{t,\tau}^{(n)} \equiv \ln\left(\frac{P_t^{(n)}}{P_{t-\tau}^{(n)}}\right). \tag{3.183}$$

The first explicit factor, in addition to a constant, is the compounded return C^M of a broad stock index like the S&P 500. The second factor is the difference SmB ("small minus big") between the compounded returns of a small-cap stock index and the compounded returns of a large-cap stock index; the third factor is the difference HmL ("high minus low") between the compounded returns of a large book-to-market-value stock index and the compounded returns of a small book-to-market-value stock index. Therefore, this *market-size-type* three-factor linear model reads:

$$\begin{aligned} C_{t,\tau}^{(n)} \equiv \mathrm{E}\left\{C_{t,\tau}^{(n)}\right\} &+ \beta^{(n)}\left(C_{t,\tau}^{M} - \mathrm{E}\left\{C_{t,\tau}^{M}\right\}\right) \\ &+\gamma^{(n)}\left(SmB_{t,\tau} - \mathrm{E}\left\{SmB_{t,\tau}\right\}\right) \\ &+\zeta^{(n)}\left(HmL_{t,\tau} - \mathrm{E}\left\{HmL_{t,\tau}\right\}\right) + U_{t,\tau}^{(n)}, \end{aligned} \tag{3.184}$$

where n ranges through all the stocks considered. From (3.127), the regression coefficients (β, γ, ζ) are defined in terms of the cross-covariances among factors and invariants: due to the "square-root" property (3.76), these coefficients do not depend on the estimation interval τ.

Hidden factors and principal component analysis

One of the most widely used applications of hidden factor dimension reduction stems from the principal component analysis of the yield curve. We detail every step of this analysis in our case study, see Section 3.5.2.

Hidden factors and the arbitrage pricing theory model

A notable example of the idiosyncratic approach to hidden factors linear models (3.164) is provided by the *Arbitrage Pricing Theory (APT)* of Ross (1976). Like the CAPM, this is a factor model for the linear returns of the stocks in a broad index such as the S&P 500:

$$\mathbf{L} \equiv \mathrm{E}\left\{\mathbf{L}\right\} + \mathbf{BF}\left(\mathbf{L}\right) + \mathbf{U}. \tag{3.185}$$

The APT superimposes a restriction on the distribution of the linear returns, namely:

$$\mathrm{E}\left\{\mathbf{L}\right\} = \xi_0 \mathbf{1} + \mathbf{B}\boldsymbol{\xi}, \tag{3.186}$$

where $\mathbf{1}$ is an N-dimensional vector of ones, ξ_0 is a constant and $\boldsymbol{\xi}$ is a K-dimensional vector of risk premia. See Ingersoll (1987) and Connor and Korajczyk (1995) for an introduction to the APT.

3.4.5 A useful routine

In the context of dimension reduction, a challenging problem that often arises is the *selection* of the best K in a pool of N potential candidates to perform a given task. This is a combinatorial problem. The pool of candidates can be indexed by the first N integers:

$$I_N \equiv \{1, \ldots, N\}; \tag{3.187}$$

we have to consider all the possible combinations of K elements from the pool of candidates:

$$I_K \equiv \{n_1, \ldots, n_K\} ; \tag{3.188}$$

and we must select the best combination I_K^* among all the above combinations.

For example, consider reducing the dimension by means of an explicit factor model as in Section 3.4.1. There exists a pool of N potential explicit factors:

$$\mathbf{F}_{I_N} \equiv (F_1, \ldots, F_N)' , \tag{3.189}$$

but eventually we only consider K among the N potential factors:

$$\mathbf{F}_{I_K^*} \equiv \left(F_{n_1^*}, \ldots, F_{n_K^*}\right)' . \tag{3.190}$$

As another example, consider an allocation problem in a market of N securities, where the final portfolio is constrained to contain a number K of these securities. This dimension-reduction problem is known as *portfolio replication*, namely replicating with as few as K securities a portfolio that should ideally contain N securities.

The best combination I_K^* is defined as the one that maximizes a given objective $\mathcal{O}$:

$$I_K^* = \underset{I_K \subset I_N}{\operatorname{argmax}} \, \mathcal{O}(I_K) . \tag{3.191}$$

For instance, in the case of regression dimension reduction the objective is represented by the generalized r-square:

$$\mathcal{O}(I_K) \equiv R^2 \left\{ \mathbf{X}, \widetilde{\mathbf{X}}(I_K) \right\} , \tag{3.192}$$

see (3.120). In this expression $\widetilde{\mathbf{X}}$ follows from (3.121) and reads:

$$\widetilde{\mathbf{X}}(I_K) \equiv \mathrm{E}\left\{\mathbf{X}\mathbf{F}_{I_K}'\right\} \mathrm{E}\left\{\mathbf{F}_{I_K}\mathbf{F}_{I_K}'\right\}^{-1} \mathbf{F}_{I_K} . \tag{3.193}$$

An alternative specification of the objective is provided for instance by the Akaike criterion, see Parzen, Tanabe, and Kitagawa (1998).

In the case of the PCA approach to dimension reduction the selection problem does not exist, because the PCA factors are naturally sorted in decreasing order of importance, i.e. $I_K^* \equiv (1, \ldots, K)$.

In a portfolio replication problem, the objective is minimizing the tracking error:

$$\mathcal{O}(I_K) \equiv -\,\mathrm{TE}\left(\boldsymbol{\alpha}(I_K)\right) , \tag{3.194}$$

see (6.179) later in the text.

Combinatorial problems are computationally very challenging. Indeed, the optimization (3.191) implies evaluating the objective $\binom{N}{K}$ times. Furthermore, the number K is often a decision variable. In other words, the optimal number K is only decided after evaluating the trade-offs of the dimension reduction process, i.e. after computing the following function:

$$K \mapsto \mathcal{O}\left(I_K^*\right), \quad K = 1, \ldots, N. \tag{3.195}$$

For instance, in portfolio replication problems, the ideal number K of securities in the final portfolio is evaluated according to the trade-off between the quality of the replication and the transaction costs.

Computing (3.195) implies evaluating the objective the following number of times:

$$\sum_{K=1}^{N} \binom{N}{K} = 2^N - 1. \tag{3.196}$$

This number is exorbitant precisely when a dimension reduction is most needed, namely when N is large. Here we propose a routine which evaluates the objective only the following number of times:

$$\sum_{K=1}^{N} K = \frac{N\left(N+1\right)}{2}. \tag{3.197}$$

The routine proceeds as follows:

Step 0. Set $K \equiv N$, and consider the initial set $I_K \equiv \{1, \ldots, N\}$

Step 1. Consider the K sets obtained from I_K by dropping the generic k-th element:

$$I_K^k \equiv \{n_1, \ldots, n_{k-1}, n_{k+1}, \ldots n_K\}, \quad k = 1, \ldots, K. \tag{3.198}$$

Step 2. Evaluate the above sets:

$$k \mapsto v_K^k \equiv \mathcal{O}\left(I_K^k\right), \quad k = 1, \ldots, K. \tag{3.199}$$

Step 3. Determine the worst element in I_K:

$$k^* \equiv \underset{k \in \{1, \ldots, K\}}{\operatorname{argmax}} \left\{v_K^k\right\}. \tag{3.200}$$

Step 4. Drop the worst element in I_K:

$$I_{K-1} \equiv I_K^{k^*}. \tag{3.201}$$

Step 5. If $K = 2$ stop. Otherwise set $K \equiv K - 1$ and go to Step 1.

Although this routine yields suboptimal results, in practice it proved very close to optimal in a variety of applications. In other words, the function

$$K \mapsto v_K^{k^*}, \quad K = 1, \ldots, N. \tag{3.202}$$

is in general a very good approximation of (3.195).

3.5 Case study: modeling the swap market

In this section we discuss how to model the swap market. Swaps are very liquid securities and many new contracts are traded every day. A *v-swap* $(E - t)$-*forward* is a contract whose value at the generic time t reads:

$$P_t^{(E,v)} \equiv s\rho \sum_{k=1}^{v/\rho} Z_t^{(E_k)} + Z_t^{(E+v)} - Z_t^{(E)}. \tag{3.203}$$

In this formula s is the agreed upon *fixed rate* expressed in annualized terms: at inception t_0 this rate is typically set in such a way that the value of the contract zero, i.e. it is set as the $(E - t_0)$-into-v forward par swap rate defined in (3.57); ρ is a fixed time-interval of the order of a few months; the generic term $E_k \equiv E + k\rho$ is one *fixed-leg payment date*; $Z_t^{(E)}$ is the price of a zero-coupon bond with maturity E. The pricing formula (3.203) originates from the structure of the contract, according to which agreed upon fixed payments are swapped against floating payments that depend on the current levels of interest rates, see Rebonato (1998) and Brigo and Mercurio (2001). Nevertheless, we can take (3.203) as the definition of a security.

In this case study the investment decision is taken at $T \equiv$ January 1st 2000 and we plan to invest in an "eight-year swap two-years forward", i.e. a swap that starts $(E - T) \equiv$ two years from the investment date on $E \equiv$ January 1st 2002 and ends $v \equiv$ eight years later on $E + v \equiv$ January 1st 2010. The fixed payments occur every $\rho \equiv$ three months. Therefore, this contract is determined by the price of thirty-three zero-coupon bonds.

We assume that the investment horizon is $\tau \equiv$ two months. Our aim is to determine the distribution of $P_{T+\tau}^{(E,v)}$. To do this, we dispose of the daily database of all the zero-coupon bond prices for the past ten years.

3.5.1 The market invariants

Everyday, many new forward swap contracts are issued with new starting and ending dates. Therefore, the swap market is completely priced by the set of all the zero-coupon bond prices for virtually all the maturities on a daily basis up to around thirty years in the future:

$$Z_t^{(E)} \text{ such that } E = t + 1d, t + 2d, \ldots, t + 30y. \tag{3.204}$$

The first step to model a market is to determine its invariants. We have seen in Section 3.1.2 that the natural invariants for the fixed-income market are the changes in yield to maturity:

$$X_{t,\widetilde{\tau}}^{(v)} \equiv Y_t^{(v)} - Y_{t-\widetilde{\tau}}^{(v)}. \tag{3.205}$$

In this expression $\widetilde{\tau}$ is the estimation interval and v denotes a specific time to maturity in the *yield curve*, which is the plot of the yield to maturity as a function of the respective time to maturity:

$$v \mapsto Y_t^{(v)} \equiv -\frac{1}{v} \ln \left(Z_t^{(t+v)} \right), \quad v = 1d, 2d, \ldots, 30y. \tag{3.206}$$

If we were to invest in several swap contracts, due to the large number (3.204) of bonds involved in the swap market, we would need to model the joint distribution of the changes in yield to maturity for the whole yield curve. Nevertheless, even for our example of one swap contract we still need to model a big portion of the swap curve, namely the sector between two and ten years.

In our example we have access to the database of the prices of the zero-coupon bonds (3.204) every day for the past ten years. Equivalently, we have access to the whole yield curve (3.206) every day for the past ten years:

$$v \mapsto y_t^{(v)} \equiv -\frac{1}{v} \ln \left(z_t^{(t+v)} \right), \quad \begin{cases} v = 1d, 2d, \ldots, 30y \\ t = T - 10y, \ldots, T - 1d, T. \end{cases} \tag{3.207}$$

The lower case letters in this expression denote the realizations in the past of the respective random variables (3.206), which we denote with upper case letters.

We remark that in reality the zero-coupon bonds are not traded in the swap market and therefore their price is not directly available. Instead (3.203) represents the set of implicit equations, one for each swap contract, that define the prices of the underlying zero coupon bonds. The process of determining the zero-coupon prices from the prices of the swap contracts is called *bootstrapping*, see James and Webber (2000). This operation is performed by standard software packages.

To determine the distribution of the changes in the yield curve (3.205) we choose an estimation interval $\widetilde{\tau} \equiv$ one week, which presents a reasonable trade-off between the number of observations in the database and the reliability of the data with respect to the investment horizon. Indeed, the number of weekly observations from a ten-year sample exceeds five hundred. If we chose an estimation interval $\widetilde{\tau}$ equal to the investment horizon τ of two months, the number of observations in the dataset would be too small, i.e. about sixty observations for ten years of data. On the other hand, an estimation interval as short as, say, one day might give rise to spurious data and would not be suitable to extrapolate the distribution of the invariants at the investment horizon.

3.5.2 Dimension reduction

Consider the weekly invariants (3.205) relative to the section of the yield curve that prices our eight year swap two years forward:

$$X^{(v)} \equiv Y_t^{(v)} - Y_{t-\widetilde{\tau}}^{(v)}, \quad v = 2y, 2y + 1d, \ldots, 10y. \tag{3.208}$$

To ease the notation in this expression we dropped in the left hand side the specification of the estimation interval $\widetilde{\tau}$, which is fixed, and the dependence on time t, because the distribution of the invariants does not depend on t.

The invariants (3.208) constitute a set of a few thousand random variables. Therefore we need to reduce the dimension of the market invariants. In view of the principal component analysis approach to dimension reduction we focus on the covariance matrix of the weekly invariants:

$$C(v,p) \equiv \operatorname{Cov}\left\{X^{(v)}, X^{(v+p)}\right\}. \tag{3.209}$$

The following intuitive relations can be checked with the data.

The covariance matrix is a smooth function of the times to maturity in both directions. For example, the covariance of the three-year rate with the five-year rate is very close to both the covariance of the three-year rate with the five-year-plus-one-day rate and to the covariance of the three-year-plus-one-day rate with the five-year rate. Therefore:

$$C(v, p + dv) \approx C(v + dv, p), \tag{3.210}$$

which means that C is a smooth function of its arguments.

The diagonal elements of the covariance matrix, i.e. the variances of the rate changes at the different maturities, are approximately similar. For example, if borrowing money for three years becomes all of a sudden more expensive, so does borrowing money for ten years, and the change is approximately similar. Therefore:

$$C(v, 0) \approx C(v + \tau, 0). \tag{3.211}$$

In particular, the correlation matrix is approximately proportional to the covariance matrix.

The correlation of equally spaced times to maturity is approximately the same. For example, the correlation of the one-year rate with the two-year rate is approximately similar to the correlation of the four-year rate with the five-year rate. Therefore:

$$C(v, p) \approx C(v + \tau, p). \tag{3.212}$$

The correlation matrix decreases away from the diagonal. For example, the correlation of the one-year rate with the two-year rate is higher than the correlation of the one-year rate with the five-year rate.

From the above properties we derive that the covariance matrix, in addition to being symmetric and positive, has the following approximate structure:

$$C(v, p) \approx h(p), \tag{3.213}$$

where h is a smooth, positive and decreasing function that is symmetrical around the origin:

$$h(p) = h(-p). \tag{3.214}$$

A matrix with this structure is called a *Toeplitz matrix*, see Figure 3.16. Therefore, the covariance matrix is a symmetric and positive smooth Toeplitz matrix that decays to zero away from the diagonal.

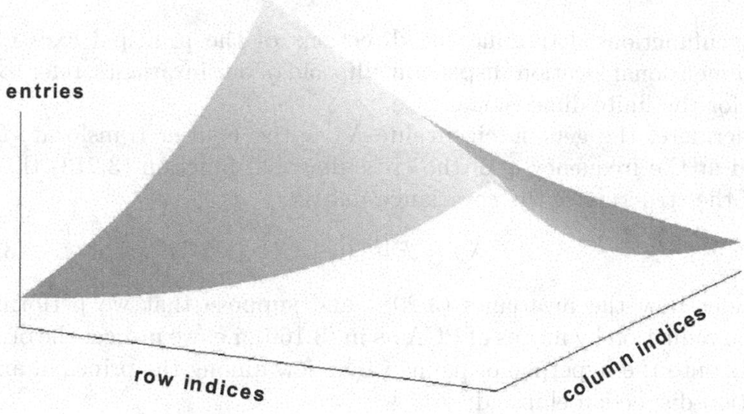

Fig. 3.16. Toeplitz matrix

The continuum limit

Although our ultimate purpose is to reduce the dimension of the swap market invariants, to gain more insight into the structure of randomness of this market we start looking in the opposite direction, namely the infinite-dimensional, continuum limit of the yield curve.

Indeed, the set of possible times to maturity is so dense, i.e. every day from two to ten years, that we can consider this parameter as a continuum: $v \in [2, 10]$. Therefore we consider the yield curve $Y_t^{(v)}$ as a stochastic object parametrized by two continuous indices: time t and time to maturity v.

We now perform the principal component analysis (PCA) of the yield curve and then perform the PCA dimension reduction discussed in Section 3.4.2. We recall from (3.149) that the PCA decomposition of the covariance matrix of the invariants reads:

$$\mathrm{Cov}\{\mathbf{X}\}\,\mathbf{e}^{(n)} = \lambda_n \mathbf{e}^{(n)}, \tag{3.215}$$

for each eigenvector $\mathbf{e}^{(n)}$ and each eigenvalue λ_n, $n = 1, \ldots, N$.

By means of the analogies in Table B.4, Table B.11 and Table B.20, in the continuum (3.215) becomes the following spectral equation:

$$\int_{\mathbb{R}} \mathrm{Cov}\left\{X^{(v)}, X^{(p)}\right\} e^{(\omega)}(p)\, dp = \lambda_\omega e^{(\omega)}(v), \tag{3.216}$$

where λ_ω is the generic eigenvalue and $e^{(\omega)}$ is the *eigenfunction* relative to that eigenvalue.

We prove in Appendix www.3.6 that the generic eigenfunction of a Toeplitz covariance matrix must be an oscillating function with frequency ω, modulo a multiplicative factor:

$$e^{(\omega)}(\upsilon) \equiv e^{i\omega\upsilon}, \quad \omega \in [0, +\infty). \tag{3.217}$$

These eigenfunctions determine the directions of the principal axes of the infinite-dimensional location-dispersion ellipsoid of our invariants, refer to Figure 3.14 for the finite-dimensional case.

Furthermore, the generic eigenvalue λ_ω is the Fourier transform $(B.34)$ evaluated at the frequency ω of the cross-diagonal function (3.213) that determines the structure of the covariance matrix:

$$\lambda_\omega = \mathcal{F}[h](\omega). \tag{3.218}$$

Consider now the invariants (3.208) and suppose that we perform the dimension reduction by means of PCA as in (3.160), i.e. we project the original invariants onto the hyperplane spanned by a few among the principal axes of the location-dispersion ellipsoid:

$$X^{(\upsilon)} \mapsto \widetilde{X}^{(\upsilon)}, \quad \upsilon \in [2, 10]. \tag{3.219}$$

This corresponds to selecting a subset Ω of all the possible frequencies. To evaluate the quality of the PCA dimension reduction, we compute the generalized r-square (3.162), which in this context reads:

$$R^2\left\{X, \widetilde{X}\right\} \equiv \frac{\int_\Omega \lambda_\omega d\omega}{\int_0 \lambda_\omega d\omega}. \tag{3.220}$$

In order to select the best frequencies Ω we need to assign a parametric form to the cross-diagonal function (3.213) that determines the structure of the covariance matrix. To this purpose, we introduce the concept of *string*, or *random field*, see James and Webber (2000): a string is a stochastic object parametrized by two continuous indices. In our case, the yield curve is a string, and the market invariants, i.e. the changes in yield to maturity (3.208), become the discrete increment of a random field along the time dimension. A result in Kennedy (1997) states that, under fairly general conditions, the covariance of a family of invariants that stem from a random field has the following structure:

$$\text{Cov}\left\{X^{(\upsilon)}, X^{(\upsilon+p)}\right\} = \sigma^2 e^{-\mu\upsilon} e^{-\gamma|p|}, \tag{3.221}$$

where $\mu \geq 0$ and $\gamma \geq \mu/2$.

Indeed, this is the structure of the covariance matrix that we have derived in (3.213), where $\mu \approx 0$. In other words, the covariance of the weekly changes in yield to maturity in the continuum limit has the following cross-diagonal functional form:

$$h(p) = \sigma^2 \exp(-\gamma|p|). \tag{3.222}$$

Substituting this expression in (3.218) we obtain the explicit form of the eigenvalues:[4]

---

[4] The reader might notice that this principal component analysis is the spectral analysis of a one-dimensional Ornstein-Uhlenbeck process.

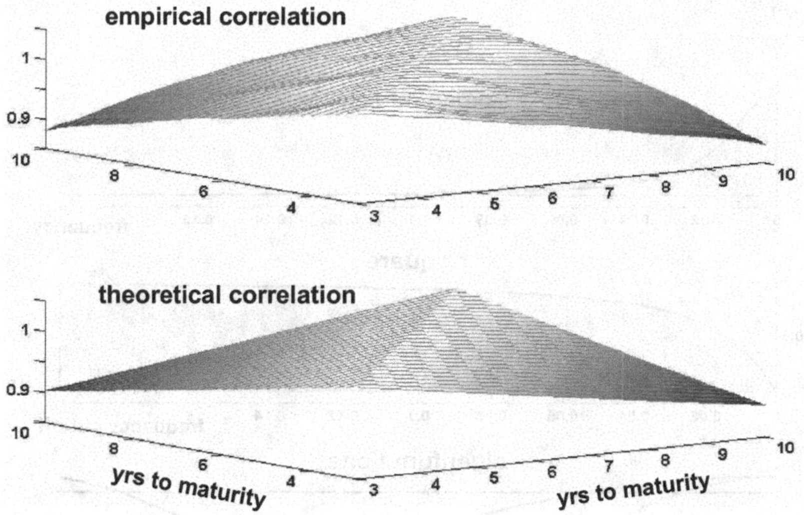

Fig. 3.17. Correlations among changes in interest rates

$$\lambda_\omega = \frac{2\sigma^2}{\sqrt{\gamma^2}} \left(1 + \frac{\omega^2}{\gamma^2}\right)^{-1}, \tag{3.223}$$

see Appendix www.3.6. Notice that the eigenvalues decrease with the frequency. Indeed, (3.223) is the *Lorentz function*, which is proportional to the probability density function of a Cauchy distribution centered in the origin $\omega \equiv 0$ with dispersion parameter γ, see (1.79). In other words, the set of preferred frequencies reads:

$$\Omega \equiv [0, \overline{\omega}], \tag{3.224}$$

for some cut-off value $\overline{\omega}$.

To choose the proper cut-off, we fit the theoretical expression of the correlation matrix, which we obtain from (3.222), to the empirical correlation matrix:

$$\mathrm{Cor}\left\{X^{(v)}, X^{(v+p)}\right\} = \exp\left(-\gamma\,|p|\right). \tag{3.225}$$

In the swap market, measuring time to maturity in years, we obtain for γ the numerical value $\gamma \approx 0.0147$. Such a low value of the parameter corresponds to highly correlated changes in yield, see Figure 3.17. Since the parameter γ in (3.225) is small, i.e. the changes in yield at different times to maturity are highly correlated, the eigenvalues (3.223) decrease sharply to zero as the respective frequency moves away from the origin. We plot this profile in the top portion of Figure 3.18. This situation is typical of the Fourier transform: when a function decays slowly to infinity, its Fourier transform decays fast, and viceversa, as we see for instance in (B.37). At an extreme, the Fourier transform of the Dirac delta centered in zero (B.38) is a constant, i.e. a flat function.

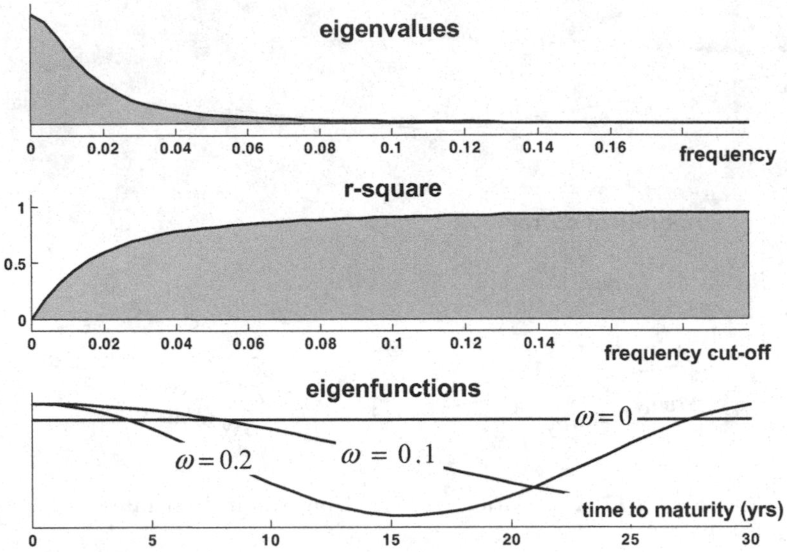

Fig. 3.18. Swap curve PCA: the continuum limit

Therefore, we expect the lowest frequencies to recover almost all of the randomness in the swap market. To quantify this more precisely, we compute the generalized r-square (3.220) obtained by considering only the lowest frequencies. A simple integration of (3.223) yields the following analytical expression:

$$R^2 \left\{ X, \widetilde{X} \right\} \equiv \frac{\int_0^{\overline{\omega}} \lambda_\omega d\omega}{\int_0^{+\infty} \lambda_\omega d\omega} = \frac{2}{\pi} \arctan \left(\frac{\overline{\omega}}{\gamma} \right). \tag{3.226}$$

In the middle portion of Figure 3.18 we display the generalized r-square (3.226) as a function of the cut-off frequency $\overline{\omega}$: the bulk of randomness is recovered by frequencies lower than $\overline{\omega} \approx 0.2$.

In the bottom portion of Figure 3.18 we see that this frequency corresponds to eigenfunctions that complete an oscillation over a thirty-year period. This is the span of time-to-maturities covered by the swap market.

To summarize, we draw the following lesson from the continuum-limit of the swap market. Since changes in interest rates are highly correlated, we can reduce the dimension of randomness in the swap market by considering a limited number of directions of randomness, i.e. those directions defined by the eigenfunctions that oscillate less than once within the set of time to maturities considered.

The discrete and finite case

Given the high correlation among changes in yield at adjacent times to maturity, instead of considering a continuous yield curve, we can safely consider time to maturities one year apart and implement the PCA dimension reduction on this discrete set, see Litterman and Scheinkman (1991).

In our example, this step shrinks the dimension from infinity to nine:

$$\mathbf{X} \equiv \left(X^{(2y)}, \ldots, X^{(10y)} \right)'. \tag{3.227}$$

First of all we estimate the 9×9 covariance matrix as in Chapter 4. Then we perform the PCA decomposition (3.149) of the covariance matrix:

$$\operatorname{Cov}\{\mathbf{X}\} = \mathbf{E}\boldsymbol{\Lambda}\mathbf{E}'. \tag{3.228}$$

In this expression $\boldsymbol{\Lambda}$ is the diagonal matrix of the $N \equiv 9$ eigenvalues of the covariance sorted in decreasing order:

$$\boldsymbol{\Lambda} \equiv \operatorname{diag}(\lambda_1, \ldots, \lambda_9); \tag{3.229}$$

and the matrix $\mathbf{E}$ is the juxtaposition of the respective eigenvectors and represents a rotation:

$$\mathbf{E} \equiv \left(\mathbf{e}^{(1)}, \ldots, \mathbf{e}^{(9)} \right). \tag{3.230}$$

At this point we are ready to reduce the dimension further. First we *define* all the potential $N \equiv 9$ factors:

$$\mathbf{F} \equiv \mathbf{E}' (\mathbf{X} - \mathrm{E}\{\mathbf{X}\}). \tag{3.231}$$

Then we decide the number K of factors to consider for the dimension reduction, see (3.159).

In the top portion of Figure 3.19 we plot the eigenvalues (3.229). The first few eigenvalues overwhelmingly dominate the others. This becomes more clear in the middle portion of Figure 3.19, where we draw the generalized r-square of the first K factors (3.162) as a function of K. We see that the first three factors account for 99% of the total randomness. Therefore we set $K \equiv 3$, i.e. we consider as factors the first three entries of (3.231).

Since the invariants (3.208) are the changes in yield to maturity:

$$\mathbf{X} \equiv \mathbf{Y}_t - \mathbf{Y}_{t-\widetilde{\tau}}, \tag{3.232}$$

and since the expected value of the yield curve is approximately the previous realization of the yield curve:

$$\mathrm{E}\{\mathbf{Y}_t\} \approx \mathbf{Y}_{t-\widetilde{\tau}} \Leftrightarrow \mathrm{E}\{\mathbf{X}\} \approx \mathbf{0}, \tag{3.233}$$

recovering the invariants as in (3.160) corresponds to recovering the following yield curve at time t from the yield curve realized at time $t - \widetilde{\tau}$:

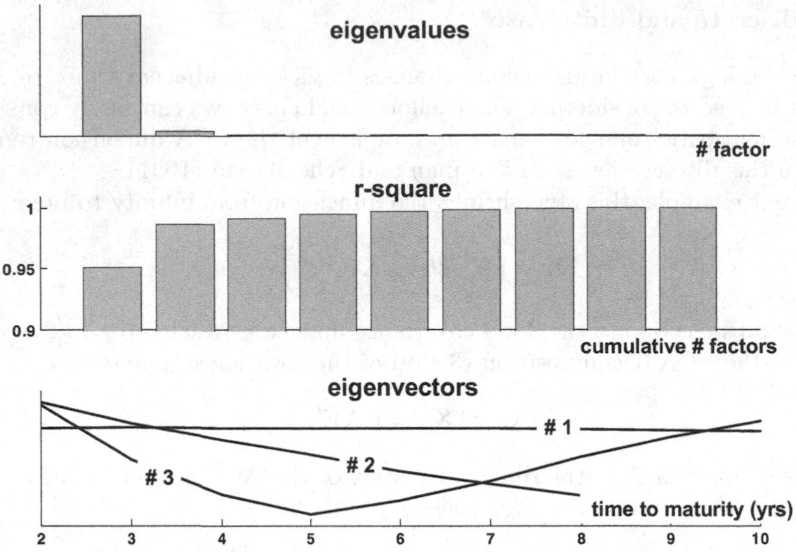

Fig. 3.19. Swap curve PCA: the discrete case

$$\widetilde{\mathbf{Y}}_t \equiv \mathbf{Y}_{t-\widetilde{\tau}} + \mathbf{e}^{(1)} F_1 + \mathbf{e}^{(2)} F_2 + \mathbf{e}^{(3)} F_3. \tag{3.234}$$

Consider the first factor. From (3.163) the square root of the first eigenvalue is the standard deviation of the first factor:

$$\mathrm{Sd}\{F_1\} = \sqrt{\lambda_1}. \tag{3.235}$$

Therefore a one-standard-deviation event in the first factor moves the curve as follows:

$$\mathbf{Y}_t \mapsto \mathbf{Y}_{t+\widetilde{\tau}} \equiv \mathbf{Y}_t \pm \sqrt{\lambda_1} \mathbf{e}^{(1)}, \tag{3.236}$$

where $\mathbf{e}^{(1)}$ is the first eigenvector. The plot of this eigenvector in the bottom portion of Figure 3.19 shows that the first eigenvector is approximately a positive constant. In a geometrical interpretation, the longest principal axis of the location-dispersion ellipsoid pierces the positive octant, see Figure 3.20. This happens because, as in (3.222), the elements of the covariance matrix of the changes in yield to maturity are positive, and thus the Perron-Frobenius theorem applies, see Appendix A.5.

Therefore the one-standard-deviation event in the first factor (3.236) corresponds to a *parallel shift* of the curve. In the top portion of Figure 3.21 we plot a three-standard-deviation parallel shift. From the estimation we obtain that this corresponds to the following change:

$$3\sqrt{\lambda_1} e_k^{(1)} \approx 3 \times 42 \times .33 \approx 42 \text{ b.p.}, \tag{3.237}$$

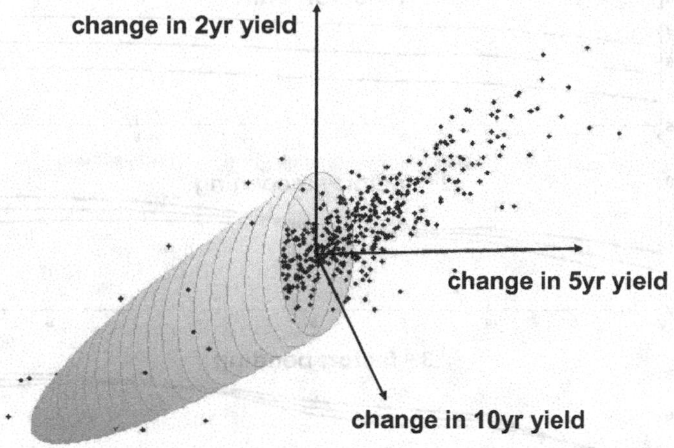

Fig. 3.20. Swap curve PCA: location-dispersion ellipsoid fitted to observations

where "b.p." stands for *basis point*. The basis point is a unit of measure for interest rates which is equal to 1/10000: in other words, an $r\%$ interest rate is equal to $100r$ b.p..

Consider now the second factor. Similarly to the first eigenvalue, from (3.163) the second eigenvalue is the variance of the second factor. Therefore a one-standard-deviation event in the second factor moves the curve as follows:

$$\mathbf{Y}_t \mapsto \mathbf{Y}_{t+\widetilde{\tau}} \equiv \mathbf{Y}_t \pm \sqrt{\lambda_2}\mathbf{e}^{(2)}, \tag{3.238}$$

where $\mathbf{e}^{(2)}$ is the second eigenvector, which we display in the bottom portion of Figure 3.19. This eigenvector is a decreasing line: thus a one-standard-deviation event in the second factor corresponds to a *steepening/flattening* of the curve. We remark that $\lambda_2 \ll \lambda_1$, whereas the entries of $\mathbf{e}^{(2)}$ are of the same order as those of $\mathbf{e}^{(1)}$ (because both vectors have length equal to one by construction). Therefore a one-standard-deviation event in the second factor has a much smaller influence on the yield curve than a one-standard-deviation event in the first factor. In the middle portion of Figure 3.21 we plot a three-standard-deviation steepening/flattening.

Finally, consider the third factor. From (3.163) the third eigenvalue is the variance of the third factor. Therefore a one-standard-deviation event in the third factor moves the curve as follows:

$$\mathbf{Y}_t \mapsto \mathbf{Y}_{t+\widetilde{\tau}} \equiv \mathbf{Y}_t \pm \sqrt{\lambda_3}\mathbf{e}^{(3)}, \tag{3.239}$$

where $\mathbf{e}^{(3)}$ is the third eigenvector, which we display in the bottom portion of Figure 3.19. This eigenvector is hump-shaped: therefore a one-standard-

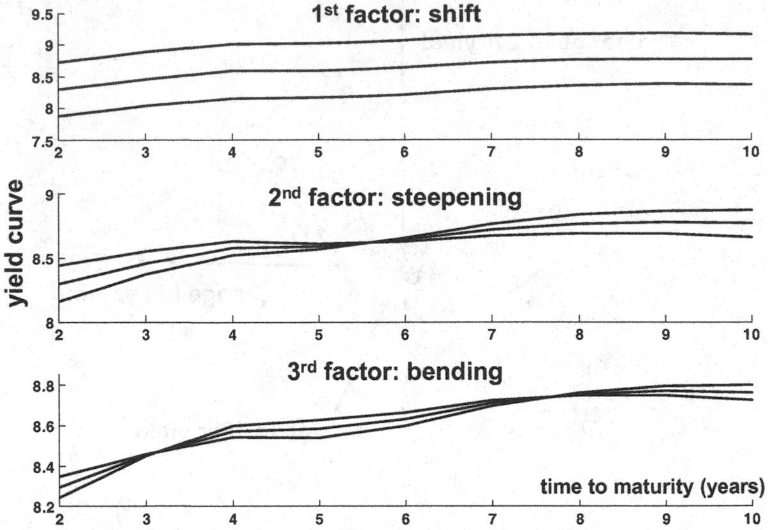

Fig. 3.21. Three-standard-deviation effects of PCA factors on swap curve

deviation event in the third factor corresponds to a *curvature effect* on the yield curve. We remark that $\lambda_3 \ll \lambda_2 \ll \lambda_1$, whereas the entries of $\mathbf{e}^{(3)}$ are of the same order as those of $\mathbf{e}^{(1)}$ and $\mathbf{e}^{(2)}$. Therefore a one-standard-deviation event in the third factor has a much smaller influence on the yield curve than a one-standard-deviation event in either the first or the second factor. In the bottom portion of Figure 3.21 we plot a three-standard-deviation curvature.

Looking back, we have managed to reduce the dimension of the two-to-ten-year section of the swap market from an infinite number of factors, as summarized in Figure 3.18, to three factors only, as summarized in Figure 3.19 with an accuracy of 99%. The reader is invited to ponder on the many analogies between these two figures.

3.5.3 The invariants at the investment horizon

We have reduced above the sources of randomness in the swap market to only three hidden factors that explain 99% of the total yield curve changes. Here we model the distribution of these three factors for the estimation period, which in our example is one week, and project it to the investment horizon, which in our case is two month.

First, we estimate the joint distribution of the three factors which from (3.231) and (3.233) are defined as follows:

$$(F_1, F_2, F_3) \equiv \mathbf{X}' \left(\mathbf{e}^{(1)}, \mathbf{e}^{(2)}, \mathbf{e}^{(3)} \right). \tag{3.240}$$

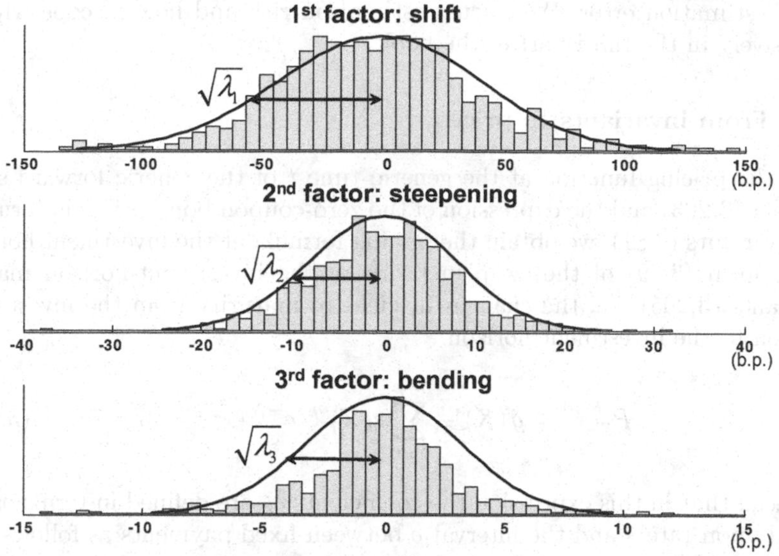

Fig. 3.22. Marginal distribution of swap curve PCA factors

In Figure 3.22 we plot the histogram of the observed values of each factor, which is a proxy for their respective marginal distributions. We model the joint distribution of the factors as a normal distribution:

$$\mathbf{F}_{t,\widetilde{\tau}} \sim \mathrm{N}\left(\mathbf{0}, \operatorname{diag}\left(\lambda_1, \lambda_2, \lambda_3\right)\right), \tag{3.241}$$

where in the notation we stressed that the distribution of the factors refers to the estimation interval $\widetilde{\tau}$, which in our case is one week.

To project the distribution of the invariants to the investment horizon τ we make use of (3.64). Performing the same steps as in Example (3.74) we obtain that the factors at the investment horizon are normally distributed with the following parameters:

$$\mathbf{F}_{T+\tau,\tau} \sim \mathrm{N}\left(\mathbf{0}, \frac{\tau}{\widetilde{\tau}} \operatorname{diag}\left(\lambda_1, \lambda_2, \lambda_3\right)\right). \tag{3.242}$$

In our case the investment horizon is two months ahead, and thus $\tau/\widetilde{\tau} \approx 8$.

We stress that here we are neglecting estimation risk. In other words, the distribution at the investment horizon is given precisely by (3.242) *if* the estimation-horizon distribution is precisely (3.241). Nevertheless, in the first place we used here a very rough estimation/fitting process: we will discuss the estimation of the market invariants in detail in Chapter 4. Secondly, no matter how good the estimate, an estimate is only an approximation to reality, and thus the distribution at the investment horizon cannot be precise. In fact, the farther in the future the investment horizon, the larger the effect of the estimation error. We discuss estimation risk and how to cope with it extensively in the third part of the book.

3.5.4 From invariants to prices

From the pricing function at the generic time t of the generic forward swap contract (3.203) and the expression of the zero-coupon bond prices in terms of the invariants (3.81), we obtain the pricing formula at the investment horizon in the form (3.79) of the swap in terms of the investment-horizon market invariants (3.205), i.e. the changes in yield to maturity from the investment decision to the investment horizon:

$$P_{T+\tau}^{(E,\upsilon)} = g(\mathbf{X}) \equiv \sum_{k=0}^{\upsilon/\rho} c_k Z_T^{(\upsilon_k)} e^{-X_{T+\tau,\tau}^{(\upsilon_k)} \upsilon_k}. \tag{3.243}$$

We recall that in this expression the coefficients c_k are defined in terms of the agreed upon rate s and the interval ρ between fixed payments as follows:

$$c_0 \equiv -1, \quad c_{\upsilon/\rho} \equiv 1 + s\rho, \quad c_k \equiv s\rho, \quad k = 1, \ldots, \frac{\upsilon}{\rho} - 1; \tag{3.244}$$

and that the set of times to maturities read:

$$\upsilon_k \equiv E + k\rho - (T + \tau), \qquad k = 0, \ldots, \frac{\upsilon}{\rho}. \tag{3.245}$$

We stress that these are the times to maturity at the investment horizon, *not* at the time the investment decision is made.

To compute the distribution of the value of the forward swap contract at the investment horizon (3.243) we can take two routes.

On the one hand, since (3.243) is a sum of log-distributions, by means of (3.93) we can compute all the cross-moments of the terms in the sum (3.243) and from these and (2.93) we can compute all the moments of the distribution of the swap contract.

Alternatively, we can obtain quick and intuitive results by means of a series of approximations. Here we choose this second option.

To evaluate the goodness of the approximations to come, we also compute the exact distribution numerically: we simulate a large number of invariant scenarios by means of the three-factor model (3.242) and we apply the exact pricing formula (3.243) for all the above scenarios. In the bottom portion of Figure 3.23 we plot the histogram of the simulation, which represents the profile of the probability density function of the distribution of (3.243). We also plot the value at the time the investment decision is made, in view of evaluating the risk/reward profile of our investment.

- **Approximation 1: one factor**

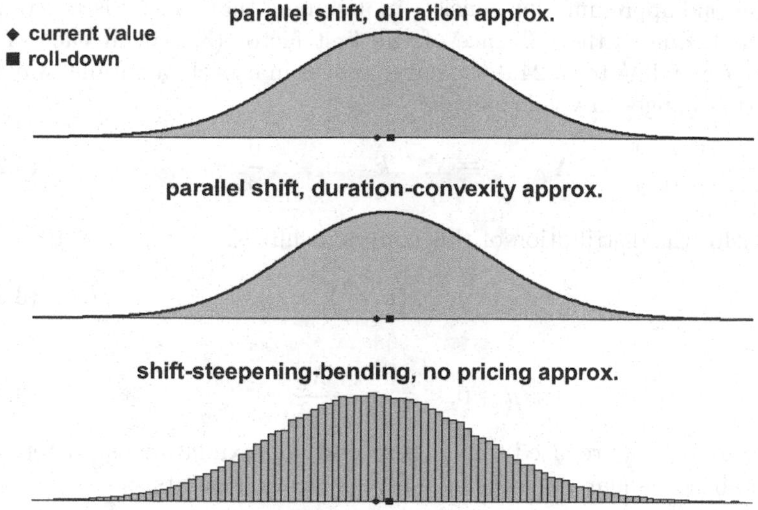

Fig. 3.23. Swap price distribution at the investment horizon

In view of performing an approximation, we do not need to consider three factors in the dimension reduction process discussed in Section 3.5.2. As we see in Figure 3.19 the first factor already explains 95% of the randomness in the swap market. Therefore, we focus on one factor only which from (3.242) has the following distribution:

$$F_{T+\tau,\tau} \sim \mathrm{N}\left(0, \frac{\tau}{\widetilde{\tau}}\lambda_1\right). \tag{3.246}$$

- **Approximation 2: parallel shift**

Without loss of generality, we can always rescale the first factor loading by a positive constant as follows:

$$\mathbf{e}^{(1)} \mapsto \alpha\mathbf{e}^{(1)}, \tag{3.247}$$

as long as we rescale the first factor accordingly, along with the distribution (3.246), in such a way that the effect of the factor on the curve (3.236) remains unaltered:

$$F \mapsto \frac{F}{\alpha} \quad \Rightarrow \quad \lambda \mapsto \frac{\lambda}{\alpha^2}. \tag{3.248}$$

We see in Figure 3.19 that the first eigenvector is almost "flat". Therefore, we can choose α in such a way that

$$\alpha\mathbf{e}^{(1)} \approx \mathbf{1}, \tag{3.249}$$

where $\mathbf{1}$ is a vector of ones.

The second approximation consists in assuming that (3.249) is exact. This implies that, due to the influence of the first factor (3.236), at each of the nodes corresponding to (3.245) the curve moves in exactly a parallel way and thus all the changes in yield coincide:

$$X_{T+\tau,\tau}^{(v_k)} \equiv X, \quad k = 0, \ldots, \frac{v}{\rho}. \tag{3.250}$$

From (3.246) the distribution of this common shift is:

$$X \sim \mathrm{N}\left(\mu, \sigma^2\right), \tag{3.251}$$

where

$$\mu \equiv 0, \quad \sigma^2 \equiv \frac{\tau}{\widetilde{\tau}} \frac{\lambda_1}{\alpha^2}. \tag{3.252}$$

From Figure 3.19 we realized that this second approximation can deteriorate the generalized r-square at most by a few percentage points.

- **Approximation 3: Taylor expansion**

We can consider the swap contract as a derivative product and perform as in (3.108) a Taylor expansion approximation of the exact pricing function (3.243) around zero, which from (3.233) represents the expected value of the invariants.

The order zero coefficient in the Taylor expansion of the value of the swap at the investment horizon (3.243) is called the *roll-down*. The roll-down is the value of the swap at the investment horizon if the invariant is zero, i.e. if the yield curve remains unchanged:

$$\mathrm{RD}\left(T, \tau\right) \equiv g\left(\mathbf{0}\right) = \sum_{k=0}^{v/\rho} c_k Z_T^{(v_k)}. \tag{3.253}$$

The (opposite of the) first-order coefficient in the Taylor expansion of the value of the swap at the investment horizon (3.243) is called the *present value of a basis point* (PVBP):

$$\mathrm{PVBP}\left(T, \tau\right) \equiv -\sum_{k=0}^{v/\rho} \partial_k g|_{\mathbf{X}=\mathbf{0}} = \sum_{k=0}^{v/\rho} v_k c_k Z_T^{(v_k)}. \tag{3.254}$$

The PVBP is a weighted sum of the times to maturity at the investment horizon (3.245) of the zero-coupon bonds involved in the pricing formula of the swap contract. We mention that practitioners in the (very similar) bond market consider the first-order term normalized by the roll-down, which is called the *duration* of the bond and has the dimensions of time.

The second-order coefficient in the Taylor expansion of the value of the swap at the investment horizon (3.243) is called the *convexity adjustment*:

$$\mathrm{Conv}\,(T,\tau) \equiv \sum_{k,j=0}^{v/\rho} \partial_{jk} g|_{\mathbf{X}=\mathbf{0}} = \sum_{k=0}^{v/\rho} c_k v_k^2 Z_T^{(v_k)}. \tag{3.255}$$

In standard contracts the convexity is typically positive: the the only negative term in the sum (3.255) is the one corresponding to $k \equiv 0$ and is typically outweighed by the other terms.

From the definitions of roll-down, PVBP and convexity, the value at the investment horizon of the swap (3.243) and the relation (3.250) we obtain the following second-order Taylor approximation:

$$P_{T+\tau}^{(E,v)} \approx \mathrm{RD} - \mathrm{PVBP}\, X + \frac{1}{2}\,\mathrm{Conv}\, X^2 + \cdots. \tag{3.256}$$

If we stop at the order zero in (3.256) we obtain a value, the roll-down, which is different than the value of the swap at the time the investment decision is made, see Figure 3.23. This difference is known as the *slide* of the contract. Some traders try to "cheat the curve", investing based on the (dis)advantages of the roll-down.

If we stop at the first order in (3.256) the distribution at the investment horizon of our swap contract is linear in the invariant. From (3.251) we obtain its distribution:

$$P_{T+\tau}^{(E,v)} \sim \mathrm{N}\left(\mathrm{RD}, \sigma^2\, \mathrm{PVBP}^2\right). \tag{3.257}$$

In the top plot of Figure 3.23 we display the probability density function of (3.257) as well as the current value and the roll-down.

If we stop at the second order in (3.256), rearranging the terms we can rewrite the swap value at the investment horizon as follows:

$$P_{T+\tau}^{(E,v)} = c + W^2, \tag{3.258}$$

where c is a constant defined as follows:

$$c \equiv \mathrm{RD} - \frac{1}{2}\frac{\mathrm{PVBP}^2}{\mathrm{Conv}}, \tag{3.259}$$

and W is the following normal random variable:

$$W \equiv \sqrt{\frac{\mathrm{Conv}}{2}} X - \frac{\mathrm{PVBP}}{\sqrt{2\,\mathrm{Conv}}}. \tag{3.260}$$

Therefore the second-order approximation of the swap contract has a shifted non-central gamma distribution with one degree of freedom and the following non-centrality and scale parameters:

$$P_{T+\tau}^{(E,v)} \sim \mathrm{Ga}\left(1, \sqrt{\frac{\mathrm{Conv}}{2}}\mu - \frac{\mathrm{PVBP}}{\sqrt{2\,\mathrm{Conv}}}, \frac{\mathrm{Conv}}{2}\sigma^2\right), \tag{3.261}$$

see (1.107). It is convenient to represent this distribution in terms of its characteristic function, which reads:

$$\phi_P(\omega) = \left(1 - i\omega \operatorname{Conv} \sigma^2\right)^{-\frac{1}{2}} e^{-\frac{1}{2}\frac{(\operatorname{Conv}\mu - \mathrm{PVBP})^2 \sigma^2}{1 - i\omega \operatorname{Conv}\sigma^2}} \tag{3.262}$$
$$e^{i\omega\left(\mathrm{RD} - \mathrm{PVBP}\,\mu + \frac{1}{2}\operatorname{Conv}\mu^2\right)}.$$

This is a specific instance of (5.30), a result which we prove and discuss later in a more general context.

In the middle plot of Figure 3.23 we display the probability density function of (3.258) as well as the current value and the roll-down. From a comparison of the three plots in Figure 3.23 we see that the simple parallel shift/duration approximation provides very satisfactory results.

Part II

Classical asset allocation

4

Estimating the distribution of the market invariants

In this chapter we discuss how to estimate the distribution of the market invariants from empirical observations.

In Section 4.1 we define the concept of estimator, which is simply a function of current information that yields a number, the estimate. Such a general definition includes estimators that perform poorly, i.e. functions that yield an estimate which has little in common with the real distribution of the market invariants. Therefore we discuss optimality criteria to evaluate an estimator.

After defining estimators and how to evaluate them, we need to actually construct estimators for the market invariants. Nevertheless, constructing estimators by maximizing the above optimality criteria is not possible. First of all, the search of the best estimator among all possible functions of current information is not feasible. Secondly the optimality criteria rarely yield a univocal answer. In other words, an estimator might perform better than another one in given circumstances, and worse in different circumstances. Therefore, we construct estimators from general intuitive principles, making sure later that their performance is acceptable, and possibly improving them with marginal corrections. In this spirit, we proceed as follows.

In Section 4.2 we introduce nonparametric estimators. These estimators are based on the law of large numbers. Therefore, they perform well when the number of empirical observations in the time series of the market invariants is large, see Figure 4.1. When this is the case, nonparametric estimators are very flexible, in that they yield sensible estimates no matter the underlying true distribution of the market invariants. In particular we discuss the sample quantile, the sample mean, the sample covariance and the ordinary least square estimate of the regression factor loadings in an explicit factor model, stressing the geometrical properties of these estimators. We conclude with an overview of kernel estimators.

When the number of observations is not very large, nonparametric estimators are no longer suitable. Therefore we take a parametric approach, by assuming that the true distribution of the market invariants belongs to a restricted class of potential distributions. In Section 4.3 we discuss maximum

A. Meucci, *Risk and Asset Allocation*, Springer Finance,

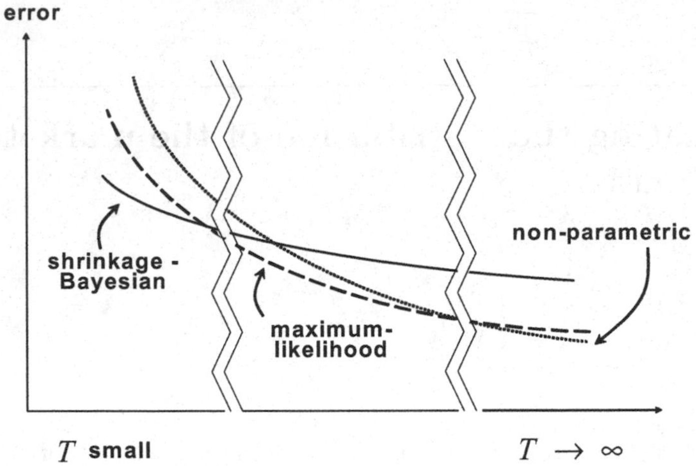

Fig. 4.1. Performance of different types of estimators

likelihood estimators, which are built in such a way that the past observations of the market invariants become the most likely outcomes of the estimated parametric distribution. We compute the maximum likelihood estimators of location, dispersion, and factor loadings under the assumption that the market invariants are elliptically distributed: this shows the intrinsic outlier-rejection mechanism of maximum likelihood estimators. Then we study thoroughly the normal case: as it turns out, the main driver of the performance of the maximum likelihood estimators is the overall level of correlation among the market invariants, as summarized by the condition number.

In some applications the number of observations is so scanty, and the result of the estimate so unreliable, that it is advisable to average the final estimates with fixed, yet potentially wrong, values: this way we obtain shrinkage estimators, see Figure 4.1. In Section 4.4 we discuss the shrinkage estimators for the location parameters, the dispersion parameters and the factor loadings of a linear model.

In Section 4.5 we discuss robust estimation. Indeed, the parametric approach dramatically restricts the set of potential distributions for the market invariants. Robust estimation provides a set of techniques to evaluate and possibly fix the consequences of not having included the true, unknown distribution among the set of potential distributions. We discuss classical measures of robustness, such as the sensitivity curve and the jackknife, and develop the more general concept of influence function. Then we evaluate the robustness of the estimators previously introduced in this chapter and show how to build robust M-estimators. In particular, we discuss M-estimators of the location parameters, the dispersion parameters and the factor loadings of a linear model.

In Section 4.6 we conclude with a series of practical tips to improve the estimation of the distribution of the market invariants in specific situations. Among other issues, we discuss outliers detection, which we tackle by means of high breakdown estimators such as the minimum volume ellipsoid and the minimum covariance determinant; missing data, which we tackle by means of the EM algorithm; and weighted estimation techniques such as the exponential smoothing, which accounts for the higher reliability of more recent data with respect to data farther back in the past.

4.1 Estimators

Before introducing the concept of estimator, we review our working assumptions, which we set forth in Section 3.1.

The randomness in the market is driven by the market invariants. The invariants are random variables that refer to a specific estimation-horizon $\widetilde{\tau}$ and are *independent and identically distributed (i.i.d.)* across time. The generic invariant $\mathbf{X}_{t,\widetilde{\tau}}$ becomes known at the respective time t, which is part of the set of equally spaced estimation dates:

$$t \in \mathcal{D}_{\widetilde{t},\widetilde{\tau}} \equiv \left\{\widetilde{t}, \widetilde{t}+\widetilde{\tau}, \widetilde{t}+2\widetilde{\tau}, \ldots\right\}. \tag{4.1}$$

For example, we have seen in Section 3.1.1 that the invariants in the equity market are the compounded returns. In other words, for a stock that at the generic time t trades at the price P_t, the following set of random variables are independent and identically distributed across time, as t varies in (4.1):

$$X_{t,\widetilde{\tau}} \equiv \ln\left(\frac{P_t}{P_{t-\widetilde{\tau}}}\right), \quad t \in \mathcal{D}_{\widetilde{t},\widetilde{\tau}}. \tag{4.2}$$

Furthermore, these variables become known at time t.

Notice that once the time origin $\widetilde{t}$ and the time interval $\widetilde{\tau}$ have been fixed, we can measure time in units of $\widetilde{\tau}$ and set the origin in $\widetilde{t}-\widetilde{\tau}$. This way, without loss of generality, we can always reduce the estimation dates to the set of positive integers:

$$t \in \mathcal{D}_{\widetilde{t},\widetilde{\tau}} \equiv \{1, 2, 3, \ldots\}. \tag{4.3}$$

We will use this more convenient notation throughout this chapter.

In this notation the market invariants of our example, namely the compounded returns (4.2), read:

$$X_t \equiv \ln\left(\frac{P_t}{P_{t-1}}\right), \quad t = 1, 2, \ldots. \tag{4.4}$$

Since the invariants are independent and identically distributed across time, from (2.44) we obtain that their across-time joint distribution, as represented by their probability density function, factors as follows:

$$f_{\mathbf{X}_1,\mathbf{X}_2,\ldots}(\mathbf{x}_1, \mathbf{x}_2, \ldots) = f_{\mathbf{X}}(\mathbf{x}_1) f_{\mathbf{X}}(\mathbf{x}_2) \cdots, \tag{4.5}$$

where we stress that the single-period joint distribution of the invariants $f_{\mathbf{X}}$ does not depend on the time index. Therefore all the information about the invariants is contained in one single-period multivariate distribution.

In (4.5) we chose to represent the distribution of the invariants in terms of their probability density function $f_{\mathbf{X}}$. Equivalently, we might find it more convenient to represent the distribution of the invariants in terms of either the cumulative distribution function $F_{\mathbf{X}}$ or the characteristic function $\phi_{\mathbf{X}}$, see Figure 2.2. The factorization (4.5) holds for any representation, as we see from (2.46) and (2.48).

4.1.1 Definition

Our aim is to infer the single-period distribution of the invariants. More precisely, we aim at inferring the "truth", as represented by a generic number S of features of the distribution of the market invariants. These features can be expressed as an S-dimensional vector of functionals of the probability density function (or of the cumulative distribution function, or of the characteristic function):

$$\mathbf{G}[f_{\mathbf{X}}] \equiv \text{"unknown truth"}. \tag{4.6}$$

For example, if we are interested in the expected value of the compounded return on a stock (4.4), the "unknown truth" is the following one-dimensional functional:

$$G[f_X] \equiv \int_{-\infty}^{+\infty} x f_X(x)\, dx, \tag{4.7}$$

where f_X is the unknown probability density function of any compounded return X_t, and does not depend on the specific time t.

The current time is T. We base inference on the information i_T about the invariants available at the time T when the investment decision is made. This information is represented by the time series of all the past realizations of the invariants:

$$i_T \equiv \{\mathbf{x}_1, \ldots, \mathbf{x}_T\}, \tag{4.8}$$

where the lower-case notation stresses the fact that the once random variables $\mathbf{X}_t$ have become observable numbers. An *estimator* is a vector-valued function that associates a vector in $\mathbb{R}^S$, i.e. a set of S numbers, with available information:

$$\boxed{\textbf{estimator: information } i_T \mapsto \textbf{number } \widehat{\mathbf{G}}} \tag{4.9}$$

For example the following is an estimator:

$$\widehat{G}\left[i_T\right] \equiv \frac{1}{T}\sum_{t=1}^{T} x_t. \tag{4.10}$$

Notice that the definition of estimator (4.9) is not related to the goal of estimation (4.6). Again, an estimator is simply a function of currently available information.

For example, the following is a function of information and thus it is an estimator:

$$\widehat{G}\left[i_T\right] \equiv x_1 x_T. \tag{4.11}$$

Similarly, for strange that it might sound, the following is also an estimator:

$$\widehat{G}\left[i_T\right] \equiv 3. \tag{4.12}$$

4.1.2 Evaluation

Although the definition of estimator is very general, an estimator serves its purpose only if its value is close to the true, unknown value (4.6) that we are interested in:

$$\widehat{\mathbf{G}}\left[i_T\right] \approx \mathbf{G}\left[f_{\mathbf{X}}\right]. \tag{4.13}$$

To make this statement precise, we need a criterion to evaluate estimators. In order to evaluate an estimator, the main requirement is its *replicability*: an estimator is good not only if the result of the estimation is close to the true unknown value, but also if this does not happen by chance.

For example, the estimator (4.12) could yield by chance the true, unknown parameter if this happens to be equal to 3, much like the hands of a broken watch happen to display the correct time twice a day.

To tackle replicability, notice that the available information (4.8), namely the time series of the market invariants, is the realization of a set of random variables:

$$I_T \equiv \left\{\mathbf{X}_1, \ldots, \mathbf{X}_T\right\}. \tag{4.14}$$

In a different scenario, the realization of this set of variables would have assumed a different value i'_T and therefore the outcome of the estimate would

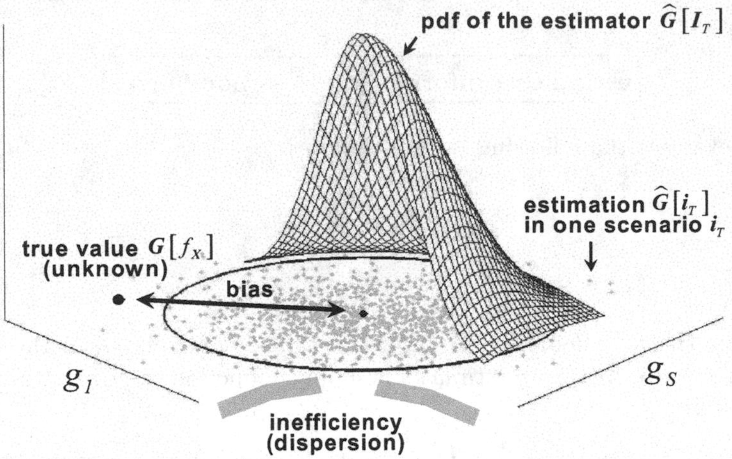

Fig. 4.2. Estimation: replicability, bias and inefficiency

have been a different number $\widehat{\mathbf{G}}[i'_T]$, see Figure 4.2. Therefore the estimator (4.9), as a function of the random variable I_T instead of the specific occurrence i_T, becomes a (multivariate) *random variable*:

$$\widehat{\mathbf{G}}[i_T] \mapsto \widehat{\mathbf{G}}[I_T]. \tag{4.15}$$

The distribution of the information (4.14) is fully determined by the true, unknown distribution $f_{\mathbf{X}}$ of the market invariants through (4.5). Therefore, the distribution of the estimator (4.15) is also determined by the true, unknown distribution $f_{\mathbf{X}}$ of the market invariants, see Figure 4.2.

For example, if the invariants (4.4) are normally distributed with the following unknown parameters:

$$X_t \sim \mathrm{N}(\mu, \sigma^2), \tag{4.16}$$

then the estimator (4.10) is normally distributed with the following parameters:

$$\widehat{G}[I_T] \equiv \frac{1}{T}\sum_{t=1}^{T} X_t \sim \mathrm{N}\left(\mu, \frac{\sigma^2}{T}\right), \tag{4.17}$$

where μ and σ^2 are unknown.

The distribution associated with an estimator is at least as important as the specific outcome $\widehat{\mathbf{G}}[i_T]$ of the estimation process: an estimator is suitable, i.e. (4.13) holds, if the distribution of the multivariate random variable $\widehat{\mathbf{G}}[I_T]$

is highly concentrated around the true unknown value $\mathbf{G}[f_{\mathbf{X}}]$. For instance, this is not the case in Figure 4.2.

Suppose we use the estimator (4.10) to estimate (4.7), i.e. the expected value of the invariants. From (4.16) this reads:

$$G[f_X] = \mu. \tag{4.18}$$

Therefore the distribution of the estimator (4.17) is centered around the true unknown value μ and the concentration of this distribution is of the order of $\sigma/\sqrt{T}$.

Nevertheless, evaluating a multivariate distribution can be complex. To summarize the goodness of an estimator into a univariate distribution we introduce the *loss*:

$$\text{Loss}\left(\widehat{\mathbf{G}}, \mathbf{G}\right) \equiv \left\|\widehat{\mathbf{G}}[I_T] - \mathbf{G}[f_{\mathbf{X}}]\right\|^2, \tag{4.19}$$

where $\|\cdot\|$ denotes a norm, see (A.7). For reasons to become clear in a moment, it is common to induce the norm from a quadratic form, i.e. a symmetric and positive $S \times S$ matrix $\mathbf{Q}$ such that the following relation holds true:

$$\|\mathbf{v}\|^2 \equiv \mathbf{v}'\mathbf{Q}\mathbf{v}. \tag{4.20}$$

Since the loss is the square of a norm, from (A.7) the loss is zero only for those outcomes where the estimator $\widehat{\mathbf{G}}$ yields an estimate that is equal to the true value to be estimated, and is strictly positive otherwise. Therefore, the estimator is good if the distribution of the loss is tightly squeezed above the value of zero.

In our example, from (4.17) and (4.18) we obtain:

$$\widehat{G}[I_T] - G[f_X] \sim \text{N}\left(0, \frac{\sigma^2}{T}\right). \tag{4.21}$$

We can summarize the goodness of this estimator with the quadratic loss induced by $Q \equiv 1$ in (4.20). Then from (1.106) we obtain the distribution of the loss, which is the following central gamma with one degree of freedom:

$$\text{Loss}\left(\widehat{G}, G\right) \equiv \left(\widehat{G} - G\right)^2 \sim \text{Ga}\left(1, \frac{\sigma^2}{T}\right). \tag{4.22}$$

In the presence of a large number of observations, or when the underlying market is not too volatile, this loss is a random variable tightly squeezed above the value of zero.

Even evaluating the shape of a univariate distribution can be complex, see Chapter 5. To further summarize the analysis of the goodness of an estimator we consider the expected value of the loss: the higher the expected value, the worse the performance of the estimator. Since the loss is a square distance, we consider the square root of the expectation of the loss. The *error*[1] is the average distance between the outcome of the estimation process and the true value to be estimated over all the possible scenarios:

$$\operatorname{Err}\left(\widehat{\mathbf{G}}, \mathbf{G}\right) \equiv \sqrt{\mathrm{E}\left\{\left\|\widehat{\mathbf{G}}\left(I_T\right) - \mathbf{G}\left[f_{\mathbf{X}}\right]\right\|^2\right\}}. \tag{4.23}$$

In our example, from (4.22) and (1.113) the error reads:

$$\operatorname{Err}\left(\widehat{G}, G\right) = \frac{\sigma}{\sqrt{T}}. \tag{4.24}$$

As expected, the larger the number of observations in the time series and the lower the volatility of the market, the lower the estimation error.

The definition (4.19)-(4.20) of the loss in terms of a square norm and the definition (4.23) of the error as the square root of its expected value are not the only viable choices. Nevertheless, the above definitions are particularly intuitive because they allow to decompose the error into bias and inefficiency.

The *bias* measures the distance between the "center" of the distribution of the estimator and the true unknown parameter to estimate:

$$\operatorname{Bias}^2\left[\widehat{\mathbf{G}}, \mathbf{G}\right] \equiv \left\|\mathrm{E}\left\{\widehat{\mathbf{G}}\left[I_T\right]\right\} - \mathbf{G}\left[f_{\mathbf{X}}\right]\right\|^2, \tag{4.25}$$

see Figure 4.2.

The *inefficiency* is a measure of the dispersion of the estimator, and as such it does not depend on the true unknown value:

$$\operatorname{Inef}^2\left[\widehat{\mathbf{G}}\right] \equiv \mathrm{E}\left\{\left\|\widehat{\mathbf{G}}\left[I_T\right] - \mathrm{E}\left\{\widehat{\mathbf{G}}\left[I_T\right]\right\}\right\|^2\right\}, \tag{4.26}$$

see Figure 4.2.

It is easy to check that in terms of bias and inefficiency the error (4.23) factors as follows:

$$\operatorname{Err}^2\left[\widehat{\mathbf{G}}, \mathbf{G}\right] = \operatorname{Bias}^2\left[\widehat{\mathbf{G}}, \mathbf{G}\right] + \operatorname{Inef}^2\left[\widehat{\mathbf{G}}\right]. \tag{4.27}$$

In these terms, the statement that the replicability distribution of a good estimator is highly peaked around the true value can be rephrased as follows: a good estimator is very efficient and displays little bias.

---

[1] The error is called *risk* in the statistical literature. We prefer to reserve this term for financial risk

In our example, from (4.17) and (4.18) we obtain the bias:

$$\text{Bias}\left[\widehat{G}, G\right] = \left|\text{E}\left\{\widehat{G}\left[I_T\right]\right\} - G\left[f_X\right]\right| = |\mu - \mu| = 0. \tag{4.28}$$

In other words, the estimator is centered around the true, unknown value μ.

From (4.17) we obtain the inefficiency:

$$\text{Inef}\left[\widehat{G}\right] = \text{Sd}\left\{\widehat{G}\left[I_T\right]\right\} = \frac{\sigma}{\sqrt{T}}. \tag{4.29}$$

In other words, the estimator has a dispersion of the order of $\sigma/\sqrt{T}$.

Comparing with (4.24) we see that the factorization (4.27) holds.

Notice that the definitions of loss and error are scale dependent: for example if the true value $\mathbf{G}$ has the dimension of money and we measure it in US dollars, the error is about one hundred times smaller than if we measure it in Japanese yen. To make the evaluation scale independent we can normalize the loss and the error by the length of the true value, if this length is not zero. Therefore at times we consider the percentage loss, which is a random variable:

$$\text{PLoss}\left(\widehat{\mathbf{G}}, \mathbf{G}\right) \equiv \frac{\left\|\widehat{\mathbf{G}}\left(I_T\right) - \mathbf{G}\left[f_{\mathbf{X}}\right]\right\|^2}{\left\|\mathbf{G}\left[f_{\mathbf{X}}\right]\right\|^2}; \tag{4.30}$$

and the percentage error, which is a scale-independent number:

$$\text{PErr}\left(\widehat{\mathbf{G}}, \mathbf{G}\right) \equiv \frac{\sqrt{\text{E}\left\{\left\|\widehat{\mathbf{G}}\left(I_T\right) - \mathbf{G}\left[f_{\mathbf{X}}\right]\right\|^2\right\}}}{\left\|\mathbf{G}\left[f_{\mathbf{X}}\right]\right\|}. \tag{4.31}$$

An estimator is suitable if its percentage error is much smaller than one.

At this point we face a major problem: the distribution of the loss, and thus the error of an estimator, depends on the underlying true distribution of the market invariants $f_{\mathbf{X}}$. If this distribution were known, we would not need an estimator in the first place.

In our example, from (4.24) the error of the estimator (4.10) depends on the standard deviation σ of the unknown distribution of the invariants (4.16): this estimator is good if the invariants are not too volatile.

Similarly, the estimator (4.12) gives rise to a deterministic loss, which is equal to the error and reads:

$$\text{Err}\left(\widehat{G}, G\right) = |\mu - 3|. \tag{4.32}$$

This estimator is suitable if the expected value of the invariants happens to lie in the neighborhood of the value $\mu \equiv 3$.

Nevertheless, neither μ nor σ are known parameters.

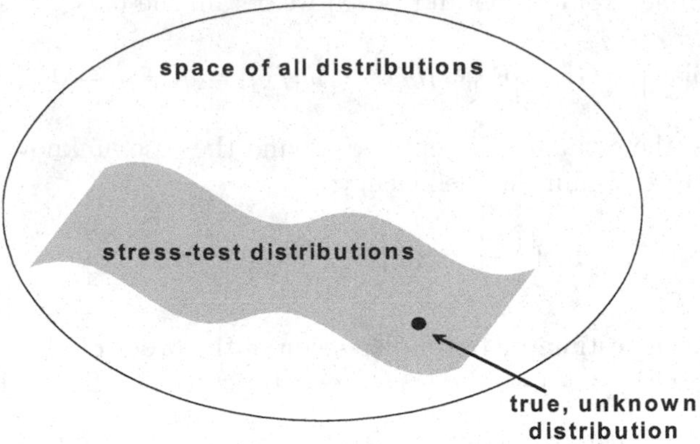

Fig. 4.3. Evaluation of estimators: choice of stress-test distributions

Therefore in order to evaluate an estimator we have to proceed as follows.

First we consider, among all the possible distributions of the market invariants, a subset of stress test distributions that is large enough to contain the true, unknown distribution, see Figure 4.3.

Then we make sure that the estimator is suitable, i.e. its distribution is peaked around the true unknown value to be estimated for all the distributions in the stress test set, see Figure 4.4.

In general an estimator performs well with some stress test distributions and performs poorly with other stress test distributions, see Figure 4.4. Consequently, in choosing the set of stress test distributions we face the following dichotomy: on the one hand, the stress test set should be as broad as possible, in such a way to encompass the true, unknown distribution; on the other hand, the stress test set should be as narrow as possible, in such a way that estimators can be built which display small errors for all the stress test distributions.

4.2 Nonparametric estimators

Assume that the number of observations T in the time series i_T is very large. The nonparametric approach is based on the following intuitive result, well known to practitioners: under fairly general conditions, sample averages computed over the whole time series approximate the expectation computed with the true distribution, and the approximation improves with the number of observations in the time series.

This result is known as the *law of large numbers (LLN)*, which we represent as follows:

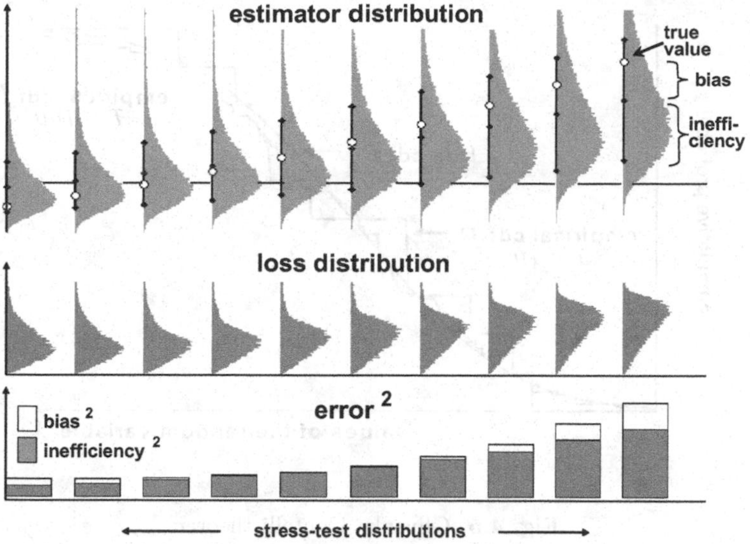

Fig. 4.4. Evaluation of estimators: loss and error

$$\frac{1}{T}\sum_{t=1}^{T}\{\text{past}\} \underset{T\rightarrow\infty}{\approx} \mathrm{E}\{\text{future}\}. \tag{4.33}$$

The Law of Large Numbers implies the *Glivenko-Cantelli theorem*. This theorem states that the empirical distribution (2.239) of a set of independent and identically distributed variables, as represented for example by its cumulative distribution function, tends[2] to the true distribution as the number of observations goes to infinity, see Figure 4.5:

$$\lim_{T\rightarrow\infty} F_{i_T}(\mathbf{x}) = F_{\mathbf{X}}(\mathbf{x}). \tag{4.34}$$

Expression (4.34) suggests how to define the estimator of a generic functional $\mathbf{G}[f_{\mathbf{X}}]$ of the true, yet unknown, distribution of the market invariants. Indeed, we only need to replace in the functional $\mathbf{G}[f_{\mathbf{X}}]$ the true, unknown probability density function $f_{\mathbf{X}}$ with the empirical probability density function (2.240), which we report here:

$$f_{i_T}(\mathbf{x}) \equiv \frac{1}{T}\sum_{t=1}^{T}\delta^{(\mathbf{x}_t)}(\mathbf{x}), \tag{4.35}$$

where δ is the Dirac delta (B.17). In other words we define the estimator of $\mathbf{G}[f_{\mathbf{X}}]$ as follows:

[2] One should specify the topology for the limits in the law of large numbers and in the Glivenko-Cantelli theorem, see e.g. Shirayaev (1989) for details. Here we choose a heuristic approach.

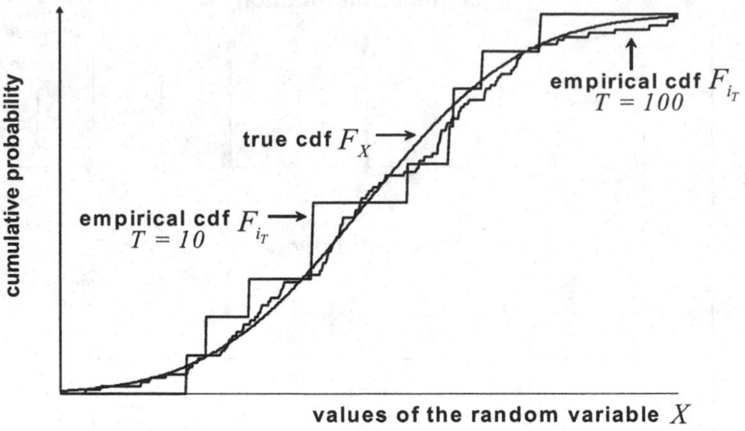

Fig. 4.5. Glivenko-Cantelli theorem

$$\widehat{\mathbf{G}}\left[i_T\right] \equiv \mathbf{G}\left[f_{i_T}\right]. \tag{4.36}$$

To test the goodness of this estimator we should compute its replicability, i.e. the distribution of $\widehat{G}\left[I_T\right]$ as in (4.15), for all possible distributions. This is an impossible task. Nevertheless, under fairly general conditions, when the number of observations T is very large the *central limit theorem (CLT)* states that the estimator is approximately normally distributed:

$$\widehat{\mathbf{G}}\left[I_T\right] \sim \mathrm{N}\left(\mathbf{G}\left[f_{\mathbf{X}}\right], \frac{\mathbf{A}}{T}\right), \tag{4.37}$$

where $\mathbf{A}$ is a suitable symmetric and positive matrix [3]. The above approximation becomes exact only in the limit of an infinite number of observations T in the time series: although this limit is never attained in practice, for a large enough number of observations the nonparametric approach yields benchmark estimators that can subsequently be refined.

We now use the nonparametric approach to estimate the features of the distribution of the market invariants that are most interesting in view of financial applications.

[3] The matrix $\mathbf{A}$ is defined in terms of the influence function (4.185) as follows:

$$A_{jk} \equiv \int_{\mathbb{R}^N} \mathrm{IF}\left(\mathbf{x}, f_{\mathbf{X}}, G_j\right) \mathrm{IF}\left(\mathbf{x}, f_{\mathbf{X}}, G_k\right) f_{\mathbf{X}}\left(\mathbf{x}\right) d\mathbf{x},$$

see Huber (1981).

4.2.1 Location, dispersion and hidden factors

If the invariants X_t are univariate random variables, we can use as location parameter the generic quantile q_p, which is defined implicitly in terms of the probability density function f_X of the invariant as follows:

$$\int_{-\infty}^{q_p[f_X]} f_X(x)\, dx \equiv p, \tag{4.38}$$

see (1.18). By applying (4.36) to the definition of quantile, we obtain the respective estimator $\widehat{q}_p$. This is the *sample quantile* (1.124):

$$\widehat{q}_p[i_T] \equiv x_{[pT]:T}, \tag{4.39}$$

where $[\cdot]$ denotes the integer part. In particular, for $p \equiv 1/2$ this expression becomes the *sample median*.

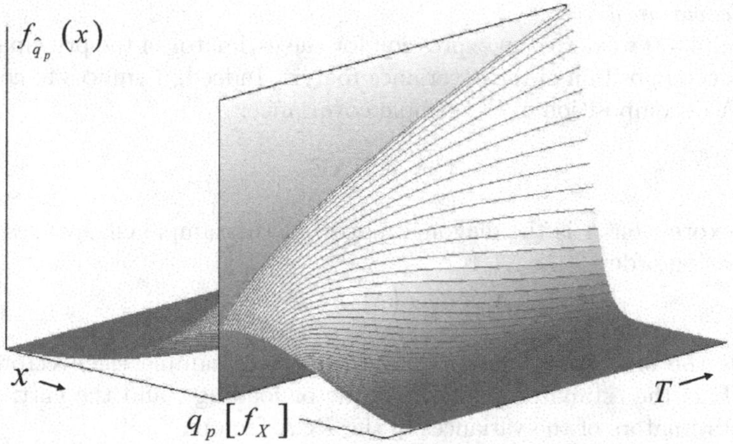

Fig. 4.6. Sample quantile: evaluation

To evaluate this estimator, we consider it as a random variable as in (4.15). From (2.248) the probability density function of the estimator $\widehat{q}_p$ reads:

$$f_{\widehat{q}_p}(x) = \frac{T!\,[F_X(x)]^{[pT]-1}\,[1 - F_X(x)]^{T-[pT]}\, f_X(x)}{([pT]-1)!\,(T-[pT])!}. \tag{4.40}$$

From (2.253) this density is concentrated around the quantile q_p and from (2.252) the quality of the estimator improves as the sample size T increases, see Figure 4.6.

Similarly, to estimate the dispersion of the univariate invariants X_t we can use the *sample interquantile range*, derived by applying (4.36) to (1.37).

In the multivariate case, we can rely on the expected value of the invariant $\mathbf{X}$ as parameter of location. We derive the nonparametric estimator of the expected value by applying (4.36) to the definition (2.54) of expected value. This is the expected value of the empirical distribution (2.244), i.e. the *sample mean*:

$$\widehat{\mathrm{E}}\left[i_T\right] \equiv \int_{\mathbb{R}^N} \mathbf{x} f_{i_T}\left(\mathbf{x}\right) d\mathbf{x} = \frac{1}{T}\sum_{t=1}^{T} \mathbf{x}_t. \tag{4.41}$$

Similarly, as a multivariate parameter of dispersion we choose the covariance matrix. By applying (4.36) to the definition (2.67) of covariance we derive the respective nonparametric estimator:

$$\widehat{\mathrm{Cov}}\left[i_T\right] = \frac{1}{T}\sum_{t=1}^{T} \left(\mathbf{x}_t - \widehat{\mathrm{E}}\left[i_T\right]\right)\left(\mathbf{x}_t - \widehat{\mathrm{E}}\left[i_T\right]\right)'. \tag{4.42}$$

This is the covariance matrix of the empirical distribution (2.245), i.e. the *sample covariance*.

From (4.42) we derive an expression for the estimator of the principal component decomposition of the covariance matrix. Indeed, it suffices to compute the PCA decomposition of the sample covariance:

$$\widehat{\mathrm{Cov}} \equiv \widehat{\mathbf{E}}\widehat{\mathbf{\Lambda}}\widehat{\mathbf{E}}'. \tag{4.43}$$

In this expression $\widehat{\mathbf{\Lambda}}$ is the diagonal matrix of the sample eigenvalues sorted in decreasing order:

$$\widehat{\mathbf{\Lambda}} \equiv \mathrm{diag}\left(\widehat{\lambda}_1, \ldots, \widehat{\lambda}_N\right); \tag{4.44}$$

and $\widehat{\mathbf{E}}$ is the orthogonal matrix of the respective sample eigenvectors. The matrix $\widehat{\mathbf{E}}$ is the estimator of the PCA factor loadings, and the entries of $\widehat{\mathbf{\Lambda}}$ are the estimators of the variances of the PCA factors.

We do not evaluate here the performance of the estimators (4.41), (4.42), (4.43) and (4.44) on a set of stress test distributions, because the same estimators reappear in a different context in Section 4.3.

The sample mean and the sample covariance display an interesting geometrical interpretation. To introduce this property, consider a generic N-dimensional vector $\boldsymbol{\mu}$ and a generic $N \times N$ scatter matrix $\mathbf{\Sigma}$, i.e. a symmetric and positive matrix. Consider the ellipsoid (A.73) defined by these two parameters, see Figure 4.7:

$$\mathcal{E}_{\boldsymbol{\mu},\mathbf{\Sigma}} \equiv \left\{\mathbf{x} \in \mathbb{R}^N \text{ such that } \left(\mathbf{x} - \boldsymbol{\mu}\right)' \mathbf{\Sigma}^{-1}\left(\mathbf{x} - \boldsymbol{\mu}\right) \leq 1\right\}. \tag{4.45}$$

Consider now the set of the Mahalanobis distances from $\boldsymbol{\mu}$ through the metric $\mathbf{\Sigma}$ of each observation $\mathbf{x}_t$ in the time series of the invariants:

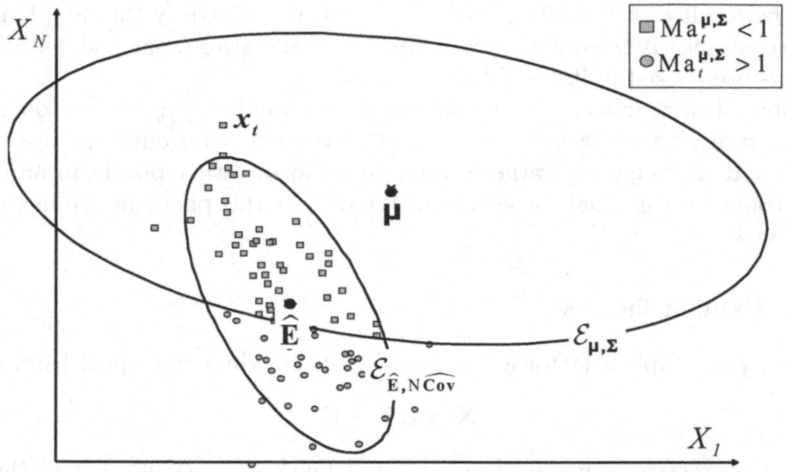

Fig. 4.7. Sample mean and sample covariance: geometric properties

$$\text{Ma}_t^{\boldsymbol{\mu},\boldsymbol{\Sigma}} \equiv \text{Ma}\left(\mathbf{x}_t, \boldsymbol{\mu}, \boldsymbol{\Sigma}\right) \equiv \sqrt{\left(\mathbf{x}_t - \boldsymbol{\mu}\right)' \boldsymbol{\Sigma}^{-1} \left(\mathbf{x}_t - \boldsymbol{\mu}\right)}. \tag{4.46}$$

The Mahalanobis distance is the "radius" of the ellipsoid concentric to $\mathcal{E}_{\boldsymbol{\mu},\boldsymbol{\Sigma}}$ that crosses the observation $\mathbf{x}_t$. In particular, if $\text{Ma}_t^{\boldsymbol{\mu},\boldsymbol{\Sigma}}$ is one, then the observation $\mathbf{x}_t$ lies on the ellipsoid (4.45). Consider the average of the square distances:

$$\overline{r^2}\left(\boldsymbol{\mu}, \boldsymbol{\Sigma}\right) \equiv \frac{1}{T} \sum_{t=1}^{T} \left(\text{Ma}_t^{\boldsymbol{\mu},\boldsymbol{\Sigma}}\right)^2. \tag{4.47}$$

If this number is close to one, the ellipsoid passes through the cloud of observations.

The sample mean and sample covariance represent the choices of location and scatter parameter respectively that give rise to the smallest ellipsoid among all those that pass through the cloud of observations, see Figure 4.7. More formally, we prove in Appendix www.4.1 the following result:

$$\left(\widehat{\text{E}}, N\widehat{\text{Cov}}\right) = \underset{(\boldsymbol{\mu},\boldsymbol{\Sigma})\in\mathcal{C}}{\operatorname{argmin}} \left[\text{Vol}\left\{\mathcal{E}_{\boldsymbol{\mu},\boldsymbol{\Sigma}}\right\}\right], \tag{4.48}$$

where the set of constraints $\mathcal{C}$ imposes that $\boldsymbol{\Sigma}$ be symmetric and positive and that the average Mahalanobis distance be one:

$$\overline{r^2}\left(\boldsymbol{\mu}, \boldsymbol{\Sigma}\right) \equiv 1. \tag{4.49}$$

In other words, the set of constraints $\mathcal{C}$ imposes that the respective ellipsoid (4.45) passes trough the cloud of observations, see Figure 4.7.

The result (4.48) is intuitive: the ellipsoid generated by the sample mean and covariance is the one that best fits the observations, since all the observations are packed in its neighborhood.

Nevertheless, in some circumstances the ellipsoid $\mathcal{E}_{\widehat{\mathrm{E}},N\widehat{\mathrm{Cov}}}$ "tries too hard" to embrace all the observations: if an observation is an outlier, the sample mean and the sample covariance tend to perform rather poorly in an effort to account for this single observation. We discuss this phenomenon further in Section 4.5.

4.2.2 Explicit factors

Consider the explicit factor affine model (3.119), which we report here:

$$\mathbf{X} = \mathbf{BF} + \mathbf{U}. \tag{4.50}$$

Since we observe both the N-dimensional market invariants $\mathbf{X}$ and the K-dimensional explicit factors $\mathbf{F}$, the available information (4.8) consists of the time series of both the invariants and the factors:

$$i_T \equiv \{\mathbf{x}_1, \mathbf{f}_1, \ldots, \mathbf{x}_T, \mathbf{f}_T\}. \tag{4.51}$$

By applying (4.36) to the definition of the regression factor loadings (3.121) we obtain the nonparametric estimator of the regression factor loadings of the explicit factor affine model:

$$\widehat{\mathbf{B}}[i_T] \equiv \left(\sum_t \mathbf{x}_t \mathbf{f}_t'\right)\left(\sum_t \mathbf{f}_t \mathbf{f}_t'\right)^{-1}. \tag{4.52}$$

This matrix represents the *ordinary least square (OLS)* estimator of the regression factor loadings.

The name is due to a geometric property of the OLS coefficients, which we sketch in Figure 4.8. Indeed, as we show in Appendix www.4.1, the OLS estimator $\widehat{\mathbf{B}}$ provides the best fit to the observations, in the sense that it minimizes the sum of the square distances between the original observations $\mathbf{x}_t$ and the recovered values $\widehat{\mathbf{B}}\mathbf{f}_t$:

$$\widehat{\mathbf{B}} = \underset{\mathbf{B}}{\operatorname{argmin}} \sum_t \|\mathbf{x}_t - \mathbf{B}\mathbf{f}_t\|^2, \tag{4.53}$$

where $\|\cdot\|$ is the standard norm (A.6).

By applying (4.36) to the covariance of the residuals (3.129) we obtain the respective nonparametric estimator:

$$\widehat{\mathrm{Cov}}[i_T] \equiv \frac{1}{T}\sum_t \left(\mathbf{x}_t - \widehat{\mathbf{B}}\mathbf{f}_t\right)\left(\mathbf{x}_t - \widehat{\mathbf{B}}\mathbf{f}_t\right)'. \tag{4.54}$$

This is the *ordinary least square (OLS)* estimator of the covariance of the residuals.

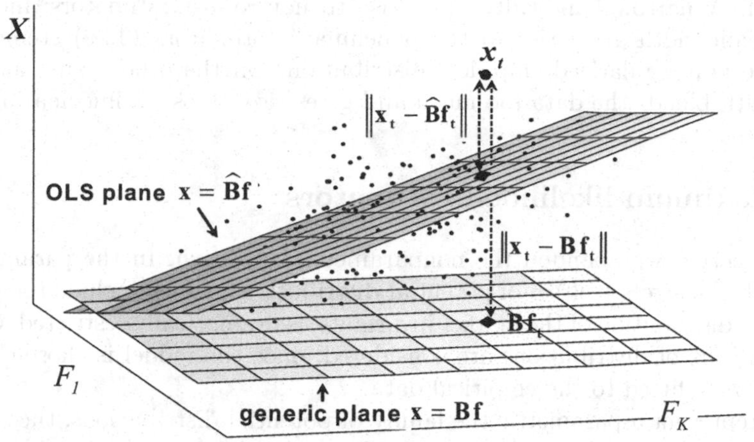

Fig. 4.8. OLS estimates of factor loadings: geometric properties

4.2.3 Kernel estimators

Here we briefly put into perspective a nonparametric approach to estimation that is becoming very popular in financial applications, see Campbell, Lo, and MacKinlay (1997).

The nonparametric estimators defined by the recipe (4.36) are very sensitive to the input data and thus are not robust, in a sense to be discussed precisely in Section 4.5. Intuitively, this happens because the empirical probability density function (4.35), is a sum of Dirac deltas, which are not regular, smooth functions.

One way to solve this problem consists in replacing the empirical distribution with a regularized, smoother distribution by means of the convolution, see (*B*.54). In other words, we replace the empirical probability density function as follows:

$$f_{i_T} \mapsto f_{i_T;\epsilon} \equiv \frac{1}{T} \sum_{t=1}^{T} \frac{1}{(2\pi)^{\frac{N}{2}} \epsilon^N} e^{-\frac{1}{2\epsilon^2}(\mathbf{x}-\mathbf{x}_t)'(\mathbf{x}-\mathbf{x}_t)}. \tag{4.55}$$

The outcome of this operation is a smoother empirical probability density function such as the one sketched in Figure 2.18. In this context, the Gaussian exponential, or any other smoothing function, takes the name of *kernel*, and the width ϵ of the regularizing function takes on the name of *bandwidth*.

Once the probability density function has been smoothened, we can define new nonparametric estimators that replace (4.36) as follows:

$$\widehat{\mathbf{G}}[i_T] \equiv \mathbf{G}[f_{i_T;\epsilon}]. \tag{4.56}$$

The bandwidth of the kernel must be chosen according to the following trade-off. A narrow bandwidth gives rise to non-robust estimators: indeed, a null bandwidth gives rise to the benchmark estimators (4.36) stemming from the non-regularized empirical distribution. On the other hand, a wide bandwidth blends the data too much and gives rise to loss of information.

4.3 Maximum likelihood estimators

In this section we abandon the nonparametric approach. In the parametric approach the stress test set of potential distributions, which include the true, unknown distribution of the market invariants, is dramatically restricted. Only a few models of distributions are considered: once the model is chosen, it is subsequently fitted to the empirical data.

We represent a parametric the family of potential distributions, the stress test distributions, in terms of their probability density function $f_{\boldsymbol{\theta}}$, where $\boldsymbol{\theta}$ is an S-dimensional parameter that fully determines the distribution and that ranges in a given set $\boldsymbol{\Theta}$, see Figure 4.9 and compare with Figure 4.3.

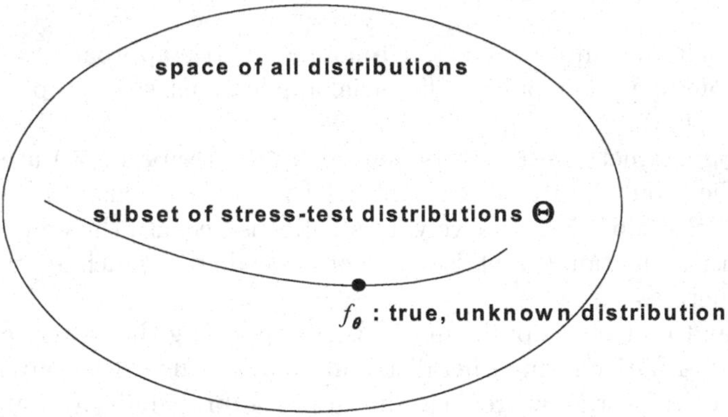

Fig. 4.9. Parametric approach to estimation

For example, from empirical observations it might seem reasonable to model a given market invariant by means of the lognormal distribution as follows:

$$X_t \sim \mathrm{LogN}\left(\theta, 1\right). \tag{4.57}$$

In this case the distribution's parameters are one-dimensional; the distribution's probability density function reads:

$$f_{\theta}(x) \equiv \frac{1}{\sqrt{2\pi}x} e^{-\frac{1}{2}(\ln x - \theta)^2}; \tag{4.58}$$

and the parameter space Θ is the real line $\mathbb{R}$.

Since the distribution of the invariants is completely determined by the parameters $\boldsymbol{\theta}$, estimating the distribution corresponds to determining these parameters. In other words, the estimation process (4.9) consists of determining some function of the available information $\widehat{\boldsymbol{\theta}}[i_T]$ that is close to the true, unknown parameters.

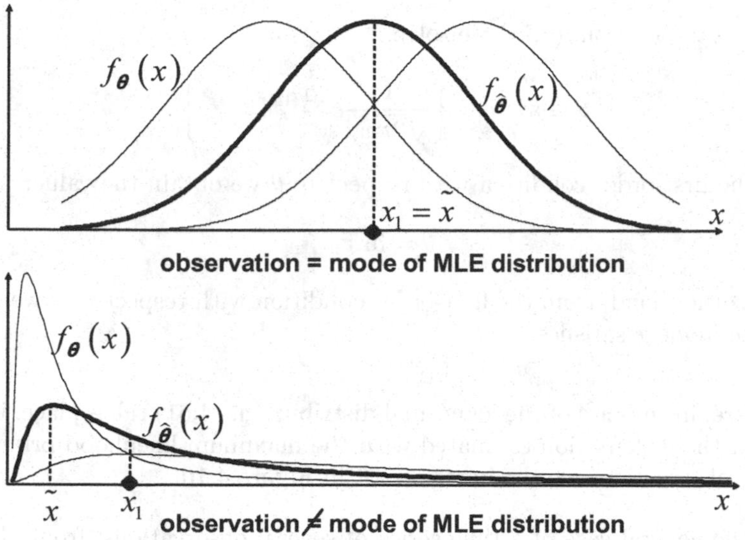

Fig. 4.10. Maximum likelihood estimator as mode

The *maximum likelihood principle* provides a method to determine an estimator which is related to the intuitive concept of mode. We recall that the mode $\widetilde{\mathbf{x}}$ of a distribution $f_{\mathbf{X}}$ is the value that corresponds to the peak of the distribution, i.e. the largest value of the probability density function:

$$\widetilde{\mathbf{x}} \equiv \underset{\mathbf{x} \in \mathbb{R}^N}{\operatorname{argmax}} f_{\mathbf{X}}(\mathbf{x}). \tag{4.59}$$

Suppose that only one observation $\mathbf{x}_1$ is available. Most likely, this observation lies in a region where the probability density function is comparatively large, i.e. near the the mode. Therefore, once we assume that the distribution that generated that observation belongs to a specific parametric family $f_{\mathbf{X}} \equiv f_{\boldsymbol{\theta}}$, the most intuitive value for the parameter $\boldsymbol{\theta}$ is the value $\widehat{\boldsymbol{\theta}}$ that makes the pdf in that point the largest, see the top plot in Figure 4.10.

In other words, according to the maximum likelihood principle we define the estimator $\widehat{\boldsymbol{\theta}}$ as follows:

$$\widehat{\boldsymbol{\theta}} \equiv \underset{\boldsymbol{\theta} \in \boldsymbol{\Theta}}{\operatorname{argmax}} f_{\boldsymbol{\theta}} (\mathbf{x}_1) . \tag{4.60}$$

Notice that, although the maximum likelihood estimator draws on the concept of mode, the observation $\mathbf{x}_1$ does not necessarily turn out to be the mode of the distribution $f_{\widehat{\boldsymbol{\theta}}}$:

$$\mathbf{x}_1 \neq \underset{\mathbf{x} \in \mathbb{R}^N}{\operatorname{argmax}} f_{\widehat{\boldsymbol{\theta}}} (\mathbf{x}) , \tag{4.61}$$

see the bottom plot in Figure 4.10.

For example, from (4.58) we solve:

$$\widehat{\theta} \equiv \underset{\theta \in \mathbb{R}}{\operatorname{argmax}} \left\{ \frac{1}{\sqrt{2\pi} x_1} e^{-\frac{1}{2}(\ln x_1 - \theta)^2} \right\} . \tag{4.62}$$

From the first-order condition with respect to θ we obtain the value:

$$\widehat{\theta} = \ln x_1 . \tag{4.63}$$

On the other hand, from the first-order condition with respect to x we obtain that the mode $\widehat{x}$ satisfies:

$$\ln \widehat{x} = \widehat{\theta} - 1 . \tag{4.64}$$

Therefore, in the case of the lognormal distribution, (4.61) takes place, i.e. the mode of the distribution estimated with the maximum likelihood principle is not the observation, see the bottom plot in Figure 4.10.

In the general case of a time series of several observations, from (4.5) we obtain the joint probability density function of the time series, which is the product of the single-period probability density functions:

$$f_{\boldsymbol{\theta}} (i_T) \equiv f_{\boldsymbol{\theta}} (\mathbf{x}_1) \cdots f_{\boldsymbol{\theta}} (\mathbf{x}_T) . \tag{4.65}$$

Expression (4.65) is also called the *likelihood function* of the time series. Now we can apply the maximum likelihood principle (4.60) to the whole time series. Therefore the *maximum likelihood estimator (MLE)* of the parameters $\boldsymbol{\theta}$ is defined as follows:

$$\begin{aligned} \widehat{\boldsymbol{\theta}} [i_T] &\equiv \underset{\boldsymbol{\theta} \in \boldsymbol{\Theta}}{\operatorname{argmax}} f_{\boldsymbol{\theta}} (i_T) \\ &= \underset{\boldsymbol{\theta} \in \boldsymbol{\Theta}}{\operatorname{argmax}} \sum_{t=1}^{T} \ln f_{\boldsymbol{\theta}} (\mathbf{x}_t) . \end{aligned} \tag{4.66}$$

For example, in the case of lognormal invariants, from (4.58) we solve:

$$\widehat{\theta} \equiv \underset{\theta \in \Theta}{\operatorname{argmax}} \left\{ -\sum_{t=1}^{T} \frac{1}{2} \left(\ln x_t - \theta \right)^2 \right\}. \tag{4.67}$$

The first-order condition reads:

$$0 = \frac{1}{T} \sum_{t=1}^{T} \left(\ln x_t - \widehat{\theta} \right), \tag{4.68}$$

which implies the following expression for the maximum likelihood estimate of the parameter:

$$\widehat{\theta} = \frac{1}{T} \sum_{t=1}^{T} \ln x_t. \tag{4.69}$$

The maximum likelihood estimator displays a few appealing properties.

For instance, the *invariance property*, which states that the MLE of a function of the parameters is that function applied to the MLE of the parameters:

$$\widehat{g(\boldsymbol{\theta})} = g\left(\widehat{\boldsymbol{\theta}}\right). \tag{4.70}$$

This property follows from the definition (4.66).

Furthermore, similarly to the nonparametric approach (4.37), the maximum likelihood principle provides good estimators in the limit case of a very large number of observations T in the time series i_T, as sketched in Figure 4.1. Indeed, the following relation holds in approximation, and the approximation becomes exact as T tends to infinity:

$$\widehat{\boldsymbol{\theta}}\left[I_T\right] \sim \mathrm{N}\left(\boldsymbol{\theta}, \frac{\boldsymbol{\Gamma}}{T}\right). \tag{4.71}$$

In this expression $\boldsymbol{\Gamma}$ is a symmetric and positive matrix called the *Fisher information matrix*:

$$\boldsymbol{\Gamma} \equiv \operatorname{Cov}\left\{ \frac{\partial \ln\left(f_{\boldsymbol{\theta}}\left(\mathbf{X}\right)\right)}{\partial \boldsymbol{\theta}} \right\}, \tag{4.72}$$

see e.g. Haerdle and Simar (2003).

The *Cramer-Rao lower bound* theorem states that the inefficiency of the maximum likelihood estimator, as represented by (4.72), is the smallest possible achievable with an unbiased estimator, and from (4.71) we see that the MLE becomes unbiased in the limit of many observations.

Nevertheless, we introduced the parametric approach to estimation in order to build estimators that perform well in the realistic case of a finite number of observations of market invariants. Therefore below we evaluate the maximum likelihood estimators of parametric models that are apt to describe the market invariants.

4.3.1 Location, dispersion and hidden factors

In Chapter 3 we saw that the market invariants are quite symmetrical. Therefore, in this section we construct and evaluate maximum-likelihood estimators under the assumption that the N-dimensional invariants $\mathbf{X}$ are elliptically distributed:

$$\mathbf{X} \sim \mathrm{El}\left(\boldsymbol{\mu}, \boldsymbol{\Sigma}, g\right), \tag{4.73}$$

where $\boldsymbol{\mu}$ is the N-dimensional location parameter, $\boldsymbol{\Sigma}$ is the $N \times N$ dispersion matrix and g is the probability density generator, see (2.268). In other words, the probability density function of the invariants invariants $\mathbf{X}$ is of the form:

$$f_{\boldsymbol{\theta}}\left(\mathbf{x}\right) \equiv \frac{1}{\sqrt{|\boldsymbol{\Sigma}|}} g\left(\mathrm{Ma}^2\left(\mathbf{x}, \boldsymbol{\mu}, \boldsymbol{\Sigma}\right)\right), \tag{4.74}$$

where $\mathrm{Ma}\left(\mathbf{x}, \boldsymbol{\mu}, \boldsymbol{\Sigma}\right)$ is the *Mahalanobis distance* of the point $\mathbf{x}$ from the point $\boldsymbol{\mu}$ through the metric $\boldsymbol{\Sigma}$:

$$\mathrm{Ma}\left(\mathbf{x}, \boldsymbol{\mu}, \boldsymbol{\Sigma}\right) \equiv \sqrt{\left(\mathbf{x} - \boldsymbol{\mu}\right)' \boldsymbol{\Sigma}^{-1} \left(\mathbf{x} - \boldsymbol{\mu}\right)}, \tag{4.75}$$

see (2.61).

Under the assumption (4.73) the parameters $\boldsymbol{\theta} \equiv \left(\boldsymbol{\mu}, \boldsymbol{\Sigma}\right)$ completely determine the distribution of the market invariants[4]. These parameters span the set:

$$\boldsymbol{\Theta} \equiv \left\{\boldsymbol{\mu} \in \mathbb{R}^N, \boldsymbol{\Sigma} \in \mathbb{R}^{N \times N}, \boldsymbol{\Sigma} \succeq \mathbf{0}\right\}, \tag{4.76}$$

where $\succeq \mathbf{0}$ denotes symmetric and positive.

In this context, estimating the distribution of the market invariants means estimating from currently available information i_T the parameters $\left(\boldsymbol{\mu}, \boldsymbol{\Sigma}\right)$. In Appendix www.4.2 we prove that the MLE estimators $\widehat{\boldsymbol{\mu}}\left[i_T\right]$ and $\widehat{\boldsymbol{\Sigma}}\left[i_T\right]$ are the solutions to the following joint set of implicit equations:

$$\widehat{\boldsymbol{\mu}} = \sum_{t=1}^{T} \frac{w\left(\mathrm{Ma}^2\left(\mathbf{x}_t, \widehat{\boldsymbol{\mu}}, \widehat{\boldsymbol{\Sigma}}\right)\right)}{\sum_{s=1}^{T} w\left(\mathrm{Ma}^2\left(\mathbf{x}_s, \widehat{\boldsymbol{\mu}}, \widehat{\boldsymbol{\Sigma}}\right)\right)} \mathbf{x}_t \tag{4.77}$$

$$\widehat{\boldsymbol{\Sigma}} = \frac{1}{T} \sum_{t=1}^{T} \left(\mathbf{x}_t - \widehat{\boldsymbol{\mu}}\right)\left(\mathbf{x}_t - \widehat{\boldsymbol{\mu}}\right)' w\left(\mathrm{Ma}^2\left(\mathbf{x}_t, \widehat{\boldsymbol{\mu}}, \widehat{\boldsymbol{\Sigma}}\right)\right), \tag{4.78}$$

where the function w is defined as follows in terms of the probability density generator:

$$w\left(z\right) \equiv -2\frac{g'\left(z\right)}{g\left(z\right)}. \tag{4.79}$$

Notice that defining the following weights:

---

[4] We assume known the specific density generator, otherwise we would obtain a *semiparametric* model.

$$w_t \equiv w\left(\mathrm{Ma}^2\left(\mathbf{x}_t, \widehat{\boldsymbol{\mu}}, \widehat{\boldsymbol{\Sigma}}\right)\right), \tag{4.80}$$

we can interpret the maximum likelihood estimators of location and dispersion (4.77) and (4.78) as weighted sums:

$$\widehat{\boldsymbol{\mu}} = \sum_{t=1}^{T} \frac{w_t}{\sum_{s=1}^{T} w_s} \mathbf{x}_t \tag{4.81}$$

$$\widehat{\boldsymbol{\Sigma}} = \frac{1}{T} \sum_{t=1}^{T} w_t \left(\mathbf{x}_t - \widehat{\boldsymbol{\mu}}\right)\left(\mathbf{x}_t - \widehat{\boldsymbol{\mu}}\right)'. \tag{4.82}$$

Each observation is weighted according to its Mahalanobis distance from the ML estimator of location through the metric defined by the ML estimator of dispersion.

For example, assume that the market invariants are Cauchy distributed, see (2.208). In this case the density generator reads:

$$g^{\mathrm{Ca}}(z) = \frac{\Gamma\left(\frac{1+N}{2}\right)}{\Gamma\left(\frac{1}{2}\right)(\pi)^{\frac{N}{2}}} (1+z)^{-\frac{1+N}{2}}, \tag{4.83}$$

where Γ is the gamma function (B.80). Therefore the weights (4.80) become:

$$w_t = \frac{N+1}{1 + \mathrm{Ma}^2\left(\mathbf{x}_t, \widehat{\boldsymbol{\mu}}, \widehat{\boldsymbol{\Sigma}}\right)}. \tag{4.84}$$

This is a decreasing function of the Mahalanobis distance: the maximum likelihood estimators of location and dispersion of a set of Cauchy-distributed invariants tend to neglect outliers.

This result is intuitive: we recall that the Cauchy distribution is fat-tailed, see Figure 1.9. Therefore extreme observations, i.e. observations with large Mahalanobis distance, are quite frequent. These extreme observations might distort the estimation, which is why the maximum likelihood estimator tends to taper their influence in the estimation process.

After solving (4.77)-(4.79) for $\widehat{\boldsymbol{\Sigma}}$ we can derive the expression for the maximum likelihood estimator of the principal component factor model. Indeed, it suffices to compute the PCA decomposition of the estimator:

$$\widehat{\boldsymbol{\Sigma}}\left[i_T\right] \equiv \widehat{\mathbf{E}}\widehat{\boldsymbol{\Lambda}}\widehat{\mathbf{E}}', \tag{4.85}$$

where $\widehat{\boldsymbol{\Lambda}}$ is the diagonal matrix of the eigenvalues in decreasing order and $\widehat{\mathbf{E}}$ is the orthogonal matrix of the respective eigenvectors. Then $\widehat{\mathbf{E}}$ becomes the MLE estimator of the hidden factor loadings and $\widehat{\boldsymbol{\Lambda}}$ becomes the estimator of the dispersion of the hidden factors.

To evaluate the performance of the maximum likelihood estimators of location and dispersion $\widehat{\boldsymbol{\mu}}$ and $\widehat{\boldsymbol{\Sigma}}$ we should determine the distribution of $\widehat{\boldsymbol{\mu}}$ and $\widehat{\boldsymbol{\Sigma}}$ when in (4.77)-(4.79) the market invariants are considered as random variables as in (4.15).

Unfortunately, in the generic elliptical case it is not possible to convert the implicit equations (4.77)-(4.79) into explicit functional expressions of current information. Therefore we must solve for the estimators numerically and resort to simulations to evaluate their performance, unless the invariants are normally distributed. We discuss the normal case in detail in Section 4.3.3.

4.3.2 Explicit factors

Consider the explicit factor affine model (3.119), which we report here:

$$\mathbf{X} = \mathbf{BF} + \mathbf{U}. \tag{4.86}$$

Since we observe both the N-dimensional invariants $\mathbf{X}$ and the K-dimensional factors $\mathbf{F}$, the available information (4.8) is the time series of both the invariants and the factors:

$$i_T \equiv \{\mathbf{x}_1, \mathbf{f}_1, \ldots, \mathbf{x}_T, \mathbf{f}_T\}. \tag{4.87}$$

To implement the maximum likelihood approach we could model the $(N+K)$-dimensional joint distribution of invariants and factors by means of some parametric distribution $f_{\boldsymbol{\theta}}$ and then maximize the likelihood over the parameters $\boldsymbol{\theta}$ and the factor loadings $\mathbf{B}$.

Nevertheless, most explicit factor models serve the purpose of stress testing the behavior of the invariants under assumptions on the future realization of the factors. For example, practitioners ask themselves such questions as what happens to a given stock if the market goes up, say, 2%. Therefore, it is more convenient to model the N-dimensional distribution of the perturbations $f_{\boldsymbol{\theta}|\mathbf{f}} \equiv f_{\mathbf{U}|\mathbf{f}}$ *conditional* on knowledge of the factors and model the *conditional* distribution of the invariants accordingly:

$$\mathbf{X}|\mathbf{f} = \mathbf{Bf} + \mathbf{U}|\mathbf{f}. \tag{4.88}$$

Under the above assumptions the *conditional* distribution $f_{\mathbf{X}|\mathbf{f}}$ of the invariants becomes a parametric function $f_{\boldsymbol{\theta},\mathbf{B}}$ of the parameters $\boldsymbol{\theta}$ of the perturbations and the factor loadings $\mathbf{B}$. Therefore we can apply the maximum likelihood principle (4.66) to the conditional distribution of the invariants, determining the maximum likelihood estimator $\widehat{\boldsymbol{\theta}}$ of the distribution of the perturbations and the maximum likelihood estimator of the factor loadings $\widehat{\mathbf{B}}$:

$$\left(\widehat{\boldsymbol{\theta}}, \widehat{\mathbf{B}}\right) \equiv \underset{\boldsymbol{\theta}\in\boldsymbol{\Theta},\mathbf{B}}{\operatorname{argmax}}\, f_{\boldsymbol{\theta},\mathbf{B}}\left(i_T\right). \tag{4.89}$$

In Chapter 3 we saw that the market invariants are quite symmetrical. Therefore, we construct the maximum-likelihood estimators under the assumption that the conditional distribution of the perturbations be an N-dimensional elliptical random variable:

$$\mathbf{U}_t|\mathbf{f}_t \sim \mathrm{El}\left(\mathbf{0}, \mathbf{\Sigma}, g\right). \tag{4.90}$$

In other words we assume that the perturbations are centered in zero; that $\mathbf{\Sigma}$ is their $N \times N$ dispersion matrix and that g is their probability density generator.

From (2.270) the invariants are elliptically distributed with the same generator:

$$\mathbf{X}_t|\mathbf{f}_t \sim \mathrm{El}\left(\mathbf{B}\mathbf{f}_t, \mathbf{\Sigma}, g\right). \tag{4.91}$$

In this context the parameters to be estimated are $\mathbf{B}$ and $\mathbf{\Sigma}$.

In Appendix www.4.2 we show that the MLE estimators of these parameters solve the following set of joint implicit equations:

$$\widehat{\mathbf{B}} = \left[\sum_{t=1}^{T} w\left(\mathrm{Ma}^2\left(\mathbf{x}_t, \widehat{\mathbf{B}}\mathbf{f}_t, \widehat{\mathbf{\Sigma}}\right)\right)\mathbf{x}_t\mathbf{f}_t'\right] \left[\sum_{t=1}^{T} w\left(\mathrm{Ma}^2\left(\mathbf{x}_t, \widehat{\mathbf{B}}\mathbf{f}_t, \widehat{\mathbf{\Sigma}}\right)\right)\mathbf{f}_t\mathbf{f}_t'\right]^{-1} \tag{4.92}$$

and

$$\widehat{\mathbf{\Sigma}} = \frac{1}{T}\sum_{t=1}^{T} w\left(\mathrm{Ma}^2\left(\mathbf{x}_t, \widehat{\mathbf{B}}\mathbf{f}_t, \widehat{\mathbf{\Sigma}}\right)\right)\left(\mathbf{x}_t - \widehat{\mathbf{B}}\mathbf{f}_t\right)\left(\mathbf{x}_t - \widehat{\mathbf{B}}\mathbf{f}_t\right)', \tag{4.93}$$

where the function w is defined in terms of the probability density generator:

$$w\left(z\right) \equiv -2\frac{g'\left(z\right)}{g\left(z\right)}. \tag{4.94}$$

In the generic elliptical case, the implicit equations (4.92)-(4.94) must be solved numerically and the evaluation of the estimators must be performed by means of simulations. On the other hand, in the specific normal case the above implicit equations can be solved analytically. We discuss the normal explicit factor model at the end of Section 4.3.3.

4.3.3 The normal case

In the special case where the market invariants are normally distributed the analysis of the maximum likelihood estimators of location, dispersion and explicit factors can be performed analytically. This analysis provides insight into the more general case.

Location, dispersion and hidden factors

Assume that the market invariants are normally distributed:

$$\mathbf{X} \sim \mathrm{N}(\boldsymbol{\mu}, \boldsymbol{\Sigma}). \tag{4.95}$$

In the normal case the location parameter $\boldsymbol{\mu}$ is the expected value of the distribution and the dispersion parameter $\boldsymbol{\Sigma}$ is its covariance matrix.

The normal distribution is a special case of elliptical distribution, which corresponds to the following choice of the density generator:

$$g^{\mathrm{N}}(z) \equiv \frac{e^{-\frac{z}{2}}}{(2\pi)^{\frac{N}{2}}}, \tag{4.96}$$

see (2.264). It is immediate to check that in the normal case the weights (4.79) are constant:

$$w(z) \equiv 1. \tag{4.97}$$

To interpret this result, we compare it with the respective result for the Cauchy distribution. The normal distribution is very thin-tailed and therefore extreme observations are rare. If an observation is far from the location parameter, the reason must be due to a large dispersion matrix: therefore, unlike (4.84), the maximum likelihood estimator gives full weight to that observation, in such a way to effectively modify the estimation and lead to a larger estimate of the dispersion matrix.

From (4.77) we obtain the explicit expression of the estimator of location in terms of current information:

$$\widehat{\boldsymbol{\mu}}[i_T] = \frac{1}{T}\sum_{t=1}^{T} \mathbf{x}_t. \tag{4.98}$$

Similarly, from (4.78) we obtain the explicit expression of the estimator of dispersion in terms of current information:

$$\widehat{\boldsymbol{\Sigma}}[i_T] = \frac{1}{T}\sum_{t=1}^{T} (\mathbf{x}_t - \widehat{\boldsymbol{\mu}})(\mathbf{x}_t - \widehat{\boldsymbol{\mu}})'. \tag{4.99}$$

These estimators are the sample mean (4.41) and the sample covariance (4.42) respectively. It is reassuring that two completely different methods yield the same estimators for both location and dispersion. This supports our statement that the sample mean and the sample covariance are the benchmark estimators of location and dispersion respectively.

To evaluate the goodness of the sample mean and of the sample covariance under the normal hypothesis we proceed as in (4.15), computing the joint distribution of the following random variables:

$$\widehat{\boldsymbol{\mu}}[I_T] \equiv \frac{1}{T}\sum_{t=1}^{T} \mathbf{X}_t \tag{4.100}$$

$$\widehat{\boldsymbol{\Sigma}}[I_T] \equiv \frac{1}{T}\sum_{t=1}^{T} (\mathbf{X}_t - \widehat{\boldsymbol{\mu}})(\mathbf{X}_t - \widehat{\boldsymbol{\mu}})'. \tag{4.101}$$

In Appendix www.4.3 we prove the following results. The sample mean is normally distributed:

$$\widehat{\boldsymbol{\mu}}\left[I_T\right] \sim \mathrm{N}\left(\boldsymbol{\mu}, \frac{\boldsymbol{\Sigma}}{T}\right). \tag{4.102}$$

The distribution of the sample covariance is related to the Wishart-distribution (2.223) by the following expression:

$$T\widehat{\boldsymbol{\Sigma}}\left[I_T\right] \sim \mathrm{W}\left(T-1, \boldsymbol{\Sigma}\right). \tag{4.103}$$

Furthermore, (4.102) and (4.103) are independent of each other.

- Component-wise evaluation

From the above expressions we can evaluate component-wise the error (4.23) of the sample estimators, using the standard quadratic form $Q \equiv 1$ in (4.20) and decomposing the error into bias and inefficiency as in (4.27).

For the sample mean, from (4.102) we obtain:

$$\mathrm{Bias}\left(\widehat{\mu}_i, \mu_i\right) = 0 \tag{4.104}$$

$$\mathrm{Inef}\left(\widehat{\mu}_i\right) = \sqrt{\frac{\Sigma_{ii}}{T}}. \tag{4.105}$$

This shows that the sample mean is unbiased and that its inefficiency shrinks to zero as the number of observations grows to infinity.

As for the estimator of the sample covariance, from (4.103) and (2.227)-(2.228) we obtain:

$$\mathrm{Bias}\left(\widehat{\Sigma}_{mn}, \Sigma_{mn}\right) = \frac{1}{T}\left|\Sigma_{mn}\right| \tag{4.106}$$

$$\mathrm{Inef}\left(\widehat{\Sigma}_{mn}\right) = \sqrt{\frac{T-1}{T^2}}\sqrt{\Sigma_{mm}\Sigma_{nn} + \Sigma_{mn}^2}. \tag{4.107}$$

As expected, bias and inefficiency shrink to zero as the number of observations grows to infinity.

Formulas (4.104)-(4.107) provide the measure of performance for each of the entries of the estimators separately. It is nonetheless interesting to obtain a global measure of performance. Since the sample mean $\widehat{\boldsymbol{\mu}}$ and the sample covariance $\widehat{\boldsymbol{\Sigma}}$ are independent, we evaluate them separately.

- Evaluation of sample mean

To evaluate the sample mean (4.100), we consider the loss (4.19) induced by the quadratic form $\mathbf{Q} \equiv \mathbf{I}_N$. In other words, the loss is the following random variable:

$$\mathrm{Loss}\left(\widehat{\boldsymbol{\mu}}, \boldsymbol{\mu}\right) \equiv \left[\widehat{\boldsymbol{\mu}}\left[I_T\right] - \boldsymbol{\mu}\right]' \left[\widehat{\boldsymbol{\mu}}\left[I_T\right] - \boldsymbol{\mu}\right]. \tag{4.108}$$

We then summarize the information contained in the loss by means of the error (4.23). We prove in Appendix www.4.3 that the error reads:

$$\text{Err}^2(\widehat{\boldsymbol{\mu}}, \boldsymbol{\mu}) = \frac{1}{T}\operatorname{tr}(\boldsymbol{\Sigma}). \tag{4.109}$$

The whole error is due to inefficiency, as the sample estimator is unbiased:

$$\text{Inef}^2(\widehat{\boldsymbol{\mu}}) = \frac{1}{T}\operatorname{tr}(\boldsymbol{\Sigma}) \tag{4.110}$$

$$\text{Bias}^2(\widehat{\boldsymbol{\mu}}, \boldsymbol{\mu}) = 0. \tag{4.111}$$

As expected, the error decreases as the number of observations grows to infinity. Furthermore, it is an increasing function of the average variance: intuitively, more volatile invariants give rise to larger estimation errors.

To gain further insight into the estimation error of the sample mean, we consider the PCA decomposition (A.70) of the scatter matrix:

$$\boldsymbol{\Sigma} \equiv \mathbf{E}\boldsymbol{\Lambda}\mathbf{E}'. \tag{4.112}$$

In this expression $\boldsymbol{\Lambda}$ is the diagonal matrix of the eigenvalues of $\boldsymbol{\Sigma}$ sorted in decreasing order:

$$\boldsymbol{\Lambda} \equiv \operatorname{diag}(\lambda_1, \ldots, \lambda_N); \tag{4.113}$$

and $\mathbf{E}$ is the juxtaposition of the respective orthogonal eigenvectors. From the PCA decomposition the following identity follows:

$$\operatorname{tr}[\boldsymbol{\Sigma}] = \operatorname{tr}[\boldsymbol{\Lambda}]. \tag{4.114}$$

Therefore, the estimation error of sample mean (4.109), along with its factorization in terms of bias and inefficiency, is completely determined by the eigenvalues of $\boldsymbol{\Sigma}$. To interpret this result geometrically, consider the ellipsoid $\mathcal{E}_{\boldsymbol{\mu},\boldsymbol{\Sigma}}$ determined by the market parameters as described in (A.73), which is also the location-dispersion ellipsoid of the invariants (2.75). Since the eigenvalues represent the (square of) the length of the principal axes of the ellipsoid, the estimation error of the sample mean is completely determined by the shape of the location-dispersion ellipsoid of the invariants, and not by its location or orientation.

In particular, a key parameter is the *condition number* or the *condition ratio* defined as the ratio between the smallest and the largest eigenvalue:

$$\text{CN}\{\mathbf{X}\} \equiv \frac{\lambda_N}{\lambda_1}. \tag{4.115}$$

The condition number ranges in the interval $[0, 1]$. When the condition number is close to one the invariants $\mathbf{X}$ are *well-conditioned* and the location-dispersion ellipsoid that represents the invariants resembles a sphere. When the condition number is close to zero the invariants $\mathbf{X}$ are *ill-conditioned*: the ellipsoid is elongated, shaped like a cigar, since the actual dimension of risk is less than the number of invariants. This is the case in highly correlated markets, such as the swap market, see Figure 3.20.

To capture the effect of the shape of the location-dispersion ellipsoid on the estimation error, we keep the location $\boldsymbol{\mu}$ constant and we let the scatter matrix $\boldsymbol{\Sigma}$ vary as follows:

$$\boldsymbol{\Sigma} \equiv \begin{pmatrix} 1 & \theta & \cdots & \theta \\ \theta & 1 & \ddots & \vdots \\ \vdots & \ddots & \ddots & \theta \\ \theta & \cdots & \theta & 1 \end{pmatrix}, \quad \theta \in (0,1). \tag{4.116}$$

The parameter θ represents the overall level of correlation among the invariants: as the correlation varies between zero and one, the condition number varies between one and zero.

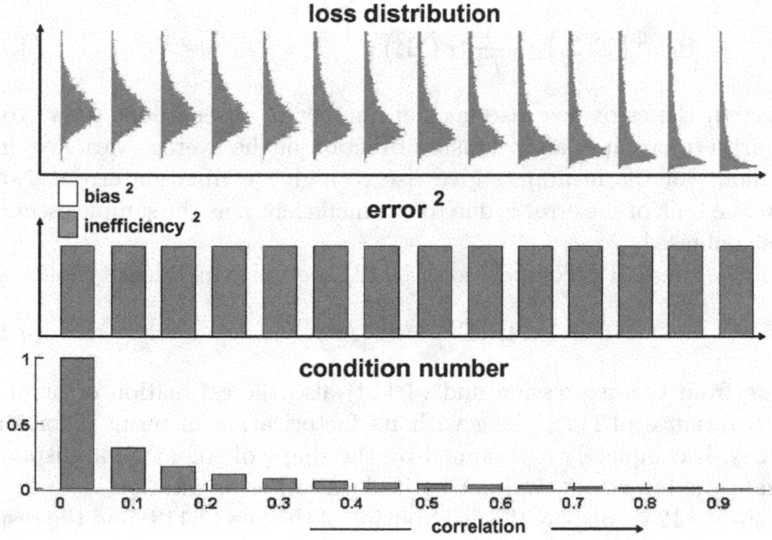

Fig. 4.11. Sample mean: evaluation

In Figure 4.11 we display the distribution of the loss (4.108) and the respective error (4.109) as the market parameters vary according to (4.116). Notice how the distribution of the loss varies, although the inefficiency and thus the error remain constant.

- Evaluation of sample covariance

To evaluate the sample covariance (4.101) we introduce the *Frobenius quadratic form* for a generic symmetric matrix $\mathbf{S}$:

$$\|\mathbf{S}\|^2 \equiv \operatorname{tr}\left[\mathbf{S}^2\right]. \tag{4.117}$$

This corresponds to the choice $\mathbf{Q} \equiv \mathbf{I}_{N^2}$ in (4.20) acting on vec $(\mathbf{S})$, the stacked columns of $\mathbf{S}$. Accordingly, the loss (4.19) becomes the following random variable:

$$\operatorname{Loss}\left(\widehat{\boldsymbol{\Sigma}}, \boldsymbol{\Sigma}\right) \equiv \operatorname{tr}\left[\left(\widehat{\boldsymbol{\Sigma}}\left[I_T\right] - \boldsymbol{\Sigma}\right)^2\right]. \tag{4.118}$$

In Appendix www.4.3 we show that the estimation error (4.23) relative to this loss reads:

$$\operatorname{Err}^2\left(\widehat{\boldsymbol{\Sigma}}, \boldsymbol{\Sigma}\right) = \frac{1}{T}\left[\operatorname{tr}\left(\boldsymbol{\Sigma}^2\right) + \left(1 - \frac{1}{T}\right)\left[\operatorname{tr}\left(\boldsymbol{\Sigma}\right)\right]^2\right]. \tag{4.119}$$

The error factors as follows into bias and inefficiency:

$$\operatorname{Inef}^2\left(\widehat{\boldsymbol{\Sigma}}\right) = \frac{1}{T}\left(1 - \frac{1}{T}\right)\left[\operatorname{tr}\left(\boldsymbol{\Sigma}^2\right) + \left[\operatorname{tr}\left(\boldsymbol{\Sigma}\right)\right]^2\right] \tag{4.120}$$

$$\operatorname{Bias}^2\left(\widehat{\boldsymbol{\Sigma}}, \boldsymbol{\Sigma}\right) = \frac{1}{T^2}\operatorname{tr}\left(\boldsymbol{\Sigma}^2\right). \tag{4.121}$$

As expected, the error decreases as the number of observations grows to infinity. Furthermore, it is an increasing function of the average variance: intuitively, more volatile invariants give rise to higher estimation errors. Notice also that the bulk of the error is due to the inefficiency, as the sample estimator is almost unbiased.

From the spectral decomposition (4.112) the following identity follows:

$$\operatorname{tr}\left[\boldsymbol{\Sigma}^2\right] = \operatorname{tr}\left[\boldsymbol{\Lambda}^2\right]. \tag{4.122}$$

Therefore from this expression and (4.114) also the estimation error of the sample covariance (4.119), along with its factorization in terms of bias and inefficiency, is completely determined by the shape of the location-dispersion ellipsoid of the invariants, and not by its location or orientation.

In Figure 4.12 we display the distribution of the loss (4.118) and the respective error (4.119) as the market parameters vary according to (4.116). Notice in the top plot that for high correlations the peak of the distribution of the loss is close to zero, although its dispersion increases dramatically. Indeed, we see in the middle plot how the inefficiency increases with the correlation of the market invariants.

Explicit factors

Consider the particular case of the conditional linear factor model (4.90) where the perturbations are normally distributed:

$$\mathbf{U}_t | \mathbf{f}_t \sim \mathrm{N}\left(\mathbf{0}, \boldsymbol{\Sigma}\right). \tag{4.123}$$

From the expression (2.264) of the density generator:

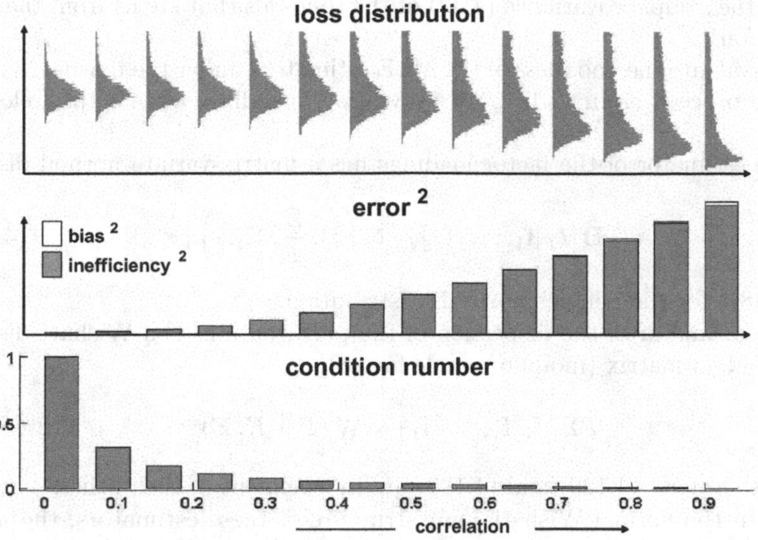

Fig. 4.12. Sample covariance: evaluation

$$g^{\mathrm{N}}(z) \equiv (2\pi)^{-\frac{N}{2}} e^{-\frac{z}{2}}, \tag{4.124}$$

we obtain that the weights (4.94) are constant:

$$w(z) \equiv 1. \tag{4.125}$$

Therefore (4.92) yields the explicit expression of the estimator of the factor loadings in terms of current information:

$$\widehat{\mathbf{B}}[i_T] = \widehat{\mathbf{\Sigma}}_{XF}[i_T]\, \widehat{\mathbf{\Sigma}}_F^{-1}[i_T], \tag{4.126}$$

where

$$\widehat{\mathbf{\Sigma}}_{XF}[i_T] \equiv \frac{1}{T}\sum_{t=1}^{T} \mathbf{x}_t \mathbf{f}_t', \quad \widehat{\mathbf{\Sigma}}_F[i_T] \equiv \frac{1}{T}\sum_{t=1}^{T} \mathbf{f}_t \mathbf{f}_t'. \tag{4.127}$$

This is the ordinary least squares estimator (4.52) of the regression factor loadings. It is reassuring that two completely different methods yield the same estimator for the factor loadings. This supports our statement that the OLS estimator is the benchmark estimator for the factor loadings.

On the other hand (4.93) yields the explicit expression of the estimator of the dispersion of the perturbations in terms of current information:

$$\widehat{\mathbf{\Sigma}}[i_T] = \frac{1}{T}\sum_{t=1}^{T} \left(\mathbf{x}_t - \widehat{\mathbf{B}}[i_T]\, \mathbf{f}_t\right)\left(\mathbf{x}_t - \widehat{\mathbf{B}}[i_T]\, \mathbf{f}_t\right)'. \tag{4.128}$$

This is the sample covariance (4.42) of the residuals that stems from the OLS estimation.

To evaluate the goodness of the MLE estimators under the normal hypothesis, we proceed as in (4.15). We prove in Appendix www.4.4 the following results.

The estimator of the factor loadings has a matrix-variate normal distribution:

$$\widehat{\mathbf{B}}\left[I_T|\mathbf{f}_1,\ldots,\mathbf{f}_T\right] \sim \mathrm{N}\left(\mathbf{B},\frac{\boldsymbol{\Sigma}}{T},\widehat{\boldsymbol{\Sigma}}_F^{-1}\right), \tag{4.129}$$

see (2.181) for the definition of this distribution.

The estimator of the dispersion of the perturbations is a Wishart distributed random matrix (modulo a scale factor):

$$T\widehat{\boldsymbol{\Sigma}}\left[I_T|\mathbf{f}_1,\ldots,\mathbf{f}_T\right] \sim \mathrm{W}\left(T-K,\boldsymbol{\Sigma}\right). \tag{4.130}$$

Furthermore, (4.129) and (4.130) are independent of each other.

Given the normal-Wishart joint structure of these estimators, the maximum likelihood estimators of the factor loadings and of the dispersion of the perturbations can be evaluated by exactly the same methodology used for (4.102) and (4.103) respectively.

4.4 Shrinkage estimators

We have discussed in Section 4.2 the benchmark estimators of location and dispersion of the generic market invariants $\mathbf{X}$, namely the sample mean and sample covariance respectively, and the benchmark estimators of the explicit factor models, namely the OLS regression coefficients. These estimators perform well in the limit case of an infinite number of observations, see Figure 4.1. We have also seen in Section 4.3 that when the underlying distribution of the invariants is normal these estimators satisfy the maximum likelihood principle.

Nevertheless, when the sample is very short, the error associated with the benchmark estimators is quite large.

An estimator is *admissible* if it is not systematically outperformed, i.e. if there does not exist another estimator which displays less error for all the stress-test distributions considered in the evaluation of that estimator, see Figure 4.9. The benchmark estimators are not admissible. Indeed, although the maximum likelihood principle is an intuitive recipe with many palatable features, it does not guarantee that the ensuing estimators be optimal.

In particular, the bulk of the error of the benchmark estimators is due to their inefficiency, whereas their bias is quite limited, see Figures 4.11 and 4.12. A key feature of the underlying distribution of the invariants $\mathbf{X}$ that deeply affects the efficiency of the benchmark estimators is the condition number

(4.115), namely the ratio between the smallest and the largest eigenvalues of the unknown underlying scatter matrix:

$$\mathrm{CN}\{\mathbf{X}\} \equiv \frac{\lambda_N}{\lambda_1}. \tag{4.131}$$

We see below that the benchmark estimators are very inefficient when the condition number is close to one, i.e. when the invariants are well-diversified and display little correlation with each other.

In order to fix the inefficiency of the benchmark estimators we consider estimators that are very efficient, although they display a large bias, namely constant estimators. Then we blend the benchmark estimators with the constant estimators by means of weighted averages. Such estimators are called *shrinkage estimators*, because the benchmark estimators are shrunk towards the target constant estimators.

As we see below, the gain in efficiency of the shrinkage estimators with respect to the original benchmark estimators more than compensates for the increase in bias, and thus the overall error of the shrinkage estimators is reduced.

4.4.1 Location

Assume that the market invariants are normally distributed with the following parameters:

$$\mathbf{X}_t \sim \mathrm{N}(\boldsymbol{\mu}, \boldsymbol{\Sigma}). \tag{4.132}$$

Consider the standard definition (4.108) of loss of a generic estimator of location $\widehat{\boldsymbol{\mu}}$ with respect to the true unknown location parameter $\boldsymbol{\mu}$ of the invariants:

$$\mathrm{Loss}(\widehat{\boldsymbol{\mu}}, \boldsymbol{\mu}) \equiv (\widehat{\boldsymbol{\mu}}[I_T] - \boldsymbol{\mu})' (\widehat{\boldsymbol{\mu}}[I_T] - \boldsymbol{\mu}); \tag{4.133}$$

and the respective definition of error:

$$\mathrm{Err}^2(\widehat{\boldsymbol{\mu}}, \boldsymbol{\mu}) \equiv \mathrm{E}\{(\widehat{\boldsymbol{\mu}}[I_T] - \boldsymbol{\mu})' (\widehat{\boldsymbol{\mu}}[I_T] - \boldsymbol{\mu})\}. \tag{4.134}$$

Consider the benchmark estimator of location (4.98) of the market invariants, namely the sample mean:

$$\widehat{\boldsymbol{\mu}}[i_T] \equiv \frac{1}{T} \sum_{t=1}^{T} \mathbf{x}_t. \tag{4.135}$$

From (4.109) the error (4.134) of the sample mean reads:

$$\mathrm{Err}^2(\widehat{\boldsymbol{\mu}}, \boldsymbol{\mu}) = \frac{1}{T} \mathrm{tr}(\boldsymbol{\Sigma}). \tag{4.136}$$

In a pathbreaking publication, Stein (1955) proved that the sample mean is not an admissible estimator. In other words, when the dimensions N of the vector of invariants $\mathbf{X}$ is larger than one, there exists an estimator of location $\widehat{\boldsymbol{\mu}}^S$ such that:

$$\operatorname{Err}^2\left(\widehat{\boldsymbol{\mu}}^S, \boldsymbol{\mu}\right) < \frac{1}{T}\operatorname{tr}(\boldsymbol{\Sigma}), \tag{4.137}$$

no matter the values of the underlying parameters in (4.132). The hypotheses in the original work were somewhat more restrictive than (4.132). Here we discuss the more general case, see also Lehmann and Casella (1998).

First of all, from (4.111) we see that we cannot improve on the sample mean's bias, as the whole error is due to the estimator's inefficiency (4.110). In other words, the sample mean is properly centered around the true, unknown value, but it is too dispersed, see Figure 4.2.

To reduce the error of the estimator we must reduce its inefficiency, although this might cost something in terms of bias. The most efficient estimator is a constant estimator, i.e., an estimator such as (4.12), which with any information associates the same fixed value. Indeed, constant estimators display zero inefficiency, although their bias is very large.

Therefore we consider weighted averages of the sample estimator with a constant estimator of location $\mathbf{b}$, i.e. *any* fixed N-dimensional vector. This way we obtain the *James-Stein shrinkage estimators* of location:

$$\widehat{\boldsymbol{\mu}}^S \equiv (1-\alpha)\,\widehat{\boldsymbol{\mu}} + \alpha\mathbf{b}. \tag{4.138}$$

We show in Appendix www.4.5 that an optimal choice for the weight α in this expression is the following:

$$\alpha \equiv \frac{1}{T}\frac{N\overline{\lambda} - 2\lambda_1}{(\widehat{\boldsymbol{\mu}} - \mathbf{b})'(\widehat{\boldsymbol{\mu}} - \mathbf{b})}, \tag{4.139}$$

where λ_1 is the largest among the N eigenvalues of $\boldsymbol{\Sigma}$ and $\overline{\lambda}$ is the average of the eigenvalues.

By means of *Stein's lemma* we prove in Appendix www.4.5 that the shrinkage estimator (4.138)-(4.139) performs better than the sample mean, i.e. it satisfies (4.137). In real applications the true underlying covariance matrix $\boldsymbol{\Sigma}$ is not known, and thus we cannot compute its eigenvalues. Therefore we replace it with an estimate $\boldsymbol{\Sigma} \mapsto \widehat{\boldsymbol{\Sigma}}$. Furthermore, to obtain more sensible results and to interpret α as a weight, we impose the additional constraint that α be comprised in the interval $(0, 1)$.

As intuition suggests, the optimal amount of shrinkage (4.139) vanishes as the amount of observations T increases.

Furthermore, the optimal shrinkage weight (4.139) is largest in well conditioned market, i.e. when the condition number (4.131) is one. Indeed, in the limit case of full correlation among the invariants, the multivariate setting becomes a one-dimensional setting, in which case the sample estimate is no longer inadmissible.

On the other hand, in the case of extremely well conditioned markets all the eigenvalues are equal to the common value $\overline{\lambda}$ and the optimal shrinkage weight reads:

$$\alpha \equiv \frac{N-2}{T} \frac{\overline{\lambda}}{(\widehat{\boldsymbol{\mu}} - \mathbf{b})' (\widehat{\boldsymbol{\mu}} - \mathbf{b})}. \tag{4.140}$$

Notice in particular that, as intuition suggests, shrinking toward the target $\mathbf{b}$ becomes particularly effective when the number of observations T is low with respect to the dimension of the invariants N and, since $\overline{\lambda}$ is the average variance of the invariants, when the markets are very volatile.

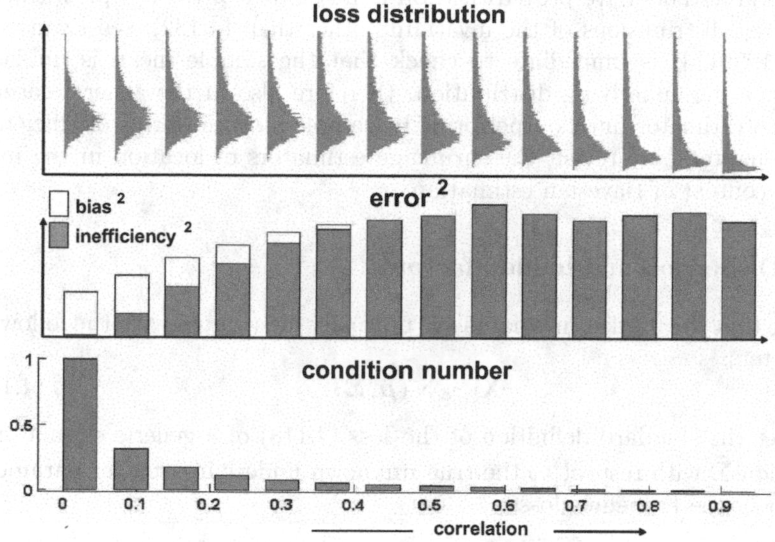

Fig. 4.13. Shrinkage estimator of mean: evaluation

In Figure 4.13 we display the distribution of the loss (4.133) of the shrinkage estimator (4.138)-(4.139) and the respective error (4.134) as the market parameters vary according to (4.116), along with the ensuing condition number. As expected, in well-conditioned markets the amount of shrinkage is maximal. Indeed, the bias is large, whereas the sample mean, which corresponds to a null shrinkage, is unbiased. Nevertheless, the overall error is reduced with respect to the sample mean, compare Figure 4.13 with Figure 4.11.

Shrinking the sample mean towards a constant vector $\mathbf{b}$ is not the only option to improve the estimation. Another possibility consists in shrinking the sample mean towards a scenario-dependent target vector, such as the grand mean. This corresponds to replacing the constant vector $\mathbf{b}$ in (4.138) as follows:

$$\mathbf{b} \mapsto \frac{\mathbf{1}'\widehat{\boldsymbol{\mu}}}{N}\mathbf{1}, \tag{4.141}$$

where $\mathbf{1}$ is an N-dimensional vector of ones.

Another choice of scenario-dependent target is the volatility-weighted grand mean, see Jorion (1986). This corresponds to replacing the constant vector $\mathbf{b}$ in (4.138) as follows:

$$\mathbf{b} \mapsto \frac{\mathbf{1}'\widehat{\boldsymbol{\Sigma}}^{-1}\widehat{\boldsymbol{\mu}}}{\mathbf{1}'\widehat{\boldsymbol{\Sigma}}^{-1}\mathbf{1}}\mathbf{1}. \tag{4.142}$$

where $\widehat{\boldsymbol{\Sigma}}$ is an estimator of the scatter matrix of the invariants.

Several authors have proved the non-admissibility of the sample mean for underlying distributions of the invariants other than (4.132), see Evans and Stark (1996). It is immediate to check that the sample mean is unbiased no matter the underlying distribution, therefore also in the general case an improved estimator must outperform the sample mean in terms of efficiency.

In Chapter 7 we revisit the shrinkage estimators of location in the more general context of Bayesian estimation.

4.4.2 Dispersion and hidden factors

Assume that the market invariants are normally distributed with the following parameters:

$$\mathbf{X}_t \sim \mathrm{N}(\boldsymbol{\mu}, \boldsymbol{\Sigma}). \tag{4.143}$$

Consider the standard definition of the loss (4.118) of a generic estimator of dispersion $\widehat{\boldsymbol{\Sigma}}$ with respect to the true unknown underlying scatter parameter $\boldsymbol{\Sigma}$, namely the Frobenius loss:

$$\operatorname{Loss}\left(\widehat{\boldsymbol{\Sigma}}, \boldsymbol{\Sigma}\right) \equiv \operatorname{tr}\left[\left(\widehat{\boldsymbol{\Sigma}}[I_T] - \boldsymbol{\Sigma}\right)^2\right]; \tag{4.144}$$

and the respective definition of error:

$$\operatorname{Err}^2\left(\widehat{\boldsymbol{\Sigma}}, \boldsymbol{\Sigma}\right) \equiv \mathrm{E}\left\{\operatorname{tr}\left[\left(\widehat{\boldsymbol{\Sigma}}[I_T] - \boldsymbol{\Sigma}\right)^2\right]\right\}. \tag{4.145}$$

Consider the benchmark estimator of dispersion (4.99), namely the sample covariance:

$$\widehat{\boldsymbol{\Sigma}}[i_T] \equiv \frac{1}{T}\sum_{t=1}^{T}[\mathbf{x}_t - \widehat{\boldsymbol{\mu}}[i_T]][\mathbf{x}_t - \widehat{\boldsymbol{\mu}}[i_T]]', \tag{4.146}$$

where $\widehat{\boldsymbol{\mu}}$ is the sample mean (4.98).

From (4.119) the error (4.145) of the sample covariance reads:

$$\operatorname{Err}^2\left(\widehat{\boldsymbol{\Sigma}}, \boldsymbol{\Sigma}\right) = \frac{1}{T}\left[\operatorname{tr}(\boldsymbol{\Sigma}^2) + \left(1 - \frac{1}{T}\right)[\operatorname{tr}(\boldsymbol{\Sigma})]^2\right]. \tag{4.147}$$

This is not the minimum error achievable and thus it is possible to define an estimator of dispersion that performs better than the sample covariance.

In order to determine this better estimator we analyze further the error (4.147). Consider as in (4.112) the principal component decomposition of the true unknown scatter matrix:

$$\boldsymbol{\Sigma} \equiv \mathbf{E}\boldsymbol{\Lambda}\mathbf{E}'. \tag{4.148}$$

In this expression $\boldsymbol{\Lambda}$ is the diagonal matrix of the eigenvalues of $\boldsymbol{\Sigma}$ sorted in decreasing order:

$$\boldsymbol{\Lambda} \equiv \operatorname{diag}(\lambda_1, \ldots, \lambda_N); \tag{4.149}$$

and the matrix $\mathbf{E}$ is the juxtaposition of the respective orthogonal eigenvectors. Using the identities (4.114) and (4.122), the percentage error (4.31) reads in this context:

$$\mathrm{PErr}^2\left(\widehat{\boldsymbol{\Sigma}}, \boldsymbol{\Sigma}\right) = \frac{1}{T}\left(1 + \left(1 - \frac{1}{T}\right)\frac{\left(\sum_{n=1}^N \lambda_n\right)^2}{\sum_{n=1}^N \lambda_n^2}\right). \tag{4.150}$$

In this expression we can assume without loss of generality that the eigenvalues

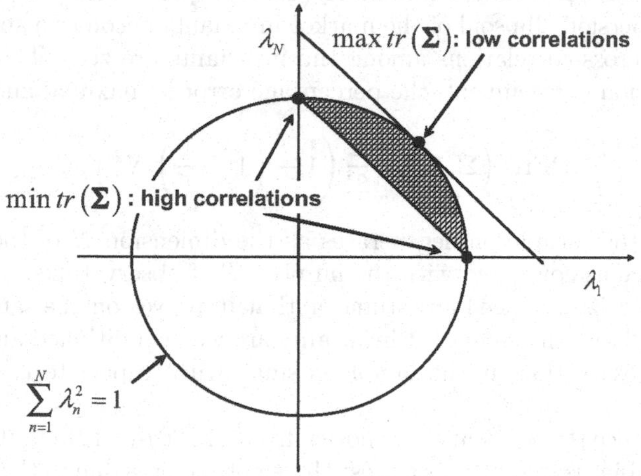

Fig. 4.14. Bounds on the error of the sample covariance matrix

lie on the unit sphere, see Figure 4.14. The sum in the numerator is the trace of $\boldsymbol{\Sigma}$. The different values γ that the trace can assume are represented by the family of hyperplanes (a line in the figure) with equation $\sum_{n=1}^N \lambda_n = \gamma$. Since the eigenvalues are constrained on the unit sphere and must be positive, the trace can only span the patterned volume of the hypersphere in Figure 4.14.

The minimum trace corresponds to the following corner solution[5]:

[5] There are actually N solutions, but we only consider one, since we sort the eigenvalues in decreasing order.

$$\lambda_1 = 1, \lambda_2 = \cdots = \lambda_N = 0 \Leftrightarrow \operatorname{tr}(\boldsymbol{\Sigma}) = 1, \tag{4.151}$$

which gives rise to a condition number (4.131) equal to zero. In this situation the ellipsoid $\mathcal{E}_{\boldsymbol{\mu},\boldsymbol{\Sigma}}$ determined by the market parameters as described in (A.73), which is also the location-dispersion ellipsoid of the invariants (2.75), is squeezed into a line. In other words, there exists only one actual dimension of risk, as all the invariants can be expressed as functions of one specific invariant. This is approximately the case in the swap market, as we see in Figure 3.20. In this environment of high correlations the percentage estimation error is minimal, and reads:

$$\operatorname{PErr}^2\left(\widehat{\boldsymbol{\Sigma}}, \boldsymbol{\Sigma}\right) = \frac{1}{T}\left(2 - \frac{1}{T}\right). \tag{4.152}$$

On the other hand, the maximum trace corresponds to the following combination:

$$\lambda_1 = \cdots = \lambda_N = \frac{1}{\sqrt{N}} \Leftrightarrow \operatorname{tr}(\boldsymbol{\Sigma}) = \sqrt{N}, \tag{4.153}$$

which gives rise to a condition number (4.131) equal to one. In this case the location-dispersion ellipsoid of the market invariants becomes a sphere, which means that cross-correlations among the invariants are zero. Therefore, in a zero-correlation environment, the percentage error is maximal and reads:

$$\operatorname{PErr}^2\left(\widehat{\boldsymbol{\Sigma}}, \boldsymbol{\Sigma}\right) = \frac{1}{T}\left(1 + \left(1 - \frac{1}{T}\right) N^2\right). \tag{4.154}$$

Notice that the estimation degenerates as the dimension N of the invariants becomes large as compared with the number T of observations.

To summarize, we need an estimator that improves on the sample covariance especially when the market invariants are well conditioned and when the number of observations in the sample is small with respect to the number of invariants.

To introduce this estimator, we notice from (4.120)-(4.121) that the sample covariance's bias is minimal, as almost the whole error is due to the estimator's inefficiency. In other words, the sample covariance is properly centered around the true, unknown value, but it is too dispersed, see Figure 4.2.

The sample covariance is inefficient because the estimation process tends to scatter the sample eigenvalues $\widehat{\boldsymbol{\Lambda}}$ away from the mean value $\overline{\lambda}$ of the true unknown eigenvalues. Indeed, Ledoit and Wolf (2004) prove the following general result:

$$\mathrm{E}\left\{\sum_{n=1}^{N}\left(\widehat{\lambda}_n - \overline{\lambda}\right)^2\right\} = \sum_{n=1}^{N}\left(\lambda_n - \overline{\lambda}\right)^2 + \operatorname{Err}^2\left(\widehat{\boldsymbol{\Sigma}}, \boldsymbol{\Sigma}\right). \tag{4.155}$$

Geometrically, the estimation process squeezes and stretches the location-dispersion ellipsoid $\mathcal{E}_{\boldsymbol{\mu},\boldsymbol{\Sigma}}$ of the market invariants.

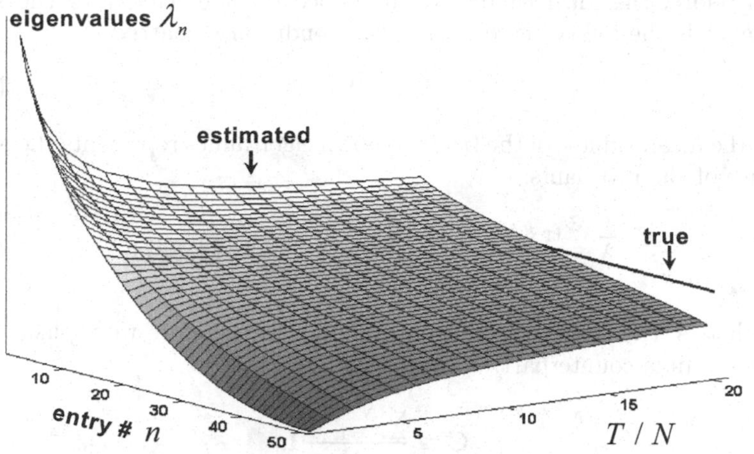

Fig. 4.15. Scattering of sample eigenvalues

Since the estimation error is large when the number of observations T is small, the scattering effect is larger when T is small with respect to the number of invariants. We plot this phenomenon in Figure 4.15 for the case of $N \equiv 50$ market invariants. As we show in Appendix www.4.6, in the extreme case where the number of observations T is lower than the number of invariants N, the last sample eigenvalues become null and thus the sample covariance becomes singular.

Furthermore, the scattering of the eigenvalues of the sample covariance is more pronounced for those invariants whose location-dispersion ellipsoid is close to a sphere. This result is intuitive: comparatively speaking, a sphere gets squeezed and stretched more than an elongated ellipsoid, which is already elongated to begin with. This result is also consistent with (4.152) and (4.154).

We can summarize the cause of the inefficiency of the sample covariance in terms of the condition number (4.131). Indeed, the estimation process worsens the condition number of the market invariants:

$$\widehat{\mathrm{CN}}\{\mathbf{X}\} \equiv \frac{\widehat{\lambda}_N}{\widehat{\lambda}_1} < \frac{\lambda_N}{\lambda_1} \equiv \mathrm{CN}\{\mathbf{X}\}. \tag{4.156}$$

To reduce the error of the sample covariance we must reduce its inefficiency by averaging it with an efficient and well conditioned estimator of dispersion. On the one hand, the most efficient estimator is a constant estimator, i.e. an estimator such as (4.12), which with any information associates a given fixed value: indeed, constant estimators display zero inefficiency, although their bias is very large. On the other hand, the best-conditioned matrices are multiples of the identity, in which case the condition number is one.

Therefore, the ideal candidate to reduce the inefficiency of the sample covariance is the following constant, well conditioned matrix:

$$\mathbf{C} \equiv \overline{\lambda} \mathbf{I}_N, \tag{4.157}$$

where the mean value $\overline{\lambda}$ of the true unknown eigenvalues represents the average variance of the invariants:

$$\overline{\lambda} \equiv \frac{\operatorname{tr}\{\mathbf{\Lambda}\}}{N} = \frac{\operatorname{tr}\{\mathbf{\Sigma}\}}{N} = \frac{1}{N}\sum_{n=1}^{N} \operatorname{Var}\{X_n\}. \tag{4.158}$$

Nevertheless, the true eigenvalues are unknown, therefore we replace (4.157) with its sample counterpart:

$$\widehat{\mathbf{C}} \equiv \frac{\sum_{n=1}^{N} \widehat{\lambda}_n}{N} \mathbf{I}. \tag{4.159}$$

At this point, following Ledoit and Wolf (2004) we define the *shrinkage estimator of dispersion* as the weighted average of the sample covariance and the target matrix:

$$\widehat{\mathbf{\Sigma}}^S \equiv (1-\alpha)\,\widehat{\mathbf{\Sigma}} + \alpha \widehat{\mathbf{C}}. \tag{4.160}$$

The optimal shrinkage weight in this expression is defined as follows:

$$\alpha \equiv \frac{1}{T} \frac{\frac{1}{T}\sum_{t=1}^{T} \operatorname{tr}\left\{\left(\mathbf{x}_t \mathbf{x}_t' - \widehat{\mathbf{\Sigma}}\right)^2\right\}}{\operatorname{tr}\left\{\left(\widehat{\mathbf{\Sigma}} - \widehat{\mathbf{C}}\right)^2\right\}}, \tag{4.161}$$

if $\alpha < 1$, and 1 otherwise.

The shrinkage estimator (4.160) is indeed better conditioned than the sample covariance:

$$\frac{\widehat{\lambda}_N^S}{\widehat{\lambda}_1^S} > \frac{\widehat{\lambda}_N}{\widehat{\lambda}_1}, \tag{4.162}$$

see Appendix www.4.6. Thus the ensuing error (4.145) is less than for the sample covariance.

As intuition suggests, the optimal amount of shrinkage (4.161) vanishes as the amount of observations T increases.

Furthermore, the optimal shrinkage weight is largest when the condition number of the market invariants is close to one. Indeed, in this case the denominator in (4.161) becomes very small. This is consistent with the fact that the percentage error is maximal in well-condition markets, see (4.154).

In Figure 4.16 we display the distribution of the loss (4.144) of the shrinkage estimator (4.160) and the respective error (4.145) as the market parameters vary according to (4.116), along with the ensuing condition number. Notice that shrinking towards a multiple of the identity matrix introduces a

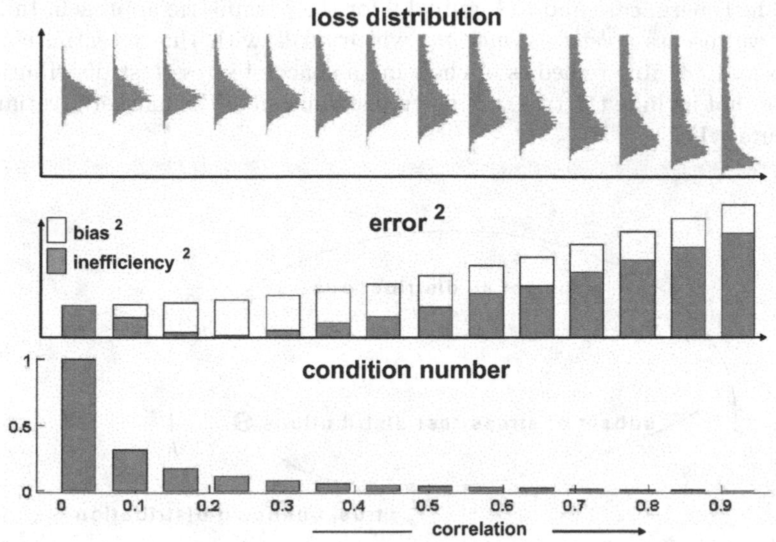

Fig. 4.16. Shrinkage estimator of covariance: evaluation

bias that was not present in the case of the sample covariance, see Figure 4.12. Nevertheless, the overall error is reduced by the shrinkage process.

In Chapter 7 we revisit the shrinkage estimators of dispersion in the more general context of Bayesian estimation.

4.4.3 Explicit factors

The benchmark estimator of the factor loadings in an explicit factor model is the ordinary least square estimator of the regression coefficients (4.126) and the estimator of the dispersion of the residuals is the respective sample covariance matrix (4.128). Like in the case of the estimators of location and dispersion, it is possible to improve on these estimators by shrinking them towards suitable targets, see Ledoit and Wolf (2003) for an application of a one-factor model to the stock market.

We discuss in Chapter 7 the shrinkage estimators of explicit-factor models in the more general context of Bayesian estimation.

4.5 Robustness

In our journey throughout the possible approaches to building estimators we have always assumed that the true, unknown distribution of the market invariants lies somewhere in the subset of stress test distributions, refer to Figure

4.3 for the general case and to Figure 4.9 for the parametric approach. In this section we discuss *robust estimation*, which deals with the potential consequences and possible remedies of choosing a space of stress test distributions that does not include the true, unknown distribution of the market invariants, see Figure 4.17.

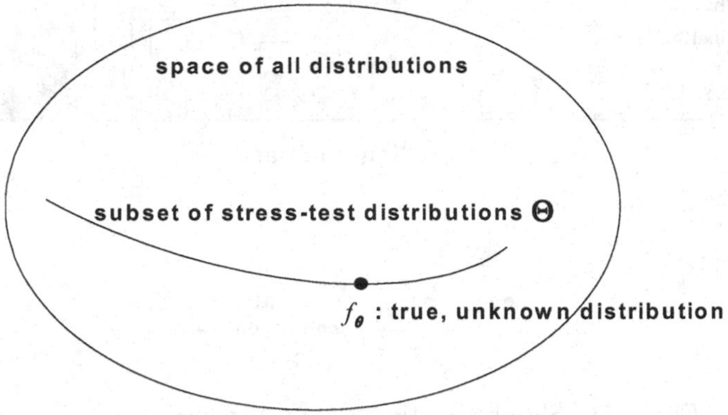

Fig. 4.17. Robust approach to estimation

To provide the intuition behind robust estimation, consider as in Figure 4.7 the location-dispersion ellipsoid (2.75) defined by the sample-mean (4.41) and sample-covariance (4.42) of a set of observations of market invariants:

$$\mathcal{E}_{\widehat{\mathrm{E}},\widehat{\mathrm{Cov}}} \equiv \left\{ \mathbf{x} \in \mathbb{R}^N \text{ such that } \left(\mathbf{x} - \widehat{\mathrm{E}}\right)' \left(\widehat{\mathrm{Cov}}\right)^{-1} \left(\mathbf{x} - \widehat{\mathrm{E}}\right) \leq 1 \right\}. \quad (4.163)$$

Then add a fake observation, an outlier, and repeat the estimation based on the enlarged sample. The new ellipsoid, which represents the new sample mean and sample covariance, is completely different, see Figure 4.18: one single observation completely disrupted the estimation.

In the above experiment we know that the extra-observation is spurious. Therefore, such an extreme sensitivity does not represent a problem. If we knew for a fact that some observations were spurious, a sensitive estimator would help us detect the unwelcome outliers. This is the subject of outlier detection, which we tackle in Section 4.6.1.

On the other hand, in many applications we do not know the true underlying distribution and, most importantly, we have no reason to believe that some observations could be spurious. Therefore we cannot trust sensitive estimators such as the sample estimators. Instead, we need to develop estimators that properly balance the trade-off between the precision and the robustness of the final results.

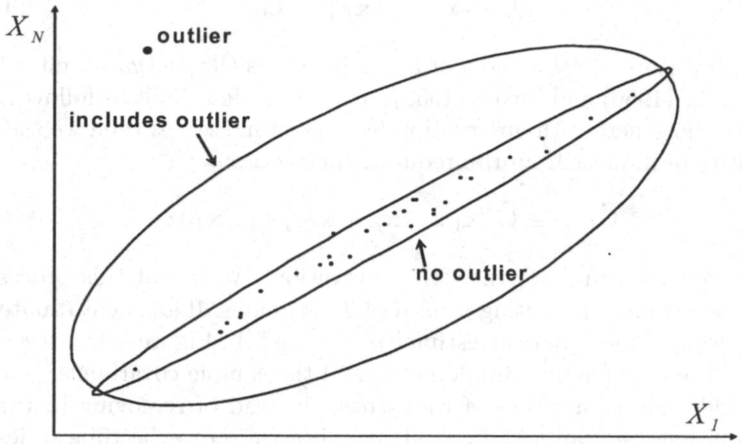

Fig. 4.18. Sample estimators: lack of robustness

In this section, first we discuss a few measures of robustness for an estimator, namely the jackknife, the sensitivity curve and, most notably, the influence function: when this is bounded, the respective estimator is robust.

Then we compute the influence function of the estimators introduced so far, namely nonparametric sample estimators and parametric maximum likelihood estimators of location, dispersion and explicit factor loadings. As it turns out, the sample estimators display an unbounded influence function and therefore they are not robust. On the other hand, the maximum likelihood estimators display a range of behaviors: for instance, MLE of normally distributed invariants are the sample estimators and therefore they are not robust. On the other hand, MLE of Cauchy-distributed invariants have bounded influence function and therefore they are robust.

Finally, we show how to build robust estimators of the main parameters of interest for asset allocation problem.

4.5.1 Measures of robustness

To tackle robustness issues, we need first of all to be able to measure the robustness of a generic estimator. First we introduce two qualitative measures, namely the jackknife and the sensitivity curve. Relying on the intuition behind these measures we introduce a tool that precisely quantifies the robustness of an estimator, namely the influence function.

Consider a generic estimator $\widehat{\mathbf{G}}$ of S features of an unknown distribution. As in (4.9), an estimator is a vector-valued function of currently available information, which is represented as in (4.8) by the time series of the past occurrences of the market invariants:

$$i_T \equiv \{\mathbf{x}_1, \ldots, \mathbf{x}_T\} \mapsto \widehat{\mathbf{G}}. \tag{4.164}$$

A first measure of robustness of an estimator is the *jackknife*, introduced by Quenouille (1956) and Tukey (1958). The jackknife is built as follows. First we remove the generic t-th observation from the time series; then we estimate the quantity of interest from the reduced time series:

$$\widehat{\mathbf{G}}_{(-t)} \equiv \widehat{\mathbf{G}}\left(\mathbf{x}_1 \ldots, \mathbf{x}_{t-1}, \mathbf{x}_{t+1}, \ldots, \mathbf{x}_T\right); \tag{4.165}$$

finally we put back in place the t-th observation. We repeat this process for all the observations, computing a total of T estimates. If all the estimates are comparable, we assess that the estimator is robust. In Figure 4.18 we see that this is not the case for the sample mean and the sample covariance.

To build another measure of robustness, instead of removing in turn all the observations, we can add an arbitrary observation to the time series and evaluate its effect on the estimate. This way we obtain the *sensitivity curve*, introduced by Tukey (1977) and defined as follows:

$$\operatorname{SC}\left(\mathbf{x}, \widehat{\mathbf{G}}\right) \equiv T\widehat{\mathbf{G}}\left(\mathbf{x}_1, \ldots, \mathbf{x}_T, \mathbf{x}\right) - T\widehat{\mathbf{G}}\left(\mathbf{x}_1, \ldots, \mathbf{x}_T\right), \tag{4.166}$$

where the normalization T is meant to make the evaluation less sensitive to the sample size. If the sensitivity curve is small for any value of the extra-observation $\mathbf{x}$, we assess that the estimator $\widehat{\mathbf{G}}$ is robust. We see in Figure 4.18 that this is not the case for the sample mean and the sample covariance.

Both jackknife and sensitivity curve are qualitative tools that can detect lack of robustness: if either measure shows that the given estimator is not robust, we should reject that estimator and search for a better one.

Nevertheless, if an estimator is not rejected, we cannot draw any conclusion on the degree of robustness of that estimator. Indeed, as far as the sensitivity curve is concerned, whatever result we obtain depends on the specific sample. On the other hand, as far as the jackknife is concerned, the sample might contain two or more outliers instead of one: in this case we might consider tests that remove more than one observation at a time, but we would not be sure where to stop.

To obtain a tool that quantifies robustness independently of the specific sample, we should move in the opposite direction, considering the marginal effect of an outlier when the sample size tends to infinity. The *influence function* can be defined heuristically as the infinite-sample limit of the sensitivity curve, see Hampel, Ronchetti, Rousseeuw, and Stahel (1986). Intuitively, the influence function quantifies the marginal effect on an estimator of an extra-observation in the limit of infinite observations.

In order to introduce this limit, we need to express the generic S-dimensional estimator as an S-dimensional functional of the empirical probability density function:

$$\widehat{\mathbf{G}} \equiv \widetilde{\mathbf{G}}\left[f_{i_T}\right], \tag{4.167}$$

where the empirical probability density function (2.240) is defined in terms of the Dirac delta (B.16) as follows:

$$f_{i_T} \equiv \frac{1}{T} \sum_{t=1}^{T} \delta^{(\mathbf{x}_t)}. \tag{4.168}$$

The sample estimators are explicit functionals of the empirical probability density function. Indeed the sample estimators aim at estimating some functional $\mathbf{G}[f_{\mathbf{X}}]$ of the unknown probability density function $f_{\mathbf{X}}$ of the market invariants. Therefore by their very definition (4.36) the functional that defines the estimator is the functional that defines the quantity of interest of the unknown distribution of the market invariants:

$$\widehat{\mathbf{G}} \equiv \mathbf{G}[f_{i_T}]. \tag{4.169}$$

This expression is clearly in the form (4.167).

For example, consider the following functional:

$$\mathbf{G}[h] \equiv \int_{\mathbb{R}^N} \mathbf{x} h(\mathbf{x})\, d\mathbf{x}, \tag{4.170}$$

where h is any function such that the integral (4.170) makes sense. This functional, when applied to the probability density function of a distribution, yields its expected value:

$$\mathbf{G}[f_{\mathbf{X}}] = \mathrm{E}\{\mathbf{X}\}. \tag{4.171}$$

Consider now the sample mean:

$$\widehat{\mathbf{G}} \equiv \frac{1}{T} \sum_{t=1}^{T} \mathbf{x}_t. \tag{4.172}$$

The sample mean is the functional (4.170) applied to the empirical pdf:

$$\widehat{\mathbf{G}} = \int_{\mathbb{R}^N} \mathbf{x} f_{i_T}(\mathbf{x})\, d\mathbf{x} \equiv \mathbf{G}[f_{i_T}]. \tag{4.173}$$

On the other hand, the maximum likelihood estimators are implicit functionals of the empirical probability density function. Consider the ML estimator $\widehat{\boldsymbol{\theta}}$ of the S-dimensional parameter $\boldsymbol{\theta}$ of a distribution $f_{\boldsymbol{\theta}}$. The ML estimator as a functional is defined implicitly by the first-order conditions on the log-likelihood. Indeed, from their definition (4.66) the ML estimators solve in quite general cases the following implicit equation:

$$\mathbf{0} = \frac{1}{T} \sum_{t=1}^{T} \boldsymbol{\psi}\left(\mathbf{x}_t, \widehat{\boldsymbol{\theta}}\right), \tag{4.174}$$

where $\boldsymbol{\psi}$ is the S-dimensional vector of first-order partial derivatives of the log-likelihood:

$$\boldsymbol{\psi}(\mathbf{x}, \boldsymbol{\theta}) \equiv \frac{\partial}{\partial \boldsymbol{\theta}} \ln (f_{\boldsymbol{\theta}}(\mathbf{x})). \tag{4.175}$$

Consider now the S-dimensional functional $\widetilde{\boldsymbol{\theta}}[h]$ defined implicitly for a generic function h in a suitable domain as follows:

$$\widetilde{\boldsymbol{\theta}}[h]: \quad \int_{\mathbb{R}^N} \boldsymbol{\psi}\left(\mathbf{x}, \widetilde{\boldsymbol{\theta}}\right) h(\mathbf{x})\, d\mathbf{x} \equiv \mathbf{0}. \tag{4.176}$$

In this notation, the ML estimator (4.174) can be written as follows:

$$\widehat{\boldsymbol{\theta}} \equiv \widetilde{\boldsymbol{\theta}}[f_{i_T}], \tag{4.177}$$

which is in the form (4.167).

For example, consider the functional $\widetilde{\theta}[h]$ defined implicitly by the following equation:

$$\widetilde{\theta}[h]: \quad 0 \equiv \int_{\mathbb{R}} \left(\ln x - \widetilde{\theta}\right) h dx. \tag{4.178}$$

Now assume as in (4.57) that there exists a lognormally distributed invariant with the following parameters:

$$X \sim \text{LogN}(\theta, 1). \tag{4.179}$$

The ML estimator of θ reads:

$$\widehat{\theta} = \frac{1}{T} \sum_{t=1}^{T} \ln x_t, \tag{4.180}$$

see (4.69). Clearly, the ML estimator of θ solves:

$$\begin{aligned} 0 &= \frac{1}{T} \sum_{t=1}^{T} \left(\ln x_t - \widehat{\theta}\right) \\ &= \int_{\mathbb{R}} \left(\ln x - \widehat{\theta}\right) \frac{1}{T} \sum_{t=1}^{T} \delta^{(x_t)}(x)\, dx \\ &= \int_{\mathbb{R}} \left(\ln x - \widehat{\theta}\right) f_{i_T}(x)\, dx. \end{aligned} \tag{4.181}$$

Therefore:

$$\widehat{\theta} = \widetilde{\theta}[f_{i_T}]. \tag{4.182}$$

Notice that the term in brackets in the integral (4.178) is the first-order derivative of the logarithm of the probability density function (4.58), as prescribed by (4.175).

Consider the sensitivity curve (4.166). Adding one observation in an arbitrary position $\mathbf{x}$ corresponds to modifying the empirical probability density function (4.168) as follows:

$$f_{i_T} \mapsto (1-\epsilon) f_{i_T} + \epsilon \delta^{(\mathbf{x})}, \tag{4.183}$$

where $\epsilon \equiv 1/(T+1)$ is the relative weight of the extra-observation and δ is the Dirac delta (B.16). Therefore in the functional notation (4.167) the sensitivity curve (4.166) reads:

$$\mathrm{SC}\left(\mathbf{x}, \widehat{\mathbf{G}}\right) \equiv \frac{1-\epsilon}{\epsilon} \left\{ \widetilde{\mathbf{G}}\left[(1-\epsilon) f_{i_T} + \epsilon \delta^{(\mathbf{x})}\right] - \widetilde{\mathbf{G}}\left[f_{i_T}\right] \right\}. \tag{4.184}$$

In the limit of an infinite number of observations T the relative weight ϵ tends to zero. Furthermore, from the Glivenko-Cantelli theorem (4.34) the empirical pdf f_{i_T} tends to the true, unknown pdf $f_{\mathbf{X}}$. Therefore the influence function of a generic S-dimensional estimator $\widehat{\mathbf{G}}$, which is the infinite-sample limit of the sensitivity curve, is defined as the following S-dimensional vector:

$$\mathrm{IF}\left(\mathbf{x}, f_{\mathbf{X}}, \widehat{\mathbf{G}}\right) \equiv \lim_{\epsilon \to 0} \frac{1}{\epsilon} \left(\widetilde{\mathbf{G}}\left[(1-\epsilon) f_{\mathbf{X}} + \epsilon \delta^{(\mathbf{x})}\right] - \widetilde{\mathbf{G}}\left[f_{\mathbf{X}}\right] \right), \tag{4.185}$$

where $\widetilde{\mathbf{G}}$ is the S-dimensional functional (4.167) that links the estimator to the empirical probability density function.

In order to use the influence function in applications, we need to define it more formally as a Gateaux derivative, which is the equivalent of a partial derivative in the world of functional analysis.

We recall that the partial derivatives of a function g defined in $\mathbb{R}^N$ at the point $\mathbf{v}$ are N numbers D, commonly denoted as follows:

$$D(n, \mathbf{v}, g) \equiv \frac{\partial g(\mathbf{v})}{\partial v_n}, \quad n = 1, \ldots, N. \tag{4.186}$$

These N numbers are such that such that whenever $\mathbf{u} \approx \mathbf{v}$ the following approximation holds:

$$g(\mathbf{u}) - g(\mathbf{v}) \approx \sum_{n=1}^{N} D(n, \mathbf{v}, g)(u_n - v_n). \tag{4.187}$$

According to Table B.4, in the world of functional analysis the vector's index n is replaced by the function's argument $\mathbf{x}$, vectors such as $\mathbf{v}$ are replaced by functions $v(\cdot)$ and sums are replaced by integrals. Furthermore, functions g are replaced with functionals G.

The *Gateaux derivative* is the partial derivative (4.187) in this new notation. In other words it is the number D such that whenever two functions are close $u \approx v$ the following approximation holds:

$$G[u] - G[v] \approx \int_{\mathbb{R}^N} D(\mathbf{x}, v, G)\, u(\mathbf{x})\, d\mathbf{x}, \tag{4.188}$$

where we used the normalization:

$$\int_{\mathbb{R}^N} D(\mathbf{x}, v, G)\, v d\mathbf{x} \equiv 0. \tag{4.189}$$

Consider an estimator $\widehat{\mathbf{G}}$ that is represented by the functional $\widetilde{\mathbf{G}}$ as in (4.167). The influence function of each entry of the estimator $\widehat{\mathbf{G}}$ for a given distribution $f_{\mathbf{X}}$ is the Gateaux derivative of the respective entry of $\widetilde{\mathbf{G}}$ in $f_{\mathbf{X}}$:

$$\mathrm{IF}\left(\mathbf{x}, f_{\mathbf{X}}, \widehat{\mathbf{G}}\right) \equiv \mathbf{D}\left(\mathbf{x}, f_{\mathbf{X}}, \widetilde{\mathbf{G}}\right). \tag{4.190}$$

Indeed, setting $u \equiv (1-\epsilon) f_{\mathbf{X}} + \epsilon \delta^{(\mathbf{x})}$ in (4.188) yields the heuristic definition (4.185).

An estimator is robust if its influence function is small, or at least bounded, as the extra observation $\mathbf{x}$ varies in the space of observations and as the distribution of the invariants $f_{\mathbf{X}}$ varies in a wide, yet bounded, range in the space of distributions.

More precisely, suppose that we are interested in some parameters $\mathbf{G}[f_{\mathbf{X}}]$ of the unknown distribution $f_{\mathbf{X}}$ of the market invariants. As usual, we make assumptions on the set of possible distributions for $f_{\mathbf{X}}$ and we build an estimator $\widehat{\mathbf{G}}$. Suppose that we choose inappropriately a family of stress test distributions that does not include the true, unknown distribution $f_{\mathbf{X}}$, i.e. we miss the target by some extent as in Figure 4.17. Under these "wrong" assumptions we develop the "wrong" estimator $\widehat{\mathbf{G}}$, which can be expressed as a functional of the empirical pdf as in (4.167). The influence function provides a measure of the damage:

$$\widehat{\mathbf{G}} - \mathbf{G}[f_{\mathbf{X}}] \approx \frac{1}{T} \sum_{t=1}^{T} \mathrm{IF}\left(\mathbf{x}_t, f_{\mathbf{X}}, \widehat{\mathbf{G}}\right), \tag{4.191}$$

where the approximation improves with the number of observations. This follows immediately by setting $u \equiv f_{i_T}$ in (4.188) and using the fact that estimators are typically *Fisher consistent*, i.e. such that:

$$\widetilde{\mathbf{G}}[f_{\mathbf{X}}] = \mathbf{G}[f_{\mathbf{X}}]. \tag{4.192}$$

Of course, we do not know the true underlying distribution $f_{\mathbf{X}}$ of the market invariants, but as long as the influence function is bounded for a wide range of underlying distributions $f_{\mathbf{X}}$, the damage is contained.

4.5.2 Robustness of previously introduced estimators

In Section 4.2 we introduced the nonparametric sample estimators $\widehat{\mathbf{G}}$ of the unknown features $\mathbf{G}[f_{\mathbf{X}}]$ of the distribution of the market invariants. The functional representation $\widetilde{\mathbf{G}}[f_{i_T}]$ of sample estimators in terms of the empirical probability density function is explicit and defined by (4.169). Therefore, the expression of the influence function of generic nonparametric estimators follows directly from the heuristic definition (4.185) of the influence function and reads:

$$\mathrm{IF}\left(\mathbf{x}, f_{\mathbf{X}}, \widehat{\mathrm{G}}\right) = \lim_{\epsilon \to 0} \frac{1}{\epsilon}\left(\mathbf{G}\left[(1-\epsilon) f_{\mathbf{X}} + \epsilon \delta^{(\mathbf{x})}\right] - \mathbf{G}\left[f_{\mathbf{X}}\right]\right). \tag{4.193}$$

In Section 4.3 we introduced the maximum likelihood estimators of the parameters $\boldsymbol{\theta}$ of the distribution $f_{\boldsymbol{\theta}}$ of the market invariants. The functional representation $\widetilde{\boldsymbol{\theta}}\left[f_{i_T}\right]$ of the maximum likelihood estimators in terms of the empirical probability density function is implicit and defined by (4.176). We prove in Appendix www.4.7 that in this case the influence function reads:

$$\mathrm{IF}\left(\mathbf{x}, f_{\mathbf{X}}, \widehat{\boldsymbol{\theta}}\right) = \mathbf{A} \left.\frac{\partial \ln f_{\boldsymbol{\theta}}(\mathbf{x})}{\partial \boldsymbol{\theta}}\right|_{\boldsymbol{\theta} \equiv \widetilde{\boldsymbol{\theta}}[f_{\mathbf{X}}]}, \tag{4.194}$$

where the constant $S \times S$ matrix $\mathbf{A}$ is defined as follows:

$$\mathbf{A} \equiv -\left[\int_{\mathbb{R}^N} \left.\frac{\partial^2 \ln f_{\boldsymbol{\theta}}(\mathbf{x})}{\partial \boldsymbol{\theta} \partial \boldsymbol{\theta}'}\right|_{\boldsymbol{\theta} \equiv \widetilde{\boldsymbol{\theta}}[f_{\mathbf{X}}]} f_{\mathbf{X}}(\mathbf{x})\, d\mathbf{x}\right]^{-1}. \tag{4.195}$$

We proceed below to apply these formulas to the sample and maximum likelihood estimators of interest for asset allocation problems.

Location and dispersion

Consider the sample estimators of location and dispersion of the market invariants $\mathbf{X}_t$, i.e. the sample mean (4.41) and the sample covariance (4.42) respectively:

$$\widehat{\mathrm{E}} \equiv \frac{1}{T} \sum_{t=1}^{T} \mathbf{x}_t \tag{4.196}$$

$$\widehat{\mathrm{Cov}} \equiv \frac{1}{T} \sum_{t=1}^{T} \left(\mathbf{x}_t - \widehat{\mathrm{E}}\right)\left(\mathbf{x}_t - \widehat{\mathrm{E}}\right)'. \tag{4.197}$$

We prove in Appendix www.4.7 that the influence function (4.193) for the sample mean reads:

$$\mathrm{IF}\left(\mathbf{x}, f_{\mathbf{X}}, \widehat{\mathrm{E}}\right) = \mathbf{x} - \mathrm{E}\{\mathbf{X}\}; \tag{4.198}$$

and the influence function (4.193) for the sample covariance reads:

$$\mathrm{IF}\left(\mathbf{x}, f_{\mathbf{X}}, \widehat{\mathrm{Cov}}\right) = (\mathbf{x} - \mathrm{E}\{\mathbf{X}\})(\mathbf{x} - \mathrm{E}\{\mathbf{X}\})' - \mathrm{Cov}\{\mathbf{X}\}. \tag{4.199}$$

Notice that the influence function of the sample estimators is not bounded. Therefore, the sample estimators are not robust: a strategically placed outlier, also known as *leverage point*, can completely distort the estimation. This is the situation depicted in Figure 4.18.

Assume now that the invariants $\mathbf{X}_t$ are elliptically distributed:

$$\mathbf{X}_t \sim \text{El}(\boldsymbol{\mu}, \boldsymbol{\Sigma}, g), \tag{4.200}$$

where $\boldsymbol{\mu}$ is the N-dimensional location parameter, $\boldsymbol{\Sigma}$ is the $N \times N$ dispersion matrix and g is the probability density generator. In other words, the probability density of the invariants $\mathbf{X}$ reads:

$$f_{\boldsymbol{\theta}}(\mathbf{x}) \equiv \frac{1}{\sqrt{|\boldsymbol{\Sigma}|}} g\left(\text{Ma}^2(\mathbf{x}, \boldsymbol{\mu}, \boldsymbol{\Sigma})\right), \tag{4.201}$$

where $\text{Ma}(\mathbf{x}, \boldsymbol{\mu}, \boldsymbol{\Sigma})$ is the Mahalanobis distance of the point $\mathbf{x}$ from the point $\boldsymbol{\mu}$ through the metric $\boldsymbol{\Sigma}$:

$$\text{Ma}^2(\mathbf{x}, \boldsymbol{\mu}, \boldsymbol{\Sigma}) \equiv (\mathbf{x} - \boldsymbol{\mu})' \boldsymbol{\Sigma}^{-1} (\mathbf{x} - \boldsymbol{\mu}). \tag{4.202}$$

In this case the parametric distribution of the market invariants is fully determined by the set of parameters is $\boldsymbol{\theta} \equiv (\boldsymbol{\mu}, \boldsymbol{\Sigma})$.

Consider the maximum likelihood estimators of the parameters $\boldsymbol{\theta}$, which are defined by the implicit equations (4.77)-(4.79) as follows:

$$\widehat{\boldsymbol{\mu}} = \sum_{t=1}^{T} \frac{w\left(\text{Ma}^2\left(\mathbf{x}_t, \widehat{\boldsymbol{\mu}}, \widehat{\boldsymbol{\Sigma}}\right)\right)}{\sum_{s=1}^{T} w\left(\text{Ma}^2\left(\mathbf{x}_s, \widehat{\boldsymbol{\mu}}, \widehat{\boldsymbol{\Sigma}}\right)\right)} \mathbf{x}_t \tag{4.203}$$

$$\widehat{\boldsymbol{\Sigma}} = \frac{1}{T} \sum_{t=1}^{T} (\mathbf{x}_t - \widehat{\boldsymbol{\mu}})(\mathbf{x}_t - \widehat{\boldsymbol{\mu}})' w\left(\text{Ma}^2\left(\mathbf{x}_t, \widehat{\boldsymbol{\mu}}, \widehat{\boldsymbol{\Sigma}}\right)\right), \tag{4.204}$$

where

$$w(z) \equiv -2\frac{g'(z)}{g(z)}. \tag{4.205}$$

These parameters can be expressed as functionals $\widetilde{\boldsymbol{\mu}}[f_{i_T}]$ and $\widetilde{\boldsymbol{\Sigma}}[f_{i_T}]$ of the empirical pdf. The functionals are defined implicitly as in (4.176) as follows:

$$\int_{\mathbb{R}^N} \boldsymbol{\psi}\left(\mathbf{x}, \widetilde{\boldsymbol{\mu}}[h], \widetilde{\boldsymbol{\Sigma}}[h]\right) h(\mathbf{x})\, d\mathbf{x} \equiv \mathbf{0}. \tag{4.206}$$

The vector-valued function $\boldsymbol{\psi}$ in this expression follows from (4.203)-(4.204) and reads:

$$\boldsymbol{\psi}(\mathbf{x}, \boldsymbol{\mu}, \boldsymbol{\Sigma}) \equiv \begin{pmatrix} w\left(\text{Ma}^2(\mathbf{x}, \boldsymbol{\mu}, \boldsymbol{\Sigma})\right)(\mathbf{x} - \boldsymbol{\mu}) \\ w\left(\text{Ma}^2_{\mathbf{x},\boldsymbol{\mu},\boldsymbol{\Sigma}}\right) \text{vec}\left[(\mathbf{x} - \boldsymbol{\mu})(\mathbf{x} - \boldsymbol{\mu})'\right] - \text{vec}[\boldsymbol{\Sigma}] \end{pmatrix}, \tag{4.207}$$

where vec is the operator (A.104) that stacks the columns of a matrix into a vector. From (4.194) and (4.175) the norm of the influence function is proportional to the norm of the above vector:

$$\left\| \text{IF}\left(\mathbf{x}, f_{\mathbf{X}}, \left(\widehat{\boldsymbol{\mu}}, \widehat{\boldsymbol{\Sigma}}\right)\right) \right\| \propto \|\boldsymbol{\psi}\|. \tag{4.208}$$

In particular, if the invariants are normally distributed the term w in (4.207) becomes $w \equiv 1$, see (4.97). Therefore the influence function is not bounded. This is not surprising, since we know from Section 4.3 that the ML estimators of location and dispersion of normally distributed invariants are the sample estimators and thus their influence function is (4.198)-(4.199). In other words, the ML estimators of location and dispersion of normally distributed invariants are not robust.

On the other hand, if the invariants are elliptically but not normally distributed the influence function displays a different behavior. Consider for example Cauchy-distributed invariants. In this case from (4.84) the term w in (4.207) becomes:

$$w(z) = \frac{N+1}{1+z}. \tag{4.209}$$

Therefore, from (4.208) and (4.202) the influence function of the location and dispersion maximum likelihood estimators becomes bounded. In other words, the ML estimators of location and dispersion of Cauchy-distributed invariants are robust.

Explicit factors

Consider an explicit factor linear model:

$$\mathbf{X}_t = \mathbf{B}\mathbf{F}_t + \mathbf{U}_t. \tag{4.210}$$

The sample estimator of the regression factor loadings are the ordinary least squares coefficients (4.52), which we report here:

$$\widehat{\mathbf{B}} \equiv \left(\sum_t \mathbf{x}_t \mathbf{f}_t'\right)\left(\sum_t \mathbf{f}_t \mathbf{f}_t'\right)^{-1}. \tag{4.211}$$

We do not discuss the sample covariance of the perturbation, which is the same as (4.197), where $\widehat{\mathbf{B}}\mathbf{f}_t$ replaces $\widehat{\mathrm{E}}$. We prove in Appendix www.4.7 that the influence function for the OLS coefficients reads:

$$\text{IF}\left((\mathbf{x}, \mathbf{f}), f_{\mathbf{X},\mathbf{F}}, \widehat{\mathbf{B}}\right) = \left(\mathbf{x}\mathbf{f}' - \mathbf{B}\mathbf{f}\mathbf{f}'\right) \mathrm{E}\left\{\mathbf{F}\mathbf{F}'\right\}^{-1}. \tag{4.212}$$

Notice that the influence function of the sample OLS coefficients is not bounded. Therefore, the OLS estimate is not robust: a strategically placed outlier, also known as *leverage point*, can completely distort the estimation. This is the situation depicted in Figure 4.18.

Consider now a parametric explicit factor model conditioned on the factors. We assume as in (4.90) that the perturbations are elliptically distributed and centered in zero. Therefore the respective conditional explicit factor model reads:

$$\mathbf{X}_t|\mathbf{f}_t \sim \mathrm{El}\left(\mathbf{B}\mathbf{f}_t, \boldsymbol{\Sigma}, g\right), \tag{4.213}$$

where $\boldsymbol{\Sigma}$ is the $N \times N$ dispersion matrix of the perturbations and g is their probability density generator. In this case the parametric distribution of the market invariants is fully determined by the set of parameters is $\boldsymbol{\theta} \equiv (\mathbf{B}, \boldsymbol{\Sigma})$.

Consider the maximum likelihood estimators of the parameters $\boldsymbol{\theta}$, which are defined by the implicit equations (4.92)-(4.94) as follows:

$$\begin{aligned} \widehat{\mathbf{B}} = & \left[\sum_{t=1}^{T} w\left(\mathrm{Ma}^2\left(\mathbf{x}_t, \widehat{\mathbf{B}}\mathbf{f}_t, \widehat{\boldsymbol{\Sigma}}\right)\right) \mathbf{x}_t \mathbf{f}_t'\right] \\ & \left[\sum_{t=1}^{T} w\left(\mathrm{Ma}^2\left(\mathbf{x}_t, \widehat{\mathbf{B}}\mathbf{f}_t, \widehat{\boldsymbol{\Sigma}}\right)\right) \mathbf{f}_t \mathbf{f}_t'\right]^{-1} \end{aligned} \tag{4.214}$$

$$\widehat{\boldsymbol{\Sigma}} = \frac{1}{T}\sum_{t=1}^{T} w\left(\mathrm{Ma}^2\left(\mathbf{x}_t, \widehat{\mathbf{B}}\mathbf{f}_t, \widehat{\boldsymbol{\Sigma}}\right)\right)\left(\mathbf{x}_t - \widehat{\mathbf{B}}\mathbf{f}_t\right)\left(\mathbf{x}_t - \widehat{\mathbf{B}}\mathbf{f}_t\right)'. \tag{4.215}$$

where

$$w(z) \equiv -2\frac{g'(z)}{g(z)}. \tag{4.216}$$

These parameters can be expressed as functionals $\widetilde{\mathbf{B}}\left[f_{i_T}\right]$ and $\widetilde{\boldsymbol{\Sigma}}\left[f_{i_T}\right]$ of the empirical pdf. The functionals are defined implicitly as in (4.176) as follows:

$$\mathbf{0} = \int_{\mathbb{R}^{N+K}} \boldsymbol{\psi}\left(\mathbf{x}, \mathbf{f}, \widetilde{\mathbf{B}}[h], \widetilde{\boldsymbol{\Sigma}}[h]\right) h(\mathbf{x}, \mathbf{f})\, d\mathbf{x} d\mathbf{f}. \tag{4.217}$$

The vector-valued function $\boldsymbol{\psi}$ in this expression follows from (4.214)-(4.215) and reads:

$$\boldsymbol{\psi}(\mathbf{x}, \mathbf{f}, \mathbf{B}, \boldsymbol{\Sigma}) \equiv \begin{pmatrix} w\left(\mathrm{Ma}^2(\mathbf{x}, \mathbf{B}\mathbf{f}, \boldsymbol{\Sigma})\right) \mathrm{vec}\left[(\mathbf{x} - \mathbf{B}\mathbf{f})\, \mathbf{f}'\right] \\ w\left(\mathrm{Ma}^2_{\mathbf{x}, \mathbf{B}\mathbf{f}, \boldsymbol{\Sigma}}\right) \mathrm{vec}\left[(\mathbf{x} - \mathbf{B}\mathbf{f})(\mathbf{x} - \mathbf{B}\mathbf{f})'\right] - \mathrm{vec}\left[\boldsymbol{\Sigma}\right] \end{pmatrix}, \tag{4.218}$$

where vec is the operator (A.104) that stacks the columns of a matrix into a vector. From (4.194) and (4.175) the norm of the influence function is proportional to the norm of the above vector:

$$\left\| \mathrm{IF}\left((\mathbf{x}, \mathbf{f}), f_{\mathbf{X}}, \left(\widehat{\boldsymbol{\mu}}, \widehat{\boldsymbol{\Sigma}}\right)\right) \right\| \propto \|\boldsymbol{\psi}\|. \tag{4.219}$$

In particular, if the perturbations are normally distributed the term w in (4.218) becomes $w \equiv 1$, see (4.125). Therefore the influence function of the regression factor loadings estimator and the perturbation dispersion estimator is not bounded. This is not surprising, since we know from Section 4.3 that the ML estimators of the regression factor loadings of normally distributed factor models are the OLS coefficients, whose influence function is (4.212). In other words, the ML estimator of the regression factor loadings in a factor model with normally distributed perturbations is not robust, and neither is the ML estimator of the perturbation dispersion.

On the other hand, if the perturbations are elliptically but not normally distributed the influence function display a different behavior. Consider for instance Cauchy-distributed perturbations. In this case as in (4.209) the term w in (4.218) becomes:

$$w(z) = \frac{N+1}{1+z}. \tag{4.220}$$

Therefore, from (4.219) and (4.202) the influence function becomes bounded. In other words, the ML estimators of the regression factor loadings and of the perturbation dispersion stemming from Cauchy-distributed perturbations are robust.

4.5.3 Robust estimators

From the above discussion we realize that robust estimators should satisfy two requirements. In the first place, since robustness questions the accuracy of the parametric assumptions on the unknown distribution of the invariants, the construction of robust estimators should be as independent as possible of these assumptions. Secondly, robust estimators should display a bounded influence function.

By forcing maximum likelihood estimators to have a bounded influence function, Maronna (1976) and Huber (1981) developed the so-called *M-estimators*, or generalized maximum likelihood estimators.

We recall that, under the assumption that the distribution of the market invariants is $f_{\boldsymbol{\theta}}$, the maximum likelihood estimators of the parameters $\boldsymbol{\theta}$ are defined as functional of the empirical distribution $\widetilde{\boldsymbol{\theta}}[f_{i_T}]$. From (4.176), this functional is defined as follows:

$$\widetilde{\boldsymbol{\theta}}[h]: \quad \int_{\mathbb{R}^N} \boldsymbol{\psi}\left(\mathbf{x}, \widetilde{\boldsymbol{\theta}}\right) h(\mathbf{x})\, d\mathbf{x} \equiv \mathbf{0}, \tag{4.221}$$

where $\boldsymbol{\psi}$ follows from the assumptions on the underlying distribution:

$$\boldsymbol{\psi}(\mathbf{x}, \boldsymbol{\theta}) \equiv \frac{\partial \ln f_{\boldsymbol{\theta}}(\mathbf{x})}{\partial \boldsymbol{\theta}}. \tag{4.222}$$

M-estimators are also defined by (4.221), but the function $\boldsymbol{\psi}(\mathbf{x}, \boldsymbol{\theta})$ is chosen exogenously. Under these more general assumptions, the influence function (4.194) becomes:

$$\mathrm{IF}\left(\mathbf{x}, f_{\mathbf{X}}, \widehat{\boldsymbol{\theta}}\right) = \mathbf{A}\boldsymbol{\psi}\left(\mathbf{x}, \widetilde{\boldsymbol{\theta}}[f_{\mathbf{X}}]\right), \tag{4.223}$$

where the $S \times S$ matrix $\mathbf{A}$ is defined as follows:

$$\mathbf{A} \equiv -\left[\int_{\mathbb{R}^N} \left.\frac{\partial \boldsymbol{\psi}'}{\partial \boldsymbol{\theta}}\right|_{\boldsymbol{\theta} \equiv \widetilde{\boldsymbol{\theta}}[f_{\mathbf{X}}]} f_{\mathbf{X}}(\mathbf{x})\, d\mathbf{x}\right]^{-1}, \tag{4.224}$$

see Appendix www.4.7.

This way the ensuing estimator $\widetilde{\boldsymbol{\theta}}[f_{i_T}]$ is independent of any assumption on the distribution of the underlying market invariants. If the function $\boldsymbol{\psi}$ is chosen appropriately, the influence function (4.223) becomes bounded. Therefore, the estimator $\widetilde{\boldsymbol{\theta}}[f_{i_T}]$ is robust.

Location and dispersion

Consider (4.207) and replace it with a vector-valued function $\boldsymbol{\psi}$ defined exogenously as follows:

$$\boldsymbol{\psi} \equiv \begin{pmatrix} \gamma\left(\mathrm{Ma}^2(\mathbf{x},\boldsymbol{\mu},\boldsymbol{\Sigma})\right)(\mathbf{x}-\boldsymbol{\mu}) \\ \zeta(\mathrm{Ma}^2_{\mathbf{x},\boldsymbol{\mu},\boldsymbol{\Sigma}})\,\mathrm{vec}\left[(\mathbf{x}-\boldsymbol{\mu})(\mathbf{x}-\boldsymbol{\mu})'\right] - \eta\left(\mathrm{Ma}^2_{\mathbf{x},\boldsymbol{\mu},\boldsymbol{\Sigma}}\right)\mathrm{vec}\left[\boldsymbol{\Sigma}\right] \end{pmatrix}, \tag{4.225}$$

where the functions γ, ζ and η satisfy some regularity criteria and are such that $\boldsymbol{\psi}$ is bounded. The ensuing estimators, which replace (4.203)-(4.205), solve the following implicit equations:

$$\widehat{\boldsymbol{\mu}} = \sum_{t=1}^{T} \frac{\gamma\left(\mathrm{Ma}^2(\mathbf{x}_t,\widehat{\boldsymbol{\mu}},\widehat{\boldsymbol{\Sigma}})\right)}{\sum_{s=1}^{T}\gamma\left(\mathrm{Ma}^2(\mathbf{x}_s,\widehat{\boldsymbol{\mu}},\widehat{\boldsymbol{\Sigma}})\right)}\mathbf{x}_t \tag{4.226}$$

$$\widehat{\boldsymbol{\Sigma}} = \sum_{t=1}^{T} \frac{\zeta\left(\mathrm{Ma}^2_{\mathbf{x}_t,\widehat{\boldsymbol{\mu}},\widehat{\boldsymbol{\Sigma}}}\right)}{\sum_{s=1}^{T}\eta\left(\mathrm{Ma}^2_{\mathbf{x}_s,\widehat{\boldsymbol{\mu}},\widehat{\boldsymbol{\Sigma}}}\right)}(\mathbf{x}_t-\widehat{\boldsymbol{\mu}})(\mathbf{x}_t-\widehat{\boldsymbol{\mu}})'. \tag{4.227}$$

Since $\boldsymbol{\psi}$ is bounded, so is the influence function and therefore these estimators are robust.

For instance, the following is a suitable choice of weights:

$$\gamma(x) \equiv \zeta^{\frac{1}{2}}(x) \equiv \eta^{\frac{1}{2}}(x) \equiv \begin{cases} 1 & \text{if } x \leq x_0 \\ \frac{a}{\sqrt{x}}e^{-\frac{(\sqrt{x}-a)^2}{2b^2}} & \text{if } x > x_0, \end{cases} \tag{4.228}$$

where $a \equiv \sqrt{N}+\sqrt{2}$. If we set $b \equiv +\infty$ we obtain the M-estimators suggested by Huber (1964). If we set $b \equiv 1.25$ we obtain the M-estimators suggested by Hampel (1973), see also Campbell (1980).

As in the case of the maximum likelihood estimators, in general the solution to the above implicit equations cannot be computed analytically. Nevertheless, for suitable choices of the functions γ and ζ such as (4.228) a recursive approach such as the following is guaranteed to converge. Further results for existence and uniqueness of the solution are provided in Huber (1981).

Step 0. Initialize $\widehat{\boldsymbol{\mu}}$ and $\widehat{\boldsymbol{\Sigma}}$ as the sample mean and sample covariance respectively.

Step 1. Compute the right hand side of (4.226) and (4.227).

Step 2. Update the left hand side of (4.226) and (4.227).

Step 3. If convergence has been reached stop, otherwise go to Step 1.

Explicit factors

It is possible to define multivariate M-estimators of the factor loadings and of the dispersion of the perturbations of an explicit factor model. The discussion proceeds exactly as above. Nevertheless, due to the larger number of parameters, convergence problems arise for the numerical routines that should yield the estimators in practice.

4.6 Practical tips

In this section we provide a few tips that turn out useful in practical estimation problems.

4.6.1 Detection of outliers

In Section 4.5.1 we introduced the tools to measure the effect on an estimate of one outlier both in the finite sample case, namely the influence curve and the jackknife, and in the infinite sample limit, namely the influence function. Another interesting question is the maximum amount of outliers that a certain estimator can sustain before breaking down: if there is a total of $T = T_G + T_O$ observations, where T_G are good data and T_O outliers, what is the highest ratio T_O/T that the estimator can sustain?

The *breakdown point* is the limit of this ratio when the number of observations tends to infinity. Obviously, the breakdown point is a positive number that cannot exceed 0.5.

For example, suppose that we are interested in estimating the location parameter of an invariant X_t.

Consider first the sample mean (4.41), which we report here:

$$\widehat{\mathrm{E}} \equiv \frac{1}{T}\sum_{t=1}^{T} x_t. \tag{4.229}$$

From (4.198) breakdown point of the sample mean is 0, as one single outlier can disrupt the estimation completely.

Consider now the sample median (4.39), which we report here:

$$\widehat{q}_{1/2} \equiv x_{[T/2]:T}, \tag{4.230}$$

where $[\cdot]$ denotes the integer part. The breakdown point of the median is 0.5. Indeed, changing the values of half the sample, i.e. all (minus one) the observations larger than $x_{[T/2]:T}$, or all (minus one) the observations smaller than $x_{[T/2]:T}$, does not affect the result of the estimation.

Estimators whose breakdown point is close to 0.5 are called *high breakdown estimators*. These estimators are useful in financial applications because they allow us to detect outliers. Indeed, time series are often fraught with suspicious data. In the case of one-dimensional variables it is relatively easy to spot these outliers by means of graphical inspection. In the multivariate case, this task becomes much more challenging.

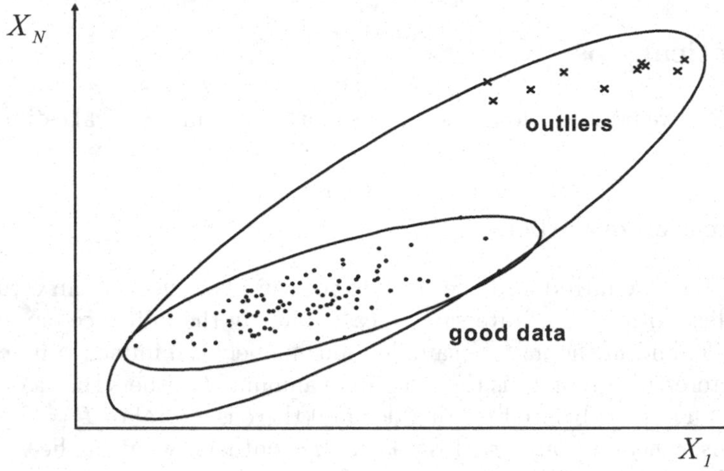

Fig. 4.19. Minimum Volume Ellipsoid

There exists a vast literature on estimators with high breakdown point, see Huber (1981) and Hampel, Ronchetti, Rousseeuw, and Stahel (1986). Here we propose two methods to build high breakdown estimators of location and dispersion: the *minimum volume ellipsoid (MVE)* and the *minimum covariance determinant (MCD)*, see Rousseeuw and Leroy (1987), Rousseeuw and VanDriessen (1999). The rationale behind these estimators rests on the assumption that the core of the good data is tightly packed, whereas the joint set of good data and outliers is much more scattered, see Figure 4.19.

Minimum volume ellipsoid

Suppose we know that T_G out of the T data are good and T_O are outliers. Due to the above rationale, the smallest ellipsoid that contains the T_G good data is the smallest among all the ellipsoids that contain any set of T_G observations.

Consider a generic location parameter $\boldsymbol{\mu}$, i.e. an N-dimensional vector, and a generic scatter matrix $\boldsymbol{\Sigma}$, i.e. a positive and symmetric $N \times N$ matrix. The parameters $(\boldsymbol{\mu},\boldsymbol{\Sigma})$ define an ellipsoid $\mathcal{E}_{\boldsymbol{\mu},\boldsymbol{\Sigma}}$ as in $(A.73)$. We can inflate this ellipsoid as follows:

$$\mathcal{E}^{q}_{\boldsymbol{\mu},\boldsymbol{\Sigma}} \equiv \left\{ \mathbf{x} \in \mathbb{R}^N \text{ such that } (\mathbf{x} - \boldsymbol{\mu})' \boldsymbol{\Sigma}^{-1} (\mathbf{x} - \boldsymbol{\mu}) \leq q^2 \right\}. \tag{4.231}$$

This locus represents a rescaled version of the original ellipsoid, where all the principal axis are multiplied by a factor q. From ($A.77$) the volume of the inflated ellipsoid reads:

$$\text{Vol}\left\{ \mathcal{E}^{q}_{\boldsymbol{\mu},\boldsymbol{\Sigma}} \right\} = \gamma_N q^N \sqrt{|\boldsymbol{\Sigma}|}, \tag{4.232}$$

where γ is the volume of the unit sphere:

$$\gamma_N \equiv \frac{\pi^{\frac{N}{2}}}{\Gamma\left(\frac{N}{2}+1\right)}. \tag{4.233}$$

Consider the set of Mahalanobis distances (2.61) of each observation from the location parameter $\boldsymbol{\mu}$ through the metric $\boldsymbol{\Sigma}$:

$$\text{Ma}^{\boldsymbol{\mu},\boldsymbol{\Sigma}}_t \equiv \text{Ma}\left(\mathbf{x}_t, \boldsymbol{\mu}, \boldsymbol{\Sigma}\right) \equiv \sqrt{(\mathbf{x}_t - \boldsymbol{\mu})' \boldsymbol{\Sigma}^{-1} (\mathbf{x}_t - \boldsymbol{\mu})}. \tag{4.234}$$

We can sort these distances in increasing order and consider the T_G-th distance:

$$q_{T_G} \equiv \text{Ma}^{\boldsymbol{\mu},\boldsymbol{\Sigma}}_{T_G:T}. \tag{4.235}$$

By construction, the ellipsoid $\mathcal{E}^{q_{T_G}}_{\boldsymbol{\mu},\boldsymbol{\Sigma}}$ contains only T_G points and from (4.232) its volume reads:

$$\text{Vol}\left\{ \mathcal{E}^{q_{T_G}}_{\boldsymbol{\mu},\boldsymbol{\Sigma}} \right\} = \gamma_N \left(\text{Ma}^{\boldsymbol{\mu},\boldsymbol{\Sigma}}_{T_G:T} \right)^N \sqrt{|\boldsymbol{\Sigma}|}. \tag{4.236}$$

Notice that the product on the right hand side of this expression does not depend on the determinant of $\boldsymbol{\Sigma}$. Therefore we can impose the constraint that the determinant of $\boldsymbol{\Sigma}$ be one.

Consequently, the parameters that give rise to the smallest ellipsoid that contains T_G observations solve the following equation:

$$\left(\widehat{\boldsymbol{\mu}}_{T_G}, \widehat{\boldsymbol{\Sigma}}_{T_G} \right) = \underset{\boldsymbol{\mu},\boldsymbol{\Sigma}\succeq\mathbf{0},|\boldsymbol{\Sigma}|=1}{\text{argmin}} \left\{ \text{Ma}^{\boldsymbol{\mu},\boldsymbol{\Sigma}}_{T_G:T} \right\}, \tag{4.237}$$

where the the notation $\boldsymbol{\Sigma} \succeq \mathbf{0}$ means that $\boldsymbol{\Sigma}$ is symmetric and positive. Once we have computed the parameters (4.237), we tag as outliers all the observation that are not contained in the ellipsoid (4.231) determined by (4.237), with the radius (4.235) implied by (4.237).

In reality we do not know a priori the true number T_G of good data. Nevertheless, if T_G is the largest set of good data, the minimum volume ellipsoid that contains $T_G + 1$ observations has a much larger volume than the minimum volume ellipsoid that contains T_G observations. Therefore, we consider the volume of the minimum volume ellipsoid as a function of the number of observations contained in the ellipsoid:

$$T_G \to \gamma_N \left(\mathrm{Ma}_{T_G:T}^{\widehat{\boldsymbol{\mu}}_{T_G}, \widehat{\boldsymbol{\Sigma}}_{T_G}} \right)^N . \tag{4.238}$$

The true number of good data is the value T_G where this function displays an abrupt jump, see Figure 4.20.

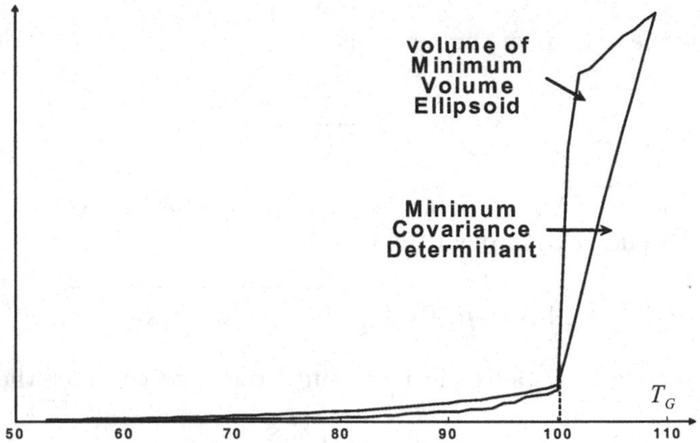

Fig. 4.20. Detection of outliers

The optimization problem (4.237) cannot be solved analytically. Numerical algorithms both deterministic and non-deterministic are available in the literature. We present below an approach that we used to generate the figures in this section.

Minimum covariance determinant

An alternative approach to detect outliers is provided by the minimum covariance determinant. This method also searches the "smallest ellipsoid". Instead of the smallest ellipsoid defined by the cloud of data we look for the smallest ellipsoid defined by the sample covariance of the data. Indeed, we recall from (4.48) that the sample covariance defines the smallest ellipsoid that fits the data in an average sense.

Suppose that we know the number of good observations T_G. Consider a generic subset of T_G observations $\{\mathbf{x}_1^*, \ldots, \mathbf{x}_{T_G}^*\}$ from the T observations in the time series i_T of the market invariants. We can compute the sample mean (4.41) and sample covariance (4.42) associated with this subset:

$$\widehat{\mathrm{E}}^*_{T_G} \equiv \frac{1}{T_G} \sum_{t=1}^{T_G} \mathbf{x}^*_t \tag{4.239}$$

$$\widehat{\mathrm{Cov}}^*_{T_G} \equiv \frac{1}{T_G} \sum_{t=1}^{T_G} \left(\mathbf{x}^*_t - \widehat{\mathrm{E}}^*_{T_G}\right) \left(\mathbf{x}^*_t - \widehat{\mathrm{E}}^*_{T_G}\right)'. \tag{4.240}$$

Consider the ellipsoid determined by these parameters:

$$\mathcal{E}^* \equiv \left\{ \mathbf{x} : \left(\mathbf{x} - \widehat{\mathrm{E}}^*_{T_G}\right)' \left(\widehat{\mathrm{Cov}}^*_{T_G}\right)^{-1} \left(\mathbf{x} - \widehat{\mathrm{E}}^*_{T_G}\right) \leq 1 \right\}. \tag{4.241}$$

From ($A.77$), the volume of $\mathcal{E}^*$ is proportional to the square root of the determinant of (4.240).

Therefore we have to determine the subset of observations that gives rise to the minimum covariance determinant:

$$\{\mathbf{x}^\times_1, \ldots, \mathbf{x}^\times_{T_G}\} = \underset{\mathbf{x}^*_1, \ldots, \mathbf{x}^*_{T_G} \in i_T}{\operatorname{argmin}} \left| \widehat{\mathrm{Cov}}^*_{T_G} \right|. \tag{4.242}$$

In reality we do not know a priori the true number T_G of good data. Nevertheless, if T_G is the largest set of good data, the minimum covariance determinant relative to $T_G + 1$ observations is much larger than the minimum covariance determinant relative to T_G observations. Therefore, we consider the minimum covariance determinant as a function of the number of observations contained in the ellipsoid:

$$T_G \to \left| \widehat{\mathrm{Cov}}^\times_{T_G} \right|. \tag{4.243}$$

The true number of good data is the value T_G where this function displays an abrupt jump, see Figure 4.20.

The optimization problem (4.242) cannot be solved exactly. We present below an approach that we used to generate the figures in this section.

Computational issues

Suppose we have a series of observations $\{\mathbf{x}_1, \ldots, \mathbf{x}_T\}$. Assume we know that $T_G \leq T$ among them are good data.

In principle we should compute the minium volume ellipsoid and the sample covariance matrix for all the possible combinations of T_G observations out of the total T observations. This number reads:

$$\binom{T}{T_G} \equiv \frac{T!}{T_G!\,(T - T_G)!}, \tag{4.244}$$

which is intractably large if T exceeds the order of the dozen. Instead, we delete the unwelcome observations one at a time from the initial set of T observations using a theoretically sub-optimal, yet for practical purposes very effective, approach.

First we build Routine A, which computes the smallest ellipsoid $\mathcal{E}_{\mathbf{m},\mathbf{S}}$ that contains a given set of observations $\{\mathbf{x}_1, \ldots, \mathbf{x}_T\}$.

Step 0. Initialize the relative weights:

$$w_t \equiv \frac{1}{T}, \quad t = 1, \ldots, T. \tag{4.245}$$

Step 1. Compute the location parameter $\mathbf{m}$ and the scatter matrix $\mathbf{S}$ as follows:

$$\mathbf{m} \equiv \frac{1}{\sum_{s=1}^{T} w_s} \sum_{t=1}^{T} w_t \mathbf{x}_t \tag{4.246}$$

$$\mathbf{S} \equiv \sum_{t=1}^{T} w_t (\mathbf{x}_t - \mathbf{m})(\mathbf{x}_t - \mathbf{m})'. \tag{4.247}$$

Notice that the weights in the scatter matrix are *not* normalized.

Step 2. Compute the square Mahalanobis distances:

$$\mathrm{Ma}_t^2 \equiv (\mathbf{x} - \mathbf{m})' \mathbf{S}^{-1} (\mathbf{x} - \mathbf{m}), \quad t = 1, \ldots, T. \tag{4.248}$$

Step 3. Update the weights: if $\mathrm{Ma}_t^2 > 1$ change the respective weight as follows:

$$w_t \mapsto w_t \, \mathrm{Ma}_t^2; \tag{4.249}$$

otherwise, leave the weight unchanged.

Step 4. If convergence has been reached, stop and define $\mathcal{E}_{\mathbf{m},\mathbf{S}}$ as in $(A.73)$, otherwise, go to Step 1.

Secondly, we build Routine B, which spots the farthest outlier in a series of observations $\{\mathbf{x}_1, \ldots, \mathbf{x}_T\}$. Define the following $T \times N$ matrix:

$$\mathbf{U} = \begin{pmatrix} \mathbf{x}_1' - \widehat{\mathrm{E}}' \\ \vdots \\ \mathbf{x}_T' - \widehat{\mathrm{E}}' \end{pmatrix}, \tag{4.250}$$

where $\widehat{\mathrm{E}}$ is the sample mean (4.41) of the data. The sample covariance matrix (4.42) can be written as follows:

$$\widehat{\mathrm{Cov}} \equiv \frac{1}{T} \mathbf{U}'\mathbf{U}. \tag{4.251}$$

We aim at finding the observation $\mathbf{x}_t$ such that if we remove it from the set $\{\mathbf{x}_1, \ldots, \mathbf{x}_T\}$ the determinant of the resulting sample covariance is reduced the most. This would mean that by dropping that observation the location-dispersion ellipsoid defined by sample mean and covariance shrinks the most, and thus that observation is the farthest outlier in the sample. To do this, we use the following result, see Poston, Wegman, Priebe, and Solka (1997):

$$\left|\mathbf{U}'_{(-t)}\mathbf{U}_{(-t)}\right| = (1-\lambda_t)\left|\mathbf{U}'\mathbf{U}\right|. \tag{4.252}$$

In this expression $\mathbf{U}_{(-t)}$ denotes the matrix (4.250) after removing the t-th row and λ_t denotes the t-th element of the diagonal of the *information matrix*:

$$\lambda_t \equiv \left(\mathbf{U}\left(\mathbf{U}'\mathbf{U}\right)^{-1}\mathbf{U}'\right)_{tt}. \tag{4.253}$$

It can be proved that

$$0 \leq \lambda_t \leq 1. \tag{4.254}$$

Therefore, the farthest outlier corresponds to the highest value of λ_t, unless $\lambda_t = 1$: in this last case, if we remove the t-th observation the sample covariance becomes singular, as is evident from (4.252).

Now we can define Routine C, which detects the outliers among the given data by means of the minimum volume ellipsoid and the minimum covariance determinant.

Step 0. Consider as data all the observations.

Step 1. Compute the sample mean and covariance $\left(\widehat{\mathrm{E}}, \widehat{\mathrm{Cov}}\right)$ of the given data and compute the determinant of the sample covariance $\left|\widehat{\mathrm{Cov}}\right|$.

Step 2. Compute with Routine A the minimum volume ellipsoid of the given data $\mathcal{E}_{\mathbf{m},\mathbf{S}}$ and compute $|\mathbf{S}|$.

Step 3. Find the farthest outlier among the data with Routine B and remove it from the data.

Step 4. If the number of data left is less than half the original number stop, otherwise go to Step 1.

The plot of $\left|\widehat{\mathrm{Cov}}\right|$ and/or $|\mathbf{S}|$ as a function of the number of observations in the dataset shows an abrupt jump when the first outlier is added to the dataset, see Figure 4.20. The respective sample covariance $\widehat{\mathrm{Cov}}$ is the minimum covariance determinant and the respective ellipsoid $\mathcal{E}_{\mathbf{m},\mathbf{S}}$ is the minimum volume ellipsoid.

4.6.2 Missing data

Sometimes some data is missing from the time series of observations. Our purpose is twofold. On the one hand, we are interested in interpolating the missing values. On the other hand we want to estimate parameters of interest regarding the market invariants, such as parameters of location or dispersion. We refer the reader to Stambaugh (1997) for a discussion of the case where some series are shorter than others. Here, we discuss the case where some observations are missing randomly from the time series.

Consider a $T \times N$ panel of observations, where T is the length of the sample and N is the number of market invariants. Each row of this matrix corresponds to a joint observation $\mathbf{x}_t$ of the invariants at a specific date. In some rows one or more entry might be missing:

$$\mathbf{x}_t \equiv \mathbf{x}_{t,\mathrm{mis}(t)} \cup \mathbf{x}_{t,\mathrm{obs}(t)}, \tag{4.255}$$

where we stressed that the set of missing and observed values depends on the specific date t. Notice that for most t the set $\mathbf{x}_{t,\mathrm{mis}(t)}$ is empty.

For example, consider a case of four market invariants and a hundred joint observations. Assume that the second entry is missing at time $t = 7$ then

$$\mathrm{mis}\,(7) \equiv \{2\}, \quad \mathrm{obs}\,(7) \equiv \{1,3,4\}. \tag{4.256}$$

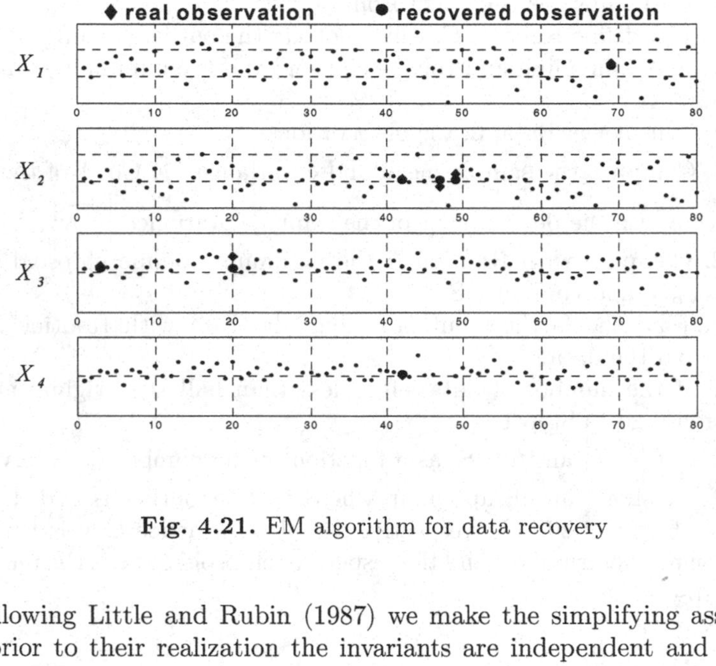

Fig. 4.21. EM algorithm for data recovery

Following Little and Rubin (1987) we make the simplifying assumption that prior to their realization the invariants are independent and normally distributed:

$$\begin{pmatrix} \mathbf{X}_{t,\mathrm{mis}(t)} \\ \mathbf{X}_{t,\mathrm{obs}(t)} \end{pmatrix} \sim \mathrm{N}\left(\begin{pmatrix} \boldsymbol{\mu}_{\mathrm{mis}(t)} \\ \boldsymbol{\mu}_{\mathrm{obs}(t)} \end{pmatrix}, \begin{pmatrix} \boldsymbol{\Sigma}_{\mathrm{mis}(t),\mathrm{mis}(t)} & \boldsymbol{\Sigma}_{\mathrm{mis}(t),\mathrm{obs}(t)} \\ \boldsymbol{\Sigma}_{\mathrm{obs}(t),\mathrm{mis}(t)} & \boldsymbol{\Sigma}_{\mathrm{obs}(t),\mathrm{obs}(t)} \end{pmatrix} \right). \tag{4.257}$$

The algorithm we propose is a specific instance of a general approach called *expectation-maximization (EM) algorithm*, see Dempster, Laird, and Rubin (1977) and also Bilmes (1998). In Figure 4.21 we recovered a few missing values with the EM algorithm.

The algorithm proceeds as follows, see Appendix www.4.8 for the proofs.

Step 0. Set $u \equiv 0$ and initialize both the location and the dispersion parameters. For all $n = 1 \ldots, N$ set:

$$\mu_n^{(u)} \equiv \frac{1}{T_n} \sum_{t \in \text{ avail. obs.}} x_{t,n} \tag{4.258}$$

$$\Sigma_{nn}^{(u)} \equiv \frac{1}{T_n} \sum_{t \in \text{ avail. obs.}} \left(x_{t,n} - \mu_n^{(u)} \right)^2 , \tag{4.259}$$

where T_n is the number of available observations for the generic n-th market invariant. For all $n, m = 1 \ldots, N$, $n \neq m$ set:

$$\Sigma_{nm}^{(u)} \equiv 0. \tag{4.260}$$

Step 1. For each $t = 1, \ldots, T$ fill in the missing entries by replacing the missing values with their expected value conditional on the observations. For the observed values we have:

$$\mathbf{x}_{t,\text{obs}(t)}^{(u)} \equiv \mathbf{x}_{t,\text{obs}(t)}; \tag{4.261}$$

and for the missing values we have:

$$\begin{aligned} \mathbf{x}_{t,\text{mis}(t)}^{(u)} &\equiv \boldsymbol{\mu}_{\text{mis}(t)}^{(u)} \\ &+ \boldsymbol{\Sigma}_{\text{mis}(t),\text{obs}(t)}^{(u)} \left(\boldsymbol{\Sigma}_{\text{obs}(t),\text{obs}(t)}^{(u)} \right)^{-1} \left(\mathbf{x}_{t,\text{obs}(t)} - \boldsymbol{\mu}_{\text{obs}(t)}^{(u)} \right). \end{aligned} \tag{4.262}$$

Step 2. For each $t = 1, \ldots, T$ compute the conditional covariance, which is zero if at least one of the invariants is observed:

$$\mathbf{C}_{t,\text{obs}(t),\text{mis}(t)}^{(u)} \equiv \mathbf{0}, \quad \mathbf{C}_{t,\text{obs}(t),\text{obs}(t)}^{(u)} \equiv \mathbf{0}, \tag{4.263}$$

and otherwise reads:

$$\begin{aligned} \mathbf{C}_{t,\text{mis}(t),\text{mis}(t)}^{(u)} &\equiv \boldsymbol{\Sigma}_{\text{mis}(t),\text{mis}(t)}^{(u)} \\ &- \boldsymbol{\Sigma}_{\text{mis}(t),\text{obs}(t)}^{(u)} \left(\boldsymbol{\Sigma}_{\text{obs}(t),\text{obs}(t)}^{(u)} \right)^{-1} \boldsymbol{\Sigma}_{\text{obs}(t),\text{mis}(t)}^{(u)}. \end{aligned} \tag{4.264}$$

Step 3. Update the estimate of the location parameter:

$$\boldsymbol{\mu}^{(u+1)} \equiv \frac{1}{T} \sum_t \mathbf{x}_t^{(u)}. \tag{4.265}$$

Step 4. Update the estimate of the dispersion parameter:

$$\boldsymbol{\Sigma}^{(u+1)} \equiv \frac{1}{T} \sum_t \left[\mathbf{C}_t^{(u)} + \left(\mathbf{x}_t^{(u)} - \boldsymbol{\mu}^{(u)} \right) \left(\mathbf{x}_t^{(u)} - \boldsymbol{\mu}^{(u)} \right)' \right]. \tag{4.266}$$

Step 5. If convergence has been reached, stop. Otherwise, set $u \equiv u + 1$ and go to Step 1.

4.6.3 Weighted estimates

We have seen in (4.167) and comments that follow that any estimator $\widehat{\mathbf{G}}$ can be represented as a functional $\widetilde{\mathbf{G}}\left[f_{i_T}\right]$ that acts on the empirical probability density function of the time series of the market invariants:

$$i_T \equiv \{\mathbf{x}_1, \ldots, \mathbf{x}_T\}. \tag{4.267}$$

In the definition of the empirical density function (4.168) and thus in the definition of the estimator $\widehat{\mathbf{G}}$ the order of the realization of the market invariants does not play a role. This is correct, since the invariants are independent and identically distributed across time, see (4.5).

Nevertheless, intuition suggests that the most recent observations should somehow play a more important role than observations farther back in the past. To account for this remark, it suffices to replace the definition of the empirical probability density function (4.168) as follows:

$$f_{i_T} \mapsto f_{i_T} \equiv \frac{1}{\sum_{s=1}^{T} w_s} \sum_{t=1}^{T} w_t \delta^{(\mathbf{x}_t)}, \tag{4.268}$$

where δ is the Dirac delta (B.17) and where the weights w_t are positive, non-decreasing functions of the time index t. We present below two notable cases.

Rolling window

A simple way to give more weight to the last observations is to assume that only the last set of observations is good at forecasting, whereas considering the previous observations might be disruptive. Therefore, we consider only the *rolling window* of the last W observations among the T in the whole time series. This corresponds to setting in (4.268) the following weights:

$$w_t \equiv 1, \text{ if } t > T - W \tag{4.269}$$

$$w_t \equiv 0, \text{ if } t \leq T - W. \tag{4.270}$$

Each time a new observation is added to the time series, we roll over the window and again we only consider the last W observations.

For example, if we are at time T, the sample mean (4.41) becomes:

$$\widehat{\mathrm{E}}_W \equiv \frac{1}{W} \sum_{t=T-W+1}^{T} \mathbf{x}_t, \tag{4.271}$$

and the sample covariance (4.42) becomes:

$$\widehat{\mathrm{Cov}}_W \equiv \frac{1}{W} \sum_{t=T-W+1}^{T} \left(\mathbf{x}_t - \widehat{\boldsymbol{\mu}}_w\right)\left(\mathbf{x}_t - \widehat{\boldsymbol{\mu}}_w\right)'. \tag{4.272}$$

To determine the most appropriate value of the rolling window one should keep in mind the specific investment horizon.

Exponential smoothing

A less dramatic approach consists in giving less and less weight to past observations in a smooth fashion. The *exponential smoothing* consists in setting in (4.268) weights that decay exponentially:

$$w_t \equiv (1-\lambda)^{T-t}, \tag{4.273}$$

where λ is a fixed *decay factor* between zero and one. Notice that the case $\lambda \equiv 0$ recovers the standard empirical pdf. If the decay factor is strictly positive, the weight of past observations in the estimate tapers at an exponential rate.

For example, if we are at time T, the sample mean (4.41) becomes:

$$\widehat{\mathrm{E}}_\lambda \equiv \frac{\lambda}{1-(1-\lambda)^T} \sum_{t=1}^{T} (1-\lambda)^{T-t} \mathbf{x}_t; \tag{4.274}$$

and the sample covariance (4.42) becomes:

$$\widehat{\mathrm{Cov}}_\lambda \equiv \frac{\lambda}{1-(1-\lambda)^T} \sum_{t=1}^{T} (1-\lambda)^{T-t} \left(\mathbf{x}_t - \widehat{\mathrm{E}}_\lambda\right) \left(\mathbf{x}_t - \widehat{\mathrm{E}}_\lambda\right)'. \tag{4.275}$$

The exponential smoothing estimate is used, among others, by RiskMetrics and Goldman Sachs, see Litterman and Winkelmann (1998). To assign a suitable value to the decay factor a possible approach is to choose a parametric form for the probability density function and then apply the maximum likelihood principle (4.66).

For example, if the invariants $\mathbf{X}_t$ are normally distributed, we determine the parameter λ in (4.274)-(4.275) by maximizing the normal log-likelihood:

$$\widetilde{\lambda} \equiv \underset{0\leq\lambda<1}{\operatorname{argmax}} \left(-\frac{T}{2} \ln \left|\widehat{\mathrm{Cov}}_\lambda\right| - \frac{1}{2} \sum_{t=1}^{T} \left(\mathbf{x}_t - \widehat{\mathrm{E}}_\lambda\right)' \widehat{\mathrm{Cov}}_\lambda^{-1} \left(\mathbf{x}_t - \widehat{\mathrm{E}}_\lambda\right) \right). \tag{4.276}$$

The exponential smoothing presents an interesting link to *GARCH* models, an acronym for Generalized AutoRegressive Conditionally Heteroskedastic models, see Engle (1982) and Bollerslev (1986). Indeed, by recursive substitution we can check that in the presence of an infinite series of observations the exponential smoothing is consistent with the following GARCH model:

$$X_t \equiv \mu + \epsilon_t, \tag{4.277}$$

where ϵ_t are random perturbations such that:

$$\mathrm{Var}\{\epsilon_t\} = \lambda \epsilon_{t-1}^2 + (1-\lambda)\,\mathrm{Var}\{\epsilon_{t-1}\}. \tag{4.278}$$

4.6.4 Overlapping data

In order for the market invariants to be independent across time it is necessary that they refer to non-overlapping time intervals as in Figure 3.11.

For example, consider the case of the equity market where the invariants are the compounded returns (3.11). Suppose that the returns are identically normally distributed and independent:

$$\begin{pmatrix} C_{t,\tau} \\ C_{t+\tau,\tau} \\ \vdots \end{pmatrix} \sim \mathrm{N}\left(\begin{pmatrix} \mu \\ \mu \\ \vdots \end{pmatrix}, \begin{pmatrix} \sigma^2 & 0 & \ddots \\ 0 & \sigma^2 & \ddots \\ \ddots & \ddots & \ddots \end{pmatrix} \right). \tag{4.279}$$

From (2.163) we immediately derive the distribution of the overlapping time series:

$$\begin{pmatrix} C_{t,2\tau} \\ C_{t+\tau,2\tau} \\ \vdots \end{pmatrix} = \begin{pmatrix} C_{t,\tau} + C_{t-\tau,\tau} \\ C_{t+\tau,\tau} + C_{t,\tau} \\ \vdots \end{pmatrix} \sim \mathrm{N}(\mathbf{m}, \mathbf{S}), \tag{4.280}$$

where

$$\mathbf{m} \equiv \begin{pmatrix} 2\mu \\ 2\mu \\ \vdots \end{pmatrix}, \quad \mathbf{S} \equiv \begin{pmatrix} 2\sigma^2 & \sigma^2 & \cdots \\ \sigma^2 & 2\sigma^2 & \ddots \\ \ddots & \ddots & \ddots \end{pmatrix}. \tag{4.281}$$

This expression shows that that the overlapping observations are not independent.

In some circumstances it is possible and even advisable to consider overlapping data, see Campbell, Lo, and MacKinlay (1997).

4.6.5 Zero-mean invariants

When a location parameter such as the expected value of a market invariant is close to null with respect to a dispersion parameter such as its standard deviation, it might be convenient to assume that the location parameter is zero, instead of estimating it. We can interpret this approach as an extreme case of shrinkage, see Section 4.4.

This approach often leads to better results, see Alexander (1998) and therefore it is often embraced by practitioners. For instance we made this assumption in (3.233) regarding the expected changes in yield in the swap market.

4.6.6 Model-implied estimation

Time-series analysis is by definition backward-looking. An alternative approach to estimation makes use of pricing models, which reflects the expectations on the market and thus is forward-looking.

Consider a parametric model $f_{\boldsymbol{\theta}}$ for the market invariants. Assume there exist pricing functions $\mathbf{F}(\boldsymbol{\theta})$ of financial products which depend on those parameters and which trade at the price $\mathbf{P}_T$ at the time the estimate is made.

In these circumstances we can compute the estimate of the parameters as the best fit to the data, i.e. as the solution of the following optimization problem:

$$\widehat{\boldsymbol{\theta}} = \underset{\boldsymbol{\theta}}{\operatorname{argmin}} \left\{ (\mathbf{P} - \mathbf{F}(\boldsymbol{\theta}))' \mathbf{Q} (\mathbf{P} - \mathbf{F}(\boldsymbol{\theta})) \right\}, \tag{4.282}$$

where $\mathbf{Q}$ is a suitably chosen symmetric and positive matrix.

Depending on the applications, some authors suggest mixed approaches, where time series analysis is used together with implied estimation.

For example, to estimate the correlation matrix of swap yield changes we can proceed as in Longstaff, Santa-Clara, and Schwartz (2001). First we estimate from (4.41) and (4.42) the sample correlation matrix:

$$\widehat{C}_{mn} \equiv \frac{\widehat{\operatorname{Cov}} \{X_m, X_n\}}{\sqrt{\widehat{\operatorname{Cov}} \{X_m, X_m\} \widehat{\operatorname{Cov}} \{X_n, X_n\}}}. \tag{4.283}$$

Then we perform the principal component decomposition (A.70) of the correlation matrix:

$$\widehat{\mathbf{C}} = \widehat{\mathbf{E}} \widehat{\boldsymbol{\Lambda}} \widehat{\mathbf{E}}', \tag{4.284}$$

where $\widehat{\boldsymbol{\Lambda}}$ is the diagonal matrix of the estimated eigenvalues and $\widehat{\mathbf{E}}$ is the orthogonal matrix of the respective estimated eigenvectors. Next, we assume that a more suitable estimate of the correlation matrix is of this form:

$$\mathbf{C} = \widehat{\mathbf{E}} \boldsymbol{\Psi} \widehat{\mathbf{E}}', \tag{4.285}$$

where $\boldsymbol{\Psi}$ is a diagonal matrix of positive entries. Finally we fit an estimate $\widetilde{\boldsymbol{\Psi}}$ from the prices of a set of swaptions which depend on the correlation through suitable pricing function.

The main problem with the model-implied approach is that the pricing functions $\mathbf{F}(\boldsymbol{\theta})$ give rise to model risk. This risk is equivalent to the risk of assuming an incorrect parametric distribution for the invariants in the derivation of maximum likelihood estimators.

5

Evaluating allocations

An allocation is a portfolio of securities in a given market. In this chapter we discuss how to evaluate an allocation for a given investment horizon, i.e. a linear combination of the prices of the securities at the investment horizon.

In Section 5.1 we introduce the investor's objectives. An objective is a feature of a given allocation on which the investor focuses his attention. For instance an objective is represented by final wealth at the horizon, or net gains, or wealth relative to some benchmark. The objective is a random variable that depends on the allocation. Although it is not possible to compute analytically the distribution of the objective in general markets, we present some approximate techniques that yield satisfactory results in most applications.

In Section 5.2 we tackle the problem of evaluating allocations, or more precisely the distribution of the objective relative to a given allocation. We do this by introducing the concept of stochastic dominance, a criterion that allows us to evaluate the distribution of the objective as a whole: when facing two allocations, i.e. the distributions of two different objectives, the investor will choose the one that is more advantageous in a global sense. Nevertheless, stochastic dominance presents a few drawbacks, most notably the fact that two generic allocations might not necessarily be comparable. In other words, the investor might not be able to rank allocations and thus make a decision regarding his investment.

As a consequence, in Section 5.3 we take a different approach. We summarize all the properties of a distribution in a single number: an index of satisfaction. If the index of satisfaction is properly defined the investor can in all circumstances choose the allocation that best suits him. Therefore we analyze a set of criteria that a proper satisfaction index should or could satisfy, such as estimability, consistency with stochastic dominance, constancy, homogeneity, translation invariance, additivity, concavity, risk aversion.

In the remainder of the chapter we discuss three broad classes of indices of satisfaction that have become popular among academics and practitioners.

A. Meucci, *Risk and Asset Allocation*, Springer Finance,

In Section 5.4 we present the first of such indices of satisfaction: the certainty-equivalent. Based on the intuitive concept of expected utility, this has been historically the benchmark criterion to assess allocations. After introducing the definition of the certainty-equivalent and discussing its general properties, we show how to build utility functions that cover a wide range of situations, including the non-standard setting of prospect theory. Then we tackle some computational issues. Indeed, the computation of the certainty-equivalent involves integrations and functional inversions, which are in general impossible to perform. Therefore we present some approximate results, such as the Arrow-Pratt expansion. Finally, we perform a second-order sensitivity analysis to determine the curvature of the certainty-equivalent. The curvature is directly linked to the investor's attitude toward diversification and it is fundamental in view of computing numerical solutions to allocation problems.

In Section 5.5 we consider another index of satisfaction, namely the quantile of the investor's objective for a given confidence level. This index is better known under the name of value at risk when the investor's objective are net gains. The quantile-based index of satisfaction has become a standard tool among practitioners after the Basel Accord enforced its use among financial institutions to monitor the riskiness of their investment policies. After introducing the definition of the quantile-based index of satisfaction and discussing its general properties, we tackle some computational issues. Approximate expressions of the quantile can be obtained with approaches such as the Cornish-Fisher expansion and extreme value theory. Finally, we perform a second-order sensitivity analysis, from which it follows that quantile-based indices of satisfaction fail to promote diversification.

In Section 5.6 we discuss a third group of measures of satisfaction: coherent indices and spectral indices, which represent a sub-class of coherent indices. These measures of satisfaction are defined axiomatically in terms of their properties, most notably the fact that by definition they promote diversification. Nevertheless, spectral indices of satisfaction can also be introduced alternatively as weighted averages of a very popular measure of risk, the expected shortfall. This representation is more intuitive and suggests how to construct coherent indices in practice. As we did for the certainty-equivalent and the quantile, we discuss the computational issues behind the spectral indices of satisfaction. Finally, we perform a second-order sensitivity analysis. In particular, from this analysis it follows that spectral measures of satisfaction are concave and thus promote diversification.

We remark that throughout the chapter all the distributions are assumed continuous and smooth, possibly after regularizing them as discussed in Appendix B.4.

5.1 Investor's objectives

Consider a market of N securities. At the time T when the investment is made the investor can purchase α_n units of the generic n-th security. These units are specific to the security: for instance, in the case of equities the units are shares, in the case of futures the units are contracts, etc. Therefore, the *allocation* is represented by the N-dimensional vector $\boldsymbol{\alpha}$.

We denote as $P_t^{(n)}$ the price at the generic time t of the generic n-th security. With the allocation $\boldsymbol{\alpha}$ the investor forms a portfolio whose value at the time the investment decision is made is:

$$w_T(\boldsymbol{\alpha}) \equiv \boldsymbol{\alpha}'\mathbf{p}_T, \tag{5.1}$$

where the lower-case notation emphasizes that the above quantities are known at the time the investment decision is made.

At the investment horizon τ the market prices of the securities are a multivariate random variable. Therefore at the investment horizon the portfolio is a one-dimensional random variable, namely the following simple function of the market prices:

$$W_{T+\tau}(\boldsymbol{\alpha}) \equiv \boldsymbol{\alpha}'\mathbf{P}_{T+\tau}. \tag{5.2}$$

The investor has one or more *objectives* Ψ, namely quantities that the investor perceives as beneficial and therefore he desires in the largest possible amounts. This is the *non-satiation principle* underlying the investor's objectives. The standard objectives are discussed below.

- **Absolute wealth**

The investor focuses on the value at the horizon of the portfolio:

$$\Psi_{\boldsymbol{\alpha}} \equiv W_{T+\tau}(\boldsymbol{\alpha}) = \boldsymbol{\alpha}'\mathbf{P}_{T+\tau}. \tag{5.3}$$

For example, personal financial planning focuses on total savings. Therefore for the private investor who makes plans on his retirement, the horizon is of the order of several years and the objective is the final absolute wealth at his investment horizon.

- **Relative wealth**

The investor is concerned with overperforming a reference portfolio, whose allocation we denote as $\boldsymbol{\beta}$. Therefore the objective is:

$$\Psi_{\boldsymbol{\alpha}} \equiv W_{T+\tau}(\boldsymbol{\alpha}) - \gamma(\boldsymbol{\alpha}) W_{T+\tau}(\boldsymbol{\beta}). \tag{5.4}$$

The function γ is a normalization factor such that at the time the investment decision is made the reference portfolio and the allocation have the same value:

$$\gamma(\boldsymbol{\alpha}) \equiv \frac{w_T(\boldsymbol{\alpha})}{w_T(\boldsymbol{\beta})}. \tag{5.5}$$

In this case the explicit expression of the objective in terms of the allocation $\boldsymbol{\alpha}$ reads:

$$\Psi_{\boldsymbol{\alpha}} \equiv \boldsymbol{\alpha}'\mathbf{K}\mathbf{P}_{T+\tau}. \tag{5.6}$$

The constant matrix $\mathbf{K}$ in this expression is defined as follows:

$$\mathbf{K} \equiv \mathbf{I}_N - \frac{\mathbf{p}_T\boldsymbol{\beta}'}{\boldsymbol{\beta}'\mathbf{p}_T}, \tag{5.7}$$

where $\mathbf{I}_N$ is the identity matrix.

For example, mutual fund managers are evaluated every year against a benchmark that defines the fund's style. Therefore for mutual fund managers the horizon is one year and the objective is relative wealth with respect to the benchmark fund.

- **Net profits**

According to *prospect theory* some investors are more concerned with changes in wealth than with the absolute value of wealth, see Kahneman and Tversky (1979). Therefore the objective becomes:

$$\Psi_{\boldsymbol{\alpha}} \equiv W_{T+\tau}(\boldsymbol{\alpha}) - w_T(\boldsymbol{\alpha}). \tag{5.8}$$

The explicit expression of the objective in terms of the allocation reads in this case:

$$\Psi_{\boldsymbol{\alpha}} \equiv \boldsymbol{\alpha}'(\mathbf{P}_{T+\tau} - \mathbf{p}_T). \tag{5.9}$$

For example, traders focus on their daily *profit and loss* (*P&L*). Therefore for a trader the investment horizon is one day and the net profits are his objective.

Notice that, in all its specifications, the objective is a linear function of the allocation and of a *market vector*:

$$\Psi_{\boldsymbol{\alpha}} = \boldsymbol{\alpha}'\mathbf{M}. \tag{5.10}$$

The market vector $\mathbf{M}$ is a simple invertible affine transformation of the market prices at the investment horizon:

$$\mathbf{M} \equiv \mathbf{a} + \mathbf{B}\mathbf{P}_{T+\tau}, \tag{5.11}$$

where $\mathbf{a}$ is a suitable conformable vector and $\mathbf{B}$ is a suitable conformable invertible matrix. Indeed, from (5.3) the market vector for the absolute wealth objective follows from the choice:

$$\mathbf{a} \equiv \mathbf{0}, \quad \mathbf{B} \equiv \mathbf{I}_N; \tag{5.12}$$

from (5.6) the market vector for the relative wealth objective follows from the choice:

$$\mathbf{a} \equiv \mathbf{0}, \quad \mathbf{B} \equiv \mathbf{K}, \tag{5.13}$$

where $\mathbf{K}$ is defined in (5.7); from (5.9) the market vector for the net profits objective follows from the choice:

$$\mathbf{a} \equiv -\mathbf{p}_T, \quad \mathbf{B} \equiv \mathbf{I}_N. \tag{5.14}$$

The distribution of $\mathbf{M}$ can be easily computed from the distribution of the security prices $\mathbf{P}_{T+\tau}$ at the investment horizon and viceversa, see Appendix www.2.4. For instance, in terms of the characteristic function we obtain:

$$\phi_{\mathbf{M}}(\boldsymbol{\omega}) = e^{i\boldsymbol{\omega}'\mathbf{a}} \phi_{\mathbf{P}}(\mathbf{B}'\boldsymbol{\omega}). \tag{5.15}$$

Therefore, with a slight abuse of terminology, we refer to both $\mathbf{M}$ and $\mathbf{P}_{T+\tau}$ as the "market vector" or simply the "market".

From (5.10) it follows that the objective as a function of the allocation is *homogeneous of first degree*:

$$\Psi_{\lambda\boldsymbol{\alpha}} = \lambda \Psi_{\boldsymbol{\alpha}}; \tag{5.16}$$

and *additive*:

$$\Psi_{\boldsymbol{\alpha}+\boldsymbol{\beta}} = \Psi_{\boldsymbol{\alpha}} + \Psi_{\boldsymbol{\beta}}. \tag{5.17}$$

These properties allow to build and compare objectives that refer to complex portfolios of securities.

If the markets were deterministic, the investor could compute the objective relative to a given allocation as a deterministic function of that allocation, and thus he would choose the allocation that gives rise to the largest value of the objective.

For example, assume that the investor's objective is final wealth, i.e. (5.3). Suppose that the market prices grew linearly:

$$\mathbf{P}_{T+t} = \operatorname{diag}(\mathbf{p}_T)\,\mathbf{h}t, \tag{5.18}$$

where $\mathbf{h}$ is a constant vector. Then from (5.12) the market vector would read:

$$\mathbf{M} \equiv \mathbf{P}_{T+\tau} = \operatorname{diag}(\mathbf{p}_T)\,\mathbf{h}\tau. \tag{5.19}$$

Consequently, the investor would allocate all his money in the asset that performs the best over the investment horizon, which corresponds to the largest entry in the vector $\mathbf{h}$.

Instead, the market prices at the investment horizon are stochastic and therefore the market vector is a random variable, and so is the investor's objective.

For example, consider normally distributed market prices:

$$\mathbf{P}_{T+\tau} \sim \mathrm{N}(\boldsymbol{\mu}, \boldsymbol{\Sigma}). \tag{5.20}$$

If the investor focuses on final wealth, from (5.12) the market vector reads:

$$\mathbf{M} \equiv \mathbf{P}_{T+\tau}. \tag{5.21}$$

Thus the objective (5.10) is normally distributed:

$$\Psi_{\boldsymbol{\alpha}} \sim \mathrm{N}\left(\mu_{\boldsymbol{\alpha}}, \sigma_{\boldsymbol{\alpha}}^2\right), \tag{5.22}$$

where

$$\mu_{\boldsymbol{\alpha}} \equiv \boldsymbol{\mu}'\boldsymbol{\alpha}, \qquad \sigma_{\boldsymbol{\alpha}}^2 \equiv \boldsymbol{\alpha}'\boldsymbol{\Sigma}\boldsymbol{\alpha}. \tag{5.23}$$

Since the objective is a random variable we need some tools to figure out in which sense a random variable is "larger" or is "better" than another one. We devote the rest of this chapter to this purpose.

We conclude this section remarking that the computation of the exact distribution of the objective $\Psi_{\boldsymbol{\alpha}} = \boldsymbol{\alpha}'\mathbf{M}$ is in general a formidable task. Indeed, the distribution of the market is easily obtained *once* the distribution of the prices is known, see (5.15). Nevertheless, the distribution of the prices is very hard to compute in general. Here we mention the *gamma approximation* of the investor's objective, a quite general approximate solution which has found a wide range of applications.

Consider the generic second-order approximation (3.108) for the prices of the securities in terms of the underlying market invariants $\mathbf{X}$, which we report here:

$$P_{T+\tau}^{(n)} \approx g^{(n)}(\mathbf{0}) + \mathbf{X}' \left.\frac{\partial g^{(n)}}{\partial \mathbf{x}}\right|_{\mathbf{x}=\mathbf{0}} + \frac{1}{2}\mathbf{X}' \left.\frac{\partial^2 g^{(n)}}{\partial \mathbf{x} \partial \mathbf{x}'}\right|_{\mathbf{x}=\mathbf{0}} \mathbf{X}, \tag{5.24}$$

where $n = 1, \ldots, N$. As we show in Appendix www.5.1, the investor's objective can be approximated by a quadratic function of the invariants:

$$\Psi_{\boldsymbol{\alpha}} \approx \Xi_{\boldsymbol{\alpha}} \equiv \theta_{\boldsymbol{\alpha}} + \boldsymbol{\Delta}_{\boldsymbol{\alpha}}'\mathbf{X} + \frac{1}{2}\mathbf{X}'\boldsymbol{\Gamma}_{\boldsymbol{\alpha}}\mathbf{X}, \tag{5.25}$$

where

$$\theta_{\boldsymbol{\alpha}} \equiv \sum_{n=1}^{N} \alpha_n a_n + \sum_{n,m=1}^{N} \alpha_n B_{nm} g^{(m)}(\mathbf{0}) \tag{5.26}$$

$$\boldsymbol{\Delta}_{\boldsymbol{\alpha}} \equiv \sum_{n,m=1}^{N} \alpha_n B_{nm} \left.\frac{\partial g^{(m)}}{\partial \mathbf{x}}\right|_{\mathbf{x}=\mathbf{0}} \tag{5.27}$$

$$\boldsymbol{\Gamma}_{\boldsymbol{\alpha}} \equiv \sum_{n,m=1}^{N} \alpha_n B_{nm} \left.\frac{\partial^2 g^{(m)}}{\partial \mathbf{x} \partial \mathbf{x}'}\right|_{\mathbf{x}=\mathbf{0}}; \tag{5.28}$$

and $\mathbf{a}$ and $\mathbf{B}$ are the coefficients (5.11) that determine the market.

In general the market invariants are sufficiently symmetric to be modeled appropriately by symmetrical distributions, such as elliptical or symmetric stable distributions, see (3.22), or (3.37), or (3.55), and comments thereafter.

Under this hypothesis it is possible to compute the distribution of the approximate objective (5.25) as represented by its characteristic function. In particular, assume that the invariants are normally distributed:

$$\mathbf{X} \sim \mathrm{N}(\boldsymbol{\mu}, \boldsymbol{\Sigma}). \tag{5.29}$$

Then we prove in Appendix www.5.1 that the characteristic function of the approximate objective (5.25) reads:

$$\begin{aligned}\phi_{\Xi_{\boldsymbol{\alpha}}}(\omega) = & \left|\mathbf{I}_K - i\omega\boldsymbol{\Gamma}_{\boldsymbol{\alpha}}\boldsymbol{\Sigma}\right|^{-\frac{1}{2}} e^{i\omega\left(\theta_{\boldsymbol{\alpha}}+\boldsymbol{\Delta}_{\boldsymbol{\alpha}}'\boldsymbol{\mu}+\frac{1}{2}\boldsymbol{\mu}'\boldsymbol{\Gamma}_{\boldsymbol{\alpha}}\boldsymbol{\mu}\right)} \\ & e^{-\frac{1}{2}[\boldsymbol{\Delta}_{\boldsymbol{\alpha}}+\boldsymbol{\Gamma}_{\boldsymbol{\alpha}}\boldsymbol{\mu}]'\boldsymbol{\Sigma}(\mathbf{I}_K-i\omega\boldsymbol{\Gamma}_{\boldsymbol{\alpha}}\boldsymbol{\Sigma})^{-1}[\boldsymbol{\Delta}_{\boldsymbol{\alpha}}+\boldsymbol{\Gamma}_{\boldsymbol{\alpha}}\boldsymbol{\mu}]},\end{aligned} \tag{5.30}$$

where the explicit dependence on the allocation $\boldsymbol{\alpha}$ is easily recovered from (5.26)-(5.28).

5.2 Stochastic dominance

In this section we present the stochastic dominance approach to assess the distribution of the investor's objective. For further references, see Ingersoll (1987), Levy (1998) and Yamai and Yoshiba (2002).

Suppose that the investor can choose between an allocation $\boldsymbol{\alpha}$ that gives rise to the objective $\Psi_{\boldsymbol{\alpha}}$ and an allocation $\boldsymbol{\beta}$ that gives rise to the objective $\Psi_{\boldsymbol{\beta}}$. All the information necessary to make a decision as to which allocation is more advantageous is contained in the joint distribution of $\Psi_{\boldsymbol{\alpha}}$ and $\Psi_{\boldsymbol{\beta}}$.

When confronted with two different objectives $\Psi_{\boldsymbol{\alpha}}$ and $\Psi_{\boldsymbol{\beta}}$, it is natural to first check whether in all possible scenarios one objective is larger than the other, see the left plot in Figure 5.1. When this happens, the objective $\Psi_{\boldsymbol{\alpha}}$, or the allocation $\boldsymbol{\alpha}$, is said to *strongly dominate* the objective $\Psi_{\boldsymbol{\beta}}$, or the allocation $\boldsymbol{\beta}$:

$$\text{strong dom.: } \Psi_{\boldsymbol{\alpha}} \geq \Psi_{\boldsymbol{\beta}} \text{ in all scenarios.} \tag{5.31}$$

In other words, strong dominance arises when the difference of the objectives relative to two allocations is a positive random variable. Therefore, an equivalent definition of strong dominance reads as follows in terms of the cumulative distribution function of the difference of the objectives:

$$\text{strong dom.: } F_{\Psi_{\boldsymbol{\alpha}}-\Psi_{\boldsymbol{\beta}}}(0) \equiv \mathbb{P}\{\Psi_{\boldsymbol{\alpha}} - \Psi_{\boldsymbol{\beta}} \leq 0\} = 0. \tag{5.32}$$

We call strong dominance also *order zero dominance*, for reasons that will become clear below.

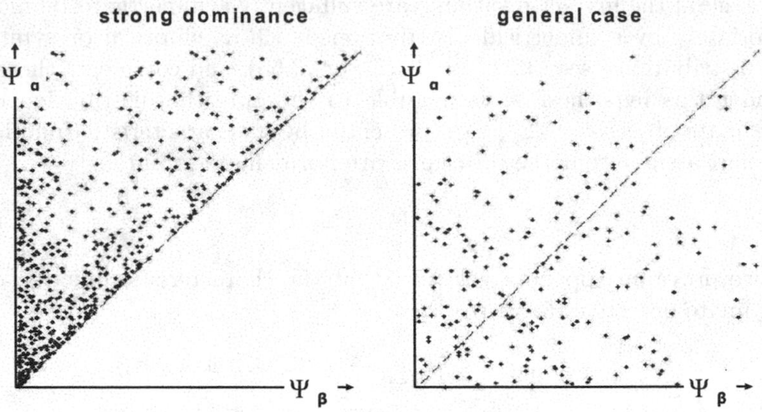

Fig. 5.1. Strong dominance

For example, suppose that the objective relative to one allocation has a chi-square distribution with two degrees of freedom:

$$\Psi_{\alpha} \sim \chi_2^2; \tag{5.33}$$

and that the objective relative to another allocation has a chi-square distribution with one degree of freedom:

$$\Psi_{\beta} \sim \chi_1^2. \tag{5.34}$$

Assume that $\Psi_{\alpha} = \Psi_{\beta} + Y$ where $Y \sim \chi_1^2$ is independent of Ψ_{β}. Then Ψ_{α} strongly dominates Ψ_{β}. With this example we generated the plot on the left in Figure 5.1.

Nevertheless, strong dominance cannot be a general criterion to evaluate allocations.

In the first place, strong dominance never takes place, for if it did, *arbitrage* opportunities, i.e. "free lunches" would arise. Instead, in general an allocation $\boldsymbol{\alpha}$ gives rise to an objective Ψ_{α} that in some scenarios is larger and in some scenarios is smaller than the objective Ψ_{β} stemming from a different allocation $\boldsymbol{\beta}$, see the plot on the right hand side in Figure 5.1.

Secondly, the definition of strong dominance relies on the joint distribution of the two objectives Ψ_{α} and Ψ_{β}, which is necessary to compute the distribution of their difference (5.32). Nevertheless, the two allocations are mutually exclusive, i.e. the investor either chooses one or the other. Therefore the interplay of the two allocations should not have an effect on the decision and thus any criterion to rank allocations should focus on comparing the marginal distributions of the objectives.

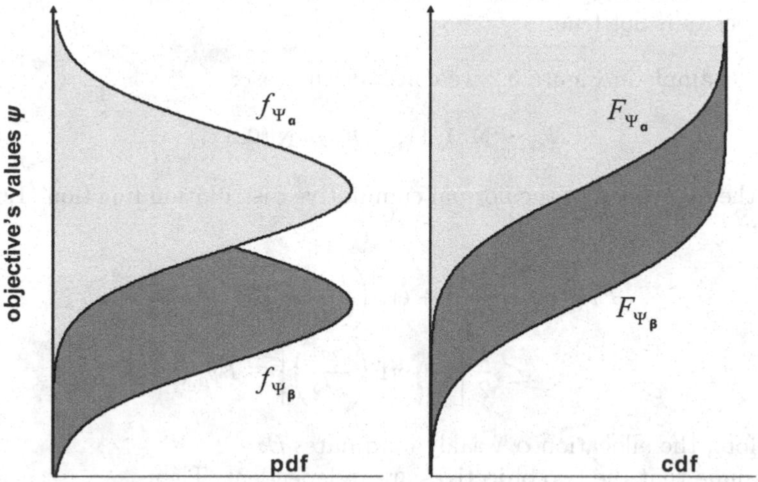

Fig. 5.2. Weak dominance

A different approach that compares two allocations only in terms of their marginal distributions is inspired from a plot of the possible values of the objectives on the vertical axis and the respective probability density functions on the horizontal axis as in Figure 5.2. We would be prone to choose an allocation $\boldsymbol{\alpha}$ over another allocation $\boldsymbol{\beta}$ if the probability density function of the ensuing objective were concentrated around larger values than for the other allocation. This condition is expressed more easily in terms of the cumulative distribution function. The objective $\Psi_{\boldsymbol{\alpha}}$, or the allocation $\boldsymbol{\alpha}$, is said to *weakly dominate* the objective $\Psi_{\boldsymbol{\beta}}$, or the allocation $\boldsymbol{\beta}$, if the following condition holds true (notice the "wrong" direction of the inequality):

$$\text{weak dom.: } F_{\Psi_{\boldsymbol{\alpha}}}(\psi) \leq F_{\Psi_{\boldsymbol{\beta}}}(\psi) \text{ for all } \psi \in (-\infty, +\infty). \tag{5.35}$$

Weak dominance is also called *first-order dominance*, for reasons to become clear below.

Comparing Figure 5.2 with Figure 1.2 we obtain a more intuitive equivalent expression for weak dominance in terms of the inverse of the cumulative distribution function, namely the quantile. The objective $\Psi_{\boldsymbol{\alpha}}$, or the allocation $\boldsymbol{\alpha}$, is said to weakly dominate the objective $\Psi_{\boldsymbol{\beta}}$, or the allocation $\boldsymbol{\beta}$, if the following condition holds true:

$$\text{weak dom: } Q_{\Psi_{\boldsymbol{\alpha}}}(p) \geq Q_{\Psi_{\boldsymbol{\beta}}}(p) \text{ for all } p \in (0,1). \tag{5.36}$$

This representation is more intuitive than (5.35), due to the "correct" direction of the inequality.

Weak dominance is not as restrictive a condition as strong dominance: if an allocation strongly dominates another one it also weakly dominates it, but the opposite is not true.

For example in Figure 5.2 we consider the case:

$$\Psi_{\boldsymbol{\alpha}} \sim \mathrm{N}(1,1), \quad \Psi_{\boldsymbol{\beta}} \sim \mathrm{N}(0,1). \tag{5.37}$$

From the expression of the normal cumulative distribution function (1.68) we obtain:

$$\begin{aligned} F_{\Psi_{\boldsymbol{\alpha}}}(\psi) &\equiv \frac{1}{2}\left[1+\operatorname{erf}\left(\frac{\psi-1}{\sqrt{2}}\right)\right] \\ &\leq \frac{1}{2}\left[1+\operatorname{erf}\left(\frac{\psi}{\sqrt{2}}\right)\right] \equiv F_{\Psi_{\boldsymbol{\beta}}}(\psi). \end{aligned} \tag{5.38}$$

Therefore, the allocation $\boldsymbol{\alpha}$ weakly dominates $\boldsymbol{\beta}$.

Assume that the two objectives are independent. Then

$$\Psi_{\boldsymbol{\alpha}} - \Psi_{\boldsymbol{\beta}} \sim \mathrm{N}(1,2), \tag{5.39}$$

or equivalently

$$F_{\Psi_{\boldsymbol{\alpha}}-\Psi_{\boldsymbol{\beta}}}(0) = \frac{1}{2}\left[1+\operatorname{erf}\left(-\frac{1}{2}\right)\right] > 0. \tag{5.40}$$

Therefore, from (5.32) the allocation $\boldsymbol{\alpha}$ does not strongly dominate the allocation $\boldsymbol{\beta}$.

A third way to express weak dominance is the following. Suppose that v is the realization of a random variable V which spans the unit interval, such as the standard uniform distribution:

$$V \sim \mathrm{U}([0,1]). \tag{5.41}$$

Applying the cumulative distribution function $F_{\Psi_{\boldsymbol{\beta}}}$ to both sides of the inequality in (5.36) we obtain:

$$\text{weak dom.: } F_{\Psi_{\boldsymbol{\beta}}}(Q_{\Psi_{\boldsymbol{\alpha}}}(V)) \geq V \text{ in all scenarios.} \tag{5.42}$$

Comparing this expression with the definition of strong dominance (5.31) we can say that the objective $\Psi_{\boldsymbol{\alpha}}$, or the allocation $\boldsymbol{\alpha}$, weakly dominates the objective $\Psi_{\boldsymbol{\beta}}$, or the allocation $\boldsymbol{\beta}$, if the distribution on the left hand side in (5.42) strongly dominates the uniform distribution on the right hand side, see Figure 5.1.

In particular, from (2.26) the grade of the objective is uniformly distributed on the unit interval: $F_{\Psi_{\boldsymbol{\alpha}}}(\Psi_{\boldsymbol{\alpha}}) \stackrel{d}{=} V$; and from (2.27) the quantile of the

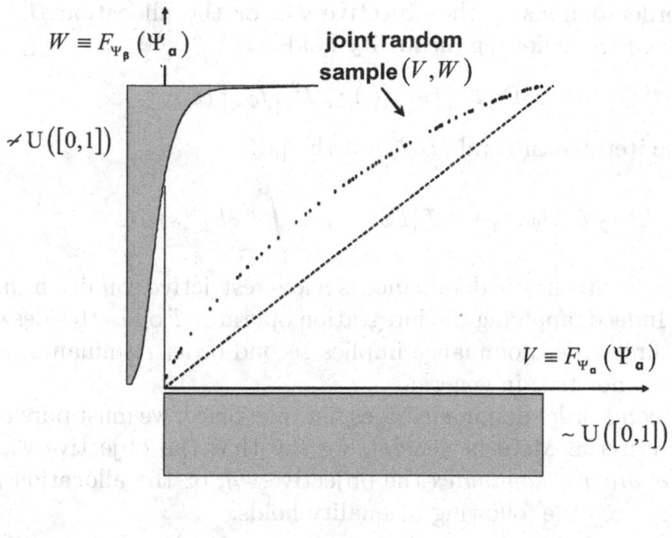

Fig. 5.3. Weak dominance in terms of strong dominance

grade is distributed as the objective: $\Psi_{\alpha} \overset{d}{=} Q_{\Psi_{\alpha}}(V)$. Therefore from (5.42) an allocation $\boldsymbol{\alpha}$ weakly dominates an allocation $\boldsymbol{\beta}$ if $F_{\Psi_{\beta}}(\Psi_{\alpha})$ strongly dominates $F_{\Psi_{\alpha}}(\Psi_{\alpha})$.

Figure 5.3 refers to our example (5.37), where the allocation $\boldsymbol{\alpha}$ weakly dominates the allocation $\boldsymbol{\beta}$. Indeed, as in Figure 5.1, all the joint outcomes of $F_{\Psi_{\alpha}}(\Psi_{\alpha})$ and $F_{\Psi_{\beta}}(\Psi_{\alpha})$ lie above the diagonal and thus $F_{\Psi_{\beta}}(\Psi_{\alpha})$ strongly dominates $F_{\Psi_{\alpha}}(\Psi_{\alpha})$.

Although weak, or first-order, dominance is not a criterion as restrictive as strong dominance, even first-order dominance hardly ever occurs. To cope with this problem we need to introduce even weaker types of dominance, such as *second-order stochastic dominance* (SSD).

The rationale behind second-order stochastic dominance is the following: we would be prone to choose the distribution Ψ_{α} over the distribution Ψ_{β} if, for any given benchmark level ψ of the objective, the events where Ψ_{α} underperforms the benchmark level are less harmful than for Ψ_{β}. In formulas, for all $\psi \in (-\infty, +\infty)$ the following inequality must hold:

$$\text{SSD: } \mathrm{E}\left\{-(\Psi_{\alpha} - \psi)^{-}\right\} \geq \mathrm{E}\left\{-(\Psi_{\beta} - \psi)^{-}\right\}, \tag{5.43}$$

where the "minus" denotes the negative part. If (5.43) holds true, Ψ_{α} is said to second-order dominate Ψ_{β}.

An equivalent formulation of second-order stochastic dominance is the following, see Ingersoll (1987) for a proof. The objective Ψ_{α}, or the allocation

$\boldsymbol{\alpha}$, second-order dominates the objective $\Psi_{\boldsymbol{\beta}}$, or the allocation $\boldsymbol{\beta}$, if for all $\psi \in (-\infty, +\infty)$ the following inequality holds:

$$\text{SSD: } \mathcal{I}^2 \left[f_{\Psi_{\alpha}} \right] (\psi) \leq \mathcal{I}^2 \left[f_{\Psi_{\beta}} \right] (\psi) , \tag{5.44}$$

where $\mathcal{I}^2$ the iterated integral $(B.27)$ of the pdf:

$$\mathcal{I}^2 \left[f_{\Psi} \right] (\psi) \equiv \mathcal{I} \left[F_{\Psi} \right] (\psi) \equiv \int_{-\infty}^{\psi} F_{\Psi} (s) \, ds. \tag{5.45}$$

Second-order stochastic dominance is a less restrictive condition than weak dominance. Indeed, applying the integration operator $\mathcal{I}$ on both sides of (5.35) we see that first-order dominance implies second-order dominance, although the opposite is not true in general.

If even second-order dominance does not take place, we must pursue weaker and weaker criteria. More in general, we say that the objective $\Psi_{\boldsymbol{\alpha}}$, or the allocation $\boldsymbol{\alpha}$, *order-q dominates* the objective $\Psi_{\boldsymbol{\beta}}$, or the allocation $\boldsymbol{\beta}$, if for all $\psi \in (-\infty, +\infty)$ the following inequality holds:

$$q\text{-dom.: } \mathcal{I}^q \left[f_{\Psi_{\alpha}} \right] (\psi) \leq \mathcal{I}^q \left[f_{\Psi_{\beta}} \right] (\psi) . \tag{5.46}$$

Notice that first-order dominance (5.35) and second-order dominance (5.44) are particular cases of (5.46).

Applying the integration operator to both sides of (5.46) we see that order q dominance implies order $(q+1)$ dominance, although the opposite is not true in general.

Recalling that we renamed strong dominance as zero-order dominance, we can write all the above implications in compact form as follows:

$$\text{0-dom.} \Rightarrow \text{1-dom.} \Rightarrow \cdots \Rightarrow q\text{-dom.} \tag{5.47}$$

Therefore in theory we only need to check that one allocation dominates another for a certain degree, as dominance for higher degrees follows. In practice the stochastic dominance approach to evaluating allocations displays major drawbacks.

First of all, the intuitive meaning behind dominance of orders higher than two is not evident.

Secondly, the computation of the generic q-th cumulative distribution is not practically feasible in most situations.

Finally, but most importantly, there is no guarantee that there exists an order q such that any two portfolios can be ranked: consequently, the investor might not be able to rank his potential investments and thus choose an allocation. Intuitively, this happens because the objective is stochastic: a deterministic variable can be represented by a point on the real line, whereas a random variable is represented by a function, such as the cumulative distribution function. Functions are infinite-dimensional vectors, i.e. points in an infinite-dimensional space, see Appendix B. In dimensions higher than one there exists no natural way to order points.

5.3 Satisfaction

The main drawback of the dominance approach to ranking two allocations $\boldsymbol{\alpha}$ and $\boldsymbol{\beta}$ is that two generic allocations might not be comparable, in the sense that neither of the respective objectives dominates the other.

To solve this problem, we summarize all the features of a given allocation $\boldsymbol{\alpha}$ into one single number $\mathcal{S}$ that indicates the respective degree of satisfaction:

$$\boldsymbol{\alpha} \mapsto \mathcal{S}(\boldsymbol{\alpha}). \tag{5.48}$$

The investor will then choose the allocation that corresponds to the highest degree of satisfaction.

For example, the expected value of the investor's objective is a number that depends on the allocation:

$$\boldsymbol{\alpha} \mapsto \mathcal{S}(\boldsymbol{\alpha}) \equiv \mathrm{E}\{\Psi_{\boldsymbol{\alpha}}\}. \tag{5.49}$$

As such, it is an index of satisfaction.

Since there exists no univocal way to summarize all the information contained in an allocation into one number, we discuss here potential features that an index of satisfaction may display, see also Frittelli and Rosazza Gianin (2002).

- **Money-equivalence**

An index of satisfaction is *money-equivalent* if it is naturally measured in units of money. This is a desirable feature, as money is "the" measure in finance.

Furthermore, money-equivalence is necessary for consistence. Indeed, consider an investor with a given objective such as absolute wealth as in (5.3) or relative wealth, as in (5.4), or net profits, as in (5.8), or possibly other specifications. In all the specifications the objective is measured in terms of money. Since in a deterministic environment the most natural index of satisfaction is the objective, it is intuitive to require that a generic index of satisfaction be measured in the same units as the objective.

For example, the expected value of the objective (5.49) has the same dimension as the objective, which is money, and thus it is a money-equivalent index of satisfaction.

The concept of money-equivalence contrasts that of *scale-invariance*, or *homogeneity of degree zero*. Scale invariant indices of satisfaction are dimensionless measures that satisfy the following relation:

$$\mathcal{S}(\lambda\boldsymbol{\alpha}) = \mathcal{S}(\boldsymbol{\alpha}), \text{ for all } \lambda > 0. \tag{5.50}$$

Scale invariant indices of satisfaction provide a tool to normalize and evaluate portfolios that differ in size.

Although in the sequel we will be concerned mainly with money-equivalent indices of satisfaction, we present the most notable scale invariant index of satisfaction, namely the *Sharpe ratio*, which is defined as follows:

$$\mathrm{SR}(\boldsymbol{\alpha}) \equiv \frac{\mathrm{E}\{\Psi_{\boldsymbol{\alpha}}\}}{\mathrm{Sd}\{\Psi_{\boldsymbol{\alpha}}\}}. \tag{5.51}$$

The rationale behind the Sharpe ratio is the following: a high standard deviation is a drawback if the expected value of the objective is positive, because it adds uncertainty to a potentially satisfactory outcome.

- **Estimability**

An index of satisfaction is *estimable* if the satisfaction associated with a generic allocation $\boldsymbol{\alpha}$ is fully determined by the marginal distribution of the investor's objective $\Psi_{\boldsymbol{\alpha}}$, which can be absolute wealth, relative wealth, net gains, etc. In other words, two allocations that give rise to two objectives with the same distribution are fully equivalent for the investor.

For example, an allocation of a thousand dollars in cash and a thousand dollars in a stock of a company quoted on the NYSE is considered fully equivalent to an investment of a thousand dollars in a currency pegged to the dollar and a thousand dollars in the same stock as quoted on the DAX.

In other words, when an index of the satisfaction is estimable, the satisfaction associated with the allocation $\boldsymbol{\alpha}$ is a functional of any of the equivalent representations of the distribution of the objective $\Psi_{\boldsymbol{\alpha}}$, namely the probability density function $f_{\Psi_{\boldsymbol{\alpha}}}$, the cumulative distribution function $F_{\Psi_{\boldsymbol{\alpha}}}$, or the characteristic function $\phi_{\Psi_{\boldsymbol{\alpha}}}$. Therefore, in order to be estimable, the simple map (5.48) must expand into the following chain of maps:

$$\boldsymbol{\alpha} \mapsto \Psi_{\boldsymbol{\alpha}} \mapsto \left(f_{\Psi_{\boldsymbol{\alpha}}}, F_{\Psi_{\boldsymbol{\alpha}}}, \phi_{\Psi_{\boldsymbol{\alpha}}}\right) \mapsto \mathcal{S}(\boldsymbol{\alpha}). \tag{5.52}$$

The concept of estimability is known in the financial literature also under the name of *law invariance*, see Kusuoka (2001).

For example, the expected value is a functional of the probability density function of the objective:

$$f_{\Psi} \mapsto \mathrm{E}\{\Psi\} \equiv \int_{\mathbb{R}} \psi f_{\Psi}(\psi)\, d\psi. \tag{5.53}$$

Therefore, the expected value is an estimable index of satisfaction:

$$\boldsymbol{\alpha} \mapsto \Psi_{\boldsymbol{\alpha}} \mapsto f_{\Psi_{\boldsymbol{\alpha}}} \mapsto \mathrm{E}\{\Psi_{\boldsymbol{\alpha}}\}. \tag{5.54}$$

- **Sensibility**

Due to the non-satiation principle underlying the investor's objective, if an objective Ψ_{α} is larger than an objective Ψ_{β} in all scenarios, then the satisfaction that the investors derive from Ψ_{α} should be greater than the satisfaction they derive from Ψ_{β}:

$$\Psi_{\alpha} \geq \Psi_{\beta} \text{ in all scenarios } \Rightarrow \mathcal{S}(\boldsymbol{\alpha}) \geq \mathcal{S}(\boldsymbol{\beta}). \tag{5.55}$$

We call this feature *sensibility* because it is the minimum requirement that any index of satisfaction needs to verify. Sensibility is also called *monotonicity* in the financial literature, see Artzner, Delbaen, Eber, and Heath (1999).

For example, the expected value of the objective (5.49) is a sensible index of satisfaction, since it trivially satisfies:

$$\Psi_{\alpha} \geq \Psi_{\beta} \text{ in all scenarios } \Rightarrow \mathrm{E}\{\Psi_{\alpha}\} \geq \mathrm{E}\{\Psi_{\beta}\}. \tag{5.56}$$

By comparing the sensibility condition (5.55) with (5.31) we notice that in order for the index of satisfaction to be sensible it must be consistent with strong dominance. In other words, if an allocation $\boldsymbol{\alpha}$ happens to strongly dominate an allocation $\boldsymbol{\beta}$, any sensible criterion should prefer the former to the latter.

Although we cannot rely on strong dominance as a criterion to compare allocations, we should always make sure that any possible criterion is consistent with strong dominance.

- **Consistence with stochastic dominance**

Sensibility stems from the intuitive non-satiation argument that the larger in a strong sense the investor's objective, the more satisfied the investor. Similarly, we can apply the non-satiation argument to weaker concepts of dominance.

For instance, if the marginal distribution of the objective Ψ_{α} of an allocation $\boldsymbol{\alpha}$ is shifted upward with respect to the marginal distribution of the objective Ψ_{β} of an allocation $\boldsymbol{\beta}$ as in Figure 5.2, then the satisfaction from Ψ_{α} should be greater than the satisfaction from Ψ_{β}. In formulas:

$$Q_{\Psi_{\alpha}}(p) \geq Q_{\Psi_{\beta}}(p) \text{ for all } p \in (0,1) \Rightarrow \mathcal{S}(\boldsymbol{\alpha}) \geq \mathcal{S}(\boldsymbol{\beta}). \tag{5.57}$$

By comparing this expression with (5.36) we realize that an index of satisfaction for which the above relation holds is consistent with weak dominance.

For example, the expected value of the objective (5.49) is consistent with weak dominance. Indeed with a change of variable we can verify the following equality:

$$\mathrm{E}\{\Psi\} \equiv \int_{-\infty}^{+\infty} \psi f_{\psi}(\psi)\, d\psi = \int_{0}^{1} Q_{\Psi}(u)\, du. \tag{5.58}$$

Therefore (5.57) is satisfied.

Consistence with weak dominance is a stronger requirement on the index of satisfaction than consistence with strong dominance: if an index of satisfaction is consistent with weak dominance it is sensible, i.e. it is consistent with strong dominance, but the opposite is not true in general.

Nevertheless, in the case of estimable indices of satisfaction the two statements are equivalent. Indeed, in Appendix www.5.2 we follow a personal communication by D. Tasche to prove:

$$\mathcal{S} \text{ estimable} + \mathcal{S} \text{ sensible } \Rightarrow \mathcal{S} \text{ weak dom. consistent.} \tag{5.59}$$

In general, an index of satisfaction $\mathcal{S}$ is consistent with q-th order dominance if, whenever an allocation $\boldsymbol{\alpha}$ dominates an allocation $\boldsymbol{\beta}$ at order q, then the satisfaction from $\boldsymbol{\alpha}$ is larger than the satisfaction from $\boldsymbol{\beta}$:

$$\Psi_{\boldsymbol{\alpha}} \; q\text{-dom. } \Psi_{\boldsymbol{\beta}} \Rightarrow \mathcal{S}(\boldsymbol{\alpha}) \geq \mathcal{S}(\boldsymbol{\beta}). \tag{5.60}$$

In particular, sensibility corresponds to consistency with zero-order dominance, see (5.55); and consistency with weak dominance corresponds to consistency with first-order dominance, see (5.57).

Given the sequence of implications (5.47) on the degrees of dominance, the reverse sequence holds for the consistency of an index of satisfaction with the degree of dominance:

$$\begin{aligned} q\text{-dom. consistence} \Rightarrow \cdots \Rightarrow &\ 1\text{-dom. consistence} \\ \Rightarrow &\ 0\text{-dom. consistence.} \end{aligned} \tag{5.61}$$

For example, the expected value of the objective (5.49) is consistent with second-order dominance. We prove this result in a broader context in Section 5.4. Therefore, it is consistent with weak dominance (first-order dominance), and therefore it is sensible, i.e. it is consistent with order-zero dominance.

- **Constancy**

If the markets were deterministic, the non-satiation principle would imply that the investor's objective, no matter whether it is absolute wealth, or relative wealth, or net profits, would serve as a suitable index of satisfaction, see p. 241.

Therefore, if there exists an allocation $\mathbf{b}$ that yields a deterministic objective $\psi_{\mathbf{b}}$, it is reasonable to require that the index coincide with the objective:

$$\Psi_{\mathbf{b}} \equiv \psi_{\mathbf{b}} \Rightarrow \mathcal{S}(\mathbf{b}) = \psi_{\mathbf{b}}. \tag{5.62}$$

This feature is called *constancy*.

For example, the expected value of the objective (5.49) is a constant index of satisfaction:

$$\Psi_{\mathbf{b}} \equiv \psi_{\mathbf{b}} \Rightarrow \mathrm{E}\left\{\Psi_{\mathbf{b}}\right\} = \psi_{\mathbf{b}}. \tag{5.63}$$

As an example of a deterministic objective, consider the case where the objective is absolute wealth and the allocation **b** is an investment in zero-coupon bonds that expire at the investment horizon.

- **Positive homogeneity**

The investor's objective is positive homogeneous of degree one, see (5.16). In other words, if we rescale the allocation by a given positive factor the objective is rescaled by the same factor:

$$\Psi_{\lambda\boldsymbol{\alpha}} = \lambda\Psi_{\boldsymbol{\alpha}}, \quad \text{for all } \lambda \geq 0. \tag{5.64}$$

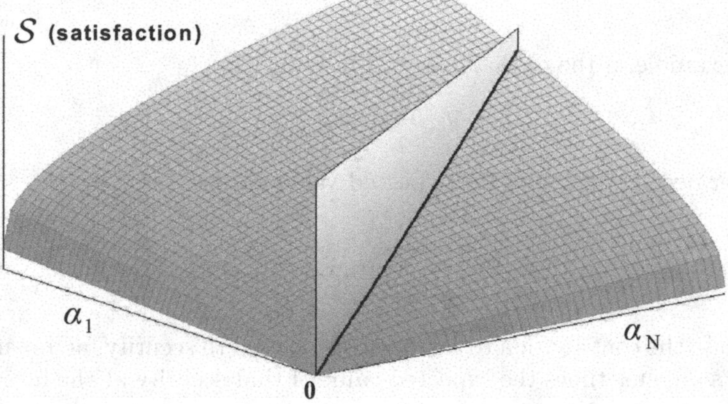

Fig. 5.4. Positive homogeneity of satisfaction index

It would be intuitive if an index of satisfaction shared the same property: loosely speaking, an index of satisfaction is homogeneous if doubling the investment makes the investor twice as happy. More precisely, a satisfaction index is *positive homogenous* (of degree one) if rescaling the allocation by a generic positive factor λ implies that satisfaction is rescaled by the same factor:

$$\mathcal{S}(\lambda\boldsymbol{\alpha}) = \lambda\mathcal{S}(\boldsymbol{\alpha}), \quad \text{for all } \lambda \geq 0. \tag{5.65}$$

It is easy to interpret positive homogeneity geometrically, see Figure 5.4. An index of satisfaction is positive homogenous if satisfaction grows linearly in any radial direction stemming from the origin of the allocation space.

For example, the expected value of the objective (5.49) is a positive homogeneous index of satisfaction. Indeed from (5.64) we obtain:

$$\mathrm{E}\{\Psi_{\lambda\boldsymbol{\alpha}}\} = \mathrm{E}\{\lambda\Psi_{\boldsymbol{\alpha}}\} = \lambda\,\mathrm{E}\{\Psi_{\boldsymbol{\alpha}}\}. \tag{5.66}$$

Positive homogeneous functions enjoy the following special property, first discovered by Euler:

$$\mathcal{S}(\boldsymbol{\alpha}) = \sum_{n=1}^{N} \alpha_n \frac{\partial \mathcal{S}(\boldsymbol{\alpha})}{\partial \alpha_n}. \tag{5.67}$$

In the case of positive homogeneous indices of satisfaction this property has a nice interpretation: satisfaction is the sum of the contributions from each security. The contribution to satisfaction from the generic n-th security in turn is the product of the amount α_n of that security times the marginal contribution to satisfaction $\partial\mathcal{S}/\partial\alpha_n$ which that security provides. Furthermore, the vector of marginal contributions to satisfaction is scale-invariant: if the allocation is multiplied by a positive factor, the marginal contribution $\partial\mathcal{S}/\partial\alpha_n$ does not change.

For example, if the objective is absolute wealth

$$\Psi_{\boldsymbol{\alpha}} \equiv \boldsymbol{\alpha}'\mathbf{P}_{T+\tau}, \tag{5.68}$$

the Euler decomposition of the expected value yields:

$$\mathrm{E}\{\Psi_{\boldsymbol{\alpha}}\} = \sum_{n=1}^{N} \alpha_n\,\mathrm{E}\left\{P_{T+\tau}^{(n)}\right\}. \tag{5.69}$$

In this case the contribution to satisfaction of the n-th security factors into the security's amount times the expected value of that security at the investment horizon.

- **Translation invariance**

The investor's objective is not only positive homogeneous, it is also additive, see (5.17): if we add two portfolios $\boldsymbol{\alpha}$ and $\boldsymbol{\beta}$ the ensuing objective is the sum of the two separate objectives:

$$\Psi_{\boldsymbol{\alpha}+\boldsymbol{\beta}} = \Psi_{\boldsymbol{\alpha}} + \Psi_{\boldsymbol{\beta}}. \tag{5.70}$$

Since the objectives are random variables the satisfaction ensuing from the sum of two random variables can be completely unrelated to the satisfaction that the investor draws from the separate portfolios.

Nevertheless, consider an allocation $\mathbf{b}$ that yields a deterministic objective $\psi_{\mathbf{b}}$. In this case the distribution of the objective relative to the joint allocation

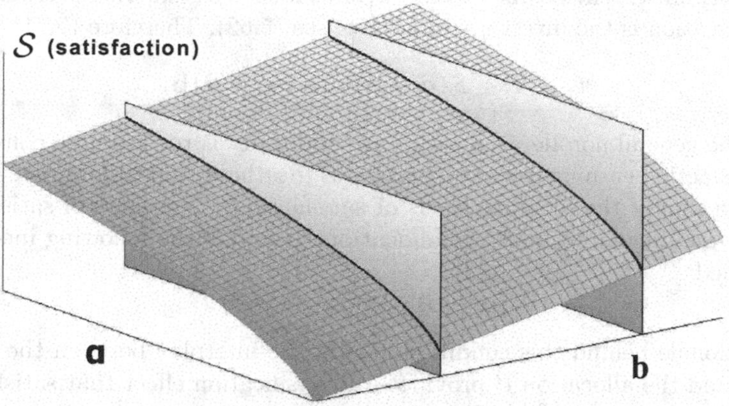

Fig. 5.5. Translation invariance of satisfaction index

$\boldsymbol{\alpha}+\mathbf{b}$ is simply a shift of the distribution of $\Psi_{\boldsymbol{\alpha}}$ by the fixed amount $\psi_{\mathbf{b}}$. If the index of satisfaction satisfies (5.62), this is the satisfaction provided by the allocation $\mathbf{b}$. Therefore, it would be intuitive if an index of satisfaction shifted by the same amount:

$$\mathcal{S}(\boldsymbol{\alpha}+\mathbf{b})=\mathcal{S}(\boldsymbol{\alpha})+\psi_{\mathbf{b}}. \tag{5.71}$$

This property is called *translation invariance*. It is easy to interpret translation invariance geometrically, see Figure 5.5.

Notice that translation invariance implies that the index of satisfaction is measured in terms of money. Without loss of generality, we can normalize the deterministic allocation to yield one unit of currency. Therefore we can restate the translation invariance property as follows:

$$\Psi_{\mathbf{b}}\equiv 1\Rightarrow\mathcal{S}(\boldsymbol{\alpha}+\lambda\mathbf{b})=\mathcal{S}(\boldsymbol{\alpha})+\lambda. \tag{5.72}$$

This expression follows from (5.71) once we take into account the positive homogeneity of the objective (5.64).

For example, the expected value of the objective (5.49) is a translation-invariant index of satisfaction. Indeed, from (5.70) and (5.66) we obtain:

$$\Psi_{\mathbf{b}}\equiv 1\Rightarrow\mathrm{E}\{\Psi_{\boldsymbol{\alpha}+\lambda\mathbf{b}}\}=\mathrm{E}\{\Psi_{\boldsymbol{\alpha}}\}+\lambda. \tag{5.73}$$

- **Sub- and super- additivity**

The translation invariance property (5.71) can be interpreted as additivity in a deterministic environment. Indeed, the index of satisfaction of a deterministic allocation is the investor's objective, see (5.62). Therefore (5.71) reads:

$$\Psi_{\mathbf{b}} \equiv 1 \Rightarrow \mathcal{S}(\boldsymbol{\alpha} + \mathbf{b}) = \mathcal{S}(\boldsymbol{\alpha}) + \mathcal{S}(\mathbf{b}). \quad (5.74)$$

In the general non-deterministic case additivity is too strong a constraint: the satisfaction ensuing from two portfolios together could be larger or smaller than the sum of the separate levels of satisfaction. A measure of satisfaction is *super-additive*[1] if for any two allocations $\boldsymbol{\alpha}$ and $\boldsymbol{\beta}$ the following inequality is satisfied:

$$\mathcal{S}(\boldsymbol{\alpha} + \boldsymbol{\beta}) \geq \mathcal{S}(\boldsymbol{\alpha}) + \mathcal{S}(\boldsymbol{\beta}). \quad (5.75)$$

The rationale behind this condition is that the interplay between the allocation $\boldsymbol{\alpha}$ and the allocation $\boldsymbol{\beta}$ provides a diversification effect that satisfies the investor more than the two allocations separately.

Similarly, a measure of satisfaction is *sub-additive* if for any two allocations $\boldsymbol{\alpha}$ and $\boldsymbol{\beta}$:

$$\mathcal{S}(\boldsymbol{\alpha} + \boldsymbol{\beta}) \leq \mathcal{S}(\boldsymbol{\alpha}) + \mathcal{S}(\boldsymbol{\beta}). \quad (5.76)$$

The rationale behind this condition is that the interplay between the allocation $\boldsymbol{\alpha}$ and the allocation $\boldsymbol{\beta}$ provides a diversification effect that the investor does not appreciate.

For example, the expected value of the objective (5.49) is an additive index of satisfaction:

$$\mathrm{E}\{\Psi_{\boldsymbol{\alpha}+\boldsymbol{\beta}}\} = \mathrm{E}\{\Psi_{\boldsymbol{\alpha}}\} + \mathrm{E}\{\Psi_{\boldsymbol{\beta}}\}. \quad (5.77)$$

This follows immediately from the additivity of the objective (5.70). Therefore the expected value is both sub- and super-additive.

- **Co-monotonic additivity**

Two allocations $\boldsymbol{\alpha}$ and $\boldsymbol{\delta}$ are *co-monotonic* if they give rise to co-monotonic objectives, i.e. such that one is a deterministic increasing function of the other, see (2.35).

For example, consider a market of two securities: a stock that trades at the price S_t and a call option on that stock with strike K that expires at the investment horizon. Consider the following two allocations:

$$\boldsymbol{\alpha} \equiv (1, 0)', \quad \boldsymbol{\delta} \equiv (0, 1)'. \quad (5.78)$$

[1] The financial literature that focuses on measures of risk rather than measures of satisfaction reverses the inequalities in the following expressions, see Artzner, Delbaen, Eber, and Heath (1999).

In words, the allocation $\boldsymbol{\alpha}$ is one share of the stock and the allocation $\boldsymbol{\delta}$ is a call option on one share of that stock. Consider an investor whose objective is final wealth. Then

$$\Psi_{\boldsymbol{\alpha}} = S_{T+\tau}, \quad \Psi_{\boldsymbol{\delta}} = \max\left(\Psi_{\boldsymbol{\alpha}} - K, 0\right). \tag{5.79}$$

The objective $\Psi_{\boldsymbol{\delta}}$ is an increasing function of the objective $\Psi_{\boldsymbol{\alpha}}$, once we regularize the call payoff as in (2.37). Thus the two allocations are co-monotonic.

A combination of co-monotonic allocations does not provide a genuine diversification effect: an extreme event in one of them is reflected in an extreme event in the other.

An index of satisfaction is *co-monotonic additive*, if it properly takes this phenomenon into account. From the remarks following (5.75), a co-monotonic additive index of satisfaction satisfies:

$$(\boldsymbol{\alpha}, \boldsymbol{\delta}) \text{ co-monotonic } \Rightarrow \mathcal{S}\left(\boldsymbol{\alpha} + \boldsymbol{\delta}\right) = \mathcal{S}\left(\boldsymbol{\alpha}\right) + \mathcal{S}\left(\boldsymbol{\delta}\right). \tag{5.80}$$

Loosely speaking, a co-monotonic additive index of satisfaction is "derivative-proof".

Since from (5.77) the expected value of the objective is an additive index of satisfaction, in particular it is co-monotonic additive.

- **Concavity/convexity**

We discussed above a few potential features of an index of satisfaction such as positive homogeneity, which refers to rescaling an allocation, and translation invariance, sub-additivity, super-additivity and co-monotonic additivity, which refer to summing allocations. These properties together help determining the level of satisfaction from a joint allocation $\lambda\boldsymbol{\alpha} + \mu\boldsymbol{\beta}$ in terms of the satisfaction $\mathcal{S}\left(\boldsymbol{\alpha}\right)$ and $\mathcal{S}\left(\boldsymbol{\beta}\right)$ from the separate portfolios $\boldsymbol{\alpha}$ and $\boldsymbol{\beta}$ and the respective amounts λ and μ of each portfolio.

In practical situations the investor is not interested in evaluating all the possible allocations spanned by two potential investments $\boldsymbol{\alpha}$ and $\boldsymbol{\beta}$. Instead, due to budget or liquidity constraints, investors typically focus on the satisfaction they draw from weighted averages of the two potential allocations, which include the two separate allocations as special cases.

An index of satisfaction is *concave* if for all $\lambda \in [0,1]$ the following inequality holds:

$$\mathcal{S}\left(\lambda\boldsymbol{\alpha} + (1-\lambda)\,\boldsymbol{\beta}\right) \geq \lambda\mathcal{S}\left(\boldsymbol{\alpha}\right) + (1-\lambda)\,\mathcal{S}\left(\boldsymbol{\beta}\right). \tag{5.81}$$

Notice that concavity is implied by the joint assumptions of positive homogeneity (5.65) and super-additivity (5.75).

Similarly, an index of satisfaction is *convex* if for all $\lambda \in [0,1]$ the following inequality holds:

$$\mathcal{S}\left(\lambda\boldsymbol{\alpha}+(1-\lambda)\boldsymbol{\beta}\right) \leq \lambda\mathcal{S}\left(\boldsymbol{\alpha}\right)+(1-\lambda)\,\mathcal{S}\left(\boldsymbol{\beta}\right). \tag{5.82}$$

It is immediate to verify that convexity is implied by the joint assumptions of positive homogeneity (5.65) and of sub-additivity (5.76).

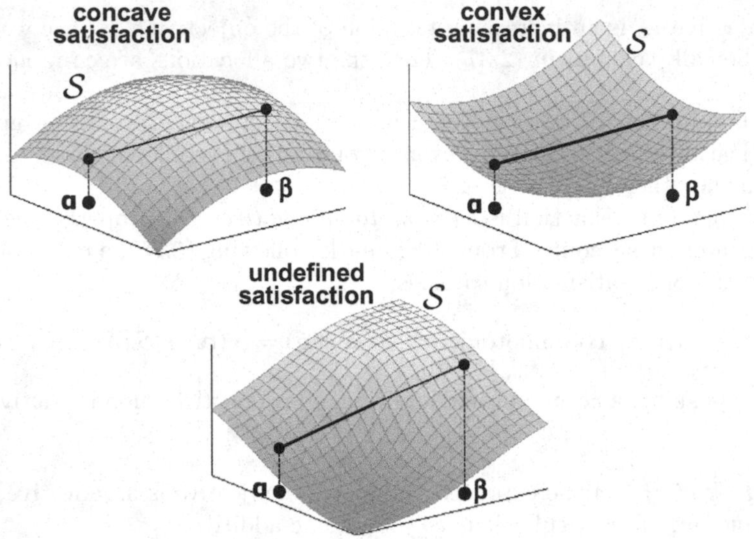

Fig. 5.6. Concavity/convexity of satisfaction index

We sketch in Figure 5.6 the geometrical interpretation of the above properties.

From a theoretical point of view, the most remarkable property of a concave index of satisfaction is the fact that such an index promotes diversification: the satisfaction derived by a diversified portfolio (the weighted average of two generic allocations) exceeds the average of the satisfaction derived by each portfolio individually. We stress that this property is independent of the market. In other words a concave index of satisfaction promotes diversification among two generic portfolios no matter their joint distribution: for example, the two portfolios might be highly positively or negatively correlated. Similarly, a convex index of satisfaction promotes concentration.

From a practical point of view, concavity is an important issue when we resort to numerical solutions to determine the best allocation, see Section 6.2.

- **Risk aversion/propensity/neutrality**

Loosely speaking, a measure of satisfaction is risk averse (risk seeking) if it rejects (welcomes) non-rewarded risk. More precisely, consider an allocation $\mathbf{b}$ that gives rise to a deterministic objective $\psi_{\mathbf{b}}$, i.e. an objective that is not a random variable. Consider now a *fair game*, i.e. an allocation $\mathbf{f}$ such that its objective $\Psi_{\mathbf{f}}$ has zero expected value.[2]

For example, a fair game is a bet on the outcome of tossing a coin: the investor wins a given amount of money if the outcome is "tail" and loses the same amount of money if the outcome is "head".

The joint allocation $\mathbf{b}+\mathbf{f}$ presents the investor with some risk, the fair game, which is not rewarded, since from (5.77) the expected value of the risky allocation is the same as the value of the risk-free allocation.

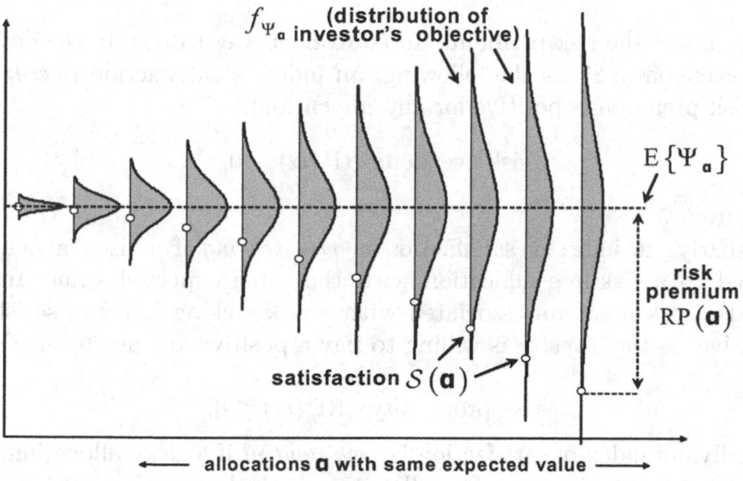

Fig. 5.7. Risk aversion and risk premium

An index of satisfaction is *risk averse* if the risk-free allocation $\mathbf{b}$ is preferred to the risky joint allocation $\mathbf{b}+\mathbf{f}$ for any level of the risk-free outcome $\psi_{\mathbf{b}}$ and any fair game $\mathbf{f}$:

$$\Psi_{\mathbf{b}} \equiv \psi_{\mathbf{b}},\ \mathrm{E}\{\Psi_{\mathbf{f}}\} \equiv 0 \ \Rightarrow\ \mathcal{S}(\mathbf{b}) \geq \mathcal{S}(\mathbf{b}+\mathbf{f}). \tag{5.83}$$

In words, the satisfaction of the risky joint allocation is less than the satisfaction of the deterministic allocation.

The *risk premium* is the dissatisfaction due to the uncertainty of a risky allocation:

$$\mathrm{RP} \equiv \mathcal{S}(\mathbf{b}) - \mathcal{S}(\mathbf{b}+\mathbf{f}). \tag{5.84}$$

[2] One can define a fair game in many different ways. We consider this definition because it is the most widely accepted in the financial literature.

If satisfaction is measured in terms of money, the risk premium is the compensation that the investor needs in order to make up for the uncertainty of his investment.

Any random variable Ψ for which the expected value is defined can be factored into the sum of a deterministic component $\mathrm{E}\{\Psi\}$ and a fair game $\Psi - \mathrm{E}\{\Psi\}$. Therefore we can define the risk premium associated with an allocation as the difference between the satisfaction arising from the expected objective and that arising from the risky allocation. In particular, if an index satisfies the constancy property (5.62) then the risk premium (5.84) associated with an allocation becomes:

$$\mathrm{RP}(\boldsymbol{\alpha}) \equiv \mathrm{E}\{\Psi_{\boldsymbol{\alpha}}\} - \mathcal{S}(\boldsymbol{\alpha}). \tag{5.85}$$

In terms of the risk premium, an equivalent way to restate the definition of risk aversion (5.83) is the following: an index of satisfaction is *risk averse* if the risk premium is positive for any allocation:

$$\text{risk aversion: } \mathrm{RP}(\boldsymbol{\alpha}) \geq 0, \tag{5.86}$$

see Figure 5.7.

Similarly, an index of satisfaction is *risk seeking* if a risky allocation is preferred to a risk-free allocation with the same expected value. In other words, the risk premium associated with a risk seeking index of satisfaction is negative, as the investor is willing to pay a positive amount to play a risky game:

$$\text{risk propensity: } \mathrm{RP}(\boldsymbol{\alpha}) \leq 0. \tag{5.87}$$

Finally, an index of satisfaction is *risk neutral* if a risky allocation is perceived as equivalent to a risk-free allocation with the same expected value. In other words, the risk premium is zero:

$$\text{risk neutrality: } \mathrm{RP}(\boldsymbol{\alpha}) \equiv 0. \tag{5.88}$$

For example, the expected value of the objective (5.49) is trivially a risk neutral index of satisfaction:

$$\mathrm{RP}(\boldsymbol{\alpha}) \equiv \mathrm{E}\{\Psi_{\boldsymbol{\alpha}}\} - \mathrm{E}\{\Psi_{\boldsymbol{\alpha}}\} \equiv 0. \tag{5.89}$$

5.4 Certainty-equivalent (expected utility)

In Section 5.3 we supported the abstract discussion on indices of satisfaction with the example (5.49) of the expected value of the investor's objective. This example in practical applications is too simplistic. In this section we discuss the first of three broad and flexible classes of indices that allow us to model the investor's satisfaction in a variety of situations.

Consider an investor with a given objective Ψ such as absolute wealth, as in (5.3), or relative wealth, as in (5.4), or net profits, as in (5.8), or possibly other specifications. Consider a generic allocation $\boldsymbol{\alpha}$ that gives rise to the objective $\Psi_{\boldsymbol{\alpha}}$. A *utility function* $u(\psi)$ describes the extent to which the investor enjoys the generic outcome $\Psi_{\boldsymbol{\alpha}} = \psi$ of the objective, in case that realization takes place.

To build an index of satisfaction we can weight the utility from every possible outcome by the probability of that outcome. In other words, we consider the expected utility from the given allocation:

$$\boldsymbol{\alpha} \mapsto \mathrm{E}\{u(\Psi_{\boldsymbol{\alpha}})\} \equiv \int_{\mathbb{R}} u(\psi)\, f_{\Psi_{\boldsymbol{\alpha}}}(\psi)\, d\psi, \tag{5.90}$$

where f_{Ψ} is the probability density function of the objective. This expression is in the form (5.48) and thus it qualifies as a potential index of satisfaction. Indeed, this is the *Von Neumann-Morgenstern* specification of expected utility as an index of satisfaction, see Varian (1992): the investor prefers an allocation that gives rise to a higher expected utility.

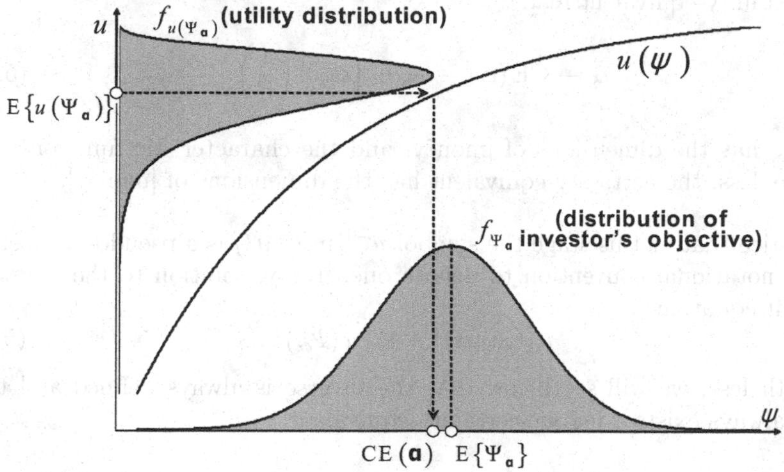

Fig. 5.8. Expected utility and certainty-equivalent

For example, consider the *exponential utility function*:

$$u(\psi) \equiv -e^{-\frac{1}{\zeta}\psi}, \tag{5.91}$$

where ζ is a constant. Since the objective has the dimensions of [money], in order to make the argument of the exponential function dimensionless, this constant must have the dimensions of [money]. The expected utility reads:

$$\mathrm{E}\left\{u\left(\Psi_{\boldsymbol{\alpha}}\right)\right\} = -\mathrm{E}\left\{e^{-\frac{1}{\zeta}\Psi_{\boldsymbol{\alpha}}}\right\} = -\phi_{\Psi_{\boldsymbol{\alpha}}}\left(\frac{i}{\zeta}\right), \tag{5.92}$$

where ϕ denotes the characteristic function (1.12) of the objective.

Nevertheless, utility cannot be measured in natural units (a meter of utility? a watt of utility?). Furthermore, for practitioners it is more intuitive to measure satisfaction in terms of money. In order to satisfy this requirement, we consider the *certainty-equivalent* of an allocation, which is the risk-free amount of money that would make the investor as satisfied as the risky allocation:

$$\boldsymbol{\alpha} \mapsto \mathrm{CE}\left(\boldsymbol{\alpha}\right) \equiv u^{-1}\left(\mathrm{E}\left\{u\left(\Psi_{\boldsymbol{\alpha}}\right)\right\}\right), \tag{5.93}$$

see Figure 5.8.

The certainty-equivalent is measured in the same units as the objective $\Psi_{\boldsymbol{\alpha}}$, which is money. Therefore, instead of (5.90), we choose the certainty-equivalent (5.93) as an index of satisfaction.

For example, consider the exponential utility function (5.91). From (5.92) the certainty-equivalent reads:

$$\boldsymbol{\alpha} \mapsto \mathrm{CE}\left(\boldsymbol{\alpha}\right) \equiv -\zeta \ln\left(\phi_{\Psi_{\boldsymbol{\alpha}}}\left(\frac{i}{\zeta}\right)\right). \tag{5.94}$$

Since ζ has the dimensions of [money] and the characteristic function is dimensionless, the certainty-equivalent has the dimensions of [money].

Notice that at this stage the symbol u^{-1} in (5.93) is a pseudo-inverse, i.e. just a notational convention to denote one, if any, solution to the following implicit equation:

$$u\left(\mathrm{CE}\left(\boldsymbol{\alpha}\right)\right) \equiv \mathrm{E}\left\{u\left(\Psi_{\boldsymbol{\alpha}}\right)\right\}. \tag{5.95}$$

Nevertheless, we will see below that the inverse is always defined and thus there always exists a unique certainty-equivalent.

5.4.1 Properties

In this section we revisit the properties of a generic index of satisfaction discussed in Section 5.3 to ascertain which are satisfied by the certainty-equivalent.

- **Money-equivalence**

By construction the certainty-equivalent (5.93) has the same dimensions as the objective and thus it is measured in terms of money. Therefore the certainty-equivalent is a money-equivalent index of satisfaction.

- **Estimability**

The certainty-equivalent is an estimable index of satisfaction. Indeed, from the formulation (5.90) we see that the expected utility of an allocation is a functional of the probability density function of the investor's objective. Since the certainty-equivalent is defined in terms of the expected utility as in (5.93), the certainty-equivalent is also a functional of the probability density function of the investor's objective. Therefore the certainty-equivalent is defined through a chain such as (5.52) and thus it is an estimable index of satisfaction:

$$\boldsymbol{\alpha} \mapsto \Psi_{\boldsymbol{\alpha}} \mapsto f_{\Psi_{\boldsymbol{\alpha}}} \mapsto \mathrm{CE}(\boldsymbol{\alpha}). \tag{5.96}$$

Depending on the situation, it might be more natural to represent the distribution of the objective equivalently in terms of the characteristic function or the cumulative distribution function.

For example, in the case of the exponential utility function (5.91) the certainty-equivalent (5.94) is defined through the following chain:

$$\boldsymbol{\alpha} \mapsto \Psi_{\boldsymbol{\alpha}} \mapsto \phi_{\Psi_{\boldsymbol{\alpha}}} \mapsto \mathrm{CE}(\boldsymbol{\alpha}) \equiv -\zeta \ln \left(\phi_{\Psi_{\boldsymbol{\alpha}}} \left(\frac{i}{\zeta} \right) \right), \tag{5.97}$$

which is in the form (5.52). Therefore the exponential certainty-equivalent is estimable.

- **Sensibility**

Due to the non-satiation principle, investors pursue the largest possible amount of their respective objectives. Therefore utility must be an increasing function of the objective. Assuming that the utility function is smooth, this corresponds to the condition that the first derivative of the utility be positive for all values in the range of the investor's objective:

$$\mathcal{D}u \geq 0, \tag{5.98}$$

where $\mathcal{D}$ is the derivative operator (B.25). This is the only, though essential, restriction that we impose on the utility function, and thus on the definition of the certainty-equivalent.

For example, consider the exponential utility function (5.91). In order for the utility to be an increasing function of the objective, we impose the constraint $\zeta > 0$.

The consequences of restricting the utility to the set of increasing functions are manifold.

In the first place, the Von Neumann-Morgenstern specification (5.90) and the certainty-equivalent specification (5.93) as indices of satisfaction become equivalent. In other words, if the utility function u is increasing, its inverse u^{-1} is well-defined and increasing. Therefore the certainty-equivalent is an increasing function of the expected utility and an allocation $\boldsymbol{\alpha}$ gives rise to a larger expected utility than an allocation $\boldsymbol{\beta}$ if and only if that allocation $\boldsymbol{\alpha}$ gives rise to a larger certainty-equivalent than $\boldsymbol{\beta}$:

$$\mathrm{E}\{u(\Psi_{\boldsymbol{\alpha}})\} \geq \mathrm{E}\{u(\Psi_{\boldsymbol{\beta}})\} \Leftrightarrow \mathrm{CE}(\boldsymbol{\alpha}) \geq \mathrm{CE}(\boldsymbol{\beta}). \tag{5.99}$$

Secondly, if the utility function is increasing, the certainty-equivalent is sensible, i.e. consistent with strong dominance. Indeed, from (5.99) and the fact that u is increasing we derive:

$$\Psi_{\boldsymbol{\alpha}} \geq \Psi_{\boldsymbol{\beta}} \text{ in all scenarios } \Rightarrow \mathrm{CE}(\boldsymbol{\alpha}) \geq \mathrm{CE}(\boldsymbol{\beta}), \tag{5.100}$$

which is the definition (5.55) of sensibility applied to the certainty-equivalent.

Finally, we can relate the utility function to the investor's *subjective probability*, see Castagnoli and LiCalzi (1996) and Bordley and LiCalzi (2000).

Suppose that the investor has an a-priori hunch as to how his investment will perform, no matter the actual investment decision. We can describe this hunch in terms of a subjective distribution of the objective Ψ, whose pdf (cdf) we denote as f_{Ψ}^{S} (F_{Ψ}^{S}).

Now consider a specific allocation $\boldsymbol{\alpha}$. We can compare this allocation with the investor's hunch by means of (5.42). In other words, we consider the variable:

$$W(V) \equiv F_{\Psi}^{\mathrm{S}}(Q_{\Psi_{\boldsymbol{\alpha}}}(V)), \tag{5.101}$$

where $V \sim \mathrm{U}([0,1])$.

If the joint outcomes of V and W plot above the diagonal as in Figure 5.3, the investor is a pessimist. Indeed in this case the objective $\boldsymbol{\alpha}$ weakly dominates the investor's hunch, which means that the allocation $\boldsymbol{\alpha}$ is better than the investor thinks. On the other hand, if the joint outcomes of V and W plot below the diagonal, the investor is an optimist, as the allocation $\boldsymbol{\alpha}$ is worse than the investor thinks.

No matter the degree of optimism of the investor, the higher the graph, the better the investment. Therefore the investor's satisfaction is an increasing function of the "aboveness" of the graph (5.101). To quantify the degree of "aboveness" of the graph, the most natural approach is to compute its expected value:

$$\begin{aligned} \mathrm{E}\{W(V)\} &\equiv \int_0^1 F_{\Psi}^{\mathrm{S}}(Q_{\Psi_{\boldsymbol{\alpha}}}(v))\,dv \\ &= \int_{\mathbb{R}} F_{\Psi}^{\mathrm{S}}(\psi)\, f_{\Psi_{\boldsymbol{\alpha}}}(\psi)\,d\psi. \end{aligned} \tag{5.102}$$

Comparing this expression with (5.90) we see that the utility function u represents the cumulative distribution function F_{Ψ}^{S} of the investor's subjective a-priori hunch on the result of his investments.

For this interpretation to make sense, the utility function must be increasing and satisfy the normalization properties (1.10) of any cumulative distribution function:

$$u(\psi_{\text{inf}}) \equiv 0, \quad u(\psi_{\text{sup}}) \equiv 1, \tag{5.103}$$

where $[\psi_{\text{inf}}, \psi_{\text{sup}}]$ is the (possibly unbounded) domain of the utility function, i.e. the range of the investor's objective.

Although (5.103) seems a very restrictive assumption on the utility, we notice that the certainty-equivalent is unaffected by positive affine transformations of the utility function. In other words, if in (5.95) we shift and stretch the investor's utility function as follows:

$$u(\psi) \mapsto a + bu(\psi), \tag{5.104}$$

where b is a positive number, the certainty-equivalent is unaffected. Therefore we can always normalize any utility function in such a way that (5.103) holds.[3]

For example, assume that the investor's objective is positive: $\psi \in [0, +\infty)$ and consider the exponential utility function (5.91). This function is equivalent to:

$$u(\psi) \equiv 1 - e^{-\frac{1}{\zeta}\psi}, \tag{5.105}$$

which satisfies (5.103) if $\zeta > 0$. The first derivative of this expression yields the probability density function of the investor's subjective view on his investments:

$$f_{\Psi}^{\text{S}}(\psi) = \frac{1}{\zeta} e^{-\frac{1}{\zeta}\psi}. \tag{5.106}$$

This is a decreasing function: in other words, an investor whose utility is exponential is a pessimist who believes that the worst scenarios are the most likely to occur.

We remark that the interpretation of the utility function u as a subjective cumulative distribution function F_{Ψ}^{S} reduces the certainty-equivalent to a quantile-based index of satisfaction like the value at risk, see Section 5.5. Indeed, denoting as Q_{Ψ}^{S} the quantile of the subjective distribution of the objective, from the definition of the certainty-equivalent (5.93) we obtain:

$$\text{CE}(\boldsymbol{\alpha}) = Q_{\Psi}^{\text{S}}(c_{\boldsymbol{\alpha}}), \tag{5.107}$$

where the confidence level reads:

$$c_{\boldsymbol{\alpha}} \equiv \text{E}\left\{F_{\Psi}^{\text{S}}(\Psi_{\boldsymbol{\alpha}})\right\}. \tag{5.108}$$

In words, the certainty-equivalent is the quantile of the investor's subjective distribution, where the confidence level is the expected subjective grade.

---

[3] If the domain $[\pi_{\text{inf}}, \pi_{\text{sup}}]$ is unbounded, the utility function should be bounded for the normalization to make sense. If this is not the case, we can overcome this problem by restricting $[\pi_{\text{inf}}, \pi_{\text{sup}}]$ to a bounded, yet arbitrarily large, domain.

- **Consistence with stochastic dominance**

To guarantee that the certainty-equivalent (5.93) is a sensible index of satisfaction we imposed the condition that the utility function be increasing. This condition also implies that the certainty-equivalent is consistent with weak dominance. In other words, if u is increasing then the following implication holds true:

$$Q_{\Psi_{\alpha}}(p) \geq Q_{\Psi_{\beta}}(p) \text{ for all } p \in (0,1) \Rightarrow \mathrm{CE}(\boldsymbol{\alpha}) \geq \mathrm{CE}(\boldsymbol{\beta}), \tag{5.109}$$

which is (5.57) in this context. This follows from (5.59).

Consistence of the certainty-equivalent with second-order dominance is guaranteed if the utility function is increasing and concave. Assuming that the utility is a smooth function, these conditions can be stated in terms of the derivative operator (B.25) as follows:

$$\mathcal{D}u \geq 0, \quad \mathcal{D}^2 u \leq 0. \tag{5.110}$$

In general, as far as consistence with higher order dominance is concerned, the certainty-equivalent is consistent with q-th order stochastic dominance if the following condition holds on the whole range of the investor's objective:

$$(-1)^k \mathcal{D}^k u \leq 0, \quad k = 1, 2, \ldots, q, \tag{5.111}$$

see Ingersoll (1987). A comparison of this condition with (5.98) and (5.110) shows that this result includes consistency with weak (first-order) and second-order stochastic dominance respectively.

- **Constancy**

If the investor's objective is deterministic, the certainty-equivalent (5.93) coincides with the objective. In other words, the certainty-equivalent satisfies the constancy requirement:

$$\Psi_{\mathbf{b}} \equiv \psi_{\mathbf{b}} \Rightarrow \mathrm{CE}(\mathbf{b}) = \psi_{\mathbf{b}}, \tag{5.112}$$

which is (5.62) in this context.

- **Positive homogeneity**

In order for the certainty-equivalent (5.93) to be a positive homogeneous index of satisfaction it has to satisfy (5.65), which in this context reads:

$$\mathrm{CE}(\lambda\boldsymbol{\alpha}) = \lambda\,\mathrm{CE}(\boldsymbol{\alpha}). \tag{5.113}$$

In Appendix www.5.3 we show that the class of utility functions that gives rise to a positive homogeneous certainty-equivalent is the power class:

$$u(\psi) \equiv \psi^{1-\frac{1}{\gamma}}, \tag{5.114}$$

where $\gamma \geq 1$. Indeed, Figure 5.4 was generated using the certainty-equivalent relative to a power utility function.

We discuss this type of utility in a more general context below, see (5.135). In particular, if the utility function is of the power class, the Euler decomposition (5.67) of the certainty-equivalent holds, see (5.152) below.

- **Translation invariance**

In order for the certainty-equivalent (5.93) to be a translation invariant index of satisfaction it has to satisfy (5.72), which in this context reads:

$$\Psi_{\mathbf{b}} \equiv 1 \Rightarrow \mathrm{CE}(\boldsymbol{\alpha} + \lambda \mathbf{b}) = \mathrm{CE}(\boldsymbol{\alpha}) + \lambda. \tag{5.115}$$

In Appendix www.5.3 we show that the class of utility functions that give rise to a translation invariant certainty-equivalent is the exponential class (5.91). Indeed, Figure 5.5 was generated using the certainty-equivalent relative to an exponential utility function. We discuss this type of utility function in a more general context below, see (5.133).

- **Super-/sub- additivity**

The certainty-equivalent (5.93) is not a super-additive index of satisfaction. For this to be the case, the certainty-equivalent should satisfy the following relation for any two allocations $\boldsymbol{\alpha}$ and $\boldsymbol{\beta}$:

$$\mathrm{CE}(\boldsymbol{\alpha} + \boldsymbol{\beta}) \geq \mathrm{CE}(\boldsymbol{\alpha}) + \mathrm{CE}(\boldsymbol{\beta}), \tag{5.116}$$

which is (5.75) in this context. The only utility function such that (5.116) is true no matter the market distribution is the linear utility, in which case the certainty-equivalent becomes the expected value of the objective:

$$u(\psi) \equiv \psi \Leftrightarrow \mathrm{CE}(\boldsymbol{\alpha}) = \mathrm{E}\{\Psi_{\boldsymbol{\alpha}}\}. \tag{5.117}$$

In this situation from (5.77) the certainty-equivalent is additive, and thus (5.116) holds as an equality.

Similarly, the certainty-equivalent is not a sub-additive index of satisfaction unless the utility is linear.

- **Co-monotonic additivity**

The certainty-equivalent (5.93) is not a co-monotonic additive index of satisfaction. In other words, if an allocation $\boldsymbol{\delta}$ gives rise to an objective that is an increasing function of the objective of another allocation $\boldsymbol{\alpha}$ as in (2.35), the certainty-equivalent of the total portfolio is not necessarily the sum of the certainty-equivalents of the separate allocations:

$$(\boldsymbol{\alpha}, \boldsymbol{\delta}) \text{ co-monotonic } \nRightarrow \mathrm{CE}(\boldsymbol{\alpha} + \boldsymbol{\delta}) = \mathrm{CE}(\boldsymbol{\alpha}) + \mathrm{CE}(\boldsymbol{\delta}). \tag{5.118}$$

This can be proved easily with a counterexample. The only utility function such that (5.118) is true no matter the market distribution is the linear utility $u(\psi) \equiv \psi$, in which case the co-monotonicity is a consequence of the additivity property (5.77).

- **Concavity/convexity**

One might think that the concavity/convexity properties of the certainty-equivalent are a straightforward consequence of the concavity/convexity properties of the utility function. This is not the case.

Intuitively, in the definition of the certainty-equivalent (5.93) the expected utility is a concave function of $\boldsymbol{\alpha}$ if and only if the utility function is concave. On the other hand, if u is concave, the inverse u^{-1} in the definition of the certainty-equivalent is convex, and the two effects tend to cancel each other.

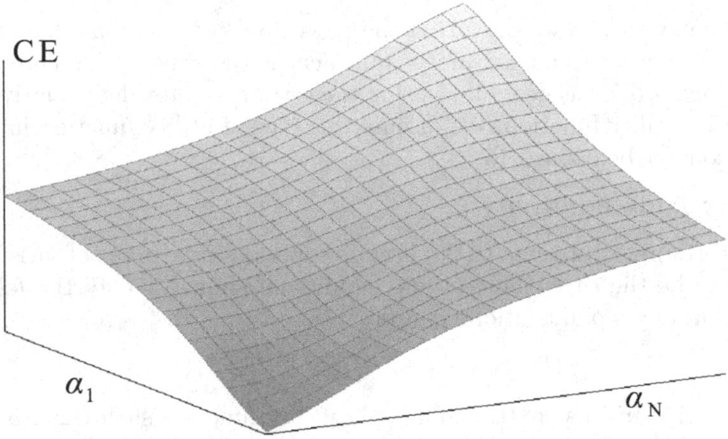

Fig. 5.9. Certainty equivalent as function of allocation

Therefore, in general the certainty-equivalent is neither concave nor convex, see Figure 5.9. We present this argument more formally in the context of the second-order sensitivity analysis in Section 5.4.4.

- **Risk aversion/propensity**

The certainty-equivalent (5.93) satisfies the constancy property, see (5.112). Therefore the risk premium associated with an allocation is given by (5.85), which in this context reads:

$$\mathrm{RP}(\boldsymbol{\alpha}) = \mathrm{E}\{\Psi_{\boldsymbol{\alpha}}\} - \mathrm{CE}(\boldsymbol{\alpha}). \tag{5.119}$$

In Appendix www.5.3 we prove that for any allocation $\boldsymbol{\alpha}$ the following result holds:

$$u \text{ concave } \Leftrightarrow \mathrm{RP}(\boldsymbol{\alpha}) \geq 0. \tag{5.120}$$

Therefore from (5.86) the certainty-equivalent is a risk averse index of satisfaction if and only if the utility function is concave. This is the situation for instance in Figure 5.8, where (5.120) is satisfied.

Similarly, the certainty-equivalent is a risk prone index of satisfaction if and only if the utility function is convex. Finally, the certainty-equivalent is a risk neutral index of satisfaction if and only if the utility function is linear.

Risk aversion is a global feature: the risk premium of a risk averse investor is positive no matter the allocation. Similarly, risk propensity is a global feature. Nevertheless, depending on their objectives and conditions, investors might display different attitudes toward risk in different situations, and thus the risk premium they require can change sign.

For instance, prospect theory asserts that the investor's objective are the net profits as in (5.8). In this context, investors tend to be cautious in their pursuit of new gains, but are unwilling to cut their losses in the hope of a recovery, see Kahneman and Tversky (1979). The ensuing utility function is S-shaped, i.e. it is concave (= risk averse) for profits and convex (= risk prone) for losses, see the fourth plot in Figure 5.10.

To better describe the investor's attitude toward risk we need a more local measure. The *Arrow-Pratt absolute risk aversion* is defined as follows:

$$\mathrm{A}(\psi) \equiv -\frac{\mathcal{D}^2 u(\psi)}{\mathcal{D} u(\psi)}, \tag{5.121}$$

where $\mathcal{D}$ is the derivative operator (B.25). In Appendix www.5.3 we show that if an allocation $\boldsymbol{\alpha}$ gives rise to an objective which is not too volatile, i.e. an objective whose distribution is highly concentrated around its expected value, the following factorization yields a good approximation of the risk premium:

$$\mathrm{RP}(\boldsymbol{\alpha}) \approx \frac{1}{2} \mathrm{A}(\mathrm{E}\{\Psi_{\boldsymbol{\alpha}}\}) \operatorname{Var}\{\Psi_{\boldsymbol{\alpha}}\}. \tag{5.122}$$

In other words, the money necessary to compensate for the riskiness of an allocation is the product of a quantity that depends on the investor's preferences, namely the Arrow-Pratt risk aversion, and a quantity that does not depend on the investor's preferences, namely the variance of the allocation, which in turn summarizes its riskiness.

The Arrow-Pratt risk aversion (5.121) is a function: as such, it is a local measure of risk aversion that depends on the expected value of the objective. Since the first derivative of the utility function is always positive, the Arrow-Pratt risk aversion is positive if and only if the second derivative of the utility function is negative, which means that the utility function is concave. In other words, from (5.120) the Arrow-Pratt risk aversion is positive if and only if the investor is locally risk averse. Similarly, the Arrow-Pratt risk aversion is negative if and only if the investor is locally risk prone.

For example, assume a prospect-theory framework where the objective are net profits as in (5.8). We use the error function (B.75) to model the investor's utility:

$$u(\psi) \equiv \operatorname{erf}\left(\frac{\psi}{\sqrt{2\eta}}\right). \tag{5.123}$$

In this expression η is a constant with the dimensions of $[\text{money}]^2$ that makes the argument of the error function dimensionless. The error function is S-shaped, with a flex point in zero, see the fourth plot in Figure 5.10.

The domain of this utility function is the whole real axis: this function describes the attitude towards risk in the case of both profits, i.e. positive values of the objective, and losses, i.e. negative values of the objective. The Arrow-Pratt risk aversion (5.121) corresponding to the error function utility (5.123) reads:

$$\mathrm{A}(\psi) = \frac{\psi}{\eta}. \tag{5.124}$$

Consider the case where η is positive. When facing net gains, i.e. when ψ is positive, the Arrow-Pratt risk aversion is positive: the investor is risk averse and seeks a compensation for the non-rewarded risk. On the other hand, when facing net losses, i.e. when ψ is negative, the Arrow-Pratt risk aversion is negative: the investor is risk prone and is willing to pay a premium to hold on to a risky allocation.

Notice that the Arrow-Pratt risk aversion is not a dimensionless number: instead, it as has the dimensions of $[\text{money}]^{-1}$. This follows from the definition (5.121) and is a necessary condition for the dimensional consistency of (5.122).

For example, in (5.124) the objective has the dimensions of [money] and the constant η has the dimensions of $[\text{money}]^2$. Therefore their ratio has the dimensions of $[\text{money}]^{-1}$.

Unlike the second derivative of the utility function, not only the sign, but also the absolute value of the Arrow-Pratt risk aversion is meaningful. Indeed, from (5.104) any utility function is defined only modulo positive affine transformations: such transformations affect the value of the first and second derivative of the utility function, but leave the Arrow-Pratt risk aversion (5.121) unaltered. Alternatively, the fact that the Arrow-Pratt risk aversion has the dimensions of $[\text{money}]^{-1}$ is proof of its importance.

5.4.2 Building utility functions

According to the certainty-equivalent approach, the attitude toward risk of an investor is described by his utility function. Therefore we should use a tailor-made utility function for each specific case. Since this is impossible, we specify the functional form of the utility function by means of parsimonious yet flexible parametrizations. We present below two methods to build parametric utility functions.

Basis

One way to build utility functions starts by specifying a *basis*. In other words, we specify a one-parameter set of (generalized) functions b that can generate a whole class of utility functions as weighted averages:

$$u(\psi) \equiv \int_{\mathbb{R}} g(\theta)\, b(\theta, \psi)\, d\theta, \tag{5.125}$$

where the weight function g is positive and sums to one:

$$g \geq 0, \quad \int_{\mathbb{R}} g(\theta)\, d\theta = 1. \tag{5.126}$$

For example, consider a basis defined in terms of the Heaviside function (B.74) as follows:

$$b(\theta, \psi) \equiv H^{(\theta)}(\psi). \tag{5.127}$$

This basis is very broad, as it can in principle generate all sensible utility functions. Indeed, any increasing function u can be expressed in terms of this basis as in (5.125), if the weight function is defined as the derivative of the utility function:

$$g \equiv \mathcal{D}u. \tag{5.128}$$

The proof follows immediately from the definition of the Heaviside step function. Also notice that the condition (5.103) guarantees that the normalization (5.126) is satisfied.

The representation of utility functions in terms of a basis is useful in two ways. In the first place, it allows us to build classes of utility functions that share the same properties.

For example, consider the min-function basis:

$$b(\theta, \psi) \equiv \min(\psi, \theta). \tag{5.129}$$

This basis can generate all concave utility functions. Therefore the min-function basis can be used to describe risk averse investors, see Gollier (2001).

Secondly, the representation of utility functions in terms of a basis provides a probabilistic interpretation of the expected utility and therefore of the certainty-equivalent:

$$\begin{aligned} \mathrm{E}\{u(\Psi)\} &= \int_{\mathbb{R}^2} b(\theta, \psi)\, g(\theta)\, f_{\Psi}(\psi)\, d\theta d\psi \\ &= \mathrm{E}\{b(\Theta, \Psi)\}. \end{aligned} \tag{5.130}$$

In other words, the expected utility is the expected value of a function of two random variables. The first random variable Θ models the investor's preferences and attitude toward risk. The second random variable Ψ, i.e. the objective, models the market. These two random variables are independent and thus fully determined by their marginal distributions. The distribution of the preferences is described by the weight function g, which due to (5.126) can be interpreted as a probability density function; the distribution of the market is described by the probability density function of the investor's objective f_{Ψ}.

Arrow-Pratt risk aversion

A different approach to building utility functions focuses on the Arrow-Pratt risk aversion.

First of all notice that the Arrow-Pratt risk aversion is an equivalent, yet more efficient, representation of the utility function. Indeed, from (5.121) the specification of the utility function yields the Arrow-Pratt risk aversion. In turn, integrating the Arrow-Pratt risk aversion, we can recover the utility function, modulo a positive affine transformation:

$$(\mathcal{I} \circ \exp \circ \mathcal{I}) [- \mathrm{A}] = a + bu, \tag{5.131}$$

where $\mathcal{I}$ is the integration operator (B.27). From (5.104), positive affine transformations are irrelevant to the determination of the certainty-equivalent: therefore, the Arrow-Pratt risk aversion contains all and only the information about the utility function that matters in determining the investor's satisfaction.

Therefore, an efficient way to define parametric forms of the investor's preferences is to specify flexible, although parsimonious, functional forms for the Arrow-Pratt risk aversion, rather than for the utility function. A possible such specification appears in LiCalzi and Sorato (2003):

$$\mathrm{A}(\psi) \equiv \frac{\psi}{\gamma\psi^2 + \zeta\psi + \eta}. \tag{5.132}$$

This specification depends on only three constant parameters[4] and yet it includes as special cases most of the parametrizations studied in the financial literature. Since Pearson (1895) discussed a similar parametrization in a different context, (5.132) is called the *Pearson specification* of the utility function.

In the special case where $\eta \equiv 0$ in (5.132) we obtain the *Hyperbolic Absolute Risk Aversion (HARA)* class of utility functions. The HARA class includes in turn a few notable parametrizations as special cases. The results below can be obtained by applying (5.131) or by checking the derivatives of the utility function in the definition (5.121) of the Arrow-Pratt risk aversion coefficient.

If $\zeta > 0$ and $\gamma \equiv 0$ the HARA class yields the *exponential utility*:

$$u(\psi) = -e^{-\frac{1}{\zeta}\psi}, \tag{5.133}$$

where $\zeta > 0$, see Figure 5.10.

[4] By adding one more parameter ξ as follows $\mathrm{A}(\psi, \xi) \equiv \mathrm{A}(\psi - \xi)$, all the ensuing utility functions are shifted along the horizontal axis by ξ.

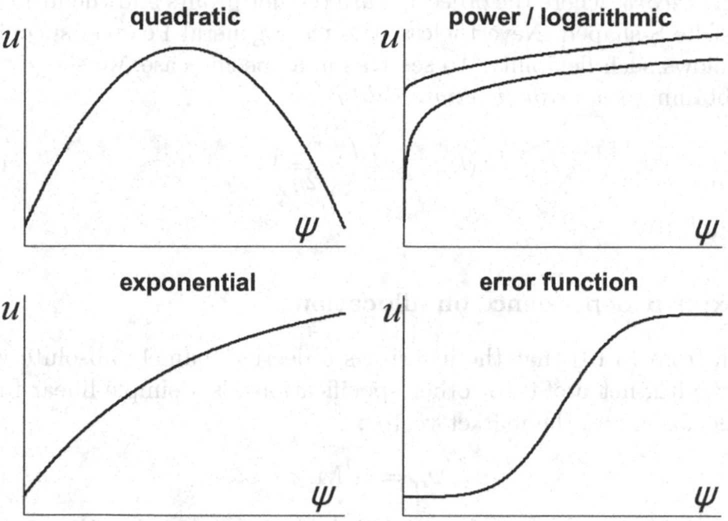

Fig. 5.10. Parametric utility functions

If $\zeta > 0$ and $\gamma \equiv -1$ the HARA class yields the *quadratic utility*:

$$u(\psi) = \psi - \frac{1}{2\zeta}\psi^2. \tag{5.134}$$

We remark that the certainty-equivalent stemming from the quadratic utility function is not sensible for $\psi > \zeta$ because in that region a larger value of the objective ψ decreases the investor's satisfaction: therefore this utility function can be used in principle only when the objective is bounded from above, see Figure 5.10.

If $\zeta \equiv 0$ and $\gamma \geq 1$ the HARA class yields the *power utility*:

$$u(\psi) \equiv \psi^{1-\frac{1}{\gamma}}, \tag{5.135}$$

see Figure 5.10. In the limit $\gamma \to 1$ the power utility (5.135) yields, modulo a positive affine transformation, the *logarithmic utility* function:

$$u(\psi) \equiv \ln \psi. \tag{5.136}$$

In the limit $\gamma \to \infty$ the power utility (5.135) becomes the *linear utility*:

$$u(\psi) \equiv \psi. \tag{5.137}$$

The utility functions in the HARA class are very flexible, but always concave. Therefore the HARA specification cannot properly model the framework of *prospect theory*, where the objective are the net profits and the utility function must be S-shaped. Nevertheless, the more general Pearson specification (5.132) allows such flexibility. To see this in a specific case, we set $\gamma \equiv 0$ and $\zeta \equiv 0$, obtaining the *error function utility*:

$$u(\psi) \equiv \operatorname{erf}\left(\frac{\psi}{\sqrt{2\eta}}\right), \tag{5.138}$$

see Figure 5.10.

5.4.3 Explicit dependence on allocation

We recall from (5.10) that the investor's objective, namely absolute wealth, relative wealth, net profits, or other specifications, is a simple linear function of the allocation and the market vector:

$$\Psi_{\boldsymbol{\alpha}} = \boldsymbol{\alpha}'\mathbf{M}. \tag{5.139}$$

Therefore the certainty-equivalent (5.93) depends on the allocation as follows:

$$\boldsymbol{\alpha} \mapsto \mathrm{CE}(\boldsymbol{\alpha}) \equiv u^{-1}\left(\mathrm{E}\left\{u\left(\boldsymbol{\alpha}'\mathbf{M}\right)\right\}\right). \tag{5.140}$$

In this section we tackle the problem of computing explicitly the certainty-equivalent of an allocation for a given distribution of the market vector $\mathbf{M}$ and a given choice of the utility function u.

For very special combinations of the distribution of the market and of the choice of the utility functions the dependence of the certainty-equivalent on the allocation can be computed analytically.

For example, we can represent the distribution of the market in terms of its characteristic function $\phi_{\mathbf{M}}$. Then the distribution of the investor's objective (5.139) is represented in terms of its characteristic function as follows:

$$\phi_{\Psi_{\boldsymbol{\alpha}}}(\omega) = \phi_{\mathbf{M}}(\omega\boldsymbol{\alpha}), \tag{5.141}$$

see Appendix www.2.4.

Suppose that the investor's satisfaction is determined by an exponential utility function (5.91). Then from (5.94) the explicit dependence of the certainty-equivalent on allocation reads:

$$\mathrm{CE}(\boldsymbol{\alpha}) = -\zeta \ln\left(\phi_{\mathbf{M}}\left(\frac{i}{\zeta}\boldsymbol{\alpha}\right)\right). \tag{5.142}$$

Consider a market which at the investment horizon is normally distributed and assume that the investor's objective is final wealth. From (5.11) and (5.12) the market vector is normally distributed:

$$\mathbf{M} \equiv \mathbf{P}_{T+\tau} \sim \mathrm{N}(\boldsymbol{\mu}, \boldsymbol{\Sigma}). \tag{5.143}$$

From (5.142) and the characteristic function of the normal distribution (2.157) we obtain:

$$\boldsymbol{\alpha} \mapsto \mathrm{CE}(\boldsymbol{\alpha}) = \boldsymbol{\alpha}'\boldsymbol{\mu} - \frac{\boldsymbol{\alpha}'\boldsymbol{\Sigma}\boldsymbol{\alpha}}{2\zeta}. \tag{5.144}$$

Notice that the expected value of the market prices $\boldsymbol{\mu}$ has the dimensions of [money] and the covariance matrix of the market prices $\boldsymbol{\Sigma}$ has the dimensions of $[\text{money}]^2$. The allocation vector $\boldsymbol{\alpha}$ is a dimensionless number. Since ζ has the dimensions of [money], so does the certainty-equivalent.

Nevertheless, for a generic utility function u and a generic market $\mathbf{M}$ the certainty-equivalent is a complex expression of the objective.

For example, consider a market for which the gamma approximation (5.24) holds, and an investor whose utility function is exponential. In this case from (5.94) and the expression of the characteristic function of the objective (5.30) we obtain:

$$\begin{aligned}\mathrm{CE}(\boldsymbol{\alpha}) = {} & \frac{\zeta}{2}\ln\left|\mathbf{I}_K + \frac{1}{\zeta}\boldsymbol{\Gamma}_{\boldsymbol{\alpha}}\boldsymbol{\Sigma}\right| + \left(\theta_{\boldsymbol{\alpha}} + \boldsymbol{\Delta}_{\boldsymbol{\alpha}}'\boldsymbol{\mu} + \frac{1}{2}\boldsymbol{\mu}'\boldsymbol{\Gamma}_{\boldsymbol{\alpha}}\boldsymbol{\mu}\right) \\ & + \frac{\zeta}{2}\left[\boldsymbol{\Delta}_{\boldsymbol{\alpha}} + \boldsymbol{\Gamma}_{\boldsymbol{\alpha}}\boldsymbol{\mu}\right]'\boldsymbol{\Sigma}\left(\mathbf{I}_K + \frac{1}{\zeta}\boldsymbol{\Gamma}_{\boldsymbol{\alpha}}\boldsymbol{\Sigma}\right)^{-1}\left[\boldsymbol{\Delta}_{\boldsymbol{\alpha}} + \boldsymbol{\Gamma}_{\boldsymbol{\alpha}}\boldsymbol{\mu}\right],\end{aligned} \tag{5.145}$$

where the explicit dependence on the allocation $\boldsymbol{\alpha}$ is given by (5.26)-(5.28).

When the explicit dependence of the certainty-equivalent on the allocation vector $\boldsymbol{\alpha}$ cannot be computed analytically, we can gain insight on this dependence by means of a first-order approximate expression, the *Arrow-Pratt approximation*:

$$\mathrm{CE}(\boldsymbol{\alpha}) \approx \mathrm{E}\{\Psi_{\boldsymbol{\alpha}}\} - \frac{\mathrm{A}(\mathrm{E}\{\Psi_{\boldsymbol{\alpha}}\})}{2}\mathrm{Var}\{\Psi_{\boldsymbol{\alpha}}\}. \tag{5.146}$$

In this expression A is the Arrow-Pratt coefficient of risk aversion (5.121), see Appendix www.5.3 for a proof. This approximation shows that high expected values of the objective are always appreciated, whereas the investor's attitude towards the variance of its objective can vary, depending on the investor's local risk aversion.

If the moments of the market vector $\mathbf{M}$ are known, (5.146) yields the (approximate) explicit dependence of the certainty-equivalent on the allocation vector $\boldsymbol{\alpha}$. This follows from (5.139) and the affine equivariance properties (2.56) and (2.71) of expected value and covariance respectively.

If the moments of the market vector $\mathbf{M}$ are not known, we can replace the objective $\Psi_{\boldsymbol{\alpha}}$ with its gamma approximation $\Xi_{\boldsymbol{\alpha}}$ as in (5.25). This corresponds to replacing the moments of $\Psi_{\boldsymbol{\alpha}}$ that appear in (5.146) with the moments of $\Xi_{\boldsymbol{\alpha}}$, which can be computed from the derivatives of the characteristic function (5.30) as illustrated in Appendix www.5.1.

In the very special case of exponential utility and normal markets the approximation (5.146) becomes exact, as we see from (5.144).

5.4.4 Sensitivity analysis

Suppose that the investor has already chosen an allocation $\boldsymbol{\alpha}$ which yields a level of satisfaction $\mathrm{CE}(\boldsymbol{\alpha})$ and that he is interested in rebalancing his portfolio marginally by means of a small change $\delta\boldsymbol{\alpha}$ in the current allocation. In this case a local analysis in terms of a Taylor expansion is useful:

$$\begin{aligned} \mathrm{CE}(\boldsymbol{\alpha}+\delta\boldsymbol{\alpha}) \approx & \; \mathrm{CE}(\boldsymbol{\alpha}) + \delta\boldsymbol{\alpha}' \frac{\partial \, \mathrm{CE}(\boldsymbol{\alpha})}{\partial \boldsymbol{\alpha}} \\ & + \frac{1}{2} \delta\boldsymbol{\alpha}' \frac{\partial^2 \, \mathrm{CE}(\boldsymbol{\alpha})}{\partial \boldsymbol{\alpha} \partial \boldsymbol{\alpha}'} \delta\boldsymbol{\alpha}. \end{aligned} \tag{5.147}$$

We prove in Appendix www.5.3 that the first-order derivatives read:

$$\frac{\partial \, \mathrm{CE}(\boldsymbol{\alpha})}{\partial \boldsymbol{\alpha}} = \frac{\mathrm{E}\{\mathcal{D}u(\boldsymbol{\alpha}'\mathbf{M})\,\mathbf{M}\}}{\mathcal{D}u(\mathrm{CE}(\boldsymbol{\alpha}))}, \tag{5.148}$$

where $\mathcal{D}$ is the derivative operator (B.25) and $\mathbf{M}$ is the random vector (5.139) that represents the market. The investor will focus on the entries of the vector (5.148) that display a large absolute value.

For example, consider a prospect theory setting where the investor's objective are the net profits. We model the investor's utility by means of the error function:

$$u(\psi) \equiv \mathrm{erf}\left(\frac{\psi}{\sqrt{2\eta}}\right). \tag{5.149}$$

Assume a normally distributed market:

$$\mathbf{P}_{T+\tau} \sim \mathrm{N}(\boldsymbol{\mu}, \boldsymbol{\Sigma}). \tag{5.150}$$

We prove in Appendix www.5.3 that the sensitivity of the certainty-equivalent to the allocation reads:

$$\frac{\partial \, \mathrm{CE}(\boldsymbol{\alpha})}{\partial \boldsymbol{\alpha}} = \gamma(\boldsymbol{\alpha}) \left[\frac{1}{\eta}\boldsymbol{\alpha}\boldsymbol{\alpha}' + \boldsymbol{\Sigma}^{-1}\right]^{-1} \boldsymbol{\Sigma}^{-1}(\boldsymbol{\mu} - \mathbf{p}_T), \tag{5.151}$$

where $\gamma(\boldsymbol{\alpha})$ is a scalar and thus equally affects all the entries of the vector of the first derivatives.

We remark that when the utility function belongs to the power class (5.114) the certainty-equivalent is positive homogeneous and thus it can be expressed in terms of the contribution to satisfaction from each security by means of the Euler decomposition (5.67). Substituting the expression of the derivative of power utility function in (5.148) we obtain:

$$u(\psi) \equiv \psi^{1-\frac{1}{\gamma}} \Rightarrow \tag{5.152}$$
$$\mathrm{CE}(\boldsymbol{\alpha}) = \sum_{n=1}^{N} \alpha_n \left[\mathrm{E}\left\{ M_n (\boldsymbol{\alpha}'\mathbf{M})^{-\frac{1}{\gamma}} \right\} (\mathrm{CE}(\boldsymbol{\alpha}))^{\frac{1}{\gamma}} \right].$$

The contribution to satisfaction from each security in turn factors into the product of the amount of that security times the marginal contribution to satisfaction from that security. The marginal contribution to satisfaction, which is the term in square brackets in (5.152), is insensitive to a rescaling of the portfolio, although it depends on the allocation.

The study of the second-order cross-derivatives provides insight on the local convexity/concavity of the certainty-equivalent. In Appendix www.5.3 we prove the following result:

$$\frac{\partial^2 \mathrm{CE}(\boldsymbol{\alpha})}{\partial \boldsymbol{\alpha} \partial \boldsymbol{\alpha}} = \frac{\mathrm{E}\left\{ \mathcal{D}^2 u(\Psi_{\boldsymbol{\alpha}}) \mathbf{M}\mathbf{M}' \right\} - \mathcal{D}^2 u(\mathrm{CE}(\boldsymbol{\alpha})) \mathbf{w}\mathbf{w}'}{\mathcal{D} u(\mathrm{CE}(\boldsymbol{\alpha}))}, \tag{5.153}$$

where $\mathcal{D}$ is the derivative operator (B.25) and the deterministic vector $\mathbf{w}$ is defined as follows:

$$\mathbf{w} \equiv \mathrm{E}\left\{ \frac{\mathcal{D} u(\Psi_{\boldsymbol{\alpha}})}{\mathcal{D} u(\mathrm{CE}(\boldsymbol{\alpha}))} \mathbf{M} \right\}. \tag{5.154}$$

In (5.153) the matrices $\mathbf{MM}'$ and $\mathbf{ww}'$are always positive, and so is $\mathcal{D}u$ by assumption. Nevertheless, even when the sign of $\mathcal{D}^2 u$ is consistently either negative or positive, we cannot be sure of the sign of the cross-derivatives (5.153). Therefore, the certainty-equivalent is neither a concave nor a convex index of satisfaction. We can see this in Figure 5.9, that refers to a prospect theory setting where utility is modeled by the error function (5.138).

5.5 Quantile (value at risk)

In Section 5.4 we discussed the certainty-equivalent, which is a very subjective index of satisfaction. Indeed, the certainty-equivalent is determined by the choice of a utility function, which is specific to each investor. In this section we discuss the second of three broad approaches to model the investor's satisfaction.

To introduce this index of satisfaction, consider a financial institution with a capital of, say, one billion dollars. The financial institution aims at investing its capital in such a way that at a given time horizon the maximum loss does not exceed, say, ten million dollars. Since in a stochastic environment there is no guarantee that the maximum loss will not be exceeded, it is more reasonable to require that the maximum loss is not exceeded within a given confidence margin of, say, ninety-five percent.

We rephrase the above situation in our notation. The investor is the financial institution, which has an initial capital w_T (one billion in the example) at the time the investment decision is made. The institution focuses on the potential loss at the investment horizon, which is the difference between the initial capital w_T and the stochastic capital at the the investment horizon $W_{T+\tau}$. This loss should not exceed the threshold $L_{\max}$ (ten million) with a confidence of at least c (ninety-five percent):

$$\mathbb{P}\{w_T - W_{T+\tau} < L_{\max}\} \geq c. \tag{5.155}$$

In other words, the investor's objective are net profits, which as in (5.8) depend on the allocation decision

$$\Psi_{\boldsymbol{\alpha}} \equiv W_{T+\tau}(\boldsymbol{\alpha}) - w_T; \tag{5.156}$$

and from the definition of quantile (1.18) the financial institution manages its investments in such a way that the lower-tail quantile of the net profits, corresponding to a confidence level $1-c$ (five percent), be above the maximum acceptable loss:

$$Q_{\Psi_{\boldsymbol{\alpha}}}(1-c) \geq -L_{\max}. \tag{5.157}$$

When the required confidence level c is high, the quantile in (5.157) is typically a negative amount, i.e. it represents a loss: the purpose of the financial institution is to make sure that its absolute value does not exceed the maximum loss $L_{\max}$.

The absolute value of the quantile of the objective, when the objective are net profits, is known among practitioners as the *value at risk* (*VaR*) with confidence c of the allocation $\boldsymbol{\alpha}$:

$$\mathrm{VaR}_c(\boldsymbol{\alpha}) \equiv -Q_{\Psi_{\boldsymbol{\alpha}}}(1-c). \tag{5.158}$$

The value at risk has become extremely popular among practitioners especially after the Basel Accord, see Crouhy, Galai, and Mark (1998).

More in general the objective $\Psi_{\boldsymbol{\alpha}}$, which depends on the allocation $\boldsymbol{\alpha}$, could be absolute wealth as in (5.3) or relative wealth, as in (5.4), or net profits, as in (5.8), or possibly other specifications. Therefore (5.157) suggests to define a *quantile-based index of satisfaction* of a given allocation $\boldsymbol{\alpha}$ for a generic investor in terms of the quantile of the investor's objective as follows:

$$\boldsymbol{\alpha} \mapsto \mathrm{Q}_c(\boldsymbol{\alpha}) \equiv Q_{\Psi_{\boldsymbol{\alpha}}}(1-c), \tag{5.159}$$

where $c \in (0,1)$ is a fixed *confidence level*. This expression is in the form (5.48) and thus it qualifies as a potential index of satisfaction. We see in Figure 5.11 the graphical interpretation of the quantile-based index of satisfaction. We also plot the interpretation of the VaR, which only applies when the investor's objective are the net profits.

5.5.1 Properties

In this section we revisit the properties of a generic index of satisfaction discussed in Section 5.3 to ascertain which are satisfied by the quantile.

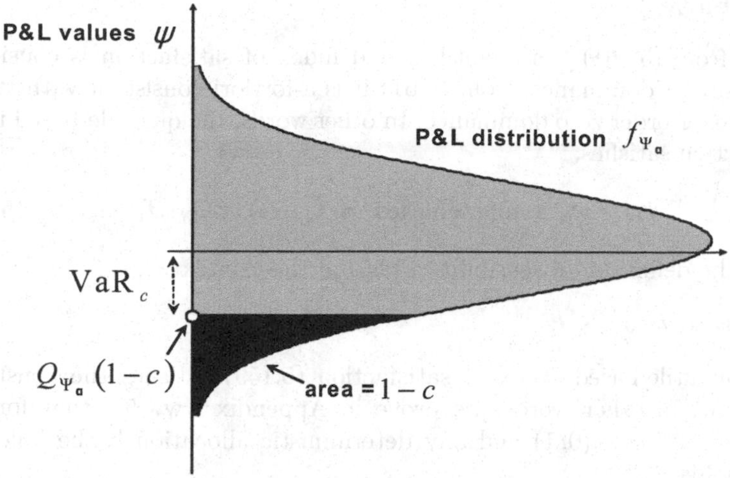

Fig. 5.11. VaR and quantile-based index of satisfaction

- **Money-equivalence**

  The quantile of the distribution of the investor's objective has the same dimensions as the objective, which is money. Therefore the quantile-based index of satisfaction (5.159) is a money-equivalent index of satisfaction.

- **Estimability**

  The quantile is the inverse of the cumulative distribution function of the objective. Therefore the quantile-based index of satisfaction is defined through a chain such as (5.52) and thus it is estimable:

$$\alpha \mapsto \Psi_\alpha \mapsto F_{\Psi_\alpha} \mapsto \mathrm{Q}_c(\alpha). \tag{5.160}$$

- **Consistence with stochastic dominance**

  The quantile-based index of satisfaction (5.159) is consistent with weak stochastic dominance. The proof is almost tautological. Indeed the quantile satisfies the following relation:

$$Q_{\Psi_\alpha}(p) \geq Q_{\Psi_\beta}(p) \text{ for all } p \in (0,1) \Rightarrow \mathrm{Q}_c(\alpha) \geq \mathrm{Q}_c(\beta), \tag{5.161}$$

which is the definition of consistence with weak dominance (5.57) in this context.

On the other hand, the quantile is not consistent with second-order dominance. The interested reader can find a counterexample in the context of value at risk in Guthoff, Pfingsten, and Wolf (1997). Since the quantile is not consistent with second-order dominance, (5.61) implies that it cannot be consistent with higher-order dominance.

- **Sensibility**

Since from (5.161) the quantile-based index of satisfaction is consistent with first-order dominance, from (5.61) it is a-fortiori consistent with strong dominance, or order zero dominance. In other words, the quantile-based index of satisfaction satisfies:

$$\Psi_{\boldsymbol{\alpha}} \geq \Psi_{\boldsymbol{\beta}} \text{ in all scenarios } \Rightarrow \mathrm{Q}_c(\boldsymbol{\alpha}) \geq \mathrm{Q}_c(\boldsymbol{\beta}), \tag{5.162}$$

which is the definition of sensibility (5.55) in this context.

- **Constancy**

The quantile-based index of satisfaction (5.159) satisfies the constancy requirement. In other words, we prove in Appendix www.5.4 that for any confidence level $c \in (0,1)$ and any deterministic allocation $\mathbf{b}$ the following relation holds:

$$\Psi_{\mathbf{b}} = \psi_{\mathbf{b}} \Rightarrow \mathrm{Q}_c(\mathbf{b}) = \psi_{\mathbf{b}}, \tag{5.163}$$

which is (5.62) in this context.

- **Positive homogeneity**

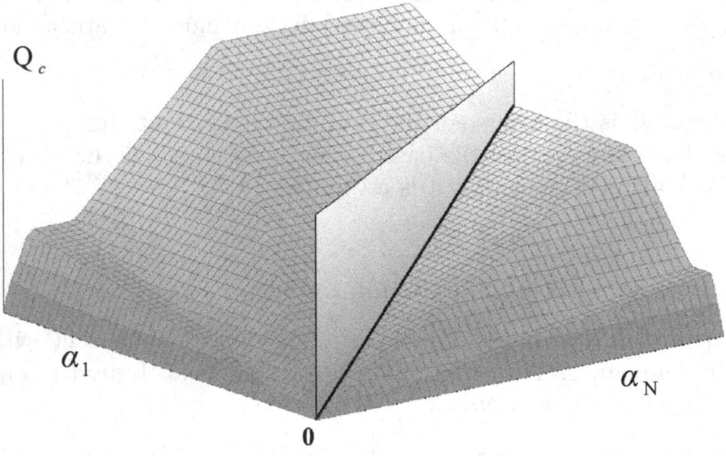

Fig. 5.12. Quantile-based satisfaction index as function of allocation

The quantile-based index of satisfaction is positive homogenous, i.e. it satisfies (5.65), which in this context reads:

$$\mathrm{Q}_c(\lambda\boldsymbol{\alpha}) = \lambda\,\mathrm{Q}_c(\boldsymbol{\alpha}), \quad \text{for all } \lambda \geq 0, \tag{5.164}$$

see Appendix www.5.4 for the proof. In other words, the quantile grows linearly in any radial direction stemming from the origin of the allocation space, see Figure 5.12 and compare with Figure 5.4. Refer to symmys.com for details on these figures.

- **Translation invariance**

The quantile-based index of satisfaction is a translation invariant index of satisfaction, i.e. it satisfies (5.72), which in this context reads:

$$\Psi_{\mathbf{b}} \equiv 1 \Rightarrow \mathrm{Q}_c(\boldsymbol{\alpha} + \lambda\mathbf{b}) = \mathrm{Q}_c(\boldsymbol{\alpha}) + \lambda, \tag{5.165}$$

see Appendix www.5.4 for the proof and Figure 5.5 for a geometrical interpretation.

- **Super-/sub- additivity**

The quantile-based index of satisfaction is not super-additive, i.e. in generic markets we have:

$$\mathrm{Q}_c(\boldsymbol{\alpha} + \boldsymbol{\beta}) \ngeq \mathrm{Q}_c(\boldsymbol{\alpha}) + \mathrm{Q}_c(\boldsymbol{\beta}). \tag{5.166}$$

Therefore (5.75) is not satisfied. This is best proved by means of counterexamples, see for instance Artzner, Delbaen, Eber, and Heath (1999) for a counterexample with discrete distributions and Tasche (2002) for a counterexample with continuous distributions, both in the context of value at risk.

Therefore the quantile-based index of satisfaction, and in particular the value at risk, fails to promote diversification. This is the main reason why alternative measures of satisfaction such as the expected shortfall were developed, refer to Section 5.6.

Similarly, the quantile-based index of satisfaction is not sub-additive.

- **Co-monotonic additivity**

The quantile-based index of satisfaction is co-monotonic additive. Indeed, we prove in Appendix www.5.4 that if an allocation $\boldsymbol{\delta}$ gives rise to an objective which is an increasing function of the objective corresponding to another allocation $\boldsymbol{\alpha}$, the satisfaction from the total portfolio is the sum of the satisfactions from the separate allocations:

$$(\boldsymbol{\alpha}, \boldsymbol{\delta}) \text{ co-monotonic } \Rightarrow \mathrm{Q}_c(\boldsymbol{\alpha} + \boldsymbol{\delta}) = \mathrm{Q}_c(\boldsymbol{\alpha}) + \mathrm{Q}_c(\boldsymbol{\delta}). \tag{5.167}$$

This is the definition of co-monotonic additivity (5.80) in this context.

In other words, the combined portfolio is not perceived as providing a diversification effect: the quantile, and in particular the value at risk, are not "fooled" by derivatives.

- **Concavity/convexity**

The quantile-based index of satisfaction (5.159) is neither a concave nor a convex function of the allocation, see Figure 5.12 and refer to `symmys.com` for details on the market behind this figure.

Indeed, the matrix of second-order cross-derivatives of the quantile with respect to the allocation is neither negative definite nor positive definite, see (5.191) and comments thereafter. In other words, quantile-based indices of satisfaction fail to promote diversification, see the discussion on p. 257. This is one of the major critiques that have been directed to the value at risk.

- **Risk aversion/propensity**

The quantile-based index of satisfaction is neither risk averse, nor risk prone, nor risk neural. Indeed, depending on the distribution of the investor's objective and on the level of confidence required, the risk premium (5.85) can assume any sign.

For example, assume that the investor's objective has a Cauchy distribution:

$$\Psi_{\boldsymbol{\alpha}} \sim \mathrm{Ca}\left(\mu, \sigma^2\right). \tag{5.168}$$

Then from (1.82) we obtain:

$$\mathrm{RP}\left(\boldsymbol{\alpha}\right) \equiv \mathrm{E}\left\{\Psi_{\boldsymbol{\alpha}}\right\} - Q_{\Psi_{\boldsymbol{\alpha}}}\left(1-c\right) = -\sigma \tan\left(\pi\left(\frac{1}{2}-c\right)\right), \tag{5.169}$$

which can be larger than, equal to, or less than zero depending on the confidence level c.

5.5.2 Explicit dependence on allocation

We recall from (5.10) that the investor's objective, namely absolute wealth, relative wealth, net profits, or other specifications, is a simple linear function of the allocation and the market vector:

$$\Psi_{\boldsymbol{\alpha}} = \boldsymbol{\alpha}'\mathbf{M}. \tag{5.170}$$

Therefore the quantile-based index of satisfaction (5.159) depends on the allocation as follows:

$$\boldsymbol{\alpha} \mapsto \mathrm{Q}_c\left(\boldsymbol{\alpha}\right) \equiv Q_{\boldsymbol{\alpha}'\mathbf{M}}\left(1-c\right). \tag{5.171}$$

In this section we tackle the problem of computing explicitly the quantile-based index of satisfaction of an allocation for a given distribution of the market $\mathbf{M}$ and a given choice of the confidence level c.

For example, consider a market which is normally distributed at the investment horizon:

$$\mathbf{P}_{T+\tau} \sim \mathrm{N}\left(\boldsymbol{\mu}, \boldsymbol{\Sigma}\right). \tag{5.172}$$

Assume that the investor's objective are the net profits as in (5.8). In this case from (5.11) and (5.14) the distribution of the objective reads:

$$\Psi_{\alpha} \equiv \alpha'\mathbf{M} \sim \mathrm{N}\left(\mu_{\alpha}, \sigma_{\alpha}^{2}\right), \tag{5.173}$$

where

$$\mu_{\alpha} \equiv \alpha'\left(\boldsymbol{\mu} - \mathbf{P}_{T}\right), \quad \sigma_{\alpha}^{2} \equiv \alpha'\boldsymbol{\Sigma}\alpha. \tag{5.174}$$

Therefore from (1.70) we have:

$$\mathrm{Q}_{c}\left(\alpha\right) = \mu_{\alpha} + \sqrt{2}\sigma_{\alpha}\,\mathrm{erf}^{-1}\left(1 - 2c\right). \tag{5.175}$$

Although the objective is a simple linear function of the allocation and the market, the quantile is in general a complex expression of the allocation. Therefore in general the explicit dependence of the index of satisfaction on the allocation cannot be computed analytically.

There exists a vast literature regarding the computation of the quantile, and in particular of the VaR, using different techniques and under different distributional assumptions for the market, see for instance the list of references at `gloriamundi.org`. We mention here two cases that play a major role in the financial literature: the gamma approximation and the extreme value theory.

Delta-gamma approximation

When the market can be described by the gamma approximation (5.24) and the market invariants are approximately normal, (5.30) yields an approximate expression for the characteristic function of the objective Ψ_{α}:

$$\begin{aligned}\phi_{\Psi_{\alpha}}\left(\omega\right) \approx \left|\mathbf{I}_{K} - i\omega\boldsymbol{\Gamma}_{\alpha}\boldsymbol{\Sigma}\right|^{-\frac{1}{2}} & e^{i\omega\left(\theta_{\alpha} + \boldsymbol{\Delta}_{\alpha}'\boldsymbol{\mu} + \frac{1}{2}\boldsymbol{\mu}'\boldsymbol{\Gamma}_{\alpha}\boldsymbol{\mu}\right)} \\ & e^{-\frac{1}{2}\left[\boldsymbol{\Delta}_{\alpha} + \boldsymbol{\Gamma}_{\alpha}\boldsymbol{\mu}\right]'\boldsymbol{\Sigma}\left(\mathbf{I}_{K} - i\omega\boldsymbol{\Gamma}_{\alpha}\boldsymbol{\Sigma}\right)^{-1}\left[\boldsymbol{\Delta}_{\alpha} + \boldsymbol{\Gamma}_{\alpha}\boldsymbol{\mu}\right]}.\end{aligned} \tag{5.176}$$

The explicit dependence of θ, $\boldsymbol{\Delta}$ and $\boldsymbol{\Gamma}$ on the allocation α is given in (5.26)-(5.28).

From the characteristic function (5.176) we can compute the probability density function of the approximate objective with a numerical inverse Fourier transform as in (1.15) and thus we can compute the quantile Q by solving numerically the following implicit equation:

$$\int_{-\infty}^{\mathrm{Q}} \mathcal{F}^{-1}\left[\phi_{\Psi_{\alpha}}\right](x)\,dx \equiv 1 - c. \tag{5.177}$$

Nevertheless, this approach does not highlight the explicit dependence of the quantile on the allocation.

To tackle this issue, we can make use of a technique developed by Cornish and Fisher (1937), which expresses the quantile of a generic random variable X in terms of its moments and the quantile of the standard normal distribution:

$$z(p) \equiv \sqrt{2}\,\text{erf}^{-1}(2p-1), \tag{5.178}$$

see (1.70). The *Cornish-Fisher expansion* is an infinite series whose terms can be easily computed up to any order by means of software packages. The first terms read:

$$Q_X(p) = \text{E}\{X\} + \text{Sd}\{X\}\left[z(p) + \frac{1}{6}\left(z^2(p)-1\right)\text{Sk}\{X\}\right] + \cdots, \tag{5.179}$$

see Kotz, Balakrishnan, and Johnson (1994).

From the characteristic function (5.176) we can recover all the moments of the investor's objective as detailed in Appendix www.5.1. Therefore we can apply the Cornish-Fisher expansion to the investor's objective Ψ_{α} to compute its generic quantile, and thus in turn the quantile-based index of satisfaction (5.159).

In particular, in Appendix www.5.4 we derive the following approximation:

$$\text{Q}_c(\alpha) \approx A_{\alpha} + B_{\alpha} z(1-c) + C_{\alpha} z^2(1-c). \tag{5.180}$$

where

$$\begin{aligned} A &\equiv \text{E}\{\Psi_{\alpha}\} - \frac{\text{E}\{\Psi_{\alpha}^3\} - 3\,\text{E}\{\Psi_{\alpha}^2\}\text{E}\{\Psi_{\alpha}\} + 2\,\text{E}\{\Psi_{\alpha}\}^3}{6\left(\text{E}\{\Psi_{\alpha}^2\} - \text{E}\{\Psi_{\alpha}\}^2\right)} \\ B &\equiv \sqrt{\text{E}\{\Psi_{\alpha}^2\} - \text{E}\{\Psi_{\alpha}\}^2} \\ C &\equiv \frac{\text{E}\{\Psi_{\alpha}^3\} - 3\,\text{E}\{\Psi_{\alpha}^2\}\text{E}\{\Psi_{\alpha}\} + 2\,\text{E}\{\Psi_{\alpha}\}^3}{6\left(\text{E}\{\Psi_{\alpha}^2\} - \text{E}\{\Psi_{\alpha}\}^2\right)}. \end{aligned} \tag{5.181}$$

Refer to Appendix www.5.1 for the explicit dependence on the allocation of the moments that appear in these coefficient.

Approximations of order higher than (5.180) can be obtained similarly.

Extreme value theory

Extreme value theory (EVT) tackles the computation of the quantile when the confidence level in (5.171) is very high, i.e. $c \sim 1$, see Embrechts, Klueppelberg, and Mikosch (1997) and references therein.

Consider the *conditional excess function* of a generic random variable X which is defined as follows:

$$L_{\widetilde{\psi}}(z) \equiv \mathbb{P}\left\{X \leq \widetilde{\psi} - z \mid X \leq \widetilde{\psi}\right\} = \frac{F_X\left(\widetilde{\psi} - z\right)}{F_X\left(\widetilde{\psi}\right)}, \tag{5.182}$$

where F_X denotes the cumulative distribution function of X. The conditional excess function describes the probability that X is less than a generic value below a given threshold, conditioned on X being less than the given threshold.

Consider now the *generalized Pareto cumulative distribution function*, defined for positive z and v and all values of ξ (possibly taking limits for $\xi = 0$) as follows:

$$G_{\xi,v}(z) \equiv 1 - \left(1 + \frac{\xi}{v} z\right)^{-1/\xi}. \tag{5.183}$$

A theorem in Pickands (1975) and Balkema and De Haan (1974) states that under fairly general conditions for very low values of the threshold $\widetilde{\psi}$ in (5.182) there exist suitable values of the parameters ξ and v such that the following approximation holds:

$$1 - L_{\widetilde{\psi}}(z) \approx G_{\xi,v}(z). \tag{5.184}$$

Substituting (5.183) in (5.184) and applying this result to the investor's objective $X \equiv \Psi_{\boldsymbol{\alpha}}$ we obtain an approximation for the cumulative distribution function of the objective for very low values of its range:

$$F_{\Psi_{\boldsymbol{\alpha}}}(\psi) \approx F_{\Psi_{\boldsymbol{\alpha}}}\left(\widetilde{\psi}\right)\left(1 + \frac{\xi(\boldsymbol{\alpha})}{v(\boldsymbol{\alpha})}\left(\widetilde{\psi} - \psi\right)\right)^{-1/\xi}. \tag{5.185}$$

Inverting this relation we obtain the approximate expression for the quantile-based index of satisfaction:

$$\mathrm{Q}_c(\boldsymbol{\alpha}) \approx \widetilde{\psi} + \frac{v(\boldsymbol{\alpha})}{\xi(\boldsymbol{\alpha})}\left[1 - \left(\frac{1-c}{F_{\Psi_{\boldsymbol{\alpha}}}\left(\widetilde{\psi}\right)}\right)^{-\xi(\boldsymbol{\alpha})}\right]. \tag{5.186}$$

Nevertheless, the applicability of this formula in this context is limited because the dependence on the allocation $\boldsymbol{\alpha}$ of the parameters v and ξ and of the threshold cdf $F_{\Psi_{\boldsymbol{\alpha}}}\left(\widetilde{\psi}\right)$ is non-trivial.

5.5.3 Sensitivity analysis

Suppose that the investor has already chosen an allocation $\boldsymbol{\alpha}$ which yields a level of satisfaction $\mathrm{Q}_c(\boldsymbol{\alpha})$ and that he is interested in rebalancing his portfolio marginally by means of a small change $\delta\boldsymbol{\alpha}$ in the current allocation. In this case a local analysis in terms of a Taylor expansion is useful:

$$\begin{aligned}\mathrm{Q}_c(\boldsymbol{\alpha} + \delta\boldsymbol{\alpha}) \approx{}& \mathrm{Q}_c(\boldsymbol{\alpha}) + \delta\boldsymbol{\alpha}' \frac{\partial\,\mathrm{Q}_c(\boldsymbol{\alpha})}{\partial\boldsymbol{\alpha}} \\ &+ \frac{1}{2}\delta\boldsymbol{\alpha}' \frac{\partial^2\,\mathrm{Q}_c(\boldsymbol{\alpha})}{\partial\boldsymbol{\alpha}\partial\boldsymbol{\alpha}'}\delta\boldsymbol{\alpha}.\end{aligned} \tag{5.187}$$

We prove in Appendix www.5.4 that the first-order derivatives read:

$$\frac{\partial\,\mathrm{Q}_c(\boldsymbol{\alpha})}{\partial\boldsymbol{\alpha}} = \mathrm{E}\left\{\mathbf{M} \middle| \boldsymbol{\alpha}'\mathbf{M} = \mathrm{Q}_c(\boldsymbol{\alpha})\right\}, \tag{5.188}$$

where $\mathbf{M}$ is the random vector (5.170) that represents the market, see Hallerbach (2003), Gourieroux, Laurent, and Scaillet (2000), Tasche (2002). The investor will focus on the entries of the vector (5.188) that display a large absolute value.

For example, in the case of normal markets from (5.175) we obtain directly the first-order derivatives:

$$\frac{\partial \, \mathrm{Q}_c(\boldsymbol{\alpha})}{\partial \boldsymbol{\alpha}} = \boldsymbol{\mu} - \mathbf{p}_T + \frac{\boldsymbol{\Sigma}\boldsymbol{\alpha}}{\sqrt{\boldsymbol{\alpha}'\boldsymbol{\Sigma}\boldsymbol{\alpha}}} \sqrt{2}\, \mathrm{erf}^{-1}(1-2c) . \tag{5.189}$$

The quantile-based index of satisfaction is positive homogeneous, see (5.164). Therefore it can be expressed in terms of the contribution to satisfaction from each security by means of the Euler decomposition (5.67), which in this context reads:

$$\mathrm{Q}_c(\boldsymbol{\alpha}) = \sum_{n=1}^{N} \alpha_n \, \mathrm{E}\{M_n | \boldsymbol{\alpha}'\mathbf{M} = \mathrm{Q}_c(\boldsymbol{\alpha})\} . \tag{5.190}$$

The contribution to satisfaction from each security in turn factors into the product of the amount of that security times the marginal contribution to satisfaction of that security. The marginal contribution to satisfaction, i.e. the conditional expectation of the market vector (5.188), is insensitive to a rescaling of the portfolio, although it depends on the allocation.

For example, consider the case of normal markets. Left-multiplying the marginal contributions to satisfaction (5.189) by the amount of each security $\boldsymbol{\alpha}'$ we obtain the quantile-based index of satisfaction (5.175). Also, a rescaling $\boldsymbol{\alpha} \mapsto \lambda\boldsymbol{\alpha}$ does not affect (5.189).

The study of the second-order cross-derivatives provides insight in the local convexity/concavity of the quantile. In Appendix www.5.4 we adapt from Gourieroux, Laurent, and Scaillet (2000) to prove the following result:

$$\begin{aligned} \frac{\partial^2 \, \mathrm{Q}_c(\boldsymbol{\alpha})}{\partial \boldsymbol{\alpha}' \partial \boldsymbol{\alpha}} = & - \left. \frac{\partial \ln f_{\Psi_{\boldsymbol{\alpha}}}(\psi)}{\partial \psi} \right|_{\psi = \mathrm{Q}_c(\boldsymbol{\alpha})} \mathrm{Cov}\{\mathbf{M} | \Psi_{\boldsymbol{\alpha}} = \mathrm{Q}_c(\boldsymbol{\alpha})\} \\ & - \left. \frac{\partial \, \mathrm{Cov}\{\mathbf{M} | \Psi_{\boldsymbol{\alpha}} = \psi\}}{\partial \psi} \right|_{\psi = \mathrm{Q}_c(\boldsymbol{\alpha})} , \end{aligned} \tag{5.191}$$

where $f_{\Psi_{\boldsymbol{\alpha}}}$ is the probability density function of the investor's objective $\Psi_{\boldsymbol{\alpha}} \equiv \boldsymbol{\alpha}'\mathbf{M}$.

For example, consider the case of normal markets. By direct derivation of (5.189) we obtain:

$$\frac{\partial^2 Q_c(\boldsymbol{\alpha})}{\partial \boldsymbol{\alpha}' \partial \boldsymbol{\alpha}} = \boldsymbol{\Sigma} \left(\mathbf{I}_N - \frac{\boldsymbol{\alpha}\boldsymbol{\alpha}'\boldsymbol{\Sigma}}{\boldsymbol{\alpha}'\boldsymbol{\Sigma}\boldsymbol{\alpha}} \right) \frac{\sqrt{2}\,\mathrm{erf}^{-1}(1-2c)}{\sqrt{\boldsymbol{\alpha}'\boldsymbol{\Sigma}\boldsymbol{\alpha}}}. \tag{5.192}$$

In the normal case the second term in (5.191) vanishes because the conditional covariance does not depend on the value on which it is conditioned.

In Appendix www.5.4 we prove that the matrix (5.192) is negative definite for high confidence levels, namely $c > 0.5$, and positive definite for low confidence levels, namely $c < 0.5$. Therefore for high confidence levels the quantile is concave and for low confidence levels the quantile is convex.

In general the first term in (5.191) is fairly easy to analyze: the conditional covariance is always positive definite and the elasticity of the marginal density is typically positive in the lower tail, which corresponds to a high level of confidence $c \sim 1$ in the quantile. Therefore for high levels of confidence the first term is negative definite and tends to make the quantile a concave index of satisfaction. Similarly, the elasticity of the marginal density is typically negative in the upper tail, which corresponds to a low level of confidence $c \sim 0$ in the quantile. Therefore for low levels of confidence the first term is positive definite and tends to make the quantile a convex index of satisfaction.

On the other hand, the sign of the second term in (5.191) is not determined. Therefore in general the quantile-based index of satisfaction is neither convex nor concave, see Figure 5.12.

5.6 Coherent indices (expected shortfall)

In Section 5.4 we discussed a first class of indices of satisfaction based on expected utility, namely the certainty-equivalent, and in Section 5.5 we introduced a second class of indices, namely the quantile-based indices of satisfaction. In this section, following the recent literature on measures of risk, we discuss a third approach to model the investor's satisfaction, namely coherent indices of satisfaction, a class of indices of satisfaction that are defined directly in terms of the properties that they are supposed to feature, see Artzner, Delbaen, Eber, and Heath (1997) and Artzner, Delbaen, Eber, and Heath (1999).

Such indices originated from the critiques directed to the value at risk for not promoting diversification. Indeed, the quantile-based index of satisfaction is not a concave function of allocation, see Figure 5.12. Therefore it fails to promote diversification. Although diversification is not necessarily a requirement for the portfolio of a private investor, it is an important requirement in the investment policy of a financial institution.

5.6.1 Properties

In this section we introduce the broad class of *coherent indices of satisfaction*, namely functions that with a generic allocation $\boldsymbol{\alpha}$ associate a level of satisfaction $\mathrm{Coh}(\boldsymbol{\alpha})$ in such a way to satisfy the properties discussed below. Since these indices are defined axiomatically in terms of their properties, we revisit the properties of a generic index of satisfaction introduced in Section 5.3 highlighting which of them define coherent indices of satisfaction and which of them are consequences of the definitions.

Consider an investor with a given objective Ψ such as absolute wealth, as in (5.3), or relative wealth, as in (5.4), or net profits, as in (5.8), or possibly other specifications. As usual, we denote as $\Psi_{\boldsymbol{\alpha}}$ the the objective relative to a generic allocation $\boldsymbol{\alpha}$, see (5.10).

- **Sensibility (definition)**

A coherent index of satisfaction must be sensible, i.e. consistent with strong dominance:

$$\Psi_{\boldsymbol{\alpha}} \geq \Psi_{\boldsymbol{\beta}} \text{ in all scenarios } \Rightarrow \mathrm{Coh}(\boldsymbol{\alpha}) \geq \mathrm{Coh}(\boldsymbol{\beta}), \tag{5.193}$$

which is (5.55) in this context.

- **Positive homogeneity (definition)**

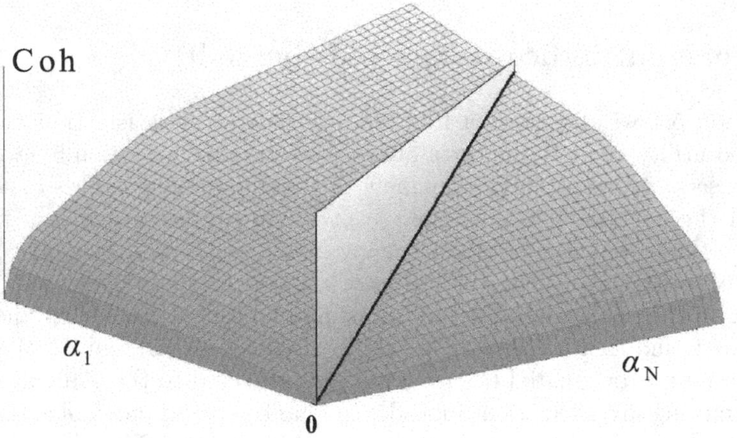

Fig. 5.13. Coherent satisfaction index as function of allocation

A coherent index of satisfaction must be positive homogeneous:

$$\text{Coh}(\lambda\boldsymbol{\alpha}) = \lambda\,\text{Coh}(\boldsymbol{\alpha}), \quad \text{for all } \lambda \geq 0, \tag{5.194}$$

which is (5.65) in this context. In other words, the coherent index grows linearly in any radial direction stemming from the origin of the allocation space, see Figure 5.13, which refers to the same market as Figure 5.12. Refer to `symmys.com` for details on these figures.

Notice that the requirement of positive homogeneity rules out the certainty-equivalent as a coherent index of satisfaction, unless the utility function is of the power class.

- **Translation invariance (definition)**

  A coherent index of satisfaction must be translation invariant:

$$\Psi_{\mathbf{b}} \equiv 1 \Rightarrow \text{Coh}(\boldsymbol{\alpha} + \lambda\mathbf{b}) = \text{Coh}(\boldsymbol{\alpha}) + \lambda, \tag{5.195}$$

which is (5.72) in this context, see Figure 5.5 for a geometrical interpretation.

Notice that the requirement of translation invariance rules out the certainty-equivalent as a coherent index of satisfaction, unless the utility function is of the exponential class.

The only intersection between the exponential class and the power class is the trivial linear utility function, in which case the certainty-equivalent becomes the expected value of the investor's objective, which is our first example of coherent index of satisfaction:

$$\text{Coh}(\boldsymbol{\alpha}) \equiv \text{CE}(\boldsymbol{\alpha}) \equiv \text{E}\{\Psi_{\boldsymbol{\alpha}}\}. \tag{5.196}$$

- **Super-additivity (definition)**

  A coherent index of satisfaction must be super-additive:

$$\text{Coh}(\boldsymbol{\alpha} + \boldsymbol{\beta}) \geq \text{Coh}(\boldsymbol{\alpha}) + \text{Coh}(\boldsymbol{\beta}), \tag{5.197}$$

which is (5.75) in this context. Notice that this requirement rules out the quantile-based index of satisfaction, see (5.166). Also, this requirement rules out the certainty-equivalent as a coherent index of satisfaction, except for the trivial case of a linear utility function (5.196).

A notable class of coherent indices of satisfaction are the one-sided moments, see Fischer (2003):

$$\text{Coh}(\boldsymbol{\alpha}) \equiv \text{E}\{\Psi_{\boldsymbol{\alpha}}\} - \gamma \left\| \min\left(0, \Psi_{\boldsymbol{\alpha}} - \text{E}\{\Psi_{\boldsymbol{\alpha}}\}\right) \right\|_{\mathbf{M};p}. \tag{5.198}$$

In this expression $\gamma \geq 0$ and $\|\cdot\|_{\mathbf{M};p}$ is the market-based expectation norm (B.57), which in terms of the market probability density function $f_{\mathbf{M}}$ reads:

$$\|g\|_{\mathbf{M};p} \equiv \left(\int |g(\mathbf{m})|^p f_{\mathbf{M}}(\mathbf{m})\, d\mathbf{m} \right)^{\frac{1}{p}} . \tag{5.199}$$

Notice that (5.198) is defined in terms of the distribution of the objective $\Psi_{\boldsymbol{\alpha}}$ and as such it is law invariant, or estimable, see also below.

The trivial case $\gamma \equiv 0$ recovers the expected value (5.196) as a coherent index of satisfaction.

The specific case $\gamma \equiv 1$ and $p \equiv 2$ gives rise to the *mean/semistandard deviation*, which practitioners use extensively.

The above four defining properties imply other features.

- **Money-equivalence (consequence of definition)**

The joint assumptions of positive homogeneity and translation invariance imply that a coherent index of satisfaction is naturally measured in terms of money.

- **Concavity (consequence of definition)**

The joint assumptions of positive homogeneity and super-additivity imply that a coherent index of satisfaction is concave, i.e. for all $\lambda \in [0,1]$ the following holds true:

$$\operatorname{Coh}(\lambda\boldsymbol{\alpha} + (1-\lambda)\boldsymbol{\beta}) \geq \lambda \operatorname{Coh}(\boldsymbol{\alpha}) + (1-\lambda)\operatorname{Coh}(\boldsymbol{\beta}), \tag{5.200}$$

which is (5.81) in this context.

In other words, coherent indices of satisfaction promote diversification by construction, see Figure 5.13. We recall that this is not the case for quantile-based indices of satisfaction such as the value at risk, see Figure 5.12, which refers to the same market as Figure 5.13. Refer to `symmys.com` for details on these figures.

The above properties of the coherent indices of satisfaction cover many but not all the potential features of a generic index of satisfaction discussed in Section 5.3.

Adding a few more intuitive requirements, Acerbi (2002) introduced the sub-class of coherent indices known as *spectral indices of satisfaction*[5]. Spectral indices of satisfaction are functions that with a generic allocation $\boldsymbol{\alpha}$ associate a level of satisfaction $\operatorname{Spc}(\boldsymbol{\alpha})$ which satisfies the properties of coherent indices of satisfaction and the additional properties discussed below. Indeed, spectral indices of satisfaction cover all the potential features of a generic index of satisfaction discussed in Section 5.3. Again, we distinguish between truly new defining features and simple consequences of the definitions.

- **Estimability (definition)**

[5] The reason for this terminology will become apparent in Section 5.6.2.

A spectral index of satisfaction is fully determined by the distribution of the investor's objective Ψ_{α}, as represented by either its probability density function, cumulative distribution function, or characteristic function. In other words, a spectral index of satisfaction is defined in terms of the following chain:

$$\alpha \mapsto \Psi_{\alpha} \mapsto \left(f_{\Psi_{\alpha}}, F_{\Psi_{\alpha}}, \phi_{\Psi_{\alpha}}\right) \mapsto \mathrm{Spc}\left(\alpha\right), \tag{5.201}$$

which is (5.52) in this context.

- **Co-monotonic additivity (definition)**

A spectral index of satisfaction is co-monotonic additive:

$$(\alpha, \delta) \text{ co-monotonic } \Rightarrow \mathrm{Spc}\left(\alpha + \delta\right) = \mathrm{Spc}\left(\alpha\right) + \mathrm{Spc}\left(\delta\right), \tag{5.202}$$

which is (5.80) in this context. In other words, like (non-coherent) quantile-based indices of satisfaction such as the value at risk, the spectral indices of satisfaction are not "fooled" by derivatives: whenever the objective relative to one allocation is an increasing function of the objective stemming from another allocation, the combined portfolio is not perceived as providing a diversification effect.

For example the one-sided moments (5.198) are estimable, but not co-monotonic additive. Therefore they give rise to coherent indices of satisfaction, but not to spectral indices of satisfaction. On the other hand, consider the expected value of the objective:

$$\mathrm{Spc}\left(\alpha\right) \equiv \mathrm{E}\left\{\Psi_{\alpha}\right\}. \tag{5.203}$$

From (5.77) the expected value is additive and thus in particular it is co-monotonic additive. Furthermore, from (5.54) it is estimable and from (5.196) it is coherent. Therefore the expected value of the investor's objective is a spectral index of satisfaction.

The above defining properties of spectral indices of satisfaction imply the following features.

- **Consistence with weak stochastic dominance (consequence of definition)**

From (5.59) sensibility and estimability imply consistence with weak stochastic dominance. Therefore the spectral indices of satisfaction are consistent with weak stochastic dominance:

$$Q_{\Psi_{\alpha}}\left(p\right) \geq Q_{\Psi_{\beta}}\left(p\right) \text{ for all } p \in (0,1) \Rightarrow \mathrm{Spc}\left(\alpha\right) \geq \mathrm{Spc}\left(\beta\right), \tag{5.204}$$

which is (5.57) in this context.

- **Constancy (consequence of definition)**

Translation invariance and homogeneity imply constancy, see Appendix www.5.2. Therefore the spectral indices of satisfaction satisfy:

$$\Psi_{\mathbf{b}} \equiv \psi_{\mathbf{b}} \Rightarrow \mathrm{Spc}\left(\mathbf{b}\right) = \psi_{\mathbf{b}}, \tag{5.205}$$

which is (5.62) in this context.

- **Risk aversion (consequence of definition)**

Spectral indices of satisfaction are risk averse. Indeed, we prove in Appendix www.5.5 that the risk premium of an allocation associated with a spectral index of satisfaction is positive: this is the definition of risk aversion (5.86).

5.6.2 Building coherent indices

In this section we discuss how to build coherent indices of satisfaction. More precisely, we focus on spectral indices. We proceed as in the case of the certainty-equivalent, see Section 5.4.2. In other words, we specify a basis for the spectral indices of satisfaction and then we obtain all possible indices as weighted averages of the basis.

To define a basis, we start from the only example of spectral index introduced so far, namely the expected value of the investor's objective (5.203). In order to build other elements for the basis of the spectral indices we generalize the expected value, which, with a change of variables, we can express as the average of all the quantiles:

$$\begin{aligned} \mathrm{E}\left\{\Psi_{\alpha}\right\} &= \int_{\mathbb{R}} \psi f_{\Psi_{\alpha}}\left(\psi\right) d\psi \\ &= \int_{0}^{1} Q_{\Psi_{\alpha}}\left(s\right) ds. \end{aligned} \tag{5.206}$$

One the one hand, it is important to start with the quantiles. Indeed, the quantile is co-monotonic additive, see (5.167), and estimable, see (5.160). These are exactly the two new features required of spectral indices of satisfaction.

On the other hand, it is important to suitably average a given range of quantiles, because the quantile per se is not super-additive, see (5.166). Super-additivity is a key feature of coherent measures of risk, and thus in particular it is a key feature of spectral indices of satisfaction.

Therefore, we define new indices of satisfaction as averages of quantiles. To make the ensuing index of satisfaction more conservative, we average the quantiles over the worst scenarios. This way we obtain the definition of *expected shortfall*[6] (*ES*):

[6] The name expected shortfall applies specifically to the case where the investor's objective are net profits, much like the value at risk in the context of quantile-based indices of satisfaction. For simplicity, we extend the terminology to a generic objective, such as absolute or relative wealth.

$$\mathrm{ES}_c(\boldsymbol{\alpha}) \equiv \frac{1}{1-c} \int_0^{1-c} Q_{\Psi_{\boldsymbol{\alpha}}}(s)\, ds, \tag{5.207}$$

where $c \in [0,1]$ is a fixed *confidence level*. The expected shortfall is indeed a generalization of the expected value, which represents the specific case $c \equiv 0$.

Notice that the expected shortfall is the expected value of the investor's objective, conditioned on the realization of the objective being less than the quantile-based index of satisfaction (5.159). Therefore the expected shortfall is the *tail conditional expectation* (*TCE*), also known as *conditional value at risk*[7] (*CVaR*):

$$\begin{aligned} \mathrm{ES}_c(\boldsymbol{\alpha}) &= \mathrm{TCE}_c(\boldsymbol{\alpha}) = \mathrm{CVaR}_c(\boldsymbol{\alpha}) \\ &\equiv \mathrm{E}\{\Psi_{\boldsymbol{\alpha}} | \Psi_{\boldsymbol{\alpha}} \leq \mathrm{Q}_c(\boldsymbol{\alpha})\}. \end{aligned} \tag{5.208}$$

We now verify that, just like the expected value, the expected shortfall satisfies the defining properties of a spectral index of satisfaction.

- **Sensibility**

  The expected shortfall is sensible:

$$\Psi_{\boldsymbol{\alpha}} \geq \Psi_{\boldsymbol{\beta}} \text{ in all scenarios } \Rightarrow \mathrm{ES}_c(\boldsymbol{\alpha}) \geq \mathrm{ES}_c(\boldsymbol{\beta}). \tag{5.209}$$

Indeed, from (5.57) and (5.207) it is consistent with first-order dominance, and thus from (5.61) it is consistent with strong dominance.

- **Positive homogeneity**

  The expected shortfall is positive homogeneous:

$$\mathrm{ES}_c(\lambda\boldsymbol{\alpha}) = \lambda\, \mathrm{ES}_c(\boldsymbol{\alpha}), \qquad \lambda \geq 0, \tag{5.210}$$

see Figure 5.13. This follows from the linearity of the integral in the definition of expected shortfall and the positive homogeneity of the quantile proved in Appendix www.5.4.

- **Translation invariance**

  The expected shortfall is translation invariant:

$$\mathrm{ES}_c(\boldsymbol{\alpha} + \lambda\mathbf{b}) = \mathrm{ES}_c(\boldsymbol{\alpha}) + \lambda, \tag{5.211}$$

see Figure 5.5 for a geometrical interpretation. This follows from the linearity of the integral in the definition of expected shortfall and the translation invariance of the quantile proved in Appendix www.5.4.

- **Super-additivity**

[7] We assume that the probability density function of the objective $f_{\Psi_{\boldsymbol{\alpha}}}$ is smooth, otherwise this is not true, see Acerbi and Tasche (2002) for a counterexample.

The expected shortfall is super-additive:

$$\mathrm{ES}_c(\boldsymbol{\alpha}+\boldsymbol{\beta}) \geq \mathrm{ES}_c(\boldsymbol{\alpha}) + \mathrm{ES}_c(\boldsymbol{\beta}). \tag{5.212}$$

For the proof of super-additivity we refer the interested reader to Acerbi and Tasche (2002).

We stress that in particular the joint assumptions of positive homogeneity and super-additivity imply that, unlike the value at risk, the expected shortfall is concave. In other words, for all $\lambda \in [0,1]$ the following holds true:

$$\mathrm{ES}_c(\lambda\boldsymbol{\alpha}+(1-\lambda)\boldsymbol{\beta}) \geq \lambda\,\mathrm{ES}_c(\boldsymbol{\alpha}) + (1-\lambda)\,\mathrm{ES}_c(\boldsymbol{\beta}). \tag{5.213}$$

We see this in Figure 5.13, which refers to a market of two securities. Compare also with Figure 5.12, which refers to the value at risk in the same market. See `symmys.com` for details on these figures.

- **Estimability**

The expected shortfall is estimable, since the quantile in the definition (5.207) of expected shortfall is the inverse of the cumulative distribution function $F_{\Psi_{\boldsymbol{\alpha}}}$ of the investor's objective. Therefore a chain-definition such as (5.201) applies:

$$\boldsymbol{\alpha} \mapsto \Psi_{\boldsymbol{\alpha}} \mapsto F_{\Psi_{\boldsymbol{\alpha}}} \mapsto \mathrm{ES}_c(\boldsymbol{\alpha}). \tag{5.214}$$

- **Co-monotonic additivity**

The expected shortfall is co-monotonic additive:

$$(\boldsymbol{\alpha},\boldsymbol{\delta}) \text{ co-monotonic } \Rightarrow \mathrm{ES}_c(\boldsymbol{\alpha}+\boldsymbol{\delta}) = \mathrm{ES}_c(\boldsymbol{\alpha}) + \mathrm{ES}_c(\boldsymbol{\delta}). \tag{5.215}$$

This follows from the linearity of the integral in the definition (5.207) of expected shortfall and the co-monotonic additivity of the quantile proved in Appendix www.5.4.

Since it satisfies the defining properties of spectral indices of satisfaction, the expected shortfall belongs to this class for any value of the confidence level c. Therefore we choose the expected shortfall as a basis to generate other spectral indices of satisfaction. In other words, we consider all weighted averages of the expected shortfall, in a way completely similar to the construction of utility functions (5.125). As we prove in Appendix www.5.5, this way we obtain the following class of spectral indices of satisfaction:

$$\mathrm{Spc}_{\varphi}(\boldsymbol{\alpha}) \equiv \int_0^1 \varphi(p)\, Q_{\Psi_{\boldsymbol{\alpha}}}(p)\, dp, \tag{5.216}$$

where the *spectrum* φ is a function that satisfies:

$$\varphi \text{ decreasing}, \quad \varphi(1) \equiv 0, \quad \int_0^1 \varphi(p)\, dp \equiv 1. \tag{5.217}$$

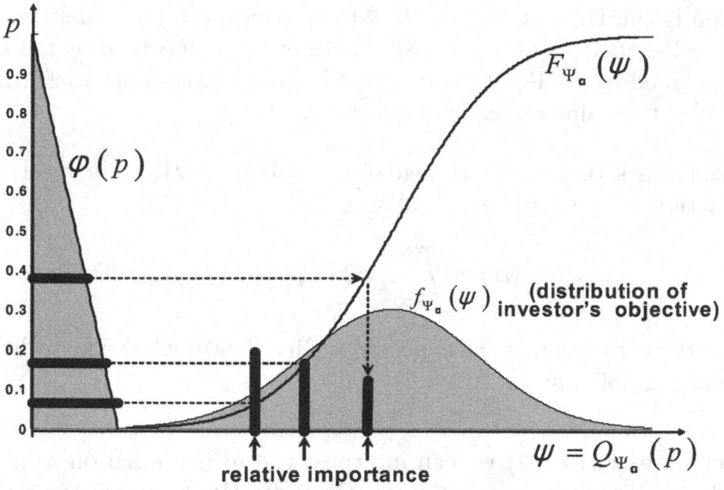

Fig. 5.14. Spectral indices of satisfaction emphasize adverse scenarios

In other words, spectral indices of satisfaction give relatively speaking more importance to the unwelcome scenarios in which the investor's objective is low, see Figure 5.14. This feature makes spectral indices of satisfaction risk averse.

As it turns out, the class (5.216) is exhaustive, as any spectral index of satisfaction can be expressed this way for a suitable choice of the spectrum φ, see Kusuoka (2001) or Tasche (2002). This clarifies why this class of indices of satisfaction is called "spectral".

For example, the expected shortfall can be represented in the form (5.216) by the following spectrum:

$$\varphi_{\mathrm{ES}_c}(p) \equiv \frac{H^{(c-1)}(-p)}{1-c}, \tag{5.218}$$

where $H^{(x)}$ is the Heaviside step function (B.74). It is easy to check that this spectrum satisfies the requirements (5.217).

The requirements (5.217) on the spectrum are essential to obtain a coherent index.

For example, also the quantile-based index of satisfaction can be represented in the form (5.216) by the following spectrum:

$$\varphi_{\mathrm{Q}_c} \equiv \delta^{(1-c)}, \tag{5.219}$$

where δ is the Dirac delta (B.16). Nevertheless, the Dirac-delta is (the limit of) a bell-shaped function (B.18). Therefore, the spectrum is not decreasing as prescribed in (5.217). Indeed, quantile-based indices of satisfaction, and in particular the value at risk, are not coherent.

We remark that with a change of variable in (5.216) any spectral index of satisfaction can be written as follows:

$$\mathrm{Spc}_{\varphi}(\boldsymbol{\alpha}) = \int_{-\infty}^{+\infty} \psi\varphi\left(F_{\Psi_{\boldsymbol{\alpha}}}(\psi)\right) f_{\Psi_{\boldsymbol{\alpha}}}(\psi)\, d\psi. \tag{5.220}$$

This is the expression of the expected utility (5.90), where the utility function is defined as follows:

$$u(\psi) \equiv \psi\varphi\left(F_{\Psi_{\boldsymbol{\alpha}}}(\psi)\right). \tag{5.221}$$

Therefore, as in (5.102) we can interpret the utility function as a subjective cumulative distribution function that reflects the investor's a-priori view on the outcome of his investment. Nevertheless, in this case the utility function depends not only on the investor's attitude toward risk, i.e. the spectral function φ, but also on the market and the allocation decision through the cumulative distribution function $F_{\Psi_{\boldsymbol{\alpha}}}$.

5.6.3 Explicit dependence on allocation

We recall from (5.10) that the investor's objective, namely absolute wealth, relative wealth, net profits, or other specifications, is a simple linear function of the allocation and the market vector:

$$\Psi_{\boldsymbol{\alpha}} = \boldsymbol{\alpha}'\mathbf{M}. \tag{5.222}$$

Therefore the spectral indices of satisfaction (5.216) depends on the allocation as follows:

$$\boldsymbol{\alpha} \mapsto \mathrm{Spc}_{\varphi}(\boldsymbol{\alpha}) \equiv \int_{0}^{1} \varphi(p)\, Q_{\boldsymbol{\alpha}'\mathbf{M}}(p)\, dp. \tag{5.223}$$

Notice that the quantile depends on the allocation and on the distribution of the market, whereas the spectrum does not.

In this section we tackle the problem of computing explicitly spectral indices of satisfaction for a given distribution of the market $\mathbf{M}$ and a given choice of the spectrum φ.

For example, consider a market that at the investment horizon is normally distributed:

$$\mathbf{P}_{T+\tau} \sim \mathrm{N}(\boldsymbol{\mu}, \boldsymbol{\Sigma}). \tag{5.224}$$

Assume that the investor's objective are the net profits as in (5.8). In this case from (5.11) and (5.14) the distribution of the objective reads:

$$\Psi_{\alpha} \equiv \alpha' \mathbf{M} \sim \mathrm{N}\left(\mu_{\alpha}, \sigma_{\alpha}^{2}\right), \tag{5.225}$$

where

$$\mu_{\alpha} \equiv \alpha' \left(\mu - \mathbf{P}_{T}\right), \qquad \sigma_{\alpha}^{2} \equiv \alpha' \Sigma \alpha. \tag{5.226}$$

From (1.70) the quantile of the investor's objective reads:

$$Q_{\alpha' \mathbf{M}}(p) = \mu_{\alpha} + \sigma_{\alpha} z(p), \tag{5.227}$$

where z is the quantile of the standard normal distribution:

$$z(p) \equiv \sqrt{2}\, \mathrm{erf}^{-1}(2p - 1). \tag{5.228}$$

Therefore in this market the generic spectral index of satisfaction reads:

$$\mathrm{Spc}_{\varphi}(\alpha) = \mu_{\alpha} + \sigma_{\alpha} \mathcal{I}[\varphi z](1), \tag{5.229}$$

where $\mathcal{I}$ denotes the integration operator (B.27). Notice that the integral does not depend on the allocation.

In generic markets the quantile of the objective in (5.223) is a complex expression of the allocation. Therefore in general the explicit dependence of a spectral index of satisfaction on allocation cannot be computed analytically.

Nevertheless, as for the quantile-based indices of satisfaction, we mention here two special quite general cases where we can compute approximate expressions for the expected shortfall: the delta-gamma approximation and extreme value theory.

Delta-gamma approximation

When the market can be described by the gamma approximation (5.24) and the market invariants are approximately normal, (5.30) yields an approximate expression for the characteristic function of the objective Ψ_{α}:

$$\begin{aligned}\phi_{\Psi_{\alpha}}(\omega) \approx & \left|\mathbf{I}_{K} - i\omega \boldsymbol{\Gamma}_{\alpha} \boldsymbol{\Sigma}\right|^{-\frac{1}{2}} e^{i\omega\left(\theta_{\alpha} + \boldsymbol{\Delta}_{\alpha}' \boldsymbol{\mu} + \frac{1}{2}\boldsymbol{\mu}' \boldsymbol{\Gamma}_{\alpha} \boldsymbol{\mu}\right)} \\ & e^{-\frac{1}{2}\left[\boldsymbol{\Delta}_{\alpha} + \boldsymbol{\Gamma}_{\alpha}\boldsymbol{\mu}\right]' \boldsymbol{\Sigma}\left(\mathbf{I}_{K} - i\omega \boldsymbol{\Gamma}_{\alpha} \boldsymbol{\Sigma}\right)^{-1}\left[\boldsymbol{\Delta}_{\alpha} + \boldsymbol{\Gamma}_{\alpha}\boldsymbol{\mu}\right]}.\end{aligned} \tag{5.230}$$

The explicit dependence of θ, $\boldsymbol{\Delta}$ and $\boldsymbol{\Gamma}$ on the allocation α is given in (5.26)-(5.28).

From the characteristic function we can compute the probability density function of the approximate objective with a numerical inverse Fourier transform as in (1.15) and then we can compute the quantile by solving numerically the following implicit equation:

$$\int_{-\infty}^{Q} \mathcal{F}^{-1}\left[\phi_{\Psi_{\alpha}}\right](x)\, dx \equiv p. \tag{5.231}$$

Finally, a third numerical integration yields the spectral index of satisfaction (5.223). Nevertheless, this approach is computationally intensive and unstable and does not highlight the explicit dependence of the spectral index on allocation on the allocation.

To tackle this issue, we can use the Cornish-Fisher expansion (5.179). In Appendix www.5.5 we show that an approximate expression of the spectral index of satisfaction in terms of the integration operator (B.27) reads:

$$\mathrm{Spc}_{\varphi}(\boldsymbol{\alpha}) \approx A_{\boldsymbol{\alpha}} + B_{\boldsymbol{\alpha}} \mathcal{I}[\varphi z](1) + C_{\boldsymbol{\alpha}} \mathcal{I}\left[\varphi z^2\right](1), \tag{5.232}$$

where the coefficients A, B, C are defined in (5.181). Notice that the integrals in this expression do not depend on the allocation: therefore they can be evaluated numerically once and for all. Higher-order approximations can be obtained similarly.

Extreme value theory

Extreme value theory does not apply to the computation of spectral measures of satisfaction in general. Nevertheless, it does apply to the computation of the most notable among the spectral measures of satisfaction, namely the expected shortfall, when the confidence level in (5.207) is very high, i.e. $c \sim 1$.

As we show in Appendix www.5.5 there exist functions of the allocation $v(\boldsymbol{\alpha})$ and $\xi(\boldsymbol{\alpha})$ such that the expected shortfall can be approximated as follows:

$$\mathrm{ES}_c(\boldsymbol{\alpha}) \approx \mathrm{Q}_c(\boldsymbol{\alpha}) - \frac{v(\boldsymbol{\alpha})}{1 - \xi(\boldsymbol{\alpha})}, \tag{5.233}$$

where the parameters v and ξ and the extreme quantile $\mathrm{Q}_c(\boldsymbol{\alpha})$ are as in (5.186). Nevertheless, the applicability of this formula in this context is limited because the explicit dependence on the allocation of the parameters v and ξ and of the quantile-based index of satisfaction Q_c is non-trivial.

5.6.4 Sensitivity analysis

Suppose that the investor has already chosen an allocation $\boldsymbol{\alpha}$ which yields a level of satisfaction $\mathrm{Spc}_{\varphi}(\boldsymbol{\alpha})$ and that he is interested in rebalancing his portfolio marginally by means of a small change $\delta\boldsymbol{\alpha}$ in the current allocation. In this case a local analysis in terms of a Taylor expansion is useful:

$$\begin{aligned} \mathrm{Spc}_{\varphi}(\boldsymbol{\alpha} + \delta\boldsymbol{\alpha}) \approx\ & \mathrm{Spc}_{\varphi}(\boldsymbol{\alpha}) + \delta\boldsymbol{\alpha}' \frac{\partial \, \mathrm{Spc}_{\varphi}(\boldsymbol{\alpha})}{\partial \boldsymbol{\alpha}} \\ & + \frac{1}{2} \delta\boldsymbol{\alpha}' \frac{\partial^2 \, \mathrm{Spc}_{\varphi}(\boldsymbol{\alpha})}{\partial \boldsymbol{\alpha} \partial \boldsymbol{\alpha}'} \delta\boldsymbol{\alpha}. \end{aligned} \tag{5.234}$$

In Appendix www.5.5 we show that the first-order derivatives read:

$$\frac{\partial \operatorname{Spc}_{\varphi}(\boldsymbol{\alpha})}{\partial \boldsymbol{\alpha}} = -\int_0^1 \mathrm{E}\{\mathbf{M}|\boldsymbol{\alpha}'\mathbf{M} \leq Q_{\boldsymbol{\alpha}'\mathbf{M}}(p)\} p\varphi'(p)\, dp, \tag{5.235}$$

where $\mathbf{M}$ is the random vector (5.222) that represents the market. The investor will focus on the entries of the vector (5.235) that display a large absolute value.

For example, in the case of normal markets, from (5.229) we obtain directly the first-order derivatives:

$$\frac{\partial \operatorname{Spc}_{\varphi}(\boldsymbol{\alpha})}{\partial \boldsymbol{\alpha}} = (\boldsymbol{\mu} - \mathbf{P}_T) + \frac{\boldsymbol{\Sigma\alpha}}{\sqrt{\boldsymbol{\alpha}'\boldsymbol{\Sigma\alpha}}} \mathcal{I}[\varphi z](1). \tag{5.236}$$

In this expression z is the quantile of the standard normal distribution (5.228).

In particular in the case of the expected shortfall the spectrum is (5.218). From (B.50) we obtain:

$$\varphi'_{\mathrm{ES}_c}(p) = -\frac{\delta^{(1-c)}(p)}{1-c}, \tag{5.237}$$

where δ is the Dirac delta (B.17). Therefore (5.235) becomes the following expression:

$$\frac{\partial \operatorname{ES}_c}{\partial \boldsymbol{\alpha}} = \mathrm{E}\{\mathbf{M}|\boldsymbol{\alpha}'\mathbf{M} \leq \mathrm{Q}_c(\boldsymbol{\alpha})\}, \tag{5.238}$$

see Tasche (1999), Hallerbach (2003), Gourieroux, Laurent, and Scaillet (2000).

We remark that since spectral indices of satisfaction are positive homogenous, they satisfy the Euler decomposition (5.67). From (5.235) this reads:

$$\operatorname{Spc}_{\varphi}(\boldsymbol{\alpha}) = \sum_{n=1}^{N} \alpha_n \left[\int_0^1 \mathrm{E}\{M_n|\Psi_{\boldsymbol{\alpha}} \leq Q_{\Psi_{\boldsymbol{\alpha}}}(p)\}[-p\varphi'(p)]\, dp\right]. \tag{5.239}$$

The contribution to satisfaction from each security in turn factors into the product of the amount of that security times the marginal contribution to satisfaction of that security. The marginal contribution to satisfaction (5.235) is insensitive to a rescaling of the portfolio, although it depends on the allocation.

The study of the second-order cross-derivatives provides insight on the local convexity/concavity of the certainty-equivalent. In Appendix www.5.5 we prove the following result:

$$\frac{\partial^2 \operatorname{Spc}_{\varphi}(\boldsymbol{\alpha})}{\partial \boldsymbol{\alpha} \partial \boldsymbol{\alpha}'} = \int_0^1 \operatorname{Cov}\{\mathbf{M}|\Psi_{\boldsymbol{\alpha}} = Q_{\Psi_{\boldsymbol{\alpha}}}(p)\} f_{\Psi_{\boldsymbol{\alpha}}}(Q_{\Psi_{\boldsymbol{\alpha}}}(p))\, \varphi'(p)\, dp, \tag{5.240}$$

where $f_{\Psi_{\boldsymbol{\alpha}}}$ is the marginal probability density function of the investor's objective $\Psi_{\boldsymbol{\alpha}} \equiv \boldsymbol{\alpha}'\mathbf{M}$.

Since any covariance matrix and any probability density function are positive, whereas the derivative of the spectrum from (5.217) is negative, the second-order cross-derivatives define a negative definite matrix. Therefore the spectral indices of satisfaction are concave, see Figure 5.13, which refers to the expected shortfall.

For example, consider the case of normal markets. By direct derivation of (5.236) we obtain:

$$\frac{\partial^2 \operatorname{Spc}_{\varphi}(\boldsymbol{\alpha})}{\partial \boldsymbol{\alpha}' \partial \boldsymbol{\alpha}} = \boldsymbol{\Sigma}\left(\mathbf{I} - \frac{\boldsymbol{\alpha}\boldsymbol{\alpha}'\boldsymbol{\Sigma}}{\boldsymbol{\alpha}'\boldsymbol{\Sigma}\boldsymbol{\alpha}}\right) \frac{\mathcal{I}[\varphi z](1)}{\sqrt{\boldsymbol{\alpha}'\boldsymbol{\Sigma}\boldsymbol{\alpha}}}. \tag{5.241}$$

In this expression z is the quantile of the standard normal distribution (5.228). Notice that the integral $\mathcal{I}[\varphi z]$ is negative, since the spectrum φ weighs the negative values of the quantile z more than the positive values. Therefore, by the same argument used in (5.192), the second derivative (5.241) is always negative definite and thus the spectral index of satisfaction is concave.

In particular, in the case of the expected shortfall, substituting (5.237) in (5.240), we obtain:

$$\frac{\partial^2 \operatorname{ES}_c(\boldsymbol{\alpha})}{\partial \boldsymbol{\alpha} \partial \boldsymbol{\alpha}'} = -\frac{f_{\Psi_{\boldsymbol{\alpha}}}(\mathrm{Q}_c(\boldsymbol{\alpha}))}{1-c} \operatorname{Cov}\{\mathbf{M} | \Psi_{\boldsymbol{\alpha}} = \mathrm{Q}_c(\boldsymbol{\alpha})\}, \tag{5.242}$$

see Rau-Bredow (2002).

6

Optimizing allocations

In this chapter we determine the optimal allocation for a generic investor in a generic market of securities.

In Section 6.1 we introduce allocation optimization by means of a fully worked-out, non-trivial leading example. First of all, one must collect the necessary inputs. The first input is the investor's profile, namely his current allocation, his market, his investment horizon, his objectives and the respective indices of satisfaction that evaluate them. The second input is information on the investor's market, namely the joint distribution of the prices at the investment horizon of the pool of securities in the investor's market, as well as information on the implementation costs associated with trading those securities. By suitably processing the above inputs we can in principle derive the most suitable allocation for a given investor.

Except in our leading example, or in trivial cases of no practical interest, the constrained optimization that yields the optimal allocation cannot be solved analytically. In order to understand which optimization problems are numerically tractable, in Section 6.2 we present an overview of results on convex optimization, with particular focus on cone programming. First-order, second-order, and semidefinite cone programming encompass a broad class of constrained optimization problems that appear in the context of asset allocation.

In Section 6.3 we discuss a two-step approach that approximates the solution to the formal general allocation optimization by means of a tractable, quadratic problem. The first step in this approach is the mean-variance optimization pioneered by Markowitz, which selects a one-parameter family of efficient allocations among all the possible combinations of assets; the second step is a simple one-dimensional search for the best among the efficient allocations, which can be performed numerically. We introduce the mean-variance framework by means of geometrical arguments and discuss how to compute the necessary inputs that feed the mean-variance optimization. We also present the mean-variance problem in the less general, yet more common, formulation in terms of returns.

A. Meucci, *Risk and Asset Allocation,* Springer Finance,

The mean-variance optimization admits analytical solutions in a broad class of cases, namely when the investment constraints are affine. In Section 6.4 we discuss these solutions, which provide insight into the properties of optimal portfolios in more general contexts. We discuss the two-fund separation theorem and study the effect of the market on the optimal allocations. Among other results, we prove wrong the common belief that a market with low correlations provides better investment opportunities than a highly correlated market.

The many advantages of the mean-variance approach at times obscure the few problems behind it. In Section 6.5 we analyze some pitfalls of this approach. We discuss the approximate nature of the mean-variance framework. We point out the nonsensical outcomes that can result from the common practice of considering the mean-variance formulation as an index of satisfaction. We highlight the conditions under which the mean-variance optimization represents a quadratic programming problem. We discuss the difference between the original mean-variance problem, which maximizes the expected value for a given level of variance, and the dual problem, which minimizes the variance for a given level of expected value. Finally, we discuss the drawbacks of presenting the mean-variance framework in terms of returns instead of prices. Indeed, returns can be used only under restrictive assumptions on the investor's preferences and constraints. Furthermore, even under those hypotheses, expressing the mean-variance problem in terms of returns leads to misinterpretations that dramatically affect the pursuit of the optimal allocation, for instance when the investment horizon is shifted in the future.

In Section 6.6 we show an application of the analytical solutions of the mean-variance problem: allocation against a benchmark. As it turns out, a benchmark is not only the explicit target of some fund managers but also the implicit target of the general investor. Indeed, we show how even so-called "total-return" investment strategies can be considered and analyzed as a special case of benchmark-driven allocation problems.

In Section 6.7 we conclude with a case study: allocation in the stock market. Unlike the leading example in Section 6.1, this case cannot be solved analytically. Therefore, we tackle it by means of the mean-variance approach. After revisiting the complete check-list of all the steps necessary to obtain the inputs of the mean-variance problem, we determine numerically the efficient allocations and then compute the optimal allocation by means of Monte Carlo simulations.

6.1 The general approach

Consider a market of N securities. We denote as $\mathbf{P}_t$ the prices at the generic time t of the N securities in the investor's market. At the time T when the investment is made, the investor can purchase α_n units of the generic n-th security. These units are specific to the security: for example in the case of equities the units are shares, in the case of futures the units are contracts, etc.

The N-dimensional vector $\boldsymbol{\alpha}$ represents the outcome of the allocation decision, which can be seen as a "black box" that processes two types of inputs: the information on the investor's profile $\mathcal{P}$ and the information i_T on the market available at the time the investment decision is made.

6.1.1 Collecting information on the investor

As far as the investor's profile $\mathcal{P}$ is concerned, information consists of knowledge of the investor's current situation and of his outlook.

The investor's current situation is summarized in his pre-existing, possibly null, portfolio $\boldsymbol{\alpha}^{(0)}$, which corresponds to his wealth, or endowment, at the time the investment decision is made:

$$w_T \equiv \mathbf{p}_T' \boldsymbol{\alpha}^{(0)}. \tag{6.1}$$

The lower-case notation for the prices at the investment date highlights the fact that these are deterministic quantities.

The investor's outlook includes first of all his choice of a market and of an investment horizon.

For example, for a private investor the market $\mathbf{P}_t$ could be a set of mutual funds and the investment horizon τ could be three years from the time the allocation decision is made.

Second, it is important to understand the investor's main objective $\Psi_{\boldsymbol{\alpha}}$, which depends on the allocation $\boldsymbol{\alpha}$. This could be final wealth as in (5.3), or relative wealth, as in (5.4), or net profits, as in (5.8), or possibly other specifications. Nevertheless, in any specification, the objective is a linear function of the allocation and of the market vector :

$$\Psi_{\boldsymbol{\alpha}} \equiv \boldsymbol{\alpha}' \mathbf{M}, \tag{6.2}$$

see (5.10). The market vector $\mathbf{M}$ is a simple invertible affine transformation of the market prices at the investment horizon:

$$\mathbf{M} \equiv \mathbf{a} + \mathbf{B} \mathbf{P}_{T+\tau}, \tag{6.3}$$

where $\mathbf{a}$ is a suitable conformable vector and $\mathbf{B}$ is a suitable conformable invertible matrix, see (5.11).

For example, assume that the investor's main objective is final wealth. Then from (5.12) we obtain $\mathbf{M} \equiv \mathbf{P}_{T+\tau}$ and therefore:

$$\Psi_{\boldsymbol{\alpha}} \equiv \boldsymbol{\alpha}' \mathbf{P}_{T+\tau}. \tag{6.4}$$

Third, we need to model the investor's attitude towards risk. This step is necessary because the markets, and thus the objective, are not deterministic. The investor's preferences are reflected in his index of satisfaction $\mathcal{S}$, see (5.48). The index of satisfaction depends on the allocation through the distribution of the investor's objective.

For instance the index of satisfaction could be the certainty-equivalent:

$$\mathcal{S}(\boldsymbol{\alpha}) \equiv \mathrm{CE}(\boldsymbol{\alpha}) \equiv u^{-1}\left(\mathrm{E}\left\{u\left(\Psi_{\boldsymbol{\alpha}}\right)\right\}\right). \tag{6.5}$$

Assume that the utility function belongs to the exponential class:

$$u(\psi) \equiv -e^{-\frac{1}{\zeta}\psi}, \tag{6.6}$$

where the risk propensity coefficient is a positive number comprised in a suitable interval:

$$\zeta \in \left[\underline{\zeta}, \overline{\zeta}\right]. \tag{6.7}$$

Under these specifications, from (5.94) we obtain the expression of the index of satisfaction:

$$\mathrm{CE}(\boldsymbol{\alpha}) = -\zeta \ln\left(\phi_{\Psi_{\boldsymbol{\alpha}}}\left(\frac{i}{\zeta}\right)\right), \tag{6.8}$$

where $\phi_{\Psi_{\boldsymbol{\alpha}}}$ is the characteristic function of the investor's objective.

In general an investor has multiple objectives. In other words, in addition to the main objective Ψ there exists a set of secondary objectives $\widetilde{\Psi}$ that the investor cares about. As for the main objective (6.2), any secondary objective is a linear function of the allocation and of its respective market vector:

$$\widetilde{\Psi}_{\boldsymbol{\alpha}} \equiv \boldsymbol{\alpha}'\widetilde{\mathbf{M}}. \tag{6.9}$$

As in (6.3), the market vector of a secondary objective is an affine transformation of the prices at the investment horizon:

$$\widetilde{\mathbf{M}} \equiv \widetilde{\mathbf{a}} + \widetilde{\mathbf{B}}\mathbf{P}_{T+\tau}. \tag{6.10}$$

A secondary objective is evaluated according to its specific index of satisfaction $\widetilde{\mathcal{S}}$, see (5.48).

In our example we assume that in addition to his level of final wealth (6.4), the investor is concerned about his net profits since the investment date. In this case from (5.14) the market vector reads $\widetilde{\mathbf{M}} \equiv \mathbf{P}_{T+\tau} - \mathbf{p}_T$ and the auxiliary objective, namely the net profits, reads:

$$\widetilde{\Psi}_{\boldsymbol{\alpha}} \equiv \boldsymbol{\alpha}'\left(\mathbf{P}_{T+\tau} - \mathbf{p}_T\right). \tag{6.11}$$

Furthermore, we assume that the investor evaluates his net profits in terms of their value at risk. In other words, other things equal the investor is happier if the value at risk of his investment is smaller. Therefore from (5.158)-(5.159) the index of satisfaction relative to the investor's net profits reads:

$$\widetilde{\mathcal{S}}(\boldsymbol{\alpha}) \equiv -\operatorname{Var}_c(\boldsymbol{\alpha}) \equiv Q_{\widetilde{\Psi}_{\boldsymbol{\alpha}}}(1-c), \tag{6.12}$$

where Q denotes the quantile of the secondary objective (6.11) and c the VaR confidence level.

The investor's current portfolio, market, investment horizon, main objective and respective index of satisfaction, as well as his secondary objectives and respective indices of satisfaction, complete the check list of the information $\mathcal{P}$ regarding the investor's profile.

6.1.2 Collecting information on the market

As far as the market is concerned, it is important to collect information about the current prices of the securities $\mathbf{p}_T$ and their future values at the investment horizon $\mathbf{P}_{T+\tau}$.

The current prices $\mathbf{p}_T$ are deterministic variables that are publicly available at time T.

The future prices $\mathbf{P}_{T+\tau}$ are a random variable: therefore information on the prices at the investment horizon corresponds to information on their distribution. We recall that the distribution of the prices is obtained as follows: first we detect from time series analysis the invariants $\mathbf{X}_{t,\widetilde{\tau}}$ behind the market prices $\mathbf{P}_t$ relative to a suitable estimation horizon $\widetilde{\tau}$, see Section 3.1; then we estimate the distribution of the invariants $\mathbf{X}_{t,\widetilde{\tau}}$ see Chapter 4; next, we project these invariants $\mathbf{X}_{t,\widetilde{\tau}}$ to the investment horizon, obtaining the distribution of $\mathbf{X}_{T+\tau,\tau}$, see Section 3.2; finally we map the distribution of the invariants $\mathbf{X}_{T+\tau,\tau}$ into the distribution of the prices at the investment horizon of the securities $\mathbf{P}_{T+\tau}$, see Section 3.3.

For example, we assume that the distribution of the prices at the investment horizon is estimated to be normal:

$$\mathbf{P}_{T+\tau} \sim \mathrm{N}(\boldsymbol{\xi}, \boldsymbol{\Phi}), \tag{6.13}$$

where $\boldsymbol{\xi}$ and $\boldsymbol{\Phi}$ are suitable values respectively of the expected value and covariance matrix of the market prices.

Furthermore, the process of switching from a generic allocation $\widetilde{\boldsymbol{\alpha}}$ to another generic allocation $\boldsymbol{\alpha}$ is costly. We denote as $\mathcal{T}(\widetilde{\boldsymbol{\alpha}}, \boldsymbol{\alpha})$ the *transaction costs* associated with this process. Transaction costs take different forms in different markets: for instance, traders face bid-ask spreads and commissions, private investors face subscription fees, etc.

For example, the investor might be charged a commission that is proportional to the number of securities transacted:

$$\mathcal{T}(\widetilde{\boldsymbol{\alpha}}, \boldsymbol{\alpha}) \equiv \mathbf{k}' |\widetilde{\boldsymbol{\alpha}} - \boldsymbol{\alpha}|, \tag{6.14}$$

where $\mathbf{k}$ is a given constant vector. In the sequel of our example we consider the simplified case of null transaction costs, i.e. $\mathbf{k} \equiv \mathbf{0}$.

The current prices of the securities in the market, the distribution of the market at the investment horizon and the details on the transaction costs complete the check list of the information i_T regarding the market.

6.1.3 Computing the optimal allocation

A generic allocation decision processes the information on the market and on the investor and outputs the amounts to invest in each security in the given market:

$$\boldsymbol{\alpha}[\cdot] : [i_T, \mathcal{P}] \mapsto \mathbb{R}^N. \tag{6.15}$$

We stress that this is the definition of a generic allocation decision, not necessarily optimal.

For example, a possible decision allocates an equal amount of the initial wealth w_T in each security in the market. This is the *equally-weighted portfolio*:

$$\boldsymbol{\alpha}[i_T, \mathcal{P}] \equiv \frac{w_T}{N} \operatorname{diag}(\mathbf{p}_T)^{-1} \mathbf{1}_N, \tag{6.16}$$

where $\mathbf{1}$ denotes a vector of ones. Notice that this decision uses very little information on the market, i.e. only the current prices of the securities, and very little information on the investor, i.e. only his initial budget.

In order to be optimal, an allocation decision ensues from carefully processing all the available information about both the investor and the market.

First of all, given the distribution of the prices at the investment horizon $\mathbf{P}_{T+\tau}$, it is possible in principle to compute explicitly the distribution of the investor's generic objective Ψ as a function of the allocation.

Indeed, the market vector $\mathbf{M}$ that correspond to the investor's objective Ψ as in (6.2) or (6.9) is a simple invertible affine transformation of the prices at the investment horizon, see (6.3) and (6.10). Therefore the distribution of the market vector $\mathbf{M}$ is easily obtained from the distribution of $\mathbf{P}_{T+\tau}$, see (5.15). Furthermore, since the objective $\Psi \equiv \boldsymbol{\alpha}'\mathbf{M}$ is a linear combination of the allocation and the market vector, we can in principle determine its distribution from the distribution of $\mathbf{M}$.

In our example, from (6.13) the market vector relative to the main objective (6.4) is normally distributed:

$$\mathbf{M} \equiv \mathbf{P}_{T+\tau} \sim \mathrm{N}(\boldsymbol{\xi}, \boldsymbol{\Phi}). \tag{6.17}$$

Therefore the investor's main objective is normally distributed with the following parameters:

$$\Psi_{\boldsymbol{\alpha}} \sim \mathrm{N}\left(\boldsymbol{\xi}'\boldsymbol{\alpha}, \boldsymbol{\alpha}'\boldsymbol{\Phi}\boldsymbol{\alpha}\right). \tag{6.18}$$

Similarly, from (6.13) the market vector relative to the secondary objective (6.11) is normally distributed:

$$\widetilde{\mathbf{M}} \equiv \mathbf{P}_{T+\tau} - \mathbf{p}_T \sim \mathrm{N}(\boldsymbol{\xi} - \mathbf{p}_T, \boldsymbol{\Phi}). \tag{6.19}$$

Therefore the investor's secondary objective is normally distributed with the following parameters:

$$\widetilde{\Psi}_{\boldsymbol{\alpha}} \sim \mathrm{N}\left((\boldsymbol{\xi} - \mathbf{p}_T)'\boldsymbol{\alpha}, \boldsymbol{\alpha}'\boldsymbol{\Phi}\boldsymbol{\alpha}\right). \tag{6.20}$$

From the distribution of the investor's primary and secondary objectives we can compute the respective indices of satisfaction.

In our example, substituting the characteristic function (1.69) of the first objective (6.18) in (6.8) we obtain the main index of satisfaction, namely the certainty-equivalent of final wealth:

$$\mathrm{CE}(\boldsymbol{\alpha}) = \boldsymbol{\xi}'\boldsymbol{\alpha} - \frac{1}{2\zeta}\boldsymbol{\alpha}'\boldsymbol{\Phi}\boldsymbol{\alpha}. \tag{6.21}$$

Similarly, substituting the quantile (1.70) of the secondary objective (6.20) in (6.12) we obtain the (opposite of the) secondary index of satisfaction, namely the value at risk:

$$\mathrm{Var}_c(\boldsymbol{\alpha}) = (\mathbf{p}_T - \boldsymbol{\xi})'\boldsymbol{\alpha} + \sqrt{2\boldsymbol{\alpha}'\boldsymbol{\Phi}\boldsymbol{\alpha}}\,\mathrm{erf}^{-1}(2c-1), \tag{6.22}$$

where erf^{-1} is the inverse of the error function (B.75).

Finally, the investor is bound by a set of investment constraints $\mathcal{C}$ that limit his feasible allocations.

One constraint that appears in different forms is the *budget constraint*, which states that the value of the initial investment cannot exceed a given budget b net of transaction costs:

$$\mathcal{C}_1: \ \mathbf{p}_T'\boldsymbol{\alpha} + \mathcal{T}\left(\boldsymbol{\alpha}^{(0)}, \boldsymbol{\alpha}\right) - b \leq 0, \tag{6.23}$$

where $\boldsymbol{\alpha}^{(0)}$ is the initial portfolio (6.1).

Notice that we defined the constraint in terms of an inequality. In most applications in the final optimal allocation this constraint turns out to be binding, i.e. it is satisfied as an equality.

In our example the transaction costs (6.14) are null, and we assume that the budget is the initial wealth. Therefore the budget constraint reads:

$$\mathcal{C}_1 : \mathbf{p}_T' \boldsymbol{\alpha} = w_T \equiv \mathbf{p}_T' \boldsymbol{\alpha}^{(0)}. \tag{6.24}$$

By means of additional constraints we can include the investor's multiple objectives in the allocation problem. Indeed, the multiple objectives are accounted for by imposing that the respective index of satisfaction $\widetilde{\mathcal{S}}$ exceed a minimum acceptable threshold $\widetilde{s}$:

$$\mathcal{C}_2 : \widetilde{s} - \widetilde{\mathcal{S}}(\boldsymbol{\alpha}) \leq 0. \tag{6.25}$$

In our example the additional objective of the investor are his net profits (6.11), which the investor monitors by means of the value at risk (6.12). An allocation is acceptable for the investor only if the respective VaR does not exceed a given *budget at risk*, i.e. a fraction γ of the initial endowment:

$$\mathcal{C}_2 : \operatorname{Var}_c(\boldsymbol{\alpha}) \leq \gamma w_T. \tag{6.26}$$

From (6.22) and (6.24) the VaR constraint reads explicitly:

$$\mathcal{C}_2 : (1-\gamma) w_T - \boldsymbol{\xi}' \boldsymbol{\alpha} + \sqrt{2 \boldsymbol{\alpha}' \boldsymbol{\Phi} \boldsymbol{\alpha}} \operatorname{erf}^{-1}(2c-1) \leq 0. \tag{6.27}$$

We denote an allocation that satisfies the given set of constraints $\mathcal{C}$ as follows:

$$\boldsymbol{\alpha} \in \mathcal{C}. \tag{6.28}$$

The set of allocations that satisfy the constraints is called the *feasible set*.

To determine the feasible set in our example, we consider the plane of coordinates:

$$e \equiv \boldsymbol{\xi}' \boldsymbol{\alpha}, \quad d \equiv \sqrt{\boldsymbol{\alpha}' \boldsymbol{\Phi} \boldsymbol{\alpha}}. \tag{6.29}$$

As we show in Figure 6.1, in this plane the budget constraint (6.24) is satisfied by all the points in the region to the right of a hyperbola:

$$d^2 \geq \frac{A}{D} e^2 - \frac{2 w_T B}{D} e + \frac{w_T^2 C}{D}, \tag{6.30}$$

where

$$\begin{array}{ll} A \equiv \mathbf{p}_T' \boldsymbol{\Phi}^{-1} \mathbf{p}_T & B \equiv \mathbf{p}_T' \boldsymbol{\Phi}^{-1} \boldsymbol{\xi} \\ C \equiv \boldsymbol{\xi}' \boldsymbol{\Phi}^{-1} \boldsymbol{\xi} & D \equiv AC - B^2, \end{array} \tag{6.31}$$

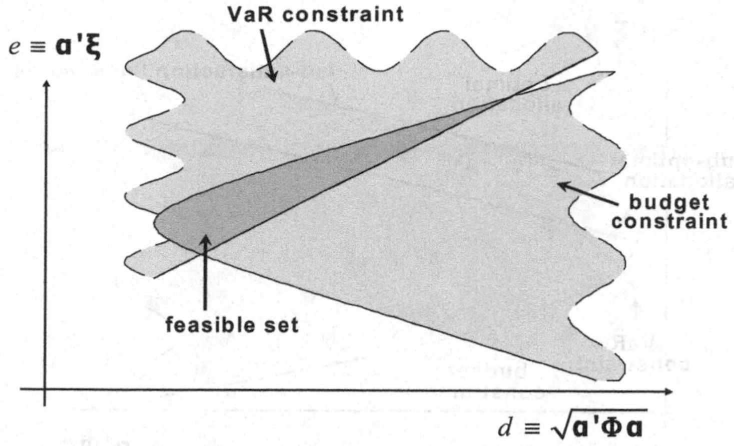

Fig. 6.1. Leading allocation example: constraints and feasible set

see Appendix www.6.1 for the proof.

On the other hand, the VaR constraint (6.27) is satisfied by all the points above a straight line:

$$e \geq (1-\gamma)\, w_T + \sqrt{2}\,\mathrm{erf}^{-1}(2c-1)\, d. \tag{6.32}$$

This follows immediately from (6.27).

The investor evaluates the potential advantages of an allocation $\boldsymbol{\alpha}$ based on his primary index of satisfaction $\mathcal{S}$, provided that the allocation is feasible. Therefore, the optimal allocation is the solution to the following maximization problem:

$$\boldsymbol{\alpha}^* \equiv \underset{\boldsymbol{\alpha}\in\mathcal{C}}{\operatorname{argmax}} \{\mathcal{S}(\boldsymbol{\alpha})\}. \tag{6.33}$$

By construction, the optimal allocation is of the form (6.15), i.e. it is a "black box" that processes both the current information on the market i_T and the investor's profile $\mathcal{P}$, and outputs a vector of amounts of each security in the market.

In our example, from (6.5), (6.24) and (6.26) the investor solves:

$$\boldsymbol{\alpha}^* \equiv \underset{\substack{\mathbf{p}_T'\boldsymbol{\alpha}=w_T \\ \mathrm{Var}_c(\boldsymbol{\alpha})\leq\gamma w_T}}{\operatorname{argmax}} \{\mathrm{CE}(\boldsymbol{\alpha})\}. \tag{6.34}$$

To compute the optimal solution we consider the plane of the following coordinates (see Figure 6.2):

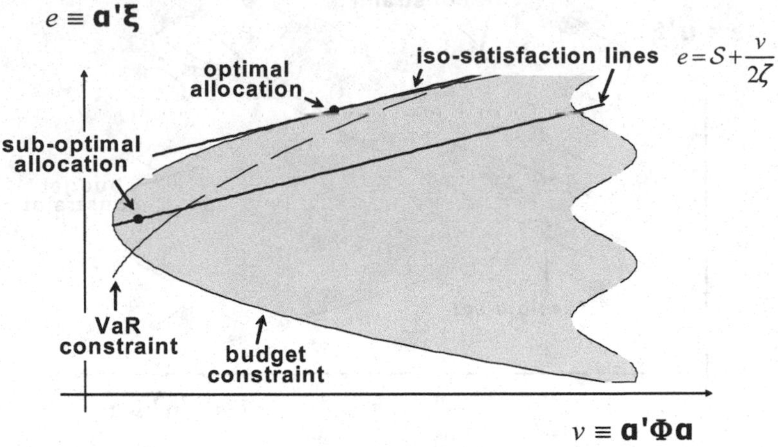

Fig. 6.2. Leading allocation example: optimal allocation

$$e \equiv \boldsymbol{\xi}'\boldsymbol{\alpha}, \quad v \equiv \boldsymbol{\alpha}'\boldsymbol{\Phi}\boldsymbol{\alpha}. \tag{6.35}$$

In these coordinates the allocations that satisfy the budget constraint plot to the right of a parabola:

$$v \geq \frac{A}{D}e^2 - \frac{2w_T B}{D}e + \frac{w_T^2 C}{D}. \tag{6.36}$$

This follows immediately from (6.30). Similarly the allocations that satisfy the VaR constraint plot to the left of a parabola:

$$v \leq \frac{1}{2\left(\operatorname{erf}^{-1}(2c-1)\right)^2}\left(e - (1-\gamma)\,w_T\right)^2. \tag{6.37}$$

This follows immediately from (6.32).

From (6.21) the iso-satisfaction contours, i.e. the allocations that give rise to the same level of satisfaction $\mathcal{S}$, plot along the following straight line:

$$e = \mathcal{S} + \frac{v}{2\zeta}. \tag{6.38}$$

The slope of this straight line is determined by the risk propensity coefficient ζ: the higher this coefficient, the flatter the line. The level of this straight line is determined by the amount of satisfaction: the larger $\mathcal{S}$, the higher the plot of the line (6.38) in the plane.

To solve for the optimal allocation we need to determine the highest iso-satisfaction line that has an intersection with the region to the right of the budget-constraint parabola (6.36), namely the tangent line to the upper branch of that parabola, see Figure 6.2. As long as the coefficient ζ is comprised between suitable limits as in (6.7), the optimal allocation automatically satisfies the VaR constraint.

We prove in Appendix www.6.1 that the optimal allocation reads:

$$\boldsymbol{\alpha}^* \equiv \zeta \boldsymbol{\Phi}^{-1} \boldsymbol{\xi} + \frac{w_T - \zeta \mathbf{p}_T' \boldsymbol{\Phi}^{-1} \boldsymbol{\xi}}{\mathbf{p}_T' \boldsymbol{\Phi}^{-1} \mathbf{p}_T} \boldsymbol{\Phi}^{-1} \mathbf{p}_T. \tag{6.39}$$

The optimal allocation is of the form (6.15). Indeed, the information on the market is summarized in the current prices and in expected values and covariances of the future prices at the investment horizon:

$$i_T \equiv (\mathbf{p}_T, \boldsymbol{\xi}, \boldsymbol{\Phi}); \tag{6.40}$$

and the investor's profile is summarized in his risk propensity and initial budget:

$$\mathcal{P} \equiv (\zeta, w_T). \tag{6.41}$$

The allocation (6.39) gives rise to the maximum level of satisfaction, given the investment constraints. We prove in Appendix www.6.1 that this level reads:

$$\begin{aligned} \mathrm{CE}(\boldsymbol{\alpha}^*) = &\frac{\zeta}{2} \boldsymbol{\xi}' \boldsymbol{\Phi}^{-1} \boldsymbol{\xi} + \frac{1}{2} \left(\frac{w_T - \zeta \mathbf{p}_T' \boldsymbol{\Phi}^{-1} \boldsymbol{\xi}}{\mathbf{p}_T' \boldsymbol{\Phi}^{-1} \mathbf{P}_T} \right) \boldsymbol{\xi}' \boldsymbol{\Phi}^{-1} \mathbf{p}_T \\ &- \frac{1}{2\zeta} \frac{\left(w_T - \zeta \mathbf{p}_T' \boldsymbol{\Phi}^{-1} \boldsymbol{\xi} \right)^2}{\mathbf{p}_T' \boldsymbol{\Phi}^{-1} \mathbf{p}_T}. \end{aligned} \tag{6.42}$$

6.2 Constrained optimization

From (6.33) we see that determining the best allocation for a given investor boils down to solving a constrained optimization problem. In this section we present a quick review of results on constrained optimization. The reader is referred to Lobo, Vandenberghe, Boyd, and Lebret (1998), Ben-Tal and Nemirovski (2001), Boyd and Vandenberghe (2004) and references therein for more on this subject.

Consider a generic constrained optimization problem:

$$\mathbf{z}^* \equiv \underset{\mathbf{z} \in \mathcal{C}}{\operatorname{argmin}}\, \mathcal{Q}(\mathbf{z}), \tag{6.43}$$

where $\mathcal{Q}$ is the objective function, and $\mathcal{C}$ is a set of constraints. Here, following the standards of the literature, we present optimization as a minimization problem. To consider maximization problems, it suffices to change the sign of the objective, turning the respective problem into a minimization.

In general it is not possible to solve (6.43) analytically. Nonetheless, even within the realm of numerical optimization, not all problems can be solved.

A broad class of constrained optimization problems that admit numerical solutions is represented by *convex programming* problems: in this framework the objective $\mathcal{Q}$ is a convex function and the feasible set determined by the constraints is the intersection of a hyperplane and a convex set. More precisely, convex programming is an optimization problem of the form:

$$\mathbf{z}^* \equiv \underset{\substack{\mathbf{z}\in\mathcal{L} \\ \mathbf{z}\in\mathcal{V}}}{\operatorname{argmin}}\, \mathcal{Q}(\mathbf{z}), \tag{6.44}$$

where $\mathcal{Q}$ is a convex function, i.e. it satisfies (5.82); $\mathcal{L}$ is a hyperplane determined by a conformable matrix $\mathbf{A}$ and a conformable vector $\mathbf{a}$:

$$\mathcal{L} \equiv \{\mathbf{z} \text{ such that } \mathbf{Az} = \mathbf{a}\}; \tag{6.45}$$

and $\mathcal{V}$ is a convex set, determined implicitly by a set of inequalities on convex functions:

$$\mathcal{V} \equiv \{\mathbf{z} \text{ such that } \mathbf{F}(\mathbf{z}) \leq \mathbf{0},\ \mathbf{F} \text{ convex}\}. \tag{6.46}$$

Although numerical solutions can be found for convex programming, these are usually computationally too expensive for the amount of variables involved in an asset allocation problem.

Nonetheless, a subclass of convex programming, called *cone programming* (CP), admits efficient numerical solutions, which are variations of *interior point algorithms*, see Nesterov and Nemirovski (1995). In conic programming the objective is linear, and the feasible set determined by the constraints is the intersection of a hyperplane and a cone. More precisely, convex programming is an optimization problem of the form:

$$\mathbf{z}^* \equiv \underset{\substack{\mathbf{z}\in\mathcal{L} \\ \mathbf{Bz}-\mathbf{b}\in\mathcal{K}}}{\operatorname{argmin}}\, \mathbf{c}'\mathbf{z}, \tag{6.47}$$

where $\mathbf{c}$ and $\mathbf{b}$ are conformable vectors; $\mathbf{B}$ is a conformable matrix; $\mathcal{L}$ is a hyperplane as in (6.45); and $\mathcal{K}$ is a *cone*, i.e. a set with the following properties, see Figure 6.3:

i. a cone is closed under positive multiplication, i.e. it extends to infinity in radial directions from the origin:

$$\mathbf{y} \in \mathcal{K}, \lambda \geq 0 \Rightarrow \lambda\mathbf{y} \in \mathcal{K}; \tag{6.48}$$

ii. a cone is closed under addition, i.e. it includes all its interior points:

$$\mathbf{y}, \widetilde{\mathbf{y}} \in \mathcal{K} \Rightarrow \mathbf{y} + \widetilde{\mathbf{y}} \in \mathcal{K}; \tag{6.49}$$

iii. a cone is "pointed", i.e. it lies on only one side of the origin. More formally, for any $\mathbf{y} \neq \mathbf{0}$ the following holds:

$$\mathbf{y} \in \mathcal{K} \Rightarrow -\mathbf{y} \notin \mathcal{K}. \tag{6.50}$$

Depending on what type of cone $\mathcal{K}$ defines the constraints in the conic programming (6.47), we obtain as special cases all the problems that currently can be solved. In particular, we distinguish three types of cones, and the respective notable classes of conic programming.

6.2.1 Positive orthants: linear programming

Consider the positive orthant of dimension M, i.e. the subset of $\mathbb{R}^M$ spanned by the positive coordinates:

$$\mathbb{R}_+^M \equiv \left\{ \mathbf{y} \in \mathbb{R}^M \text{ such that } y_1 \geq 0, \ldots, y_M \geq 0 \right\}. \tag{6.51}$$

It is easy to check that the positive orthant $\mathbb{R}_+^M$ is a cone, i.e. it satisfies (6.48)-(6.50). The ensuing conic programming (6.47) problem reads:

$$\mathbf{z}^* \equiv \underset{\substack{\mathbf{Az}=\mathbf{a} \\ \mathbf{Bz}\geq\mathbf{b}}}{\operatorname{argmin}} \, \mathbf{c}'\mathbf{z}. \tag{6.52}$$

This problem is called *linear programming* (LP), see Dantzig (1998).

6.2.2 Ice-cream cones: second-order cone programming

Consider the *ice-cream cone*, or *Lorentz cone*, of dimension M:

$$\mathbb{K}^M \equiv \left\{ \mathbf{y} \in \mathbb{R}^M \text{ such that } \left\| (y_1, \ldots, y_{M-1})' \right\| \leq y_M \right\}, \tag{6.53}$$

where $\|\cdot\|$ is the standard norm ($A.6$) in $\mathbb{R}^M$, see Figure 6.3.

It is easy to check that the ice-cream cone $\mathbb{K}^M$ is indeed a cone, i.e. it satisfies (6.48)-(6.50). Furthermore, the direct product of ice-cream cones is also cone:

$$\begin{aligned} \mathcal{K} &\equiv \mathbb{K}^{M_1} \times \cdots \times \mathbb{K}^{M_J} \\ &\equiv \left\{ \mathbf{y}_{(1)} \in \mathbb{K}^{M_1} \times \cdots \times \mathbf{y}_{(J)} \in \mathbb{K}^{M_J} \right\}. \end{aligned} \tag{6.54}$$

The ensuing conic programming problem (6.47) reads:

$$\mathbf{z}^* \equiv \underset{\mathbf{z}}{\operatorname{argmin}} \, \mathbf{c}'\mathbf{z} \tag{6.55}$$

$$\text{subject to } \begin{cases} \mathbf{Az} = \mathbf{a} \\ \left\| \mathbf{D}_{(1)}\mathbf{z} - \mathbf{q}_{(1)} \right\| \leq \mathbf{p}'_{(1)}\mathbf{z} - r_{(1)} \\ \vdots \\ \left\| \mathbf{D}_{(J)}\mathbf{z} - \mathbf{q}_{(J)} \right\| \leq \mathbf{p}'_{(J)}\mathbf{z} - r_{(J)}, \end{cases}$$

where $\mathbf{D}_{(j)}$ are conformable matrices, $\mathbf{q}_{(j)}$ are conformable vectors and $r_{(j)}$ are scalars for $j = 1, \ldots, J$. This follows by defining the matrix $\mathbf{B}$ and the vector $\mathbf{b}$ in (6.47) as below:

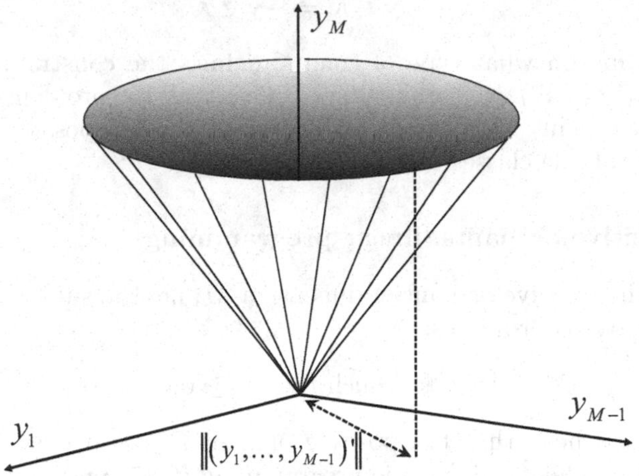

Fig. 6.3. Lorentz cone

$$\mathbf{B} \equiv \begin{pmatrix} \mathbf{D}_{(1)} \\ \mathbf{p}'_{(1)} \\ \vdots \\ \mathbf{D}_{(J)} \\ \mathbf{p}'_{(J)} \end{pmatrix}, \quad \mathbf{b} \equiv \begin{pmatrix} \mathbf{q}_{(1)} \\ r_{(1)} \\ \vdots \\ \mathbf{q}_{(J)} \\ r_{(J)} \end{pmatrix}. \tag{6.56}$$

The optimization problem (6.55) is called *second-order cone programming* (SOCP).

Second-order cone programming problems include *quadratically constrained quadratic programming* (QCQP) problems as a subclass. A generic QCQP problem reads:

$$\mathbf{z}^* \equiv \underset{\mathbf{z}}{\operatorname{argmin}} \left\{ \mathbf{z}'\mathbf{S}_{(0)}\mathbf{z} + 2\mathbf{u}'_{(0)}\mathbf{z} + v_{(0)} \right\} \tag{6.57}$$

$$\text{subject to} \begin{cases} \mathbf{A}\mathbf{z} = \mathbf{a} \\ \mathbf{z}'\mathbf{S}_{(1)}\mathbf{z} + 2\mathbf{u}'_{(1)}\mathbf{z} + v_{(1)} \leq 0 \\ \vdots \\ \mathbf{z}'\mathbf{S}_{(J)}\mathbf{z} + 2\mathbf{u}'_{(J)}\mathbf{z} + v_{(J)} \leq 0, \end{cases}$$

where $\mathbf{S}_{(j)}$ are symmetric and positive matrices, $\mathbf{u}_{(j)}$ are conformable vectors and $v_{(j)}$ are scalars for $j = 0, \ldots, J$. Consider the spectral decomposition $(A.66)$ of the matrices $\mathbf{S}_{(j)}$:

$$\mathbf{S}_{(j)} \equiv \mathbf{E}_{(j)} \boldsymbol{\Lambda}_{(j)} \mathbf{E}'_{(j)}, \tag{6.58}$$

where $\boldsymbol{\Lambda}$ is the diagonal matrix of the eigenvalues and $\mathbf{E}$ is the juxtaposition of the respective eigenvectors. As we show in Appendix www.6.2, by introducing an auxiliary variable t the QCQP problem (6.57) can be written equivalently as follows:

$$(\mathbf{z}^*, t^*) \equiv \underset{(\mathbf{z},t)}{\operatorname{argmin}}\, t \tag{6.59}$$
$$\text{s.t.} \begin{cases} \mathbf{A}\mathbf{z} = \mathbf{a} \\ \left\| \boldsymbol{\Lambda}_{(0)}^{1/2} \mathbf{E}'_{(0)} \mathbf{z} + \boldsymbol{\Lambda}_{(0)}^{-1/2} \mathbf{E}'_{(0)} \mathbf{u}_{(0)} \right\| \leq t \\ \left\| \boldsymbol{\Lambda}_{(1)}^{1/2} \mathbf{E}'_{(1)} \mathbf{z} + \boldsymbol{\Lambda}_{(1)}^{-1/2} \mathbf{E}'_{(1)} \mathbf{u}_{(1)} \right\| \leq \sqrt{\mathbf{u}_{(1)} \mathbf{S}_{(1)}^{-1} \mathbf{u}_{(1)} - v_{(1)}} \\ \vdots \\ \left\| \boldsymbol{\Lambda}_{(J)}^{1/2} \mathbf{E}'_{(J)} \mathbf{z} + \boldsymbol{\Lambda}_{(J)}^{-1/2} \mathbf{E}'_{(J)} \mathbf{u}_{(J)} \right\| \leq \sqrt{\mathbf{u}_{(J)} \mathbf{S}_{(J)}^{-1} \mathbf{u}_{(J)} - v_{(J)}}. \end{cases}$$

This problem is in the SOCP form (6.55).

Quite obviously, the QCQP problem (6.57) also includes linearly constrained quadratic programming problems (QP) and linear programming problems (LP) as special cases.

6.2.3 Semidefinite cones: semidefinite programming

Consider the following set of $M \times M$ matrices:

$$\mathbb{S}_+^M \equiv \{\mathbf{S} \succeq \mathbf{0}\}, \tag{6.60}$$

where $\mathbf{S} \succeq \mathbf{0}$ denotes symmetric and positive. It is easy to check that this set is a cone, i.e. it satisfies (6.48)-(6.50). The cone $\mathbb{S}_+^M$ is called the *semidefinite cone*. The ensuing conic programming problem (6.47) reads:

$$\mathbf{z}^* \equiv \underset{\mathbf{z}}{\operatorname{argmin}}\, \mathbf{c}'\mathbf{z} \tag{6.61}$$
$$\text{s.t.} \begin{cases} \mathbf{A}\mathbf{z} = \mathbf{a} \\ \mathbf{B}_{(1)} z_1 + \cdots + \mathbf{B}_{(N)} z_N - \mathbf{B}_{(0)} \succeq \mathbf{0}, \end{cases}$$

where $\mathbf{B}_{(j)}$ are symmetric, but not necessarily positive, matrices, for $j = 0, \ldots, N$.

The optimization problem (6.61) is called *semidefinite programming* (SDP). It is possible to show that SDP includes the SOCP problem (6.55) as a special case, see Lobo, Vandenberghe, Boyd, and Lebret (1998). Nonetheless, the computational cost to solve generic SDP problems is much higher than the cost to solve SOCP problems.

6.3 The mean-variance approach

Consider the general formalization (6.33) of an allocation optimization:

$$\boldsymbol{\alpha}^* \equiv \underset{\boldsymbol{\alpha} \in \mathcal{C}}{\operatorname{argmax}} \mathcal{S}(\boldsymbol{\alpha}). \tag{6.62}$$

In general it is not possible to determine the analytical solution of this problem. Indeed, the leading example detailed in Section 6.1 probably represents the only non-trivial combination of market, preferences and constraints that gives rise to a problem which can be solved analytically in all its steps. Therefore, we need to turn to numerical results.

Even within the domain of numerical solutions, if the primary index of satisfaction $\mathcal{S}$ targeted by the investor is not concave, or if the secondary indices of satisfaction $\widetilde{\mathcal{S}}$ that determine the constraints as in (6.25) are not concave, then the general allocation optimization problem (6.62) is not convex. Therefore, the allocation optimization problem is not as in (6.44) and thus it cannot be solved numerically. For instance, this happens when one among the primary or secondary indices of satisfaction is based on a quantile of the investor's objective (value at risk) or it based on its expected utility (certainty-equivalent), see (5.153) and (5.191). Numerical solutions cannot be computed in general for non-convex problems.

Furthermore, even if the allocation optimization problem (6.62) is convex, the computational cost of obtaining a solution is in general prohibitive: only the special class of conic programming problems can be computed efficiently, see Section 6.2. Therefore it is important to cast the general allocation optimization in this class, possibly by means of approximations.

In this section we discuss a two-step approximation to the general allocation optimization problem that is both intuitive and computationally tractable, namely the mean-variance approach. The mean-variance two-step approach is by far the most popular approach to asset allocation: it has become the guideline in all practical applications and the benchmark in all academic studies on the subject.

6.3.1 The geometry of allocation optimization

To better understand the generality and the limitations of the mean-variance approach we analyze the optimization problem (6.62) from a geometrical point of view.

In any of the formulations considered in Chapter 5, the investor's index of satisfaction is law invariant, i.e. it is a functional of the distribution of the investor's objective, see (5.52). In turn, the distribution of the investor's objective is in general univocally determined by its moments, see Appendix www.1.6. Therefore the index of satisfaction can be re-written as a function defined on the infinite-dimensional space of the moments of the distribution of the objective:

$$\mathcal{S}(\boldsymbol{\alpha}) \equiv \mathcal{H}\left(\mathrm{E}\left\{\Psi_{\boldsymbol{\alpha}}\right\}, \mathrm{CM}_2\left\{\Psi_{\boldsymbol{\alpha}}\right\}, \mathrm{CM}_3\left\{\Psi_{\boldsymbol{\alpha}}\right\}, \ldots\right). \tag{6.63}$$

In this expression CM_k denotes as in (1.48) the central moment of order k of a univariate distribution:

$$\mathrm{CM}_k \{\Psi\} \equiv \mathrm{E}\left\{(\Psi - \mathrm{E}\{\Psi\})^k\right\}. \tag{6.64}$$

We chose to represent (6.63) in terms of the central moments, but we could equivalently have chosen the raw moments (1.47). The explicit functional expression of $\mathcal{H}$ in (6.63) is determined by the specific index of satisfaction $\mathcal{S}$ adopted to model the investor's preferences. For instance, when the index of satisfaction is the certainty-equivalent, the functional expression follows from a Taylor expansion of the utility function, see (5.146). When the index of satisfaction is a quantile or a spectral index, this expression follows from the Cornish-Fisher expansion, see (5.180) and (5.232).

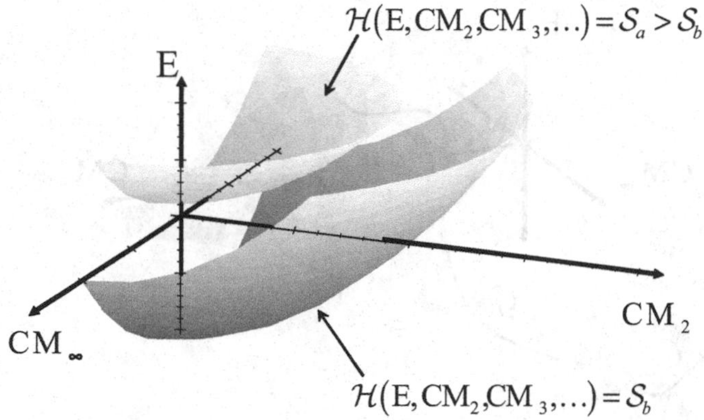

Fig. 6.4. Iso-satisfaction surfaces in the space of moments of the investor's objective

The iso-satisfaction surfaces in this space, i.e. the combinations of moments of the investor's objective that elicit an equal level of satisfaction, are defined by implicit equations as the following:

$$\mathcal{H}(\mathrm{E}, \mathrm{CM}_2, \mathrm{CM}_3, \ldots) = \mathcal{S}, \tag{6.65}$$

where $\mathcal{S}$ is a given constant level of satisfaction. Therefore, iso-satisfaction surfaces are "$(\infty - 1)$"-dimensional objects in the ∞-dimensional space of the moments of the investor's objective.

We represent this situation in Figure 6.4, where the ∞-dimensional space of moments is reduced to three dimensions and the $(\infty - 1)$-dimensional iso-satisfaction surfaces are represented by two-dimensional surfaces.

On the other hand, not all points in the space of the moments of the investor's objective correspond to an allocation. Indeed, as the allocation vector $\boldsymbol{\alpha}$ spans $\mathbb{R}^N$, the corresponding moments describe an N-dimensional surface $\mathcal{G}$ in the ∞-dimensional space of moments:

$$\mathcal{G}: \boldsymbol{\alpha} \mapsto \left(\mathrm{E}\left\{\Psi_{\boldsymbol{\alpha}}\right\}, \mathrm{CM}_2\left\{\Psi_{\boldsymbol{\alpha}}\right\}, \mathrm{CM}_3\left\{\Psi_{\boldsymbol{\alpha}}\right\}, \ldots\right). \tag{6.66}$$

In Figure 6.5 we sketch the case of $N \equiv 2$ securities: the shaded square represents $\mathbb{R}^N$ and the two-dimensional shape in the space of moments represents the combinations of moments that can be generated by an allocation.

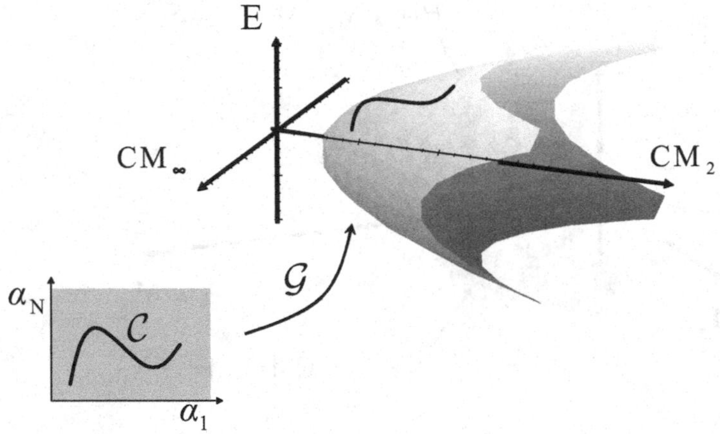

Fig. 6.5. Feasible allocations in the space of moments of the investor's objective

Finally, even within the surface (6.66), not all the allocations are viable, because the generic allocation $\boldsymbol{\alpha} \in \mathbb{R}^N$ has to satisfy a set of investment constraints, see (6.28). Therefore the feasible set becomes a subset of $\mathbb{R}^N$. This is reflected in the space of moments: the function $\mathcal{G}$ maps the feasible set into a lower-dimensional/lower-size portion of the N-dimensional surface described by (6.66).

We sketch this phenomenon in Figure 6.5 in our example of $N \equiv 2$ securities: the feasible set becomes a line, which is then mapped by $\mathcal{G}$ into the space of moments.

Solving the allocation optimization problem (6.62) corresponds to determining an iso-satisfaction surface that contains feasible points in the space

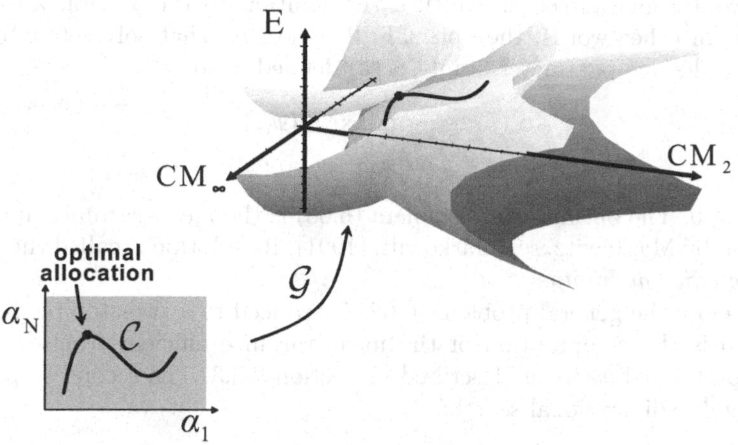

Fig. 6.6. Optimal allocation maximizes satsifaction

of moments and that corresponds to the highest possible level of satisfaction. Those feasible points corresponds to specific allocations that maximize satisfaction.

In Figure 6.6 the highest possible level of satisfaction compatible with the constraints in the space of moments corresponds to a specific allocation in the shaded square, i.e. in $\mathbb{R}^N$.

6.3.2 Dimension reduction: the mean-variance framework

In order to solve explicitly the general allocation problem (6.62) we need to determine the functional dependence (6.63) of the index of satisfaction on all the moments and the dependence of each moment on the allocation.

Suppose that we can focus on the two first moments only and neglect all the higher moments. In other words, assume that (6.63) can be approximated as follows:

$$\mathcal{S}(\boldsymbol{\alpha}) \approx \widetilde{\mathcal{H}}(\mathrm{E}\{\Psi_{\boldsymbol{\alpha}}\}, \mathrm{Var}\{\Psi_{\boldsymbol{\alpha}}\}), \tag{6.67}$$

for a suitable bivariate function $\widetilde{\mathcal{H}}$. This approximation is quite satisfactory in a wide range of applications, see Section 6.5.1. In this case the general allocation problem (6.62) is much easier to solve.

Indeed, since all the indices of satisfaction $\mathcal{S}$ discussed in Chapter 5 are consistent with weak stochastic dominance, for a given level of variance of the objective, higher expected values of the objective are always appreciated, *no matter* the functional expression of $\widetilde{\mathcal{H}}$. Therefore, if for each target value of variance of the investor's objective we pursue its maximum possible expected value, we are guaranteed to capture the solution to the general allocation problem. In other words, the optimal allocation $\boldsymbol{\alpha}^*$ that solves (6.62) must belong to the one-parameter family $\boldsymbol{\alpha}(v)$ defined as follows:

$$\boldsymbol{\alpha}(v) \equiv \underset{\substack{\boldsymbol{\alpha}\in\mathcal{C} \\ \operatorname{Var}\{\Psi_{\boldsymbol{\alpha}}\}=v}}{\operatorname{argmax}} \operatorname{E}\{\Psi_{\boldsymbol{\alpha}}\}, \tag{6.68}$$

where $v \geq 0$. The optimization problem (6.68) is the *mean-variance* approach pioneered by Markowitz, see Markowitz (1991). Its solution is called the *mean-variance efficient frontier*.

Therefore the general problem (6.62) is reduced to a two-step recipe. The first step is the computation of the mean-variance efficient frontier, which can be performed easily as described in Section 6.3.3. The second step is the following one-dimensional search:

$$\boldsymbol{\alpha}^* \equiv \boldsymbol{\alpha}(v^*) \equiv \underset{v\geq 0}{\operatorname{argmax}}\, \mathcal{S}(\boldsymbol{\alpha}(v)), \tag{6.69}$$

which can be computed numerically when analytical results are not available, see the case study in Section 6.7.

The mean-variance approach appeals intuition. The target variance v of the investor's objective $\Psi_{\boldsymbol{\alpha}}$ in the mean-variance optimization (6.68) can be interpreted as the riskiness of the solution $\boldsymbol{\alpha}(v)$: for a given level of risk v, the investor seeks the allocation that maximizes the expected value of his objective. As the risk level v spans all the positive numbers, the one-parameter family of solutions $\boldsymbol{\alpha}(v)$ describes a one-dimensional curve in the N-dimensional space of all possible allocations, and the optimal allocation $\boldsymbol{\alpha}^*$ must lie on this curve.

Making use of the Lagrangian formulation in (6.68), we can express the optimal allocation (6.69) as follows:

$$(\boldsymbol{\alpha}^*, \lambda^*) \equiv \arg\left(\max_{\boldsymbol{\alpha}\in\mathcal{C}} \min_{\lambda\in\mathbb{R}} \{\operatorname{E}\{\Psi_{\boldsymbol{\alpha}}\} - \lambda(\operatorname{Var}\{\Psi_{\boldsymbol{\alpha}}\} - v^*)\}\right). \tag{6.70}$$

The Lagrange coefficient λ^* that solves (6.70) can be interpreted as a coefficient of risk aversion. If λ^* is null the investor is risk neutral: indeed, the argument in curly brackets in (6.70) becomes the expected value. Thus the risk premium required by the investor to be exposed to market risk is null, see (5.89). On the other hand, if λ^* is positive the investor is risk averse: indeed, allocations with the same expected value but with larger variance are penalized in (6.70). Similarly, if λ^* is negative the investor is risk prone.

6.3.3 Setting up the mean-variance optimization

We recall that the investor's objective that appears in the mean-variance problem (6.68) is a linear function of the allocation and of the market vector:

$$\Psi_{\boldsymbol{\alpha}} \equiv \boldsymbol{\alpha}'\mathbf{M}, \tag{6.71}$$

see (6.2). Using the affine equivariance (2.56) and (2.71) of the expected value and the covariance respectively we obtain:

$$\mathrm{E}\{\Psi_{\boldsymbol{\alpha}}\} = \boldsymbol{\alpha}'\,\mathrm{E}\{\mathbf{M}\} \tag{6.72}$$

$$\mathrm{Var}\{\Psi_{\boldsymbol{\alpha}}\} = \boldsymbol{\alpha}'\,\mathrm{Cov}\{\mathbf{M}\}\,\boldsymbol{\alpha}. \tag{6.73}$$

Therefore we can re-express the mean-variance efficient frontier (6.68) in the following form:

$$\boldsymbol{\alpha}(v) \equiv \underset{\substack{\boldsymbol{\alpha}\in\mathcal{C} \\ \boldsymbol{\alpha}'\mathrm{Cov}\{\mathbf{M}\}\boldsymbol{\alpha}=v}}{\operatorname{argmax}} \boldsymbol{\alpha}'\,\mathrm{E}\{\mathbf{M}\}, \tag{6.74}$$

where $v \geq 0$.

In addition to the set of constraints $\mathcal{C}$, the only inputs required to compute the mean-variance efficient frontier (6.74) are the expected values of the market vector $\mathrm{E}\{\mathbf{M}\}$ and the respective covariance matrix $\mathrm{Cov}\{\mathbf{M}\}$. In order to compute these inputs, we have to follow the steps below, adapting from the discussion in Section 6.1:

Step 1. Detect the invariants $\mathbf{X}_{t,\widetilde{\tau}}$ behind the market relative to a suitable estimation horizon $\widetilde{\tau}$, see Section 3.1.

Step 2. Estimate the distribution of the invariants $\mathbf{X}_{t,\widetilde{\tau}}$, see Chapter 4.

Step 3. Project the invariants $\mathbf{X}_{t,\widetilde{\tau}}$ to the investment horizon, obtaining the distribution of $\mathbf{X}_{T+\tau,\tau}$, see Section 3.2.

Step 4. Map the distribution of the invariants $\mathbf{X}_{T+\tau,\tau}$ into the distribution of the prices at the investment horizon of the securities $\mathbf{P}_{T+\tau}$, see Section 3.3.

Step 5. Compute the expected value $\mathrm{E}\{\mathbf{P}_{T+\tau}\}$ and the covariance matrix $\mathrm{Cov}\{\mathbf{P}_{T+\tau}\}$ of the distribution of the market prices.

Step 6. Compute the inputs for the optimization (6.74), i.e. the expected value and the covariance matrix of the market vector $\mathbf{M}$. The market vector is an affine transformation of the market prices $\mathbf{M} \equiv \mathbf{a} + \mathbf{B}\mathbf{P}_{T+\tau}$, see (6.3). Therefore the inputs of the optimization follow from the affine equivariance (2.56) and (2.71) of the expected value and of the covariance matrix respectively:

$$\mathrm{E}\{\mathbf{M}\} = \mathbf{a} + \mathbf{B}\,\mathrm{E}\{\mathbf{P}_{T+\tau}\} \tag{6.75}$$

$$\mathrm{Cov}\{\mathbf{M}\} = \mathbf{B}\,\mathrm{Cov}\{\mathbf{P}_{T+\tau}\}\,\mathbf{B}'. \tag{6.76}$$

If the market is composed of equity-like and fixed-income security without derivative products, we can bypass some of the above steps. Indeed, in this case the invariants $\mathbf{X}_{t,\widetilde{\tau}}$ are the compounded returns and the changes in yield to maturity respectively. From (3.100) we obtain the expected value of the prices $\mathbf{P}_{T+\tau}$ directly from the distribution of the market invariants relative to the estimation interval:

$$\mathrm{E}\left\{P_{T+\tau}^{(n)}\right\} = e^{\boldsymbol{\gamma}'\boldsymbol{\delta}^{(n)}}\left[\phi_{\mathbf{X}_{t,\widetilde{\tau}}}\left(-i\,\mathrm{diag}\left(\boldsymbol{\varepsilon}\right)\boldsymbol{\delta}^{(n)}\right)\right]^{\frac{\tau}{\widetilde{\tau}}}, \tag{6.77}$$

where ϕ is the characteristic function of the market invariants, γ and ε are constant vectors defined in (3.84) and (3.85), and $\boldsymbol{\delta}$ is the canonical basis (A.15). Similarly, from (3.100) we obtain:

$$\mathrm{E}\left\{P_{T+\tau}^{(n)} P_{T+\tau}^{(m)}\right\} = e^{\gamma'\left(\boldsymbol{\delta}^{(n)}+\boldsymbol{\delta}^{(m)}\right)} \left[\phi_{\mathbf{X}_{t,\widetilde{\tau}}}\left(-i\,\mathrm{diag}\left(\boldsymbol{\varepsilon}\right)\left(\boldsymbol{\delta}^{(n)}+\boldsymbol{\delta}^{(m)}\right)\right)\right]^{\frac{\tau}{\widetilde{\tau}}}. \tag{6.78}$$

This expression with (6.77) in turn yields the covariance matrix:

$$\mathrm{Cov}\left\{P_{T+\tau}^{(n)}, P_{T+\tau}^{(m)}\right\} = \mathrm{E}\left\{P_{T+\tau}^{(n)} P_{T+\tau}^{(m)}\right\} - \mathrm{E}\left\{P_{T+\tau}^{(n)}\right\}\mathrm{E}\left\{P_{T+\tau}^{(m)}\right\}. \tag{6.79}$$

The inputs for the optimization (6.74) then follow from (6.75) and (6.76). For an application of these formulas, see the case study in Section 6.7.

If the constraints $\mathcal{C}$ in (6.74) are not too complex the computation of the mean-variance efficient frontier represents a quadratic programming problem which can be easily solved numerically, see Section 6.5.3.

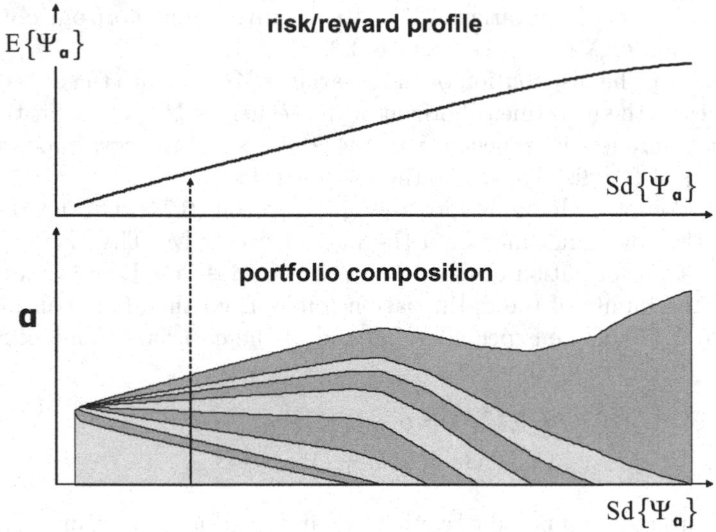

Fig. 6.7. Mean-variance efficient frontier

In Figure 6.7 we computed the solutions for the standard problem where the investor is bound by a budget constraint and a no-short-sale constraint:

$$\mathcal{C}: \boldsymbol{\alpha}' \mathbf{p}_T = w_T, \ \boldsymbol{\alpha} \geq \mathbf{0}. \tag{6.80}$$

Refer to symmys.com for the details on the market and on the computations.

If the constraints are affine, it is even possible to compute the analytical solution of the mean-variance problem, see Section 6.4 for the theory and Section 6.6 for an application.

6.3.4 Mean-variance in terms of returns

Recall from (3.10) that the linear return from the investment date T to the investment horizon τ of a security/portfolio that at time t trades at the price P_t is defined as follows:

$$L \equiv \frac{P_{T+\tau}}{P_T} - 1. \tag{6.81}$$

The mean-variance approach (6.74) is often presented and solved in terms of the returns (6.81) instead of the market vector as in (6.74). Nevertheless, this formulation presents a few drawbacks.

To present the formulation in terms of returns we need to make two restrictive assumptions. First, we assume that the investor's objective is final wealth, or equivalently that the market vector in (6.71) is represented by the prices of the securities at the investment horizon:

$$\Psi_\alpha \equiv \boldsymbol{\alpha}' \mathbf{P}_{T+\tau}. \tag{6.82}$$

Second, we assume that the investor's initial capital is not null:

$$w_T \equiv \boldsymbol{\alpha}' \mathbf{p}_T \neq 0. \tag{6.83}$$

Consider the linear return on wealth:

$$L^{\Psi_\alpha} \equiv \frac{\Psi_\alpha}{w_T} - 1. \tag{6.84}$$

As we show in Appendix www.6.6, under the assumptions (6.82) and (6.83) the mean-variance efficient frontier (6.68) can be expressed equivalently in terms of the linear return on wealth as follows:

$$\boldsymbol{\alpha}(v) = \underset{\substack{\boldsymbol{\alpha} \in \mathcal{C} \\ \operatorname{Var}\{L^{\Psi_\alpha}\} = v}}{\operatorname{argmax}} \ \mathrm{E}\{L^{\Psi_\alpha}\}, \tag{6.85}$$

where $v \geq 0$.

Consider now the *relative weights* $\mathbf{w}$ of a generic allocation:

$$\mathbf{w} \equiv \frac{\operatorname{diag}(\mathbf{p}_T)}{\boldsymbol{\alpha}' \mathbf{p}_T} \boldsymbol{\alpha}. \tag{6.86}$$

Since the current prices $\mathbf{p}_T$ are known, the relative weights $\mathbf{w}$ are a scale-independent equivalent representation of the allocation $\boldsymbol{\alpha}$.

As we show in Appendix www.6.6, we can express the linear return on wealth in terms of the linear returns (6.81) of the securities in the market and the respective relative weights:

$$L^{\Psi_\alpha} = \mathbf{w}'\mathbf{L}. \tag{6.87}$$

Therefore, using the affine equivariance properties (2.56) and (2.71) of the expected value and of the covariance matrix respectively, we can write (6.85) equivalently as follows:

$$\mathbf{w}(v) = \underset{\substack{\mathbf{w}\in\mathcal{C} \\ \mathbf{w}'\operatorname{Cov}\{\mathbf{L}\}\mathbf{w}=v}}{\operatorname{argmax}} \mathbf{w}'\operatorname{E}\{\mathbf{L}\}, \tag{6.88}$$

where $v \geq 0$. The efficient frontier in terms of the allocation vector $\boldsymbol{\alpha}(v)$ is then recovered from the relative weights (6.88) by inverting (6.86).

In order to set up the optimization in terms of linear returns and relative weights (6.88) we proceed like in the more general mean-variance case (6.74). Indeed, the inputs necessary to solve (6.88) are the expected value of the horizon-specific linear returns $\operatorname{E}\{\mathbf{L}\}$ and the respective covariance matrix $\operatorname{Cov}\{\mathbf{L}\}$. These parameters are obtained by following steps similar to those on p. 321:

Step 1. Detect the invariants $\mathbf{X}_{t,\widetilde{\tau}}$ behind the market relative to a suitable estimation horizon $\widetilde{\tau}$, see Section 3.1.

Step 2. Estimate the distribution of the invariants $\mathbf{X}_{t,\widetilde{\tau}}$, see Chapter 4.

Step 3. Project these invariants $\mathbf{X}_{t,\widetilde{\tau}}$ to the investment horizon, obtaining the distribution of $\mathbf{X}_{T+\tau,\tau}$, see Section 3.2.

Step 4. Map the distribution of the invariants $\mathbf{X}_{T+\tau,\tau}$ into the distribution of the prices at the investment horizon of the securities $\mathbf{P}_{T+\tau}$, see Section 3.3.

Step 5. Compute the expected value $\operatorname{E}\{\mathbf{P}_{T+\tau}\}$ and the covariance matrix $\operatorname{Cov}\{\mathbf{P}_{T+\tau}\}$ from the distribution of the market prices.

Step 6. Compute the inputs for the optimization (6.74), i.e. the expected value and the covariance matrix of the linear returns, from (6.81) using the affine equivariance (2.56) and (2.71) of the expected value and of the covariance matrix respectively:

$$\operatorname{E}\{\mathbf{L}\} = \operatorname{diag}(\mathbf{p}_T)^{-1}\operatorname{E}\{\mathbf{P}_{T+\tau}\} - \mathbf{1} \tag{6.89}$$

$$\operatorname{Cov}\{\mathbf{L}\} = \operatorname{diag}(\mathbf{p}_T)^{-1}\operatorname{Cov}\{\mathbf{P}_{T+\tau}\}\operatorname{diag}(\mathbf{p}_T)^{-1}. \tag{6.90}$$

If the constraints $\mathcal{C}$ in (6.88) are not too complex, the optimization problem in terms of linear returns and relative weights is quadratic and therefore it can be solved easily either analytically or numerically, just like the more general problem (6.74).

On the other hand, expressing an allocation in terms relative weights is somewhat more intuitive than expressing it in absolute terms. In other words, it is easier to interpret a statement such as "thirty percent of one's budget is invested in xyz" than "his investment consists, among others, of a thousand shares of xyz". Furthermore, in the formulation in terms of linear returns and relative weights a few expressions assume a simpler form.

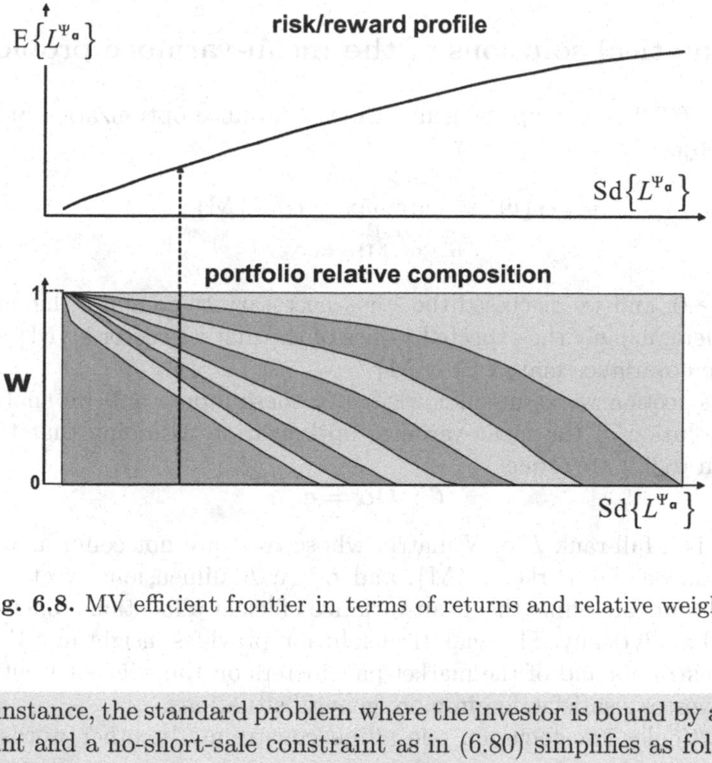

Fig. 6.8. MV efficient frontier in terms of returns and relative weights

For instance, the standard problem where the investor is bound by a budget constraint and a no-short-sale constraint as in (6.80) simplifies as follows:

$$\mathcal{C}: \ \mathbf{w}'\mathbf{1} = 1, \mathbf{w} \geq \mathbf{0}. \tag{6.91}$$

In Figure 6.8 we computed the efficient frontier under these constraints: compare with the respective plots in Figure 6.7. Refer to `symmys.com` for the details on the market and on the computations.

For the above reasons, the mean-variance framework is often presented in terms of returns and relative weights.

Nevertheless, we stress that the specification in terms of returns is not as general as the specification in terms of the investor's objective, because it applies only under the hypotheses (6.82) and (6.83). For instance, the linear returns on wealth are not defined when the initial investment is null. This prevents the analysis of *market-neutral strategies*, namely highly leveraged portfolios that pursue the largest possible final wealth by allocating zero initial net capital.

Furthermore, the formulation of the mean-variance problem in terms of returns and relative weights gives rise to misunderstandings. Indeed it makes it harder to separate the estimation process from the optimization process, see Section 6.5.4, and it gives rise to confusion when implementing allocation at different horizons, see Section 6.5.5.

6.4 Analytical solutions of the mean-variance problem

In Section 6.3.3 we set up the general mean-variance optimization of the investor's objective:

$$\boldsymbol{\alpha}(v) \equiv \underset{\substack{\boldsymbol{\alpha} \in \mathcal{C} \\ \boldsymbol{\alpha}' \operatorname{Cov}\{\mathbf{M}\}\boldsymbol{\alpha} = v}}{\operatorname{argmax}} \boldsymbol{\alpha}' \mathrm{E}\{\mathbf{M}\}, \tag{6.92}$$

where $v \geq 0$, and we discussed the steps necessary to compute the inputs of this problem, namely the expected values of the market vector $\mathrm{E}\{\mathbf{M}\}$ and the respective covariance matrix $\operatorname{Cov}\{\mathbf{M}\}$.

In this section we assume knowledge of these inputs and we analyze the explicit solution of the mean-variance optimization assuming that the constraints in (6.92) are affine:

$$\mathcal{C}: \ \mathbf{D}\boldsymbol{\alpha} = \mathbf{c}, \tag{6.93}$$

where $\mathbf{D}$ is a full-rank $K \times N$ matrix whose rows are not collinear with the expectation on the market $\mathrm{E}\{\mathbf{M}\}$, and $\mathbf{c}$ is a K-dimensional vector. When the constraints are affine the mean-variance efficient allocations (6.92) can be computed analytically. The analytical solution provides insight into the effect of the constraints and of the market parameters on the efficient frontier and on the investor's satisfaction in more general situations.

In particular, we focus on one affine constraint. In other words (6.93) becomes:

$$\mathcal{C}: \ \mathbf{d}'\boldsymbol{\alpha} = c, \tag{6.94}$$

where $\mathbf{d}$ is a generic constant vector not collinear with the expectation on the market $\mathrm{E}\{\mathbf{M}\}$ and c is a scalar. The one-dimensional case is still general enough to cover a variety of practical situations. Furthermore, in the one-dimensional case the analytical solution is very intuitive and easy to interpret geometrically. The computations and respective interpretations for the general case (6.93) follow similarly to the one-dimensional case.

The most notable example of affine constraint is the budget constraint:

$$\mathcal{C}: \ \boldsymbol{\alpha}'\mathbf{p}_T = w_T, \tag{6.95}$$

where w_T is the investor's initial capital.

6.4.1 Efficient frontier with affine constraints

In the general case where $c \neq 0$ in (6.94), the mean-variance efficient frontier (6.92) is the set of non-empty solutions to this problem:

$$\boldsymbol{\alpha}(v) \equiv \underset{\substack{\boldsymbol{\alpha}'\mathbf{d}=c \\ \mathrm{Var}\{\Psi_{\boldsymbol{\alpha}}\}=v}}{\operatorname{argmax}} \mathrm{E}\{\Psi_{\boldsymbol{\alpha}}\}, \tag{6.96}$$

where $v \geq 0$.

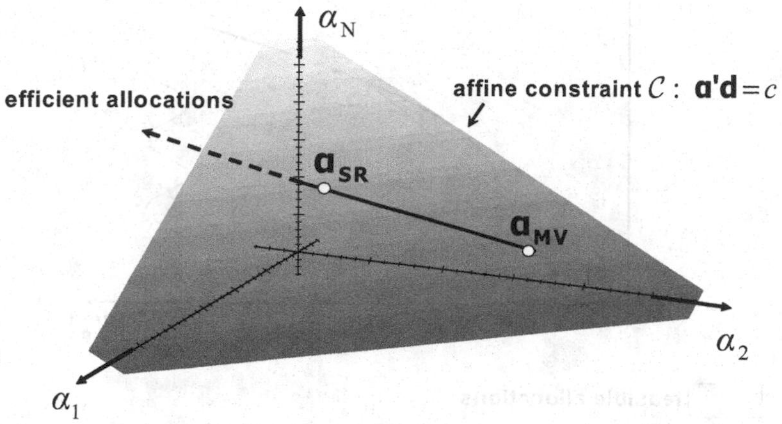

Fig. 6.9. MV efficient allocations under affine constraint: two-fund separation

In Appendix www.6.3 we prove that the above solutions are more easily parametrized in terms of the expected value of the investor's objective $e \equiv \mathrm{E}\{\Psi_{\boldsymbol{\alpha}}\}$ and read explicitly:

$$\boldsymbol{\alpha}(e) = \boldsymbol{\alpha}_{MV} + [e - \mathrm{E}\{\Psi_{\boldsymbol{\alpha}_{MV}}\}] \frac{\boldsymbol{\alpha}_{SR} - \boldsymbol{\alpha}_{MV}}{\mathrm{E}\{\Psi_{\boldsymbol{\alpha}_{SR}}\} - \mathrm{E}\{\Psi_{\boldsymbol{\alpha}_{MV}}\}}. \tag{6.97}$$

In this expression the scalar e varies in an infinite range:

$$e \in [\mathrm{E}\{\Psi_{\boldsymbol{\alpha}_{MV}}\}, \infty); \tag{6.98}$$

and the two allocations $\boldsymbol{\alpha}_{SR}$ and $\boldsymbol{\alpha}_{MV}$ are defined as follows:

$$\boldsymbol{\alpha}_{MV} \equiv \frac{c \,\mathrm{Cov}\{\mathbf{M}\}^{-1} \mathbf{d}}{\mathbf{d}' \mathrm{Cov}\{\mathbf{M}\}^{-1} \mathbf{d}}. \tag{6.99}$$

$$\boldsymbol{\alpha}_{SR} \equiv \frac{c \,\mathrm{Cov}\{\mathbf{M}\}^{-1} \mathrm{E}\{\mathbf{M}\}}{\mathbf{d}' \mathrm{Cov}\{\mathbf{M}\}^{-1} \mathrm{E}\{\mathbf{M}\}}. \tag{6.100}$$

In other words, the mean-variance efficient frontier (6.97) is a straight semi-line in the N-dimensional space of allocations that lies on the $(N-1)$-dimensional hyperplane determined by the affine constraint. This straight semi-line stems from the allocation $\boldsymbol{\alpha}_{MV}$ and passes through the allocation $\boldsymbol{\alpha}_{SR}$, see Figure 6.9.

This result is known as the *two-fund separation theorem*: a linear combination of two specific portfolios (mutual funds) suffices to generate the whole mean-variance efficient frontier.

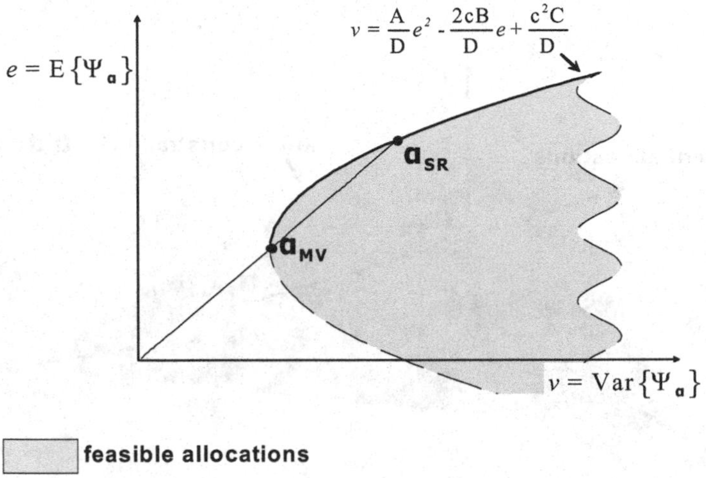

Fig. 6.10. Risk/reward profile of MV efficient allocations: expected value and variance

To evaluate the investor's satisfaction ensuing from the efficient allocations, we recall that in the mean-variance setting the investor's satisfaction by assumption only depends on the expected value and the variance of the investor's objective, see (6.67). Therefore we consider the plane of these two moments:

$$(v, e) \equiv (\mathrm{Var}\{\Psi\}, \mathrm{E}\{\Psi\}). \tag{6.101}$$

In Appendix www.6.3 we show that the feasible set in these coordinates is the region to the right of the following parabola, see Figure 6.10:

$$v = \frac{A}{D}e^2 - \frac{2cB}{D}e + \frac{c^2C}{D}, \tag{6.102}$$

where (A, B, C, D) are four scalars that do not depend on the allocations:

$$\begin{array}{ll} A \equiv \mathbf{d}' \mathrm{Cov}\{\mathbf{M}\}^{-1} \mathbf{d} & B \equiv \mathbf{d}' \mathrm{Cov}\{\mathbf{M}\}^{-1} \mathrm{E}\{\mathbf{M}\} \\ C \equiv \mathrm{E}\{\mathbf{M}\}' \mathrm{Cov}\{\mathbf{M}\}^{-1} \mathrm{E}\{\mathbf{M}\} & D \equiv AC - B^2, \end{array} \tag{6.103}$$

From (6.96), the mean-variance efficient frontier (6.97) corresponds to the allocations that give rise to the upper branch of this parabola.

The allocations $\boldsymbol{\alpha}_{SR}$ and $\boldsymbol{\alpha}_{MV}$ that generate the efficient frontier (6.97) are very special in terms of their risk/reward profile.

As we show in Appendix www.6.3, $\boldsymbol{\alpha}_{MV}$ is the allocation that displays the least possible variance. Therefore $\boldsymbol{\alpha}_{MV}$ is called the *global minimum variance portfolio*: in the risk/reward plane of Figure 6.10 the allocation $\boldsymbol{\alpha}_{MV}$ corresponds to the "belly" of the parabola (6.102).

To interpret the allocation $\boldsymbol{\alpha}_{SR}$, we recall that the Sharpe ratio (5.51) is defined as the ratio of the expected value of the investor's objective over its standard deviation:

$$\mathrm{SR}(\boldsymbol{\alpha}) \equiv \frac{\mathrm{E}\{\Psi_{\boldsymbol{\alpha}}\}}{\mathrm{Sd}\{\Psi_{\boldsymbol{\alpha}}\}}. \tag{6.104}$$

As we show in Appendix www.6.3, $\boldsymbol{\alpha}_{SR}$ is the allocation that displays the highest possible Sharpe ratio. Therefore $\boldsymbol{\alpha}_{SR}$ is called the *maximum Sharpe ratio portfolio*: in the risk/reward plane of Figure 6.10 the allocation $\boldsymbol{\alpha}_{SR}$ represents the intersection of the efficient frontier with the straight line through the origin and the minimum variance portfolio, see Appendix www.6.3.

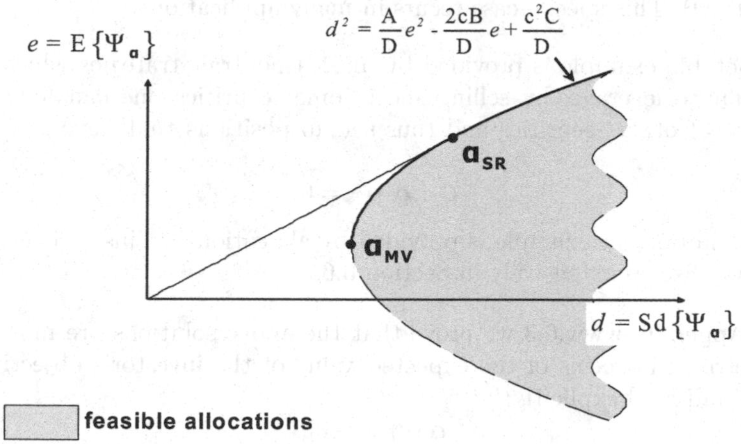

Fig. 6.11. Risk/reward profile of MV efficient allocations: expected value and standard deviation

Due to the interpretation in terms of the Sharpe ratio, it is convenient to represent the risk/reward profile of the objective also in terms of the expected value and standard deviation:

$$(d, e) \equiv (\mathrm{Sd}\{\Psi\}, \mathrm{E}\{\Psi\}). \tag{6.105}$$

In this plane the boundary of the feasible set, namely the parabola (6.102), becomes the following hyperbola, see Figure 6.11:

$$d^2 = \frac{A}{D}e^2 - \frac{2cB}{D}e + \frac{c^2C}{D}, \quad d > 0. \tag{6.106}$$

In turn, the global minimum variance portfolio is also the global minimum standard-deviation portfolio and therefore it plots as the "belly" of the hyperbola. On the other hand, from the definition of the Sharpe ratio (6.104), the maximum Sharpe ratio portfolio corresponds to the point of tangency of the hyperbola with a straight line stemming from the origin.

6.4.2 Efficient frontier with linear constraints

When $c \equiv 0$ in the affine constraint (6.94), the constraint becomes linear and the mean-variance efficient frontier (6.92) becomes the set of non-empty solutions to this problem:

$$\boldsymbol{\alpha}(v) \equiv \underset{\substack{\boldsymbol{\alpha}'\mathbf{d}=0 \\ \operatorname{Var}\{\Psi_{\boldsymbol{\alpha}}\}=v}}{\operatorname{argmax}} \operatorname{E}\{\Psi_{\boldsymbol{\alpha}}\}, \tag{6.107}$$

where $v \geq 0$. This special case recurs in many applications.

A notable example is provided by market-neutral strategies which invest with infinite leverage: by selling short some securities one can finance the purchase of other securities and thus set up positions that have zero initial value:

$$\mathcal{C}: \ \boldsymbol{\alpha}'\mathbf{p}_T = 0. \tag{6.108}$$

Another important example is provided by allocations against a benchmark, which we discuss extensively in Section 6.6.

In Appendix www.6.3 we prove that the above solutions are more easily parametrized in terms of the expected value of the investor's objective $e \equiv \operatorname{E}\{\Psi_{\boldsymbol{\alpha}}\}$ and read explicitly:

$$\boldsymbol{\alpha}(e) = e\boldsymbol{\alpha}_0, \tag{6.109}$$

where $\boldsymbol{\alpha}_0$ is a specific fixed allocation, defined in terms of the constants (6.103) as follows:

$$\boldsymbol{\alpha}_0 \equiv \operatorname{Cov}\{\mathbf{M}\}^{-1}(A\operatorname{E}\{\mathbf{M}\} - B\mathbf{d}). \tag{6.110}$$

In other words, when the investment constraint is linear, the ensuing mean-variance efficient allocations (6.109) describe a straight semi-line stemming from the origin that passes through the specific allocation $\boldsymbol{\alpha}_0$ and lies on the $(N-1)$-dimensional hyperplane determined by the constraint $\boldsymbol{\alpha}'\mathbf{d} = 0$, see Figure 6.12.

This result can be seen as a special case of (6.97) and Figure 6.9 in the limit where the constant c in (6.96) tends to zero. Indeed, in this limit the

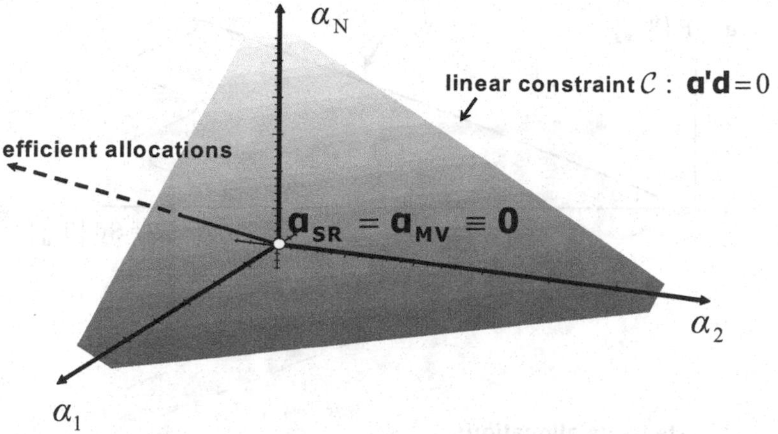

Fig. 6.12. MV efficient allocations under linear constraint

two portfolios $\boldsymbol{\alpha}_{SR}$ and $\boldsymbol{\alpha}_{MV}$, defined in (6.99) and (6.100) respectively, both shrink to zero. Nevertheless, the direction of departure of the straight semi-line (6.97) from the global minimum-variance portfolio does not depend on c and thus remains constant:

$$\boldsymbol{\alpha}_0 = \frac{\boldsymbol{\alpha}_{SR} - \boldsymbol{\alpha}_{MV}}{\mathrm{E}\{\Psi_{\boldsymbol{\alpha}_{SR}}\} - \mathrm{E}\{\Psi_{\boldsymbol{\alpha}_{MV}}\}}. \tag{6.111}$$

Therefore, as $c \to 0$ in the constraint (6.94), the straight semi-line in Figure 6.9 shifts in a parallel way towards the origin.

As in the case of a generic affine constraint, also for the special case $c \equiv 0$ in order to analyze the satisfaction ensuing from the mean-variance efficient allocations we only need to focus on the first two moments of the objective (6.101), or equivalently (6.105). In the latter coordinates, the hyperbola (6.106) which limits the feasible set in Figure 6.11 degenerates into the following locus:

$$e = \pm\sqrt{\frac{D}{A}}d, \quad d \geq 0. \tag{6.112}$$

This locus represents two straight semi-lines that stem from the origin, see Figure 6.13. The efficient frontier corresponds to the upper branch of this degenerate hyperbola, i.e. the straight line in the positive quadrant. Therefore all the efficient allocations share the same Sharpe ratio which is the highest possible in the feasible set and is equal to $\sqrt{D/A}$.

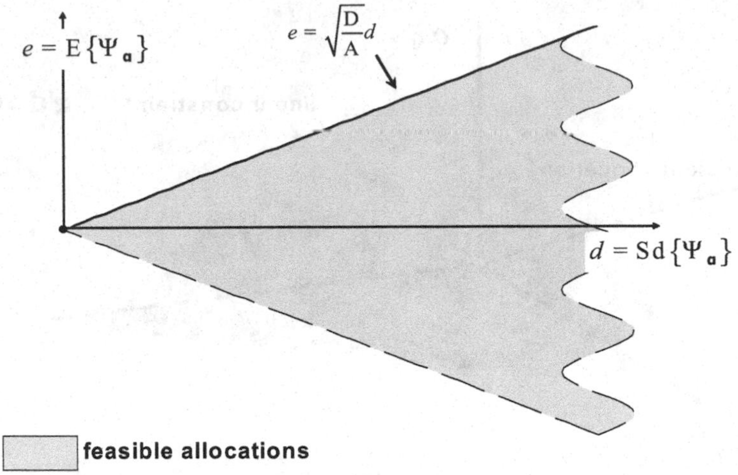

Fig. 6.13. Risk/reward profile of MV efficient allocations under linear constraint

6.4.3 Effects of correlations and other parameters

The only market parameters necessary to determine the mean-variance efficient frontier (6.92) are the expected values of the market vector $\mathrm{E}\{\mathbf{M}\}$ and the covariance matrix $\mathrm{Cov}\{\mathbf{M}\}$, which we factor into the respective standard deviations and correlations:

$$\mathrm{Cov}\{\mathbf{M}\} \equiv \mathrm{diag}(\mathrm{Sd}\{\mathbf{M}\}) \, \mathrm{Cor}\{\mathbf{M}\} \, \mathrm{diag}(\mathrm{Sd}\{\mathbf{M}\}). \tag{6.113}$$

In this section we discuss the impact of changes in these parameters on the investor's satisfaction.

In the mean-variance setting the investor's satisfaction only depends on the expected value and the variance of his objective, see (6.67), or equivalently on the expected value and the standard deviation of his objective. Therefore we analyze the effects of changes in the market parameters in the plane of these coordinates:

$$(d, e) \equiv (\mathrm{Sd}\{\Psi\}, \mathrm{E}\{\Psi\}). \tag{6.114}$$

Since all the indices of satisfaction $\mathcal{S}$ discussed in Chapter 5 are consistent with weak stochastic dominance, for a given level of standard deviation of the objective, higher expected values of the objective are always appreciated. Therefore a given market presents better investment opportunities than another market if, other things equal, the upper boundary of its feasible set in the coordinates (6.114) plots above the upper boundary of the feasible set of the other market for all values of the standard deviation.

It is immediate to determine the effect of changes in expected values $\mathrm{E}\{\mathbf{M}\}$ and standard deviations $\mathrm{Sd}\{\mathbf{M}\}$ on the feasible set. Indeed, larger expected values of the market vector shift the feasible set upward in the coordinates (6.114) and larger standard deviations shift the feasible set to the right.

To analyze the effect of the correlations, we consider the simplest case of a two-security market, which gives rise to a bivariate market vector $\mathbf{M} \equiv (M_1, M_2)'$. In this case there exists only one correlation $\rho \equiv \text{Cor}\{M_1, M_2\}$.

We consider the generic case where with a higher expected value of the investor's objective is associated a higher standard deviation. Therefore we assume without loss of generality:

$$e^{(1)} < e^{(2)}, \qquad d^{(1)} < d^{(2)}, \tag{6.115}$$

where the index $j = 1, 2$ denotes the coordinates (6.114) of a full allocation in the j-th security. The boundary of the feasible set (6.106) becomes fully determined by the value of the correlation ρ. In Figure 6.14 we show the effect of different values of the market correlation ρ on the feasible set, see Appendix www.6.4 for the analytical expressions behind these plots and the statements that follow.

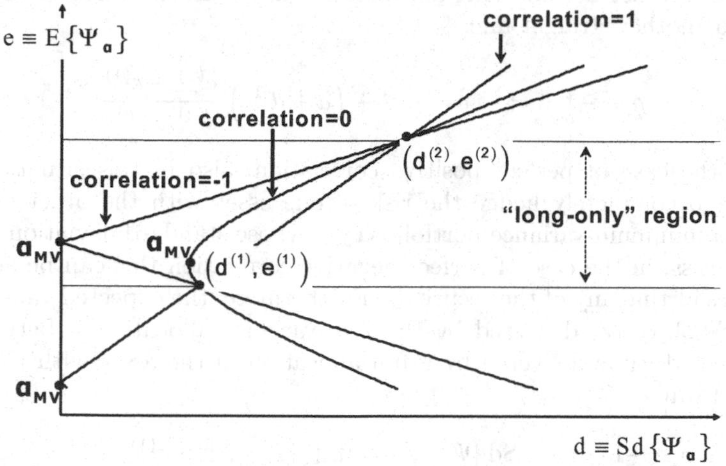

Fig. 6.14. Diversification effect of correlation

We distinguish three cases for the correlation: total correlation $\rho \equiv 1$, null correlation, $\rho \equiv 0$ and total anti-correlation $\rho \equiv -1$; and two cases for the allocation: *long-only positions*, where the amounts (α_1, α_2) of both securities in an allocation are positive, and *short positions*, where one of the two amounts α_1 or α_2 is negative.

In the case of perfect positive correlation the two securities are perceived as equivalent. The efficient frontier degenerates into a straight line that joins the coordinates of the two assets in the plane (6.114):

$$\rho \equiv 1 \quad \Rightarrow \quad e = e^{(1)} + \left(d - d^{(1)}\right) \frac{e^{(2)} - e^{(1)}}{d^{(2)} - d^{(1)}}. \tag{6.116}$$

By shorting one of the assets it is possible to completely hedge the risk of the other asset and achieve a global minimum-variance portfolio $\boldsymbol{\alpha}_{MV}$ such that the investor's objective has null standard deviation. Nevertheless, the perfect hedge comes at a price: the expected value of the investor's objective delivered by the zero-variance allocation is worse than the expected value delivered by a full allocation in the asset with the lower expected value:

$$\rho \equiv 1 \quad \Rightarrow \quad \operatorname{Sd}\{\Psi_{\boldsymbol{\alpha}_{MV}}\} = 0, \operatorname{E}\{\Psi_{\boldsymbol{\alpha}_{MV}}\} < e^{(1)}. \tag{6.117}$$

As the correlation decreases toward zero, the securities give rise to an increasingly diversified market: the diversification effect makes the expected value of the global minimum-variance portfolio to rise, although the variance of this portfolio is no longer zero. In the long-only region the efficient frontier swells upwards, providing better investment opportunities than the straight-line (6.116).

As the market become fully anti-correlated, the efficient frontier degenerates into another straight line:

$$\rho \equiv -1 \quad \Rightarrow \quad e = e^{(1)} + \left(d + d^{(1)}\right) \frac{e^{(2)} - e^{(1)}}{d^{(2)} + d^{(1)}}. \tag{6.118}$$

Like in the case of perfect positive correlation, also in this situation it is possible to completely hedge the risk of one asset with the other one, obtaining a minimum-variance portfolio $\boldsymbol{\alpha}_{MV}$ whose standard deviation is zero. Nevertheless, in the case of perfect negative correlation this can be achieved without shorting any of the securities. Furthermore, the expected value of the investor's objective delivered by the zero-variance allocation is better than the expected value delivered by a full allocation in the asset with the lower expected value:

$$\rho \equiv -1 \quad \Rightarrow \quad \operatorname{Sd}\{\Psi_{\boldsymbol{\alpha}_{MV}}\} = 0, \operatorname{E}\{\Psi_{\boldsymbol{\alpha}_{MV}}\} > e^{(1)}. \tag{6.119}$$

A comparison of (6.117) and (6.119) shows the benefits of diversification for low-variance portfolios.

Nevertheless, the perfect-correlation efficient frontier (6.116) is steeper than the perfect-anticorrelation efficient frontier (6.118). The two lines intersect at the point $\left(d^{(2)}, e^{(2)}\right)$, which corresponds to a full-investment in the riskier asset. When the investor is willing to increase his risk in pursuit of higher expected values by abandoning the long-only region, the best opportunities are provided by a highly correlated market. Therefore, contrary to a common belief, markets with low correlations do not necessarily provide better investment opportunities.

6.4.4 Effects of the market dimension

So far we have assumed that the dimension N of the market is fixed. Suppose that we allow new securities in the market. As intuition suggests, the ensuing investment opportunities can only improve. Indeed, if for a given level of standard deviation of the investor's objective it is possible to achieve a determined expected value, we can obtain the same result with a larger set of assets by simply allocating zero wealth in the new securities.

Nevertheless, from the above discussion on the effect of diversification we can guess that enlarging the market not only does not worsen, but actually substantially improves the efficient frontier. In order to verify this ansatz, we consider a market of an increasing number of securities, where we screen the effect of correlations, variances and expected values.

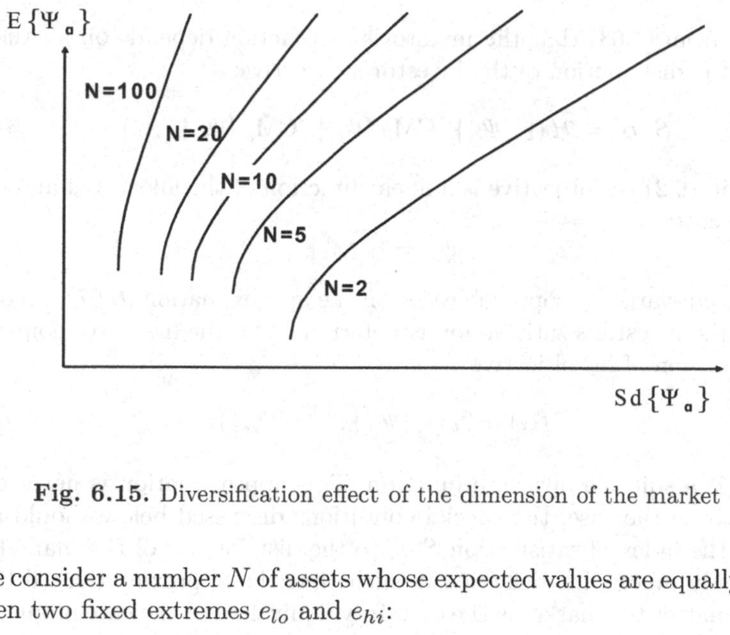

Fig. 6.15. Diversification effect of the dimension of the market

We consider a number N of assets whose expected values are equally spaced between two fixed extremes e_{lo} and e_{hi}:

$$\mathrm{E}\{\mathbf{M}\} \equiv (e_{lo}, e_{lo} + \Delta_N, \ldots, e_{hi} - \Delta_N, e_{hi})', \tag{6.120}$$

where $\Delta_N \equiv (e_{hi} - e_{lo}) / (N - 1)$. Similarly, we assume the standard deviations of these assets to be equally spaced between two fixed extremes d_{lo} and d_{hi}:

$$\mathrm{Sd}\{\mathbf{M}\} \equiv (d_{lo}, d_{lo} + \Gamma_N, \ldots, d_{hi} - \Gamma_N, d_{hi})', \tag{6.121}$$

where $\Gamma_N \equiv (d_{hi} - d_{lo}) / (N - 1)$. In order to screen out the effect of the cross-correlations, we assume zero correlation between all pairs of different entries of the market vector $\mathbf{M}$.

In Figure 6.15 we plot the efficient frontier in the above market as a function of the number N of securities. As expected, adding new assets shifts the frontier toward the upper-left region, giving rise to better investment opportunities: this effect is more pronounced when the number of assets in the market is relatively low.

6.5 Pitfalls of the mean-variance framework

In this section we discuss some common pitfalls in the interpretation and implementation of the mean-variance framework. Indeed, the very reasons that led to the success of the mean-variance approach also made it susceptible to misinterpretations.

6.5.1 MV as an approximation

We recall from (6.63) that the investor's satisfaction depends on all the moments of the distribution of the investor's objective:

$$\mathcal{S}(\boldsymbol{\alpha}) = \mathcal{H}\left(\mathrm{E}\left\{\Psi_{\boldsymbol{\alpha}}\right\}, \mathrm{CM}_2\left\{\Psi_{\boldsymbol{\alpha}}\right\}, \mathrm{CM}_3\left\{\Psi_{\boldsymbol{\alpha}}\right\}, \ldots\right), \tag{6.122}$$

where as in (6.2) the objective is a linear function of the allocation and of the market vector:

$$\Psi_{\boldsymbol{\alpha}} \equiv \boldsymbol{\alpha}'\mathbf{M}. \tag{6.123}$$

The mean-variance approach relies on the approximation (6.67), according to which the investor's satisfaction is determined by the first two moments of the distribution of his objective:

$$\mathcal{S}(\boldsymbol{\alpha}) \approx \widetilde{\mathcal{H}}\left(\mathrm{E}\left\{\Psi_{\boldsymbol{\alpha}}\right\}, \mathrm{Var}\left\{\Psi_{\boldsymbol{\alpha}}\right\}\right), \tag{6.124}$$

where $\widetilde{\mathcal{H}}$ is a suitable bivariate function. This approximation is never exact. For this to be the case, the special conditions discussed below should apply to either the index of satisfaction $\mathcal{S}$ or to the distribution of the market $\mathbf{M}$.

The only index of satisfaction $\mathcal{S}$ such that the approximation (6.124) is exact no matter the market is the certainty-equivalent in the case of quadratic utility:

$$u(\psi) = \psi - \frac{1}{2\zeta}\psi^2. \tag{6.125}$$

Indeed in this case the expected utility becomes a function of the expected value and variance of the objective, and therefore so does the certainty-equivalent.

Nevertheless the quadratic utility is not flexible enough to model the whole spectrum of the investor's preferences. Furthermore, for values of the objective such that $\psi > \zeta$ the quadratic utility becomes nonsensical, as it violates the non-satiation principle underlying the investor's objective: larger values of the objective make the investor less satisfied, see (5.134) and comments thereafter.

The only markets such that the approximation (6.124) is exact no matter the index of satisfaction are elliptically distributed markets:

$$\mathbf{M} \sim \mathrm{El}\left(\boldsymbol{\mu}, \boldsymbol{\Sigma}, g_N\right), \tag{6.126}$$

where $\boldsymbol{\mu}$ is the location parameter, $\boldsymbol{\Sigma}$ is the scatter matrix and g_N is the probability density generator for the N-dimensional case, see (2.268).

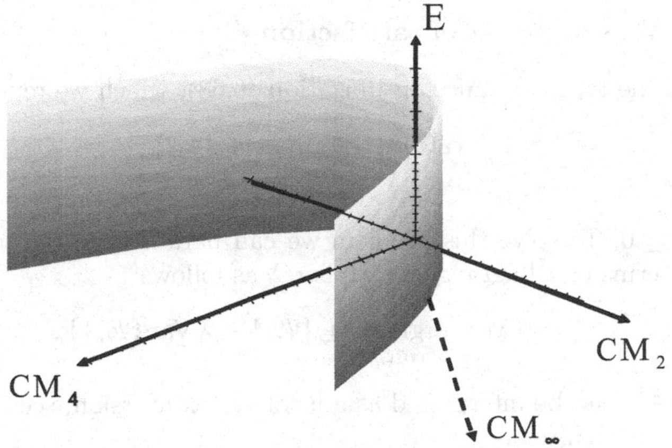

Fig. 6.16. Elliptical markets: the space of moments of the investor's objective is two-dimensional

Indeed, in this case from (2.270) and (2.276) the investor's objective is also elliptically distributed:

$$\Psi_{\alpha} \sim \mathrm{El}\left(\mathrm{E}\left\{\Psi_{\alpha}\right\}, \gamma \operatorname{Var}\left\{\Psi_{\alpha}\right\}, g_1\right), \tag{6.127}$$

where g_1 is the pdf generator for the one-dimensional case and γ is a scalar that does not depend on the allocation. In other words, if the market $\mathbf{M}$ is elliptically distributed, the infinite-dimensional space of moments is reduced to a two-dimensional manifold parametrized by expected value and variance, see Figure 6.16. As a result, also the index of satisfaction (6.122) becomes a function of expected value and variance only.

Nevertheless, the assumption that a market is elliptical is very strong. For instance, in highly asymmetric markets with derivative products the elliptical assumption cannot be accepted. Even in the absence of derivatives, the standard distribution to model prices in the stock market is the multivariate lognormal distribution, which extends to a multivariate setting the classical framework of Black and Scholes (1973). If the stock market is very volatile or the investment horizon is large, approximating the lognormal distribution in the mean-variance problem with an elliptical distribution leads to incorrect results.

Although the approximation (6.124) is never exact, it is quite accurate in many practical applications, namely when the combined effects of the distribution of the market $\mathbf{M}$ and of the functional expression of $\mathcal{H}$ in (6.122) make the relative importance of higher moments negligible. Therefore, the applicability of the approximation (6.124) must be checked on a case-by-case basis.

6.5.2 MV as an index of satisfaction

Consider the mean-variance optimization (6.68), which we report here:

$$\boldsymbol{\alpha}(v) \equiv \underset{\substack{\boldsymbol{\alpha}\in\mathcal{C} \\ \mathrm{Var}\{\Psi_{\boldsymbol{\alpha}}\}=v}}{\operatorname{argmax}} \mathrm{E}\{\Psi_{\boldsymbol{\alpha}}\}, \tag{6.128}$$

where $v \geq 0$. To solve this problem we can parametrize the set of solutions $\boldsymbol{\alpha}(v)$ in terms of a Lagrange multiplier λ as follows:

$$\boldsymbol{\alpha}(\lambda) \equiv \underset{\boldsymbol{\alpha}\in\mathcal{C}}{\operatorname{argmax}} \{\mathrm{E}\{\Psi_{\boldsymbol{\alpha}}\} - \lambda \mathrm{Var}\{\Psi_{\boldsymbol{\alpha}}\}\}, \tag{6.129}$$

where $\lambda \in \mathbb{R}$ can be interpreted as a level of risk aversion, see also (6.70) and comments thereafter.

From (6.69), the optimal allocation lies on the curve $\boldsymbol{\alpha}(\lambda)$. In other words, in order to determine the proper level of risk aversion λ^* and thus the optimal allocation $\boldsymbol{\alpha}^* \equiv \boldsymbol{\alpha}(\lambda^*)$, we perform the following one-dimensional optimization based on the investor's index of satisfaction:

$$\lambda^* \equiv \underset{\lambda\in\mathbb{R}}{\operatorname{argmax}} \mathcal{S}(\boldsymbol{\alpha}(\lambda)). \tag{6.130}$$

Consider an investor whose initial budget is one unit of currency. Assume that his objective is final wealth, and that he evaluates the riskiness of an allocation by means of a sensible index of satisfaction.

Suppose that the market consists of only two securities, that trade at the following price today:

$$p_T^{(1)} \equiv 1, \quad p_T^{(2)} \equiv 1. \tag{6.131}$$

At the investment horizon the value of the first security, which is non-stochastic, remains unaltered; the second security on the other hand has a 50% chance of doubling in value:

$$P_{T+\tau}^{(1)} \equiv 1, \quad P_{T+\tau}^{(2)} = \begin{cases} 1 \text{ (probability} = 50\%) \\ 2 \text{ (probability} = 50\%). \end{cases} \tag{6.132}$$

Taking into account the budget constraint and the no-short-sale constraint, the investor's objective is completely determined by the investment α in the risky security:

$$\Psi_{\alpha} \equiv (1-\alpha) P_{T+\tau}^{(1)} + \alpha P_{T+\tau}^{(2)}, \quad \alpha \in [0,1] . \tag{6.133}$$

From this expression and (6.132) it is immediate to compute the first two moments of the objective:

$$\mathrm{E}\{\Psi_{\alpha}\} = 1 + \frac{\alpha}{2}, \quad \mathrm{Var}\{\Psi_{\alpha}\} = \frac{\alpha^2}{4} . \tag{6.134}$$

In turn, from the first-order condition in the Lagrange formulation (6.129) we obtain the mean-variance curve:

$$\alpha(\lambda) = \frac{1}{\lambda} . \tag{6.135}$$

To compute the optimal level λ^* that gives rise to the optimal allocation $\alpha(\lambda^*)$ we do not need to specify the investor's preferences by means of a specific index of satisfaction, as long as such an index is sensible. Indeed, sensibility implies a full investment in the risky security, which strongly dominates the risk-free asset. In other words, the optimal allocation is $\alpha^* \equiv 1$ and the respective optimal value for the Lagrange multiplier reads:

$$\lambda^* \equiv 1. \tag{6.136}$$

A common misinterpretation of the Lagrangian reformulation consists in considering the level of risk aversion λ^* as a feature of the investor that is independent of the market. In other words, one is tempted to first define a pseudo-index of satisfaction as follows:

$$\mathcal{S}^*(\boldsymbol{\alpha}) \equiv \mathrm{E}\{\Psi_{\boldsymbol{\alpha}}\} - \lambda^* \mathrm{Var}\{\Psi_{\boldsymbol{\alpha}}\} ; \tag{6.137}$$

and then to solve for the optimal allocation as follows:

$$\boldsymbol{\alpha}^* \equiv \underset{\boldsymbol{\alpha} \in \mathcal{C}}{\operatorname{argmax}} \, \mathcal{S}^*(\boldsymbol{\alpha}) . \tag{6.138}$$

This is a quadratic function of the allocation, and thus an easier problem to solve than the two-step optimization (6.129)-(6.130).

Nevertheless, the definition of the pseudo-index of satisfaction (6.137) is incorrect, because it depends on the market through λ^*. In other words, the same investor displays different risk aversion coefficients λ^* when facing different markets. Therefore the pseudo-index of satisfaction (6.137) does not represent a description of the investor's preferences. Using it as if λ^* did not depend on the market might lead to nonsensical results.

Consider the previous example, where instead of the market (6.132) we have:

$$P_{T+\tau}^{(1)} \equiv 1, \quad P_{T+\tau}^{(2)} = \begin{cases} 1 \text{ (probability} = 50\%) \\ 3 \text{ (probability} = 50\%). \end{cases} \tag{6.139}$$

Then

$$\mathrm{E}\{\Psi_{\boldsymbol{\alpha}}\} = 1 + \alpha, \quad \mathrm{Var}\{\Psi_{\boldsymbol{\alpha}}\} = \alpha^2. \tag{6.140}$$

From the first-order condition in the Lagrange formulation (6.129) we obtain the mean-variance curve:

$$\alpha(\lambda) = \frac{1}{2\lambda}. \tag{6.141}$$

Since the optimal allocation is a full investment in the risky asset $\alpha^* \equiv 1$, in this market we obtain:

$$\lambda^* \equiv \frac{1}{2}. \tag{6.142}$$

Using in the mean-variance curve (6.141) the value (6.136) obtained in the previous market would give rise to a nonsensical positive allocation in the risk-free asset.

6.5.3 Quadratic programming and dual formulation

We recall that since all the indices of satisfaction $\mathcal{S}$ discussed in Chapter 5 are consistent with weak stochastic dominance, the mean-variance approach aims at maximizing the expected value of the investor's objective for a given level of variance. Therefore the mean-variance efficient allocations are the non-empty solutions of (6.68), which we report here:

$$\boldsymbol{\alpha}(v) \equiv \underset{\substack{\boldsymbol{\alpha} \in \mathcal{C} \\ \mathrm{Var}\{\Psi_{\boldsymbol{\alpha}}\} = v}}{\operatorname{argmax}} \mathrm{E}\{\Psi_{\boldsymbol{\alpha}}\}, \tag{6.143}$$

where $v \geq 0$. Notice that the variance constraint appears as an equality.

Consider the plane of coordinates $v \equiv \mathrm{Var}\{\Psi_{\boldsymbol{\alpha}}\}$ and $e \equiv \mathrm{E}\{\Psi_{\boldsymbol{\alpha}}\}$. If the upper limit of the feasible set determined by the constraints increases as we shift to the right on the horizontal axis, then (6.143) is equivalent to a problem with an inequality for the variance:

$$\boldsymbol{\alpha}(v) \equiv \underset{\substack{\boldsymbol{\alpha} \in \mathcal{C} \\ \mathrm{Var}\{\Psi_{\boldsymbol{\alpha}}\} \leq v}}{\operatorname{argmax}} \mathrm{E}\{\Psi_{\boldsymbol{\alpha}}\}, \tag{6.144}$$

where $v \geq 0$.

This is not the case in the example in Figure 6.17: the allocations on the thick line in the north-east region would not be captured by (6.144), although they are efficient according to (6.143).

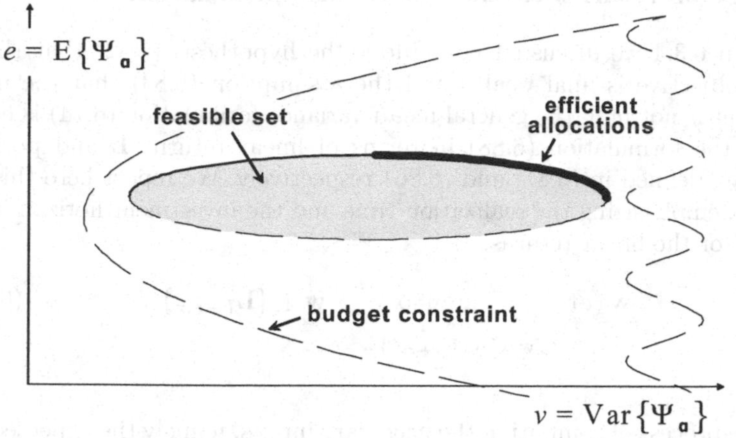

Fig. 6.17. MV efficient frontier as expected value maximization

On the other hand, we cannot rule out such allocations. For instance, in a prospect theoretical setting the investor becomes risk prone when facing losses, see the example in the shaded box on p. 269.

If the optimization with an inequality (6.144) is equivalent to the original problem (6.143) and if the investment constraints $\mathcal{C}$ are at most quadratic in the allocation, the optimization with an inequality is a quadratically constrained quadratic programming problem, see (6.57):

$$\boldsymbol{\alpha}(v) \equiv \underset{\substack{\boldsymbol{\alpha}\in\mathcal{C} \\ \boldsymbol{\alpha}' \operatorname{Cov}\{\mathbf{M}\}\boldsymbol{\alpha}\leq v}}{\operatorname{argmax}} \; \boldsymbol{\alpha}' \operatorname{E}\{\mathbf{M}\}, \tag{6.145}$$

where $v \geq 0$. Therefore this problem can be solved numerically.

At times the inequality-based mean-variance problem (6.144) is presented in its *dual formulation* as the non-empty set of the solutions to the following problem:

$$\boldsymbol{\alpha}(e) \equiv \underset{\substack{\boldsymbol{\alpha}\in\mathcal{C} \\ \operatorname{E}\{\Psi_{\boldsymbol{\alpha}}\}\geq e}}{\operatorname{argmin}} \operatorname{Var}\{\Psi_{\boldsymbol{\alpha}}\}, \tag{6.146}$$

where $e \in (-\infty, +\infty)$.

Under regularity conditions for the constraints $\mathcal{C}$ the dual formulation (6.146) is equivalent to (6.144), which in turn is equivalent to the original problem (6.143). The equivalence of these formulations must be checked on a case-by-case basis. For instance, the three formulations are equivalent when the constraints are affine, see (6.93), or for the standard no-short-sale constraint that appears in (6.80).

6.5.4 MV on returns: estimation versus optimization

In Section 6.3.4 we discussed how under the hypothesis (6.82) that the investor's objective is final wealth and the assumption (6.83) that the initial investment is not null, the general mean-variance formulation (6.74) is equivalent to the formulation (6.88) in terms of linear returns $\mathbf{L}$ and portfolio weights $\mathbf{w}$, defined in (6.81) and (6.86) respectively. We report here this formulation, emphasizing the realization time and the investment horizon in the notation for the linear returns:

$$\mathbf{w}(v) = \underset{\substack{\mathbf{w}\in\mathcal{C} \\ \mathbf{w}' \operatorname{Cov}\{\mathbf{L}_{T+\tau,\tau}\}\mathbf{w}=v}}{\operatorname{argmax}} \mathbf{w}' \operatorname{E}\{\mathbf{L}_{T+\tau,\tau}\}, \tag{6.147}$$

where $v \geq 0$.

In the process of computing the necessary inputs, namely the expected values $\operatorname{E}\{\mathbf{L}_{T+\tau,\tau}\}$ and the covariance matrix $\operatorname{Cov}\{\mathbf{L}_{T+\tau,\tau}\}$, there exists a clear distinction between estimation, which is performed on the market invariants $\mathbf{X}_{t,\widetilde{\tau}}$, and optimization, which acts on *functions* of the *projected* invariants $\mathbf{L}_{T+\tau,\tau}$, see the steps 1-6 on p. 324.

Now let us make two further assumptions. Assume that the market consists of equity-like securities as in Section 3.1.1, in which case the linear returns are market invariants:

$$\mathbf{X}_{t,\widetilde{\tau}} \equiv \mathbf{L}_{t,\widetilde{\tau}}. \tag{6.148}$$

Furthermore, assume that the investment horizon and the estimation interval coincide:

$$\tau \equiv \widetilde{\tau}. \tag{6.149}$$

Under the above combined assumptions it is possible to bypass many of the steps that lead to the inputs $\operatorname{E}\{\mathbf{L}_{T+\tau,\tau}\}$ and $\operatorname{Cov}\{\mathbf{L}_{T+\tau,\tau}\}$. Indeed, instead of estimating the whole distribution of the invariants $\mathbf{L}_{t,\widetilde{\tau}}$ as in Step 2 on p. 324, we estimate directly only its expected value $\operatorname{E}\{\mathbf{L}_{t,\widetilde{\tau}}\}$ and its covariance matrix $\operatorname{Cov}\{\mathbf{L}_{t,\widetilde{\tau}}\}$. Since by assumption the investment horizon is the estimation interval *and* since $\mathbf{L}_{t,\widetilde{\tau}}$ are invariants, the following holds:

$$\operatorname{E}\{\mathbf{L}_{T+\tau,\tau}\} = \operatorname{E}\{\mathbf{L}_{t,\widetilde{\tau}}\}, \quad \operatorname{Cov}\{\mathbf{L}_{T+\tau,\tau}\} = \operatorname{Cov}\{\mathbf{L}_{t,\widetilde{\tau}}\}. \tag{6.150}$$

Therefore we can skip Step 3, Step 4, Step 5 and Step 6 and plug (6.150) directly in the mean-variance problem (6.147).

We stress that the above shortcut is not viable in general. For instance, in the fixed-income market the linear returns are not market invariants. Instead, the market invariants are the changes in yield to maturity, see Section 3.1.2. Therefore in order to perform the mean-variance analysis in the fixed-income market in terms of relative weights and linear returns we need to go through all the steps 1-6 on p. 324. Nonetheless, one is dangerously tempted to estimate the returns as if they were invariants and proceed with the shortcut (6.150).

6.5.5 MV on returns: investment at different horizons

Another misunderstanding regarding the mean-variance framework occurs when the investment horizon τ is shifted farther in the future and the mean-variance optimization is formulated in terms of returns, see also Meucci (2001).

As in (6.147) we make the assumptions that the investor's objective is final wealth and that the initial investment is not zero, in such a way that the general mean-variance formulation (6.74) is equivalent to the formulation in term of linear returns and portfolio weights.

As in (6.148) we consider the case where the market consists of equity-like securities, in which case the linear returns are market invariants.

Nevertheless, unlike (6.149), we consider an investment horizon that is different, typically longer, than the estimation interval:

$$\tau > \widetilde{\tau}. \tag{6.151}$$

In this case the shortcut (6.150) does not apply. Instead, we need to project the distribution of the invariants to the investment horizon and then compute the quantities of interest $\mathrm{E}\{\mathbf{L}_{T+\tau,\tau}\}$ and $\mathrm{Cov}\{\mathbf{L}_{T+\tau,\tau}\}$ as described in the steps 1-6 on p. 324. As we see below, only when the market is not too volatile and both the investment horizon and the estimation interval are short is the shortcut (6.150) approximately correct, see also Meucci (2004).

Since we are dealing with equity-like securities, the projection of the invariants into the moments of the linear returns takes a simpler form than in the more general case discussed in Section 6.3.4. This is the same argument that leads to (6.78) and (6.79) in the mean-variance formulation in terms of prices. Therefore it applies also to the fixed-income market. Here we present this argument explicitly in the case of equity-like securities.

We recall from (3.11) that for a generic security or portfolio that is worth P_t at time t, the τ-horizon compounded return at time t is defined as follows:

$$C_{t,\tau} \equiv \ln\left(\frac{P_t}{P_{t-\tau}}\right). \tag{6.152}$$

Therefore the linear returns (6.81) are the following function of the compounded returns:

$$1 + L_{t,\tau} \equiv e^{C_{t,\tau}}. \tag{6.153}$$

From the above equality we obtain the following relation for the expected value of the linear returns:

$$\begin{aligned} \mathrm{E}\left\{1 + L^{(n)}_{T+\tau,\tau}\right\} &= \mathrm{E}\left\{e^{C^{(n)}_{T+\tau,\tau}}\right\} \\ &= \phi_{\mathbf{C}_{T+\tau,\tau}}\left(-i\boldsymbol{\delta}^{(n)}\right). \end{aligned} \tag{6.154}$$

In this expression $\phi_{\mathbf{C}_{T+\tau,\tau}}$ is the joint characteristic function of the compounded returns relative to the investment horizon and $\boldsymbol{\delta}^{(n)}$ is the n-th element canonical basis ($A.15$), i.e. it is a vector of zeros, except for the n-th entry, which is one. Similarly, from (6.153) we obtain:

$$\mathrm{E}\left\{\left(1+L_{T+\tau,\tau}^{(m)}\right)\left(1+L_{T+\tau,\tau}^{(n)}\right)\right\}=\mathrm{E}\left\{e^{C_{T+\tau,\tau}^{(m)}+C_{T+\tau,\tau}^{(n)}}\right\} \tag{6.155}$$
$$=\phi_{\mathbf{C}_{T+\tau,\tau}}\left(-i\left(\boldsymbol{\delta}^{(m)}+\boldsymbol{\delta}^{(n)}\right)\right).$$

From these expressions in turn we immediately obtain the desired quantities:

$$\mathrm{E}\left\{L_{T+\tau,\tau}^{(n)}\right\}=\phi_{\mathbf{C}_{T+\tau,\tau}}\left(-i\boldsymbol{\delta}^{(n)}\right)-1 \tag{6.156}$$

and

$$\operatorname{Cov}\left\{L_{T+\tau,\tau}^{(m)},L_{T+\tau,\tau}^{(n)}\right\}=\operatorname{Cov}\left\{1+L_{T+\tau,\tau}^{(m)},1+L_{T+\tau,\tau}^{(n)}\right\} \tag{6.157}$$
$$=\phi_{\mathbf{C}_{T+\tau,\tau}}\left(-i\left(\boldsymbol{\delta}^{(m)}+\boldsymbol{\delta}^{(n)}\right)\right)$$
$$-\phi_{\mathbf{C}_{T+\tau,\tau}}\left(-i\boldsymbol{\delta}^{(m)}\right)\phi_{\mathbf{C}_{T+\tau,\tau}}\left(-i\boldsymbol{\delta}^{(n)}\right).$$

Therefore, in order to compute the inputs of the mean-variance optimization (6.156) and (6.157) we need to derive the expression of the characteristic function $\phi_{\mathbf{C}_{T+\tau,\tau}}$ from the distribution of the market invariants (6.148). In order to do this, we notice that if the linear returns $\mathbf{L}_{t,\tilde{\tau}}$ are market invariants, so are the compounded returns $\mathbf{C}_{t,\tilde{\tau}}$. For the compounded returns the simple projection formula (3.64) holds, which in this context reads:

$$\phi_{\mathbf{C}_{T+\tau,\tau}}=\left(\phi_{\mathbf{C}_{t,\tilde{\tau}}}\right)^{\frac{\tau}{\tilde{\tau}}}. \tag{6.158}$$

Notice that this formula does not hold for the linear returns, whose projection formula relies on the much more complex expression (3.78).

Substituting (6.158) into (6.156) and (6.157) we obtain the desired inputs of the mean-variance problem directly in terms of the distribution of the market invariants:

$$\mathrm{E}\left\{L_{T+\tau,\tau}^{(n)}\right\}=\left[\phi_{\mathbf{C}_{t,\tilde{\tau}}}\left(-i\boldsymbol{\delta}^{(n)}\right)\right]^{\frac{\tau}{\tilde{\tau}}}-1 \tag{6.159}$$

and

$$\operatorname{Cov}\left\{L_{T+\tau,\tau}^{(m)},L_{T+\tau,\tau}^{(n)}\right\}=\left[\phi_{\mathbf{C}_{t,\tilde{\tau}}}\left(-i\left(\boldsymbol{\delta}^{(m)}+\boldsymbol{\delta}^{(n)}\right)\right)\right]^{\frac{\tau}{\tilde{\tau}}} \tag{6.160}$$
$$-\left[\phi_{\mathbf{C}_{t,\tilde{\tau}}}\left(-i\boldsymbol{\delta}^{(m)}\right)\right]^{\frac{\tau}{\tilde{\tau}}}\left[\phi_{\mathbf{C}_{t,\tilde{\tau}}}\left(-i\boldsymbol{\delta}^{(n)}\right)\right]^{\frac{\tau}{\tilde{\tau}}}.$$

For instance, assuming as in Black and Scholes (1973) that the compounded returns are normally distributed, from (2.157) we obtain their characteristic function:

$$\phi_{\mathbf{C}_{t,\tilde{\tau}}}(\boldsymbol{\omega})=e^{i\boldsymbol{\omega}'\boldsymbol{\mu}-\frac{1}{2}\boldsymbol{\omega}'\boldsymbol{\Sigma}\boldsymbol{\omega}}. \tag{6.161}$$

Therefore from (6.159) the expected values read:

$$\mathrm{E}\left\{L^{(n)}_{T+\tau,\tau}\right\} = e^{\frac{\tau}{\tilde{\tau}}\left(\mu_n+\frac{1}{2}\Sigma_{nn}\right)} - 1; \tag{6.162}$$

and from (6.160) the covariances read:

$$\mathrm{Cov}\left\{L^{(m)}_{T+\tau,\tau}, L^{(n)}_{T+\tau,\tau}\right\} = e^{\frac{\tau}{\tilde{\tau}}\left(\mu_m+\mu_n+\frac{1}{2}\Sigma_{mm}+\frac{1}{2}\Sigma_{nn}\right)}\left(e^{\frac{\tau}{\tilde{\tau}}\Sigma_{mn}} - 1\right). \tag{6.163}$$

The reader is invited to consider the limit of the expressions (6.162) and (6.163) when the market volatility is low and the investment horizon τ is short.

Instead of using the correct formulas (6.159) and (6.160) in the mean-variance optimization (6.147), some practitioners replace the linear returns with compounded returns. In other words, they *define* the mean-variance efficient frontier as follows:

$$\widetilde{\mathbf{w}}(v) \equiv \underset{\substack{\mathbf{w}\in\mathcal{C} \\ \mathbf{w}'\,\mathrm{Cov}\{\mathbf{C}_{T+\tau,\tau}\}\mathbf{w}=v}}{\operatorname{argmax}} \mathbf{w}'\,\mathrm{E}\left\{\mathbf{C}_{T+\tau,\tau}\right\}, \tag{6.164}$$

where $v \geq 0$. In this formulation, the "square-root rule" (3.75) and (3.76), which is a consequence of (6.158), applies:

$$\mathrm{E}\left\{\mathbf{C}_{T+\tau,\tau}\right\} = \frac{\tau}{\tilde{\tau}}\mathrm{E}\left\{\mathbf{C}_{t,\tilde{\tau}}\right\}, \quad \mathrm{Cov}\left\{\mathbf{C}_{T+\tau,\tau}\right\} = \frac{\tau}{\tilde{\tau}}\mathrm{Cov}\left\{\mathbf{C}_{t,\tilde{\tau}}\right\}. \tag{6.165}$$

Therefore, it suffices to estimate the expected values and the covariance matrix of the compounded return for a given estimation interval $\tilde{\tau}$, and use the results for any investment horizon τ.

Nevertheless, the definition (6.164) of the mean-variance problem is incorrect.

In the first place, unlike the formulation in terms of linear returns (6.147), this formulation is not equivalent to the general mean-variance problem (6.74), because the identity (6.87) does not hold for the compounded returns:

$$C^{\Psi_\alpha}_{T+\tau,\tau} \neq \mathbf{w}'\mathbf{C}_{T+\tau,\tau}. \tag{6.166}$$

More in general, the quantity $\mathbf{w}'\mathbf{C}_{T+\tau,\tau}$ does not represent any feature of the investor's portfolio, not only it does not represent its compounded return. Therefore also the quantities $\mathbf{w}'\,\mathrm{E}\{\mathbf{C}\}$ and $\mathbf{w}'\,\mathrm{Cov}\{\mathbf{C}_{T+\tau,\tau}\}\,\mathbf{w}$ that appear in (6.164) are not related to the investor's portfolio.

Secondly, from the square-root rule (6.165) it follows that the mean-variance efficient allocations (6.164) do not depend on the investment horizon. This is incorrect and counterintuitive.

Notice that in the case of a short investment horizon τ and a not-too-volatile market, a first-order Taylor expansion shows that the linear and the compounded returns are approximately the same:

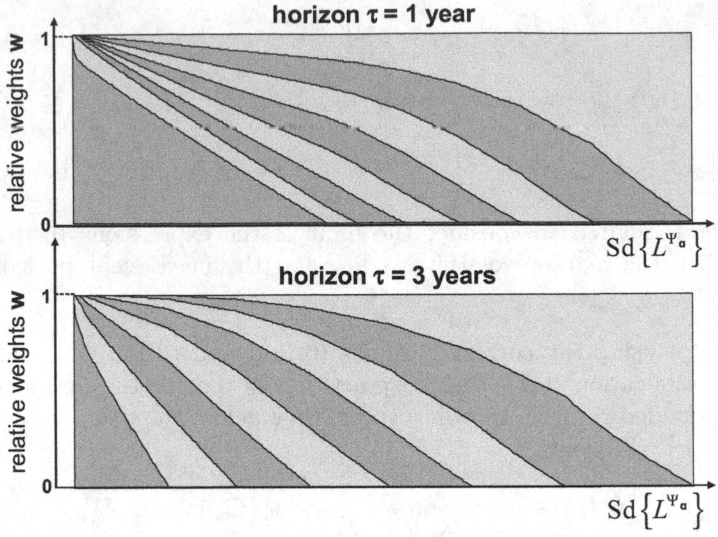

Fig. 6.18. MV efficient allocations at different investment horizons

$$L_{T+\tau,\tau} \equiv \frac{P_{T+\tau}}{P_T} - 1 \approx \ln\left(\frac{P_{T+\tau}}{P_T}\right) \equiv C_{T+\tau,\tau}. \tag{6.167}$$

In this case the correct mean-variance efficient allocations obtained from the formulation in terms of the linear returns (6.147) are approximately equal to the mean-variance efficient allocations obtained from the formulation in terms of the compounded returns (6.164), and therefore they are approximately independent of the investment horizon τ:

$$\begin{aligned} \mathbf{w}(v) \equiv & \underset{\substack{\mathbf{w}\in\mathcal{C} \\ \mathbf{w}'\,\mathrm{Cov}\{\mathbf{L}_{T+\tau,\tau}\}\mathbf{w}=v}}{\operatorname{argmax}} \mathbf{w}'\,\mathrm{E}\{\mathbf{L}_{T+\tau,\tau}\} \\ \approx & \underset{\substack{\mathbf{w}\in\mathcal{C} \\ \mathbf{w}'\,\mathrm{Cov}\{\mathbf{C}_{T+\tau,\tau}\}\mathbf{w}=v}}{\operatorname{argmax}} \mathbf{w}'\,\mathrm{E}\{\mathbf{C}_{T+\tau,\tau}\} \\ = & \underset{\substack{\mathbf{w}\in\mathcal{C} \\ \mathbf{w}'\,\mathrm{Cov}\left\{\mathbf{C}_{T,\widetilde{\tau}}\right\}\mathbf{w}=\frac{\widetilde{\tau}}{\tau}v}}{\operatorname{argmax}} \frac{\tau}{\widetilde{\tau}}\mathbf{w}'\,\mathrm{E}\{\mathbf{C}_{t,\widetilde{\tau}}\} \\ = & \,\widetilde{\mathbf{w}}(s), \end{aligned} \tag{6.168}$$

where $s \equiv v\widetilde{\tau}/\tau \geq 0$. Indeed, in the limit case of a dynamic setting, where the investor can rebalance continuously his portfolio and thus the investment horizon tends to zero, the formulation in terms of compounded returns becomes correct, and an equality holds in (6.166), see Merton (1992).

Nevertheless, for longer investment horizons and more volatile markets the first-order Taylor approximation (6.167) is not accurate. Indeed, the efficient frontiers relative to different investment horizons are different.

We see this Figure 6.18, where we plot the efficient combination of eight securities at a horizon of one and three years respectively under the normal assumption (6.161). Refer to `symmys.com` for the details on this market of securities and on the computations that generated the plots.

6.6 Total-return versus benchmark allocation

In this section we present an application of the analytical solutions of the mean-variance problem discussed in Section 6.4. We analyze two standard allocation strategies in the mutual fund industry: total-return allocation and benchmark allocation. As it turns out, benchmark allocation is the implicit strategy of the generic investor.

In the total-return strategy the investor's objective is final wealth at the investment horizon, see (5.3):

$$\Psi_{\boldsymbol{\alpha}} \equiv \boldsymbol{\alpha}' \mathbf{P}_{T+\tau}. \tag{6.169}$$

In the benchmark strategy the investor's objective is to overperform a benchmark whose allocation is $\widetilde{\boldsymbol{\beta}}$. In this case, the investor's objective is the *overperformance*, see (5.4):

$$\Phi_{\boldsymbol{\alpha}} \equiv \boldsymbol{\alpha}' \mathbf{P}_{T+\tau} - \gamma \widetilde{\boldsymbol{\beta}}' \mathbf{P}_{T+\tau}, \tag{6.170}$$

where the normalization scalar is meant to make the comparison between the portfolio and the benchmark fair:

$$\gamma \equiv \frac{\boldsymbol{\alpha}' \mathbf{p}_T}{\widetilde{\boldsymbol{\beta}}' \mathbf{p}_T}. \tag{6.171}$$

In both the total-return and the benchmark strategies the investor is bound by the same budget constraint:

$$\mathcal{C}: \; \boldsymbol{\alpha}' \mathbf{p}_T = w > 0. \tag{6.172}$$

In order to cast the total-return allocation problem in the mean-variance framework, we assume as in (6.67) that the investor's satisfaction only depends on the first two moments of his objective, namely final wealth:

$$\mathcal{S}(\boldsymbol{\alpha}) \approx \widetilde{\mathcal{H}}\left(\mathrm{E}\left\{\Psi_{\boldsymbol{\alpha}}\right\}, \mathrm{Var}\left\{\Psi_{\boldsymbol{\alpha}}\right\}\right), \tag{6.173}$$

where $\widetilde{\mathcal{H}}$ is a suitable bivariate function. Given the constraint (6.172), the mean-variance efficient frontier solves an affine constraint problem of the form (6.96), which in this context reads:

$$\widetilde{\boldsymbol{\alpha}}(v) = \underset{\substack{\boldsymbol{\alpha}'\mathbf{p}_T = w \\ \mathrm{Var}\{\Psi_{\boldsymbol{\alpha}}\}=v}}{\operatorname{argmax}} \mathrm{E}\{\Psi_{\boldsymbol{\alpha}}\}, \tag{6.174}$$

where $v \geq 0$.

The non-empty solutions of this optimization, namely the total-return efficient frontier, follow from (6.97). They represent a straight semi-line parametrized by the expected value of final wealth $e \equiv \mathrm{E}\{\Psi_{\widetilde{\boldsymbol{\alpha}}}\}$, which we report here:

$$\widetilde{\boldsymbol{\alpha}} = \boldsymbol{\alpha}_{MV} + [e - \mathrm{E}\{\Psi_{\boldsymbol{\alpha}_{MV}}\}] \frac{\boldsymbol{\alpha}_{SR} - \boldsymbol{\alpha}_{MV}}{\mathrm{E}\{\Psi_{\boldsymbol{\alpha}_{SR}}\} - \mathrm{E}\{\Psi_{\boldsymbol{\alpha}_{MV}}\}}, \tag{6.175}$$

where $e \in [\mathrm{E}\{\Psi_{\boldsymbol{\alpha}_M}\}, +\infty)$. The global minimum variance portfolio and the maximum Sharpe ratio portfolio in this expression follow from (6.99) and (6.100) respectively and read in this context as follows:

$$\boldsymbol{\alpha}_{MV} \equiv \frac{w \,\mathrm{Cov}\{\mathbf{P}_{T+\tau}\}^{-1} \mathbf{p}_T}{\mathbf{p}_T' \,\mathrm{Cov}\{\mathbf{P}_{T+\tau}\}^{-1} \mathbf{p}_T} \tag{6.176}$$

$$\boldsymbol{\alpha}_{SR} \equiv \frac{w \,\mathrm{Cov}\{\mathbf{P}_{T+\tau}\}^{-1} \mathrm{E}\{\mathbf{P}_{T+\tau}\}}{\mathbf{p}_T' \,\mathrm{Cov}\{\mathbf{P}_{T+\tau}\}^{-1} \mathrm{E}\{\mathbf{P}_{T+\tau}\}}. \tag{6.177}$$

In order to cast the benchmark allocation problem in the mean-variance framework, we first introduce some jargon used by practitioners. The expected value of the investor's objective (6.170) is called *expected overperformance*, which we denote as follows:

$$\mathrm{EOP}(\boldsymbol{\alpha}) \equiv \mathrm{E}\{\Phi_{\boldsymbol{\alpha}}\}. \tag{6.178}$$

The standard deviation of the investor's objective is called the *tracking error*[1], which we denote as follows:

$$\mathrm{TE}(\boldsymbol{\alpha}) \equiv \mathrm{Sd}\{\Phi_{\boldsymbol{\alpha}}\}. \tag{6.179}$$

The Sharpe ratio, i.e. the ratio of the above two parameters, is called the *information ratio*, which we denote as follows:

$$\mathrm{IR}(\boldsymbol{\alpha}) \equiv \frac{\mathrm{EOP}(\boldsymbol{\alpha})}{\mathrm{TE}(\boldsymbol{\alpha})}. \tag{6.180}$$

---

[1] Some authors define the tracking error differently: Roll (1992) defines it as the overperformance:

$$\mathrm{TE}(\boldsymbol{\alpha}) \equiv \Phi_{\boldsymbol{\alpha}};$$

Leibowitz, Bader, and Kogelman (1996) define it as follows:

$$\mathrm{TE}\{\boldsymbol{\alpha}\} \equiv \sqrt{\mathrm{E}\{\Phi_{\boldsymbol{\alpha}}^2\}}.$$

As in (6.67), we assume that the satisfaction of an investor whose purpose is to overperform a benchmark only depends on the first two moments of his objective:

$$\mathcal{S}(\boldsymbol{\alpha}) \approx \widetilde{\mathcal{K}}\left(\mathrm{EOP}(\boldsymbol{\alpha}), \mathrm{TE}^2(\boldsymbol{\alpha})\right), \tag{6.181}$$

where $\widetilde{\mathcal{K}}$ is a suitable bivariate function. Therefore the mean-variance efficient frontier of the benchmark strategy solves an affine constraint problem of the form (6.96), which in the newly introduced notation reads:

$$\widehat{\boldsymbol{\alpha}}(u) = \underset{\substack{\boldsymbol{\alpha}'\mathbf{p}_T = w \\ \mathrm{TE}^2(\boldsymbol{\alpha}) = u}}{\operatorname{argmax}} \ \mathrm{EOP}(\boldsymbol{\alpha}), \tag{6.182}$$

where $u \geq 0$.

To solve this problem we could cast the benchmark-relative objective in the form $\Phi_{\boldsymbol{\alpha}} \equiv \boldsymbol{\alpha}'\mathbf{M}$ for the market vector defined in (5.11) and (5.13) and then write the solution in the form (6.97). Nevertheless, we gain more insight into the differences and similarities between total-return allocation and benchmark allocation if we re-formulate the benchmark problem in terms of relative bets, which represent the difference between the allocation chosen by the investor and the allocation of the benchmark.

First of all we define the normalized benchmark allocation as follows:

$$\boldsymbol{\beta} \equiv \frac{w}{\mathbf{p}_T'\widetilde{\boldsymbol{\beta}}}\widetilde{\boldsymbol{\beta}}. \tag{6.183}$$

It is immediate to check that the normalized benchmark is a rescaled version of the original benchmark which has the same value as the investor's portfolio at the time the investment is made. The relative bets are defined as the vector $\boldsymbol{\rho}$ such that:

$$\boldsymbol{\alpha} \equiv \boldsymbol{\beta} + \boldsymbol{\rho}. \tag{6.184}$$

Since the benchmark allocation $\boldsymbol{\beta}$ is fixed, it is equivalent to determine and express the efficient frontier in terms of the allocations $\boldsymbol{\alpha}$ or in terms of the relative bets $\boldsymbol{\rho}$.

In terms of the relative bets, it is easy to check that the benchmark-relative objective (6.170) takes the form of a total-return objective:

$$\Phi_{\boldsymbol{\alpha}} = (\boldsymbol{\alpha} - \boldsymbol{\beta})' \mathbf{P}_{T+\tau} = \boldsymbol{\rho}'\mathbf{P}_{T+\tau} \equiv \Psi_{\boldsymbol{\rho}}, \tag{6.185}$$

compare with (6.169). Furthermore, the budget constraint (6.172) simplifies to a linear constraint:

$$\mathcal{C}: \ \boldsymbol{\rho}'\mathbf{p}_T = (\boldsymbol{\alpha} - \boldsymbol{\beta})' \mathbf{p}_T = 0. \tag{6.186}$$

Therefore the efficient frontier of the benchmark strategy (6.182) can be written in terms of the relative bets that solve the following problem:

$$\widehat{\boldsymbol{\rho}}(u) = \underset{\substack{\boldsymbol{\rho}'\mathbf{p}_T = 0 \\ \mathrm{Var}\{\Psi_{\boldsymbol{\rho}}\} = u}}{\operatorname{argmax}} \ \mathrm{E}\{\Psi_{\boldsymbol{\rho}}\}, \tag{6.187}$$

where $u \geq 0$. This linear constraint problem is of the from (6.107). The non-empty solutions to this optimization, namely the benchmark-relative efficient frontier, follow from (6.109) and (6.111). They represent a straight semi-line parametrized by the expected value of the overperformance $p \equiv \mathrm{E}\{\Psi_{\widehat{\rho}}\}$:

$$\widehat{\rho} = p \frac{\alpha_{SR} - \alpha_{MV}}{\mathrm{E}\{\Psi_{\alpha_{SR}}\} - \mathrm{E}\{\Psi_{\alpha_{MV}}\}}, \tag{6.188}$$

where $p \geq 0$.

From the definition of the relative bets (6.184) and the definition of the overperformance (6.185) we recover the efficient frontier of the benchmark strategy. This is a straight line parameterized by the expected final wealth of the benchmark-relative efficient allocations:

$$e \equiv \mathrm{E}\{\Psi_{\widehat{\alpha}}\} \in [\mathrm{E}\{\Psi_{\beta}\}, +\infty), \tag{6.189}$$

and reads explicitly:

$$\widehat{\alpha} = \beta + [e - \mathrm{E}\{\Psi_{\beta}\}] \frac{\alpha_{SR} - \alpha_{MV}}{\mathrm{E}\{\Psi_{\alpha_{SR}}\} - \mathrm{E}\{\Psi_{\alpha_{MV}}\}}. \tag{6.190}$$

A comparison of this expression with (6.175) shows that the total-return allocations $\widetilde{\alpha}$ can be interpreted as benchmark-relative allocations $\widehat{\alpha}$, where the benchmark is represented by the global minimum-variance portfolio.

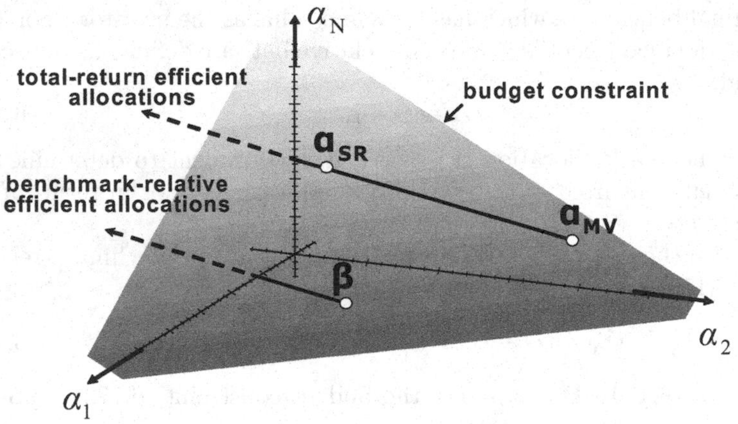

Fig. 6.19. Total-return vs. benchmark-relative MV efficient allocations

Geometrically, we can interpret these results as in Figure 6.19. In the N-dimensional space of allocations, the mean-variance efficient frontiers of both the total-return strategy and the benchmark-relative strategy are straight semi-lines that lie in the $(N-1)$-dimensional hyperplane determined by the budget constraint.

The direction of departure of the efficient straight semi-line from its starting point is determined by a two-fund separation principle. This direction of departure is the same for both the total-return strategy and the benchmark-relative strategy, no matter the specific composition of the benchmark.

On the other hand, the starting point of the efficient straight line is the investor's benchmark. In particular, in the total-return case where the investor's objective is final wealth, the starting point of the efficient straight line is the global minimum-variance portfolio.

We can now analyze the satisfaction stemming from the efficient allocations, see Roll (1992). First we consider the satisfaction of the total-return investor (6.173), which depends on the first two moments of the distribution of final wealth:

$$\begin{aligned} v &\equiv \operatorname{Var}\{\Psi_{\boldsymbol{\alpha}}\} && (6.191)\\ &= \boldsymbol{\alpha}' \operatorname{Cov}\{\mathbf{P}_{T+\tau}\} \boldsymbol{\alpha} \\ e &\equiv \operatorname{E}\{\Psi_{\boldsymbol{\alpha}}\} && (6.192)\\ &= \boldsymbol{\alpha}' \operatorname{E}\{\mathbf{P}_{T+\tau}\}. \end{aligned}$$

Since both the total-return investor and the benchmark-relative investor share the same *affine* budget constraint (6.172), their feasible set is the same. Like in Figure 6.10, the feasible set plots as the internal portion of a parabola, see Figure 6.20.

The total-return mean-variance efficient allocations (6.175) generate the upper branch of this parabola. From (6.102) the equation of this parabola reads:

$$\widetilde{\boldsymbol{\alpha}}: \quad v = \frac{A}{D}e^2 - \frac{2wB}{D}e + \frac{w^2 C}{D}, \tag{6.193}$$

where (A, B, C, D) are the four scalars (6.103) that do not depend on the allocation and which in this context read:

$$\begin{aligned} A &\equiv \mathbf{p}_T' \operatorname{Cov}\{\mathbf{P}_{T+\tau}\}^{-1} \mathbf{p}_T \\ B &\equiv \mathbf{p}_T' \operatorname{Cov}\{\mathbf{P}_{T+\tau}\}^{-1} \operatorname{E}\{\mathbf{P}_{T+\tau}\} \\ C &\equiv \operatorname{E}\{\mathbf{P}_{T+\tau}\}' \operatorname{Cov}\{\mathbf{P}_{T+\tau}\}^{-1} \operatorname{E}\{\mathbf{P}_{T+\tau}\} \\ D &\equiv AC - B^2. \end{aligned} \tag{6.194}$$

The benchmark-relative efficient allocations (6.190) are sub-optimal in these coordinates and thus do not lie on the upper branch of the parabola. As we prove in Appendix www.6.5, the benchmark-relative optimal allocations give rise to the portion of a parabola above the coordinates of the benchmark, see Figure 6.20. This parabola represents a right translation of the mean-variance efficient parabola (6.193) generated by the total-return efficient allocations. Indeed, the equation of this parabola reads:

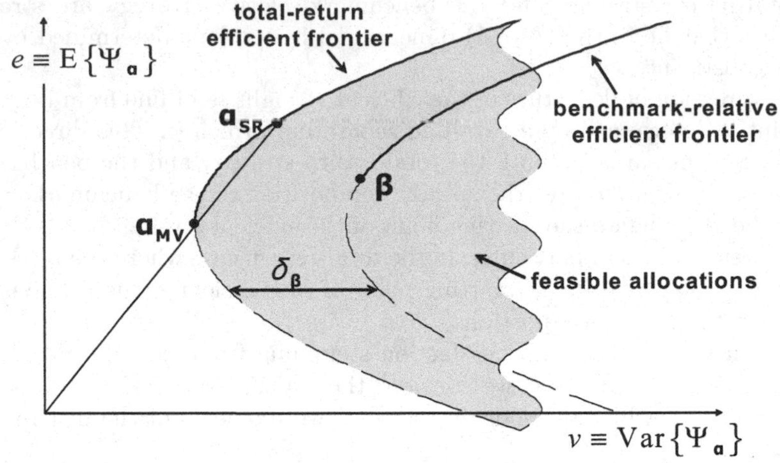

Fig. 6.20. Risk/reward profile of efficient allocations: total-return coordinates

$$\widehat{\alpha}: \quad v = \frac{A}{D}e^2 - \frac{2wB}{D}e + \frac{w^2C}{D} + \delta_{\beta}. \tag{6.195}$$

In this expression δ_{β} is a benchmark-dependent, non-negative scalar:

$$\delta_{\beta} \equiv \mathrm{Var}\{\Psi_{\beta}\} - \frac{A}{D}\mathrm{E}\{\Psi_{\beta}\}^2 + \frac{2wB}{D}\mathrm{E}\{\Psi_{\beta}\} - \frac{w^2C}{D} \geq 0, \tag{6.196}$$

As we show in Appendix www.6.5, the equality holds in (6.196) if and only if the benchmark is mean-variance efficient from a total-return point of view.

Now we discuss the satisfaction (6.181) of the investor who aims at outperforming a benchmark, which depends on the first two moments of the overperformance:

$$\begin{aligned} u &\equiv \mathrm{Var}\{\Phi_{\alpha}\} \equiv \mathrm{TE}^2(\alpha) & (6.197)\\ &= (\alpha - \beta)' \mathrm{Cov}\{\mathbf{P}_{T+\tau}\}(\alpha - \beta) \\ p &\equiv \mathrm{E}\{\Phi_{\alpha}\} \equiv \mathrm{EOP}(\alpha) & (6.198)\\ &= (\alpha - \beta)' \mathrm{E}\{\mathbf{P}_{T+\tau}\}. \end{aligned}$$

Since both the total-return investor and the benchmark-relative investor share the *linear* budget constraint (6.186), their feasible set is the same. From the discussion on p. 331, the feasible set plots as the internal region of a parabola through the origin, which represents an allocation that fully replicates the benchmark, see Figure 6.21.

The benchmark-relative efficient allocations (6.190) generate the upper branch of this parabola. From (6.112) the equation of the parabola reads:

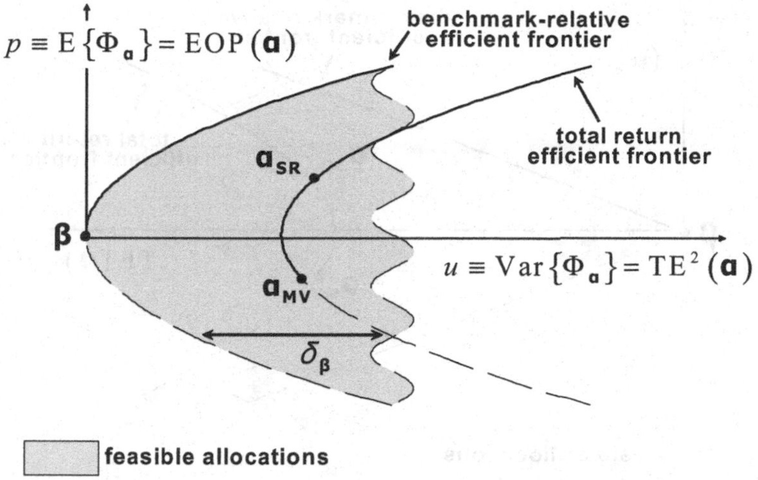

Fig. 6.21. Risk/reward profile of efficient allocations: benchmark-relative coordinates

$$\widehat{\alpha}: \quad u = \frac{A}{D}p^2, \tag{6.199}$$

where A and D are defined in (6.194).

The total-return mean-variance efficient allocations (6.175) are sub-optimal in these coordinates, and thus they do not lie on the optimal parabola. In Appendix www.6.5 we prove that the total-return mean-variance efficient allocations give rise to the portion of a parabola above the coordinates of the global minimum-variance portfolio, see Figure 6.21. This parabola represents a right translation of the parabola (6.199) generated by the benchmark-relative efficient allocations. Indeed the equation of this parabola reads:

$$\widetilde{\alpha}: \quad u = \frac{A}{D}p^2 + \delta_{\beta}. \tag{6.200}$$

In this expression δ_{β} is the same non-negative constant as (6.196), which is null if and only if the benchmark is mean-variance efficient from a total-return point of view.

Finally, it is interesting to look at the risk/reward profile of the total-return and benchmark-relative efficient allocations in the plane of the tracking error and expected overperformance plane, see Figure 6.22. Much like in (6.112), in these coordinates the boundary of the feasible set degenerates into two straight lines.

The total-return efficient allocations plot as a portion of the upper branch of a hyperbola within the feasible set. On the other hand, the benchmark-relative efficient allocations represent the straight line in the positive quadrant

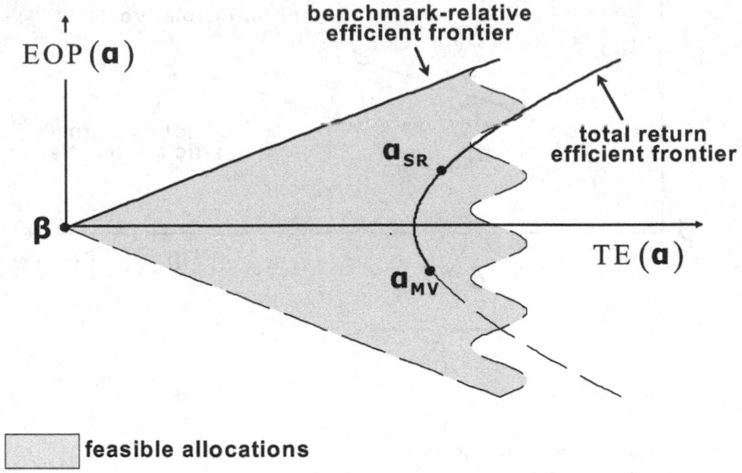

Fig. 6.22. Risk-reward profile of efficient allocations: expected overperformance and tracking error

that limits the feasible set. From (6.199) and the definitions (6.197)-(6.198), the equation of this line reads:

$$\mathrm{EOP}(\widehat{\alpha}) = \sqrt{\frac{D}{A}}\,\mathrm{TE}(\widehat{\alpha})\,. \tag{6.201}$$

As we see in Figure 6.22, such allocations give rise to the highest possible information ratio, defined in (6.180): as the investor is willing to accept a larger tracking error, the attainable expected overperformance increases linearly.

Depending on their specific index of satisfaction, some investors will abandon the benchmark, aggressively pursuing a higher expected overperformance. On the other hand, other investors will closely track the benchmark, minimizing their relative risk: this is the case for *index funds*, whose aim is to replicate the performance of a benchmark at the minimum possible cost. Refer to Section 3.4.5 for a routine to implement portfolio replication.

6.7 Case study: allocation in stocks

In this section we revisit all the steps of a real allocation problem that lead to the optimal portfolio for a given investor. Unlike the leading example in Section 6.1, this case study cannot be solved analytically. Therefore, after collecting the necessary information on the investor and on his market, we simplify the problem according to the two-step mean variance recipe, as discussed in Section 6.3 and we compute the optimal portfolio numerically.

We stress that at this stage little importance is given to the yet very important issue of estimation risk. We discuss this issue in depth in the third part of the book.

6.7.1 Collecting information on the investor

The investor starts in general with a pre-existing allocation $\boldsymbol{\alpha}^{(0)}$. In this case we assume that the investor's initial wealth is a given amount of cash w, say ten thousand dollars.

The investor determines a market. We assume that he chooses a set of $N \equiv 8$ well-diversified stocks and that he plans to re-invest any dividends. We denote as $\mathbf{P}_t$ the prices at the generic time t of one share of the stocks.

The investor determines his investment horizon τ: in this example we set τ equal to one year.

The investor specifies his objective. In this example we assume that he focuses on final wealth:

$$\Psi_{\boldsymbol{\alpha}} \equiv \boldsymbol{\alpha}' \mathbf{P}_{T+\tau}, \tag{6.202}$$

where T denotes the current time.

We model the investor's satisfaction. We assume that the investor bases his decisions according to the certainty-equivalent of his expected utility as in (5.93), where his utility function is of the power type. Therefore the investor's satisfaction reads:

$$\mathcal{S}(\boldsymbol{\alpha}) \equiv \left(\gamma \operatorname{E} \left\{ \frac{\Psi_{\boldsymbol{\alpha}}^{\gamma}}{\gamma} \right\} \right)^{\frac{1}{\gamma}}. \tag{6.203}$$

Notice that the power utility function is defined only for positive values of the investor's objective. This is consistent with the fact that prices are positive and that the investor can only hold long positions in the stocks. In our example we set the specific value $\gamma \equiv -9$ for the risk aversion parameter of the power utility functions.

6.7.2 Collecting information on the market

In order to collect information on the market we turn to data providers and we retrieve the time-series of the stock prices, which are available, say, for the past five years.

We determine the market invariants. After performing the analysis of Section 3.1, we determine that the non-overlapping compounded returns of the stocks can be modeled as independent and identically distributed across time:

$$C_{t,\widetilde{\tau}}^{(n)} \equiv \ln \left(\frac{P_t^{(n)}}{P_{t-\widetilde{\tau}}^{(n)}} \right), \tag{6.204}$$

where $n = 1, \ldots, N \equiv 8$. In our example, an estimation horizon $\widetilde{\tau}$ of one week provides a good balance in the trade-off between the number of independent observations and the homogeneity of the data.

We estimate the distribution of the invariants from currently available information. In this example, information becomes the time series of weekly compounded returns for the past five years. This is a series of approximately 250 observations. We fit the weekly compounded returns to a multivariate normal distribution:

$$\mathbf{C}_{t,\widetilde{\tau}} \sim \mathrm{N}\left(\widehat{\boldsymbol{\mu}}, \widehat{\boldsymbol{\Sigma}}\right). \tag{6.205}$$

To estimate $\widehat{\boldsymbol{\mu}}$ and $\widehat{\boldsymbol{\Sigma}}$ we first compute the sample mean and the sample covariance matrix:

$$\widehat{\mathbf{m}} \equiv \frac{1}{T}\sum_{t=1}^{T} \mathbf{c}_{t,\widetilde{\tau}}, \quad \widehat{\mathbf{S}} \equiv \frac{1}{T}\sum_{t=1}^{T} \left(\mathbf{c}_{t,\widetilde{\tau}} - \widehat{\mathbf{m}}\right)\left(\mathbf{c}_{t,\widetilde{\tau}} - \widehat{\mathbf{m}}\right)'. \tag{6.206}$$

Then as in (4.160) we shrink the covariance matrix toward a spherical estimator:

$$\widehat{\boldsymbol{\Sigma}} \equiv (1-\epsilon)\,\widehat{\mathbf{S}} + \frac{\epsilon}{N}\sum_{n=1}^{N} \widehat{S}_{nn}\mathbf{I}_N, \tag{6.207}$$

where from (4.161) the shrinkage weight reads:

$$\epsilon \equiv \frac{1}{T}\frac{\frac{1}{T}\sum_{t=1}^{T} \mathrm{tr}\left\{\left(\mathbf{c}_{t,\widetilde{\tau}}\mathbf{c}_{t,\widetilde{\tau}}' - \widehat{\mathbf{S}}\right)^2\right\}}{\mathrm{tr}\left\{\left(\widehat{\mathbf{S}} - \frac{1}{N}\sum_{n=1}^{N}\widehat{S}_{nn}\mathbf{I}_N\right)^2\right\}}. \tag{6.208}$$

Finally, as in (4.138) we shrink the sample mean towards a target vector:

$$\widehat{\boldsymbol{\mu}} \equiv (1-\gamma)\,\widehat{\mathbf{m}} + \gamma\mathbf{b}. \tag{6.209}$$

In this expression the shrinkage target follows from (4.142):

$$\mathbf{b} \equiv \frac{\mathbf{1}'\widehat{\boldsymbol{\Sigma}}^{-1}\widehat{\mathbf{m}}}{\mathbf{1}'\widehat{\boldsymbol{\Sigma}}^{-1}\mathbf{1}}\mathbf{1}; \tag{6.210}$$

and the shrinkage weight follows from (4.139) and reads in terms of the highest eigenvalue λ_1 of $\widehat{\boldsymbol{\Sigma}}$ as follows:

$$\gamma \equiv \frac{1}{T}\frac{\sum_{n=1}^{N}\widehat{\Sigma}_{nn} - 2\lambda_1}{\left(\widehat{\mathbf{m}} - \mathbf{b}\right)'\left(\widehat{\mathbf{m}} - \mathbf{b}\right)}. \tag{6.211}$$

We project the distribution of the invariants to the investment horizon by means of (3.64). As in (3.74) we obtain that compounded returns from the investment date to the investment horizon are normally distributed:

$$\mathbf{C}_{T+\tau,\tau} \sim \mathrm{N}\left(\frac{\tau}{\widetilde{\tau}}\widehat{\boldsymbol{\mu}}, \frac{\tau}{\widetilde{\tau}}\widehat{\boldsymbol{\Sigma}}\right). \tag{6.212}$$

We determine the transaction costs in the market. In this case we assume that the transaction costs grow quadratically with the number of shares transacted:

$$\mathcal{T}(\widetilde{\boldsymbol{\alpha}}, \boldsymbol{\alpha}) = \mathbf{k}'(\widetilde{\boldsymbol{\alpha}} - \boldsymbol{\alpha}) + (\widetilde{\boldsymbol{\alpha}} - \boldsymbol{\alpha})' \mathbf{D} (\widetilde{\boldsymbol{\alpha}} - \boldsymbol{\alpha}), \tag{6.213}$$

where $\mathbf{D}$ is a diagonal matrix of positive entries. The non-linear growth of the transaction costs accounts for the market impact of large stock transactions.

6.7.3 Computing the optimal allocation

First, we formulate the investor's constraints. We assume that the investor has a budget constraint:

$$\mathcal{C}_1 : \boldsymbol{\alpha}' \mathbf{p}_T \leq w - \mathcal{T}\left(\boldsymbol{\alpha}^{(0)}, \boldsymbol{\alpha}\right), \tag{6.214}$$

where w is his initial capital of ten thousand dollars, and $\mathcal{T}$ are the transaction costs (6.213). Furthermore, we assume that he can only hold long positions:

$$\mathcal{C}_2 : \boldsymbol{\alpha} \geq \mathbf{0}. \tag{6.215}$$

With the information on the investor's profile, his market and his constraints we can set up the optimization problem (6.33), which in this context is equivalent to:

$$\boldsymbol{\alpha}^* \equiv \underset{\substack{\boldsymbol{\alpha} \geq \mathbf{0} \\ \boldsymbol{\alpha}' \mathbf{p}_T \leq w - \boldsymbol{\alpha}' \mathbf{D} \boldsymbol{\alpha}}}{\operatorname{argmax}} \mathrm{E}\left\{\frac{1}{\gamma}\left(\boldsymbol{\alpha}' \operatorname{diag}(\mathbf{p}_T) e^{\mathbf{C}_{T+\tau,\tau}}\right)^{\gamma}\right\}, \tag{6.216}$$

where the distribution of $\mathbf{C}$ is provided in (6.212). Since it is not possible to determine the solution $\boldsymbol{\alpha}^*$ analytically, we resort to the mean-variance framework to restrict the search to a limited number of portfolios.

We compute the inputs of the mean-variance problem (6.74), namely the expected value and the covariance of the market prices at the investment horizon. To do this, consider the characteristic function (2.157) of the compounded returns (6.205), which reads:

$$\phi_{\mathbf{C}_{t,\widetilde{\tau}}}(\boldsymbol{\omega}) = e^{i\boldsymbol{\omega}'\widehat{\boldsymbol{\mu}} - \frac{1}{2}\boldsymbol{\omega}'\widehat{\boldsymbol{\Sigma}}\boldsymbol{\omega}}. \tag{6.217}$$

From (6.77) we obtain the expected value of the market prices:

$$\begin{aligned} \mathrm{E}\left\{P_{T+\tau}^{(n)}\right\} &= P_T^{(n)} \left[\phi_{\mathbf{C}_{t,\widetilde{\tau}}}\left(-i\boldsymbol{\delta}^{(n)}\right)\right]^{\frac{\tau}{\widetilde{\tau}}} \\ &= P_T^{(n)} e^{\frac{\tau}{\widetilde{\tau}}\left(\widehat{\mu}_n + \frac{\widehat{\Sigma}_{nn}}{2}\right)}. \end{aligned} \tag{6.218}$$

Similarly, from (6.78) we obtain:

$$\begin{aligned} \mathrm{E}\left\{P^{(m)}_{T+\tau}, P^{(n)}_{T+\tau}\right\} &= P^{(m)}_T P^{(n)}_T \left[\phi_{\mathbf{C}_{t,\tilde{\tau}}}\left(-i\boldsymbol{\delta}^{(m)} - i\boldsymbol{\delta}^{(n)}\right)\right]^{\frac{\tau}{\tilde{\tau}}} \\ &= P^{(m)}_T P^{(n)}_T e^{\frac{\tau}{\tilde{\tau}}(\widehat{\mu}_m + \widehat{\mu}_n)} e^{\frac{1}{2}\frac{\tau}{\tilde{\tau}}(\widehat{\Sigma}_{mm} + \widehat{\Sigma}_{nn} + 2\widehat{\Sigma}_{mn})}. \end{aligned} \tag{6.219}$$

Therefore from (6.79) we obtain the covariance matrix of the market:

$$\begin{aligned} \mathrm{Cov}\left\{P^{(m)}_{T+\tau}, P^{(n)}_{T+\tau}\right\} &= P^{(m)}_T P^{(n)}_T e^{\frac{\tau}{\tilde{\tau}}(\widehat{\mu}_m + \widehat{\mu}_n)} \\ &\quad e^{\frac{1}{2}\frac{\tau}{\tilde{\tau}}(\widehat{\Sigma}_{mm} + \widehat{\Sigma}_{nn})} \left(e^{\frac{\tau}{\tilde{\tau}}\widehat{\Sigma}_{mn}} - 1\right). \end{aligned} \tag{6.220}$$

With the inputs (6.218) and (6.220) we compute numerically the mean-variance efficient curve (6.74). In order to do this, we choose a significative grid of, say $I \equiv 100$ target variances $\left\{v^{(1)}, \ldots, v^{(I)}\right\}$ and solve numerically each time the following optimization:

$$\begin{aligned} \boldsymbol{\alpha}^{(i)} &\equiv \underset{\boldsymbol{\alpha}}{\operatorname{argmax}}\ \boldsymbol{\alpha}' \mathrm{E}\left\{\mathbf{P}_{T+\tau}\right\} \\ &\quad \text{subject to} \begin{cases} \boldsymbol{\alpha}' \mathrm{Cov}\left\{\mathbf{P}_{T+\tau}\right\} \boldsymbol{\alpha} \leq v^{(i)} \\ \boldsymbol{\alpha}' \mathbf{p}_T \leq w - \boldsymbol{\alpha}' \mathbf{D} \boldsymbol{\alpha} \\ \boldsymbol{\alpha} \geq \mathbf{0}. \end{cases} \end{aligned} \tag{6.221}$$

Each optimization (6.221) is a quadratically constrained linear programming problem, i.e. a subclass of (6.57). Therefore it can be efficiently solved numerically.

In the top plot of Figure 6.23 we display the risk/reward profile of the mean-variance efficient allocations (6.221) in terms of expected value and standard deviation of final wealth:

$$\mathrm{E}\left\{\Psi_{\boldsymbol{\alpha}}\right\} = \boldsymbol{\alpha}' \mathrm{E}\left\{\mathbf{P}_{T+\tau}\right\}, \quad \mathrm{Sd}\left\{\Psi_{\boldsymbol{\alpha}}\right\} = \sqrt{\boldsymbol{\alpha}' \mathrm{Cov}\left\{\mathbf{P}_{T+\tau}\right\} \boldsymbol{\alpha}}. \tag{6.222}$$

In the middle plot of Figure 6.23 we display the mean-variance efficient allocations (6.221) in terms of their relative weights:

$$\mathbf{w}^{(i)} \equiv \frac{\mathrm{diag}\left(\boldsymbol{\alpha}^{(i)}\right) \mathbf{p}_T}{\mathbf{p}_T' \boldsymbol{\alpha}^{(i)}}. \tag{6.223}$$

According to the mean-variance optimization, the optimal allocation is the portfolio (6.221) that gives rise to the higher level of satisfaction. To determine this portfolio we use Monte Carlo simulations.

We simulate a large number J of Monte Carlo market scenarios as follows:

$$_j\mathbf{P}_{T+\tau} \equiv \mathrm{diag}\left(\mathbf{p}_T\right) e^{_j\mathbf{C}}, \tag{6.224}$$

where the exponential acts component-wise and where each vector $_j\mathbf{C}$ is an independent drawing from the multivariate normal distribution (6.212) for all $j = 1, \ldots, J$. In our example we perform $J \equiv 10^5$ simulations.

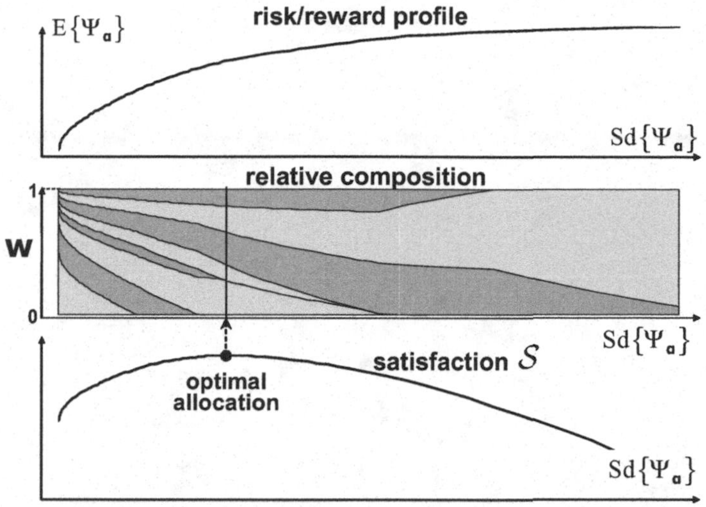

Fig. 6.23. MV approach: two-step allocation optimization

We evaluate numerically the satisfaction drawn from each of the mean-variance efficient allocations. In other words, we compute the following approximation to (6.203) for all the mean-variance efficient portfolios (6.221) in the grid:

$$\widetilde{\mathcal{S}}\left(\boldsymbol{\alpha}^{(i)}\right) \equiv \left(\frac{\gamma}{J}\sum_{j=1}^{J}\frac{\left({}_{j}\mathbf{P}'_{T+\tau}\boldsymbol{\alpha}^{(i)}\right)^{\gamma}}{\gamma}\right)^{\frac{1}{\gamma}}. \tag{6.225}$$

In the bottom plot in Figure 6.23 we display the satisfaction (6.225) ensuing from each of the allocations in the grid.

We rank the levels of satisfaction provided by the mean-variance efficient portfolios:

$$i^{*} \equiv \underset{i}{\operatorname{argmax}}\left\{\widetilde{\mathcal{S}}\left(\boldsymbol{\alpha}^{(i)}\right)\right\}. \tag{6.226}$$

Finally, we determine the optimal allocation:

$$\boldsymbol{\alpha}^{*} \equiv \boldsymbol{\alpha}^{(i^{*})}, \tag{6.227}$$

see Figure 6.23.

Part III

Accounting for estimation risk

7

Estimating the distribution of the market invariants

The classical approach to estimation discussed in the second part of the book does not account for estimation risk. To tackle this issue, in this chapter we introduce the Bayesian approach to parameter estimation.

The outcome of the Bayesian estimation process is the posterior distribution of the market parameters. This distribution explicitly acknowledges that an estimate cannot be a single number. Furthermore the posterior distribution includes within a sound statistical framework both the investor's experience, or prior knowledge, and the information from the market.

In Section 7.1 we introduce the Bayesian approach in general, showing how to blend the investor's prior and the market information to obtain the posterior distribution. Furthermore, we show how the Bayesian framework includes the classical approach to estimation in the form of "classical-equivalent" estimators. Finally, we discuss how to summarize the main features of a generic posterior distribution by means of its location-dispersion ellipsoid.

As in the classical approach in Chapter 4, we then proceed to discuss the estimation of the parameters of the market invariants that are most relevant to allocation problems, namely location, dispersion, and factor loadings.

In Section 7.2 we compute the posterior distribution of expected value and covariance matrix of the market invariants under the conjugate normal-inverse-Wishart hypothesis. Then we compute the classical-equivalent estimators of the above parameters, exploring their self-adjusting behavior. Finally we compute the joint and the marginal location-dispersion ellipsoids of expected values and covariance matrix provided by their posterior distribution.

In Section 7.3 we consider multivariate factor models. First we compute the posterior distribution of the factor loadings and of the perturbation covariance under the conjugate normal-inverse-Wishart hypothesis. Then we compute the classical-equivalent estimators of the respective parameters, exploring their self-adjusting behavior. Finally we compute the joint and the marginal location-dispersion ellipsoids of the posterior distribution of factor loadings and perturbation covariance.

A. Meucci, *Risk and Asset Allocation*, Springer Finance,

In Section 7.4 we discuss how to quantify the investor's prior knowledge of the market in practice. Indeed, in typical situations the investor does not input directly the prior market parameters: instead, he computes them under suitable assumptions from what he considers an ideal allocation.

7.1 Bayesian estimation

We recall from Chapter 4 that a classical estimator is a function that processes current information i_T and outputs an S-dimensional vector $\widehat{\boldsymbol{\theta}}$, see (4.9). Information consists as in (4.8) of a time series of T past observations of the market invariants:

$$i_T \equiv \{\mathbf{x}_1, \ldots, \mathbf{x}_T\}. \tag{7.1}$$

The output $\widehat{\boldsymbol{\theta}}$ is a *number* which is supposed to be close to the true, unknown parameter $\boldsymbol{\theta}^{\mathrm{t}}$. We can summarize the classical approach as follows:

$$\boxed{\textbf{classical estimation: } i_T \mapsto \widehat{\boldsymbol{\theta}}} \tag{7.2}$$

The Bayesian estimation process differs from the classical one in terms of both "input" and "output".

7.1.1 Bayesian posterior distribution

In the first place, in a Bayesian context an estimator does not yield a number $\widehat{\boldsymbol{\theta}}$. Instead, it yields a random variable $\boldsymbol{\theta}$, which takes values in a given range $\boldsymbol{\Theta}$. The distribution of $\boldsymbol{\theta}$ is called the *posterior distribution*, which can be represented for instance in terms of its probability density function $f_{\mathrm{po}}(\boldsymbol{\theta})$. The true, unknown parameter $\boldsymbol{\theta}^{\mathrm{t}}$ is assumed to be hidden most likely in the neighborhood of those values where the posterior distribution is more peaked, but the possibility that $\boldsymbol{\theta}^{\mathrm{t}}$ might lie in some other region of the range $\boldsymbol{\Theta}$ is also acknowledged, see Figure 7.2.

Secondly, in a Bayesian context an estimator does not depend only on backward-looking historical information i_T. Indeed, the investor/statistician typically has some prior knowledge of the unknown value $\boldsymbol{\theta}^{\mathrm{t}}$ based on his experience e_C, where C denotes the level of confidence in his experience. This experience is explicitly taken into account in the Bayesian estimation process.

Therefore we can summarize the Bayesian approach as follows:

$$\boxed{\textbf{Bayesian estimation: } i_T, e_C \mapsto f_{\mathrm{po}}(\boldsymbol{\theta})} \tag{7.3}$$

The Bayesian approach to estimation can be interpreted intuitively as follows, see Figure 7.1.

On the one hand, the purely classical estimator based on historical information i_T gives rise to a distribution of the market parameters $\boldsymbol{\theta}$ that is

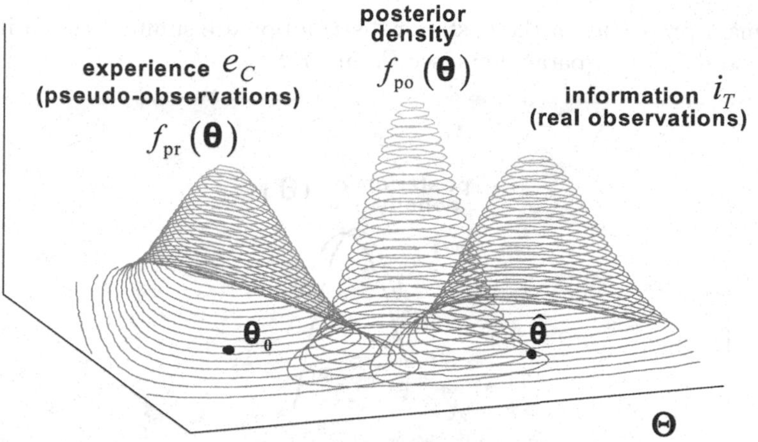

Fig. 7.1. Bayesian approach to parameter estimation

peaked around the classical estimate $\widehat{\boldsymbol{\theta}}$: the larger the number of observations T in the time series, the higher the concentration of the historical distribution around the classical estimate.

On the other hand, the investor equates his experience e_C to a number C of pseudo-observations, that only he sees, located in a "prior" value $\boldsymbol{\theta}_0$. These observations give rise to a distribution of the market parameters $\boldsymbol{\theta}$ which is called the *prior distribution*, whose probability density function we denote as $f_{pr}(\boldsymbol{\theta})$. The larger the number of these pseudo-observations, the higher the investor's confidence in his own experience and thus the more concentrated the prior distribution around $\boldsymbol{\theta}_0$.

The Bayesian posterior provides a theoretically sound way to blend the above two distributions into a third distribution, i.e. a spectrum of values and respective probabilities for the parameters $\boldsymbol{\theta}$. In particular, when the confidence C in the investor's experience is large the posterior becomes peaked around the prior value $\boldsymbol{\theta}_0$. On the other hand, when the number of observations T in the time series is large the posterior becomes peaked around the classical estimate $\widehat{\boldsymbol{\theta}}$:

$$f_{po}(\boldsymbol{\theta}) \begin{array}{l} \nearrow \ \boldsymbol{\theta}_0 \ (C \to \infty) \\ \\ \searrow \ \widehat{\boldsymbol{\theta}} \ \ (T \to \infty). \end{array} \tag{7.4}$$

7.1.2 Summarizing the posterior distribution

The main properties of the posterior distribution are summarized in its location and dispersion parameters, see Figure 7.2.

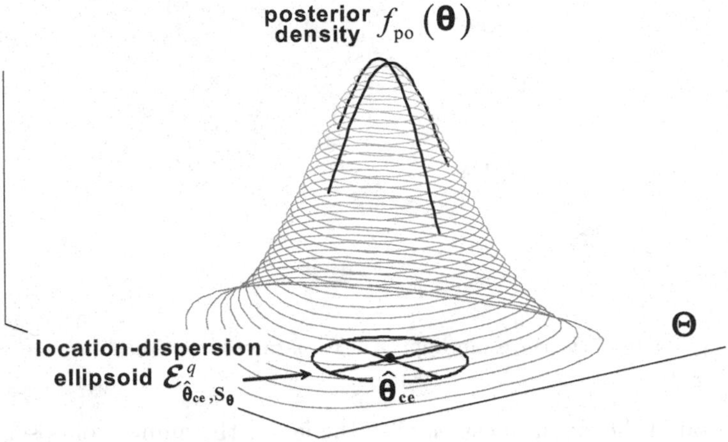

Fig. 7.2. Bayesian posterior distribution and uncertainty set

Location

The location parameter of an S-variate posterior distribution $f_{\text{po}}(\boldsymbol{\theta})$ is an S-dimensional vector $\widehat{\boldsymbol{\theta}}$, see Section 2.4. Since the posterior distribution is determined by the information i_T, in addition to the investor's experience e_C, the location parameter is a *number* that depends on information. This is the definition (7.2) of a classical estimator. Therefore a location parameter of the posterior distribution defines a *classical-equivalent estimator*.

A standard choice for the location parameter of a distribution is its expected value (2.54). Therefore, we introduce the following classical-equivalent estimator of the parameter $\boldsymbol{\theta}$:

$$\begin{aligned}\widehat{\boldsymbol{\theta}}_{\text{ce}}[i_T, e_C] &\equiv \mathrm{E}_{i_T,e_C}\{\boldsymbol{\theta}\} \\ &\equiv \int_{\boldsymbol{\Theta}} \boldsymbol{\theta} f_{\text{po}}(\boldsymbol{\theta}; i_T, e_C)\, d\boldsymbol{\theta}.\end{aligned} \tag{7.5}$$

As it turns out, this classical-equivalent estimator minimizes the estimation error defined by a quadratic loss function as in (4.19). Furthermore, under fairly general conditions, this classical-equivalent estimator is admissible, see Berger (1985).

Another standard choice for the location parameter of a distribution is its mode (2.52). Therefore, we introduce the following classical-equivalent estimator of the parameter $\boldsymbol{\theta}$:

$$\begin{aligned}\widehat{\boldsymbol{\theta}}_{\text{ce}}\left[i_T, e_C\right] &\equiv \operatorname{Mod}_{i_T, e_C}\{\boldsymbol{\theta}\} \\ &\equiv \underset{\boldsymbol{\theta} \in \boldsymbol{\Theta}}{\operatorname{argmax}}\, f_{\text{po}}\left(\boldsymbol{\theta}; i_T, e_C\right).\end{aligned} \tag{7.6}$$

This classical-equivalent estimator (7.6) based on the mode is equal to the classical-equivalent estimator (7.5) based on the expected value when the posterior is normally distributed, see (2.158). Furthermore, it yields the point of highest concentration of probability in the domain $\boldsymbol{\Theta}$ even when the moments of the posterior distribution are not defined, see Figure 7.2.

A classical-equivalent estimator is an instance of the shrinkage estimators discussed in Section 4.4. For this reason they are called *Bayes-Stein shrinkage estimators*. In the Bayesian context the shrinkage target is represented by the investor's prior experience and the extent of the shrinkage is driven by the relation between the amount of information, i.e. the length T of the time series, and the investor's confidence C in his experience, see Figure 7.1.

In particular, because of (7.4), when the investor's confidence C in his experience is high, the posterior distribution becomes extremely concentrated around the prior $\boldsymbol{\theta}_0$. Therefore the classical-equivalent estimator also shrinks to the point $\boldsymbol{\theta}_0$. Similarly, when the length T of the time series is large, the posterior distribution becomes extremely concentrated around the historical estimate $\widehat{\boldsymbol{\theta}}$. Therefore the classical-equivalent estimator also converges to the historical estimate $\widehat{\boldsymbol{\theta}}$.

Dispersion

A dispersion parameter of the S-variate posterior distribution $f_{\text{po}}(\boldsymbol{\theta})$ is a symmetric and positive $S \times S$ matrix $\mathbf{S}_{\boldsymbol{\theta}}$. Since the posterior is determined by the information i_T and by investor's experience e_C, so is the dispersion parameter.

A standard choice for the dispersion parameter of a distribution is represented by the covariance matrix (2.67), which in this context reads:

$$\begin{aligned}\mathbf{S}_{\boldsymbol{\theta}}\left[i_T, e_C\right] &\equiv \operatorname{Cov}_{i_T, e_C}\{\boldsymbol{\theta}\} \\ &\equiv \int_{\boldsymbol{\Theta}} (\boldsymbol{\theta} - \mathrm{E}\{\boldsymbol{\theta}\})(\boldsymbol{\theta} - \mathrm{E}\{\boldsymbol{\theta}\})' f_{\text{po}}\left(\boldsymbol{\theta}; i_T, e_C\right) d\boldsymbol{\theta}.\end{aligned} \tag{7.7}$$

Alternatively, we can consider the modal dispersion (2.65), which in this context reads:

$$\mathbf{S}_{\boldsymbol{\theta}}\left[i_T, e_C\right] \equiv \mathrm{MDis}_{i_T, e_C}\{\boldsymbol{\theta}\} \equiv -\left(\left.\frac{\partial^2 \ln f_{\mathrm{po}}\left(\boldsymbol{\theta}; i_T, e_C\right)}{\partial \boldsymbol{\theta} \partial \boldsymbol{\theta}'}\right|_{\boldsymbol{\theta}=\mathrm{Mod}\{\boldsymbol{\theta}\}}\right)^{-1}. \tag{7.8}$$

The modal dispersion is equal to the covariance matrix when the posterior is normally distributed:

$$\boldsymbol{\theta} \sim \mathrm{N}\left(\widehat{\boldsymbol{\theta}}_{\mathrm{ce}}, \mathbf{S}_{\boldsymbol{\theta}}\right), \tag{7.9}$$

see (2.159). Furthermore, it provides a measure of the dispersion of the parameter $\boldsymbol{\theta}$ in the range $\boldsymbol{\Theta}$ even when the moments of the posterior are not defined.

Location-dispersion ellipsoid

Together with the classical-equivalent S-dimensional vector $\widehat{\boldsymbol{\theta}}_{\mathrm{ce}}$, the $S \times S$ dispersion matrix $\mathbf{S}_{\boldsymbol{\theta}}$ defines the location-dispersion ellipsoid with radius proportional to q of the estimate of the market parameters $\boldsymbol{\theta}$:

$$\mathcal{E}^q_{\widehat{\boldsymbol{\theta}}_{\mathrm{ce}}, \mathbf{S}_{\boldsymbol{\theta}}} \equiv \left\{\boldsymbol{\theta} \text{ such that } \left(\boldsymbol{\theta} - \widehat{\boldsymbol{\theta}}_{\mathrm{ce}}\right)' \mathbf{S}_{\boldsymbol{\theta}}^{-1} \left(\boldsymbol{\theta} - \widehat{\boldsymbol{\theta}}_{\mathrm{ce}}\right) \leq q^2\right\}, \tag{7.10}$$

see Figure 7.2. Refer to Section 2.4.3 for a thorough discussion of the location-dispersion ellipsoid in a general context.

The location-dispersion ellipsoid defines naturally a self-adjusting uncertainty region for $\boldsymbol{\theta}$. Indeed, we show in Appendix www.7.1 that in the specific case (7.9) where the posterior is normally distributed the following result holds for the probability that the parameters lie within the boundaries of the ellipsoid:

$$\mathbb{P}\left\{\boldsymbol{\theta} \in \mathcal{E}^q_{\widehat{\boldsymbol{\theta}}_{\mathrm{ce}}, \mathbf{S}_{\boldsymbol{\theta}}}\right\} = F_{\chi^2_S}\left(q^2\right), \tag{7.11}$$

where $F_{\chi^2_S}$ is the cumulative distribution function of the chi-square distribution with S degrees of freedom, which is a special case of the gamma cumulative distribution function (1.111). More in general, the least upper bound (2.90) of the Chebyshev inequality applies:

$$\mathbb{P}\left\{\boldsymbol{\theta} \notin \mathcal{E}^q_{\widehat{\boldsymbol{\theta}}_{\mathrm{ce}}, \mathbf{S}_{\boldsymbol{\theta}}}\right\} \leq \frac{S}{q^2}. \tag{7.12}$$

Furthermore, because of (7.4), when the investor's confidence C in his experience is large, the posterior distribution becomes extremely concentrated around the prior $\boldsymbol{\theta}_0$. Therefore the dispersion parameter $\mathbf{S}_{\boldsymbol{\theta}}$ becomes small and the uncertainty ellipsoid (7.10) shrinks to the point $\boldsymbol{\theta}_0$, no matter the radius factor q. Similarly, when the number of observations T in the time series of the market invariants is large, the posterior distribution becomes extremely concentrated around the historical estimate $\widehat{\boldsymbol{\theta}}$. Therefore the dispersion parameter $\mathbf{S}_{\boldsymbol{\theta}}$ becomes small and the uncertainty ellipsoid (7.10) shrinks to the point $\widehat{\boldsymbol{\theta}}$, no matter the radius factor q.

The self-adjusting uncertainty region represented by the location-dispersion ellipsoid (7.10) of the posterior distribution of the parameter $\boldsymbol{\theta}$ plays an important role in robust Bayesian allocation decisions.

7.1.3 Computing the posterior distribution

To compute explicitly the posterior distribution we denote the probability density function of the market invariants by the conditional notation $f(\mathbf{x}|\boldsymbol{\theta})$. In so doing we are implicitly considering the parameters $\boldsymbol{\theta}$ as a random variable, where the true, unknown value $\boldsymbol{\theta}^{\mathrm{t}}$ is the specific instance of this random variable that is chosen by Nature.

Since the market invariants are independent and identically distributed the joint probability density function of the available information (7.1) assuming known the value of the parameters $\boldsymbol{\theta}$ is the product of the probability density functions of the invariants:

$$f_{I_T|\boldsymbol{\theta}}(i_T|\boldsymbol{\theta}) = f(\mathbf{x}_1|\boldsymbol{\theta}) \cdots f(\mathbf{x}_T|\boldsymbol{\theta}), \tag{7.13}$$

see also (4.5).

The investor has some prior knowledge of the parameters, which reflects his experience e_C and is modeled by the prior density $f_{\mathrm{pr}}(\boldsymbol{\theta})$. From the relation between the conditional and the joint probability density functions (2.40) we obtain the expression for the joint distribution of the observations and the market parameters:

$$f_{I_T,\boldsymbol{\theta}}(i_T, \boldsymbol{\theta}) = f_{I_T|\boldsymbol{\theta}}(i_T|\boldsymbol{\theta}) f_{\mathrm{pr}}(\boldsymbol{\theta}). \tag{7.14}$$

The posterior probability density function is simply the density of the parameters conditional on current information. It follows from the joint density of the observations and the parameters by applying Bayes' rule (2.43), which in this context reads:

$$f_{\mathrm{po}}(\boldsymbol{\theta}; i_T, e_C) \equiv f(\boldsymbol{\theta}|i_T) = \frac{f_{I_T,\boldsymbol{\theta}}(i_T, \boldsymbol{\theta})}{\int_{\boldsymbol{\Theta}} f_{I_T,\boldsymbol{\theta}}(i_T, \boldsymbol{\theta})\, d\boldsymbol{\theta}}. \tag{7.15}$$

By construction, the Bayes posterior distribution smoothly blends the information from the market i_T with the investor's experience e_C, which is modeled by the prior density.

Although the Bayesian approach is conceptually simple, it involves multiple integrations. Therefore, the choices of distributions that allow us to obtain analytical results is quite limited. Parametric models for the investor's prior and the market invariants that give rise to tractable posterior distributions of the market parameters are called *conjugate distributions*.

We present in Sections 7.2 and 7.3 notable conjugate models that allow us to model the markets. If analytical results are not available, one has to resort to numerical simulations, see Geweke (1999).

7.2 Location and dispersion parameters

We present here the Bayesian estimators of the location and the dispersion parameters of the market invariants under the normal hypothesis:

$$\mathbf{X}_t|\boldsymbol{\mu}, \boldsymbol{\Sigma} \sim \mathrm{N}(\boldsymbol{\mu}, \boldsymbol{\Sigma}) . \tag{7.16}$$

In this setting, the location parameter is the expected value $\boldsymbol{\mu}$ and the scatter parameter is the covariance matrix $\boldsymbol{\Sigma}$. This specification is rich and flexible enough to suitably model real problems, yet the otherwise analytically intractable computations of Bayesian analysis can be worked out completely, see also Aitchison and Dunsmore (1975).

7.2.1 Computing the posterior distribution

The Bayesian estimate of the unknown parameters is represented by the joint posterior distribution of $\boldsymbol{\mu}$ and $\boldsymbol{\Sigma}$. In order to compute this distribution we need to collect the information available and to model the investor's experience, i.e. his prior distribution.

Information from the market

The information on the market is contained in the time series (7.1) of the past realizations of the market invariants.

Since we are interested in the estimation of the location parameter $\boldsymbol{\mu}$ and of the scatter parameter $\boldsymbol{\Sigma}$, it turns out sufficient to summarize the historical information on the market into the sample estimator of location (4.98), i.e. the sample mean:

$$\widehat{\boldsymbol{\mu}} \equiv \frac{1}{T} \sum_{t=1}^{T} \mathbf{x}_t, \tag{7.17}$$

and the sample estimator of dispersion (4.99), i.e. the sample covariance:

$$\widehat{\boldsymbol{\Sigma}} \equiv \frac{1}{T} \sum_{t=1}^{T} (\mathbf{x}_t - \widehat{\boldsymbol{\mu}}) (\mathbf{x}_t - \widehat{\boldsymbol{\mu}})' . \tag{7.18}$$

Along with the number of observations T in the sample, this is all the information we need from the market. Therefore we can represent this information equivalently as follows:

$$i_T \equiv \left\{ \widehat{\boldsymbol{\mu}}, \widehat{\boldsymbol{\Sigma}}; T \right\} . \tag{7.19}$$

Prior knowledge

We model the investor's prior as a *normal-inverse-Wishart (NIW) distribution*. In other words, it is convenient to factor the joint distribution of $\boldsymbol{\mu}$ and $\boldsymbol{\Sigma}$ into the conditional distribution of $\boldsymbol{\mu}$ given $\boldsymbol{\Sigma}$ and the marginal distribution of $\boldsymbol{\Sigma}$.

We model the conditional prior on $\boldsymbol{\mu}$ given $\boldsymbol{\Sigma}$ as a normal distribution with the following parameters:

$$\boldsymbol{\mu}|\boldsymbol{\Sigma} \sim \mathrm{N}\left(\boldsymbol{\mu}_0, \frac{\boldsymbol{\Sigma}}{T_0}\right), \tag{7.20}$$

where $\boldsymbol{\mu}_0$ is an N-dimensional vector and T_0 is a positive scalar.

We model the marginal prior on $\boldsymbol{\Sigma}$ as an inverse-Wishart distribution. In other words, it is easier to model the distribution of the inverse of $\boldsymbol{\Sigma}$, which we assume Wishart-distributed with the following parameters:

$$\boldsymbol{\Sigma}^{-1} \sim \mathrm{W}\left(\nu_0, \frac{\boldsymbol{\Sigma}_0^{-1}}{\nu_0}\right), \tag{7.21}$$

where $\boldsymbol{\Sigma}_0$ is an $N \times N$ symmetric and positive matrix and ν_0 is a positive scalar. For a graphical interpretation of the prior (7.20) and (7.21) in the case $N \equiv 1$ refer to Figure 7.1.

To analyze the role played by the parameters that appear in the above distributions, we first compute the unconditional (marginal) prior on $\boldsymbol{\mu}$. As we show in Appendix www.7.5, this is a multivariate Student t distribution:

$$\boldsymbol{\mu} \sim \mathrm{St}\left(\nu_0, \boldsymbol{\mu}_0, \frac{\boldsymbol{\Sigma}_0}{T_0}\right). \tag{7.22}$$

From this expression we see that the parameter $\boldsymbol{\mu}_0$ in (7.20) reflects the investor's view on the parameter $\boldsymbol{\mu}$. Indeed, from (2.190) we obtain:

$$\mathrm{E}\left\{\boldsymbol{\mu}\right\} = \boldsymbol{\mu}_0. \tag{7.23}$$

On the other hand the parameter T_0 in (7.20) reflects his confidence in that view. Indeed, from (2.191) we obtain:

$$\mathrm{Cov}\left\{\boldsymbol{\mu}\right\} = \frac{\nu_0}{\nu_0 - 2} \frac{\boldsymbol{\Sigma}_0}{T_0}. \tag{7.24}$$

Therefore a large T_0 corresponds to little uncertainty about the view on $\boldsymbol{\mu}$.

The parameter $\boldsymbol{\Sigma}_0$ in (7.21) reflects the investor's view on the dispersion parameter $\boldsymbol{\Sigma}$. Indeed, from (2.227) we see that the prior expectation reads:

$$\mathrm{E}\left\{\boldsymbol{\Sigma}^{-1}\right\} = \boldsymbol{\Sigma}_0^{-1}. \tag{7.25}$$

On the other hand, the parameter ν_0 in (7.21) describes the investor's confidence in this view. Indeed, from (2.229) we obtain:

$$\text{Cov}\left\{\text{vec}\left[\boldsymbol{\Sigma}^{-1}\right]\right\} = \frac{1}{\nu_0}\left(\mathbf{I}_{N^2} + \mathbf{K}_{NN}\right)\left(\boldsymbol{\Sigma}_0^{-1} \otimes \boldsymbol{\Sigma}_0^{-1}\right), \tag{7.26}$$

where vec is the operator (A.104) that stacks the columns of $\boldsymbol{\Sigma}^{-1}$ into a vector, $\mathbf{I}$ is the identity matrix, $\mathbf{K}$ is the commutation matrix (A.108) and $\otimes$ is the Kronecker product (A.96). Therefore a large value ν_0 corresponds to little uncertainty about the view on $\boldsymbol{\Sigma}^{-1}$ and thus about the view on $\boldsymbol{\Sigma}$.

To summarize, the investor's experience and his confidence are described by the following prior parameters:

$$e_C \equiv \{\boldsymbol{\mu}_0, \boldsymbol{\Sigma}_0; T_0, \nu_0\}. \tag{7.27}$$

To determine the specific values of these parameters in financial applications we can use the techniques discussed in Section 7.4.

Posterior distribution

Given the above assumptions on the market, i.e. (7.16), and on the investor's experience, i.e. (7.20) and (7.21), it is possible to carry out the integration in (7.15) explicitly and compute the posterior distribution of the market parameters.

As we show in Appendix www.7.2, the posterior is, like the prior, a normal-inverse-Wishart (NIW) distribution. Indeed, recall (7.17) and (7.18), and define the following additional parameters:

$$T_1\left[i_T, e_C\right] \equiv T_0 + T \tag{7.28}$$

$$\boldsymbol{\mu}_1\left[i_T, e_C\right] \equiv \frac{1}{T_1}\left[T_0\boldsymbol{\mu}_0 + T\widehat{\boldsymbol{\mu}}\right] \tag{7.29}$$

$$\nu_1\left[i_T, e_C\right] \equiv \nu_0 + T \tag{7.30}$$

$$\boldsymbol{\Sigma}_1\left[i_T, e_C\right] \equiv \frac{1}{\nu_1}\left[\nu_0\boldsymbol{\Sigma}_0 + T\widehat{\boldsymbol{\Sigma}} + \frac{\left(\boldsymbol{\mu}_0 - \widehat{\boldsymbol{\mu}}\right)\left(\boldsymbol{\mu}_0 - \widehat{\boldsymbol{\mu}}\right)'}{\frac{1}{T} + \frac{1}{T_0}}\right]. \tag{7.31}$$

Then the posterior distribution of the location parameter conditioned on the dispersion parameter is normal:

$$\boldsymbol{\mu}|\boldsymbol{\Sigma} \sim \text{N}\left(\boldsymbol{\mu}_1, \frac{\boldsymbol{\Sigma}}{T_1}\right); \tag{7.32}$$

and the posterior distribution of the dispersion parameter is inverse-Wishart:

$$\boldsymbol{\Sigma}^{-1} \sim \text{W}\left(\nu_1, \frac{\boldsymbol{\Sigma}_1^{-1}}{\nu_1}\right). \tag{7.33}$$

Also, since both prior and posterior distributions are normal-inverse-Wishart, from (7.22) we immediately derive the unconditional posterior distribution of the location parameter:

$$\boldsymbol{\mu} \sim \text{St}\left(\nu_1, \boldsymbol{\mu}_1, \frac{\boldsymbol{\Sigma}_1}{T_1}\right). \tag{7.34}$$

7.2.2 Summarizing the posterior distribution

We can summarize the main features of the posterior distribution of $\boldsymbol{\mu}$ and $\boldsymbol{\Sigma}$ by means of its location-dispersion ellipsoid, as discussed in Section 7.1.2.

We have two options: we can consider the two separate location-dispersion ellipsoids of the marginal posterior distributions of $\boldsymbol{\mu}$ and $\boldsymbol{\Sigma}$ respectively, or we can consider the single location-dispersion ellipsoid of the joint posterior distribution of $\boldsymbol{\mu}$ and $\boldsymbol{\Sigma}$. Since both approaches find applications in allocation problems, we present both cases.

Marginal posterior distribution of the expected value $\boldsymbol{\mu}$

As far as $\boldsymbol{\mu}$ is concerned, its marginal posterior distribution is the Student t distribution (7.34).

First we compute the classical-equivalent estimator of $\boldsymbol{\mu}$, i.e. a parameter of location of the marginal posterior distribution of $\boldsymbol{\mu}$. Choosing either the expected value (7.5) or the mode (7.6) as location parameter, we obtain from (2.190) the following classical-equivalent estimator:

$$\widehat{\boldsymbol{\mu}}_{\text{ce}}\left[i_T, e_C\right] = \frac{T_0 \boldsymbol{\mu}_0 + T \widehat{\boldsymbol{\mu}}}{T_0 + T}. \tag{7.35}$$

It is easy to check that, as the number of observations T increases, this classical-equivalent estimator shrinks towards the sample mean $\widehat{\boldsymbol{\mu}}$. On the other hand, as the investor's confidence T_0 in his experience regarding $\boldsymbol{\mu}$ increases, the classical-equivalent estimator (7.35) shrinks toward the investor's view $\boldsymbol{\mu}_0$. Notice the symmetric role that the confidence level T_0 and the number of observations T play in (7.35): the confidence level T_0 can be interpreted as the number of "pseudo-observations" that would be necessary in a classical setting to support the investor's confidence about his view $\boldsymbol{\mu}_0$.

Now we turn to the dispersion parameter for $\boldsymbol{\mu}$. Choosing the covariance (7.7) as scatter parameter we obtain from (2.191) the following result:

$$\mathbf{S}_{\boldsymbol{\mu}}\left[i_T, e_C\right] = \frac{1}{T_1} \frac{\nu_1}{\nu_1 - 2} \boldsymbol{\Sigma}_1, \tag{7.36}$$

where the explicit dependence on information and experience is given in (7.28)-(7.31). It can be proved that choosing the modal dispersion (7.8) as scatter parameter the result is simply rescaled by a number close to one.

The location and dispersion parameters (7.35) and (7.36) respectively define the location-dispersion uncertainty ellipsoid (7.10) for $\boldsymbol{\mu}$ with radius proportional to q:

$$\mathcal{E}^q_{\widehat{\boldsymbol{\mu}}_{\text{ce}}, \mathbf{S}_{\boldsymbol{\mu}}} \equiv \left\{ \boldsymbol{\mu} \text{ such that } \left(\boldsymbol{\mu} - \widehat{\boldsymbol{\mu}}_{\text{ce}}\right)' \mathbf{S}_{\boldsymbol{\mu}}^{-1} \left(\boldsymbol{\mu} - \widehat{\boldsymbol{\mu}}_{\text{ce}}\right) \leq q^2 \right\}. \tag{7.37}$$

From (7.36) and the definitions (7.28)-(7.31) we observe that when either the number of observations T or the confidence in the views T_0 tends to infinity, the Bayesian setting becomes the classical setting. Indeed, in this case the uncertainty ellipsoid (7.37) shrinks to the single point $\widehat{\boldsymbol{\mu}}_{\text{ce}}$, no matter the radius factor q. In other words, the marginal posterior distribution of $\boldsymbol{\mu}$ becomes infinitely peaked around its classical-equivalent estimator.

Marginal posterior distribution of the covariance $\boldsymbol{\Sigma}$

As far as $\boldsymbol{\Sigma}$ is concerned, its marginal posterior distribution is the inverse-Wishart distribution (7.33).

First we compute the classical-equivalent estimator of $\boldsymbol{\Sigma}$, i.e. a parameter of location of the marginal posterior distribution of $\boldsymbol{\Sigma}$. Choosing the mode (7.6) as location parameter, we show in Appendix www.7.4 that the ensuing classical-equivalent estimator reads:

$$\widehat{\boldsymbol{\Sigma}}_{\text{ce}}\left[i_T, e_C\right] = \frac{1}{\nu_0 + T + N + 1}\left[\nu_0 \boldsymbol{\Sigma}_0 + T\widehat{\boldsymbol{\Sigma}} + \frac{\left(\boldsymbol{\mu}_0 - \widehat{\boldsymbol{\mu}}\right)\left(\boldsymbol{\mu}_0 - \widehat{\boldsymbol{\mu}}\right)'}{\frac{1}{T} + \frac{1}{T_0}}\right]. \tag{7.38}$$

It can be proved that choosing the expected value (7.5) as location parameter the result is simply rescaled by a number close to one.

It is easy to check that, as the number of observations T increases, the classical-equivalent estimator (7.38) shrinks towards the sample covariance $\widehat{\boldsymbol{\Sigma}}$. On the other hand, as the investor's confidence ν_0 in his experience regarding $\boldsymbol{\Sigma}$ increases, the classical-equivalent estimator (7.38) shrinks toward the investor's view $\boldsymbol{\Sigma}_0$. Notice the symmetric role that the confidence level ν_0 and the number of observations T play in (7.38): the confidence level ν_0 can be interpreted as the number of "pseudo-observations" that would be necessary in a classical setting to support the investor's confidence about his view $\boldsymbol{\Sigma}_0$.

Now we turn to the dispersion parameter for $\boldsymbol{\Sigma}$. Since $\boldsymbol{\Sigma}$ is symmetric, we disregard the redundant elements above the diagonal. In other words we only consider the vector $\operatorname{vech}[\boldsymbol{\Sigma}]$, where vech is the operator that stacks the columns of a matrix skipping the redundant entries above the diagonal. Choosing the modal dispersion (7.8) as scatter parameter, we show in Appendix www.7.4 that the dispersion of $\operatorname{vech}[\boldsymbol{\Sigma}]$ reads:

$$\mathbf{S}_{\boldsymbol{\Sigma}}\left[i_T, e_C\right] = \frac{2\nu_1^2}{\left(\nu_1 + N + 1\right)^3}\left(\mathbf{D}_N'\left(\boldsymbol{\Sigma}_1^{-1} \otimes \boldsymbol{\Sigma}_1^{-1}\right)\mathbf{D}_N\right)^{-1}, \tag{7.39}$$

where $\mathbf{D}_N$ is the duplication matrix (A.113); $\otimes$ is the Kronecker product (A.95); and the explicit dependence on information and experience is given in (7.28)-(7.31). It can be proved that choosing the covariance matrix (7.7) as scatter parameter for $\operatorname{vech}[\boldsymbol{\Sigma}]$, the result is simply rescaled by a number close to one.

The location and dispersion parameters (7.38) and (7.39) respectively define the location-dispersion uncertainty ellipsoid (7.10) for $\boldsymbol{\Sigma}$ with radius proportional to q:

$$\mathcal{E}^{q}_{\widehat{\boldsymbol{\Sigma}}_{\text{ce}},\mathbf{S}_{\boldsymbol{\Sigma}}} \equiv \left\{ \boldsymbol{\Sigma} : \ \operatorname{vech}\left[\boldsymbol{\Sigma} - \widehat{\boldsymbol{\Sigma}}_{\text{ce}}\right]' \mathbf{S}_{\boldsymbol{\Sigma}}^{-1} \operatorname{vech}\left[\boldsymbol{\Sigma} - \widehat{\boldsymbol{\Sigma}}_{\text{ce}}\right] \leq q^{2} \right\}. \tag{7.40}$$

Notice that the matrices $\boldsymbol{\Sigma}$ in this ellipsoid are always symmetric, because the vech operator only spans the non-redundant elements of a matrix. When the radius factor q is small enough, the matrices $\boldsymbol{\Sigma}$ in this ellipsoid are also positive, because positivity is a continuous property and $\widehat{\boldsymbol{\Sigma}}_{\text{ce}}$ is positive.

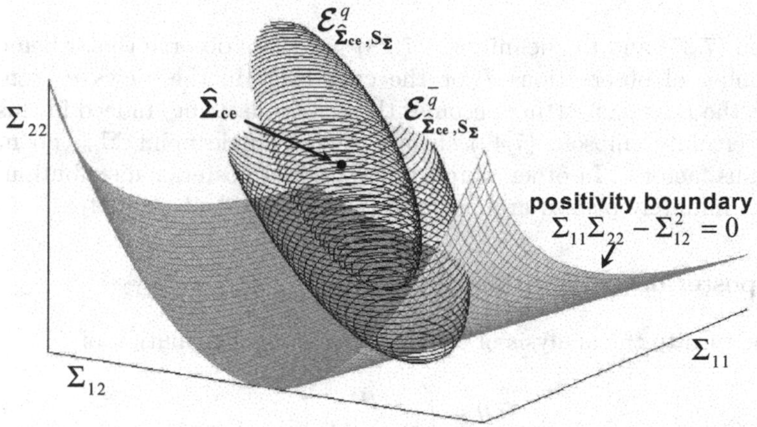

Fig. 7.3. Bayesian location-dispersion ellipsoid for covariance estimation

Consider the case of $N \equiv 2$ market invariants. In this case $\boldsymbol{\Sigma}$ is a 2×2 matrix:

$$\boldsymbol{\Sigma} \equiv \begin{pmatrix} \Sigma_{11} & \Sigma_{12} \\ \Sigma_{21} & \Sigma_{22} \end{pmatrix}. \tag{7.41}$$

The symmetry of $\boldsymbol{\Sigma}$ implies $\Sigma_{12} \equiv \Sigma_{21}$. Therefore a matrix is completely determined by the following three entries:

$$\operatorname{vech}[\boldsymbol{\Sigma}] = (\Sigma_{11}, \Sigma_{12}, \Sigma_{22})'. \tag{7.42}$$

A symmetric matrix is positive if and only if its eigenvalues are positive. In the 2×2 case, denoting as λ_1 and λ_2 the two eigenvalues, these are positive if and only if the following inequalities are satisfied:

$$\lambda_1 \lambda_2 > 0, \qquad \lambda_1 + \lambda_2 > 0. \tag{7.43}$$

On the other hand, the product of the eigenvalues is the determinant of $\boldsymbol{\Sigma}$ and the sum of the eigenvalues is the trace of $\boldsymbol{\Sigma}$, which are both invariants, see Appendix A.4. Therefore the positivity condition is equivalent to the two conditions below:

$$|\boldsymbol{\Sigma}| \equiv \Sigma_{11}\Sigma_{22} - \Sigma_{12}^2 \geq 0 \tag{7.44}$$

$$\text{tr}(\boldsymbol{\Sigma}) \equiv \Sigma_{11} + \Sigma_{22} \geq 0, \tag{7.45}$$

where the first expression follows from $(A.41)$.

In Figure 7.3 we see that when the radius factor q is small enough, every point of the ellipsoid (7.40) satisfies (7.44)-(7.45). When the radius factor becomes $\overline{q}$, a large enough scalar, the positivity condition is violated.

From (7.39) and the definitions (7.28)-(7.31) we observe that when either the number of observations T or the confidence in the views ν_0 tends to infinity, the Bayesian setting becomes the classical setting. Indeed in this case the uncertainty ellipsoid (7.40) shrinks to the single point $\widehat{\boldsymbol{\Sigma}}_{\text{ce}}$, no matter the radius factor q. In other words, the marginal posterior distribution of $\boldsymbol{\Sigma}$ becomes infinitely peaked around its classical-equivalent estimator.

Joint posterior distribution of $\boldsymbol{\mu}$ and $\boldsymbol{\Sigma}$

We now turn to the analysis of the joint posterior distribution of

$$\boldsymbol{\theta} \equiv \begin{pmatrix} \boldsymbol{\mu} \\ \text{vech}[\boldsymbol{\Omega}] \end{pmatrix}, \tag{7.46}$$

where $\boldsymbol{\Omega} \equiv \boldsymbol{\Sigma}^{-1}$. Indeed, it is much easier to parameterize the joint distribution of $\boldsymbol{\mu}$ and $\boldsymbol{\Sigma}$ in terms of the inverse of $\boldsymbol{\Sigma}$.

In Appendix www.7.3 we compute the mode (7.6) of the posterior distribution of $\boldsymbol{\theta}$, which reads:

$$\widehat{\boldsymbol{\theta}}_{\text{ce}}[i_T, e_C] \equiv \begin{pmatrix} \boldsymbol{\mu}_1 \\ \frac{\nu_1 - N}{\nu_1} \text{vech}\left[\boldsymbol{\Sigma}_1^{-1}\right] \end{pmatrix}, \tag{7.47}$$

where the explicit dependence on information and experience is given in (7.28)-(7.31).

In Appendix www.7.3 we also compute the modal dispersion (7.8) of the posterior distribution of $\boldsymbol{\theta}$, which reads:

$$\mathbf{S}_{\boldsymbol{\theta}}[i_T, e_C] = \begin{pmatrix} \mathbf{S}_{\boldsymbol{\mu}} & \mathbf{0}_{N^2 \times (N(N+1)/2)^2} \\ \mathbf{0}_{(N(N+1)/2)^2 \times N^2} & \mathbf{S}_{\boldsymbol{\Omega}} \end{pmatrix}. \tag{7.48}$$

In this expression:

$$\mathbf{S}_{\mu}\left[i_T, e_C\right] \equiv \frac{1}{T_1}\frac{\nu_1}{\nu_1 - N}\boldsymbol{\Sigma}_1 \tag{7.49}$$

$$\mathbf{S}_{\Omega}\left[i_T, e_C\right] \equiv \frac{2}{\nu_1}\frac{\nu_1 - N}{\nu_1}\left[\mathbf{D}'_N\left(\boldsymbol{\Sigma}_1 \otimes \boldsymbol{\Sigma}_1\right)\mathbf{D}_N\right]^{-1}, \tag{7.50}$$

where $\mathbf{D}_N$ is the duplication matrix (A.113); $\otimes$ is the Kronecker product (A.95); and the explicit dependence on information and experience is given in (7.28)-(7.31).

The location and dispersion parameters (7.47) and (7.48) define the joint location-dispersion uncertainty ellipsoid (7.10) with radius factor q. It is straightforward to check that all the comments regarding the self-adjusting nature of the location-dispersion ellipsoids (7.37) and (7.40) also apply to the joint location-dispersion ellipsoid.

7.3 Explicit factors

We present here the Bayesian estimators of factor loadings and perturbation dispersion in a factor model under the normal hypothesis for the market. In other words, we consider an affine explicit factor model:

$$\mathbf{X}_t = \mathbf{B}\mathbf{f}_t + \mathbf{U}_t, \tag{7.51}$$

where the factors $\mathbf{f}_t$ are known and the perturbations, conditioned on the factors, are normally distributed:

$$\mathbf{X}_t | \mathbf{f}_t, \mathbf{B}, \boldsymbol{\Sigma} \sim \mathrm{N}\left(\mathbf{B}\mathbf{f}_t, \boldsymbol{\Sigma}\right). \tag{7.52}$$

In this setting, the parameters to be determined are the factor loadings $\mathbf{B}$ and the dispersion matrix $\boldsymbol{\Sigma}$. This specification is rich and flexible enough to suitably model real problems, yet the otherwise analytically intractable computations of Bayesian analysis can be worked out completely, see also Press (1982).

7.3.1 Computing the posterior distribution

The Bayesian estimate of the unknown parameters is represented by the joint posterior distribution of $\mathbf{B}$ and $\boldsymbol{\Sigma}$. In order to compute this distribution we need to collect the available information from the market and to model the investor's experience, i.e. his prior distribution.

Information from the market

The information on the market is contained in the time series of the past joint realizations of the market invariants and the factors:

$$i_T \equiv \{\mathbf{x}_1, \mathbf{f}_1, \mathbf{x}_2, \mathbf{f}_2, \ldots, \mathbf{x}_T, \mathbf{f}_T\}. \tag{7.53}$$

Since we are interested in the estimation of the factor loadings $\mathbf{B}$ and the scatter parameter $\boldsymbol{\Sigma}$, it turns out sufficient to summarize the historical information on the market into the ordinary least squares estimator (4.126) of the factor loadings, which we report here:

$$\widehat{\mathbf{B}} \equiv \widehat{\boldsymbol{\Sigma}}_{XF} \widehat{\boldsymbol{\Sigma}}_F^{-1}, \tag{7.54}$$

where

$$\widehat{\boldsymbol{\Sigma}}_{XF} \equiv \frac{1}{T} \sum_{t=1}^{T} \mathbf{x}_t \mathbf{f}_t', \quad \widehat{\boldsymbol{\Sigma}}_F \equiv \frac{1}{T} \sum_{t=1}^{T} \mathbf{f}_t \mathbf{f}_t'; \tag{7.55}$$

and the sample covariance of the residuals (4.128), which we report here:

$$\widehat{\boldsymbol{\Sigma}} \equiv \frac{1}{T} \sum_{t=1}^{T} \left(\mathbf{x}_t - \widehat{\mathbf{B}}\mathbf{f}_t\right) \left(\mathbf{x}_t - \widehat{\mathbf{B}}\mathbf{f}_t\right)'. \tag{7.56}$$

Along with the number of observations T in the sample, this is all the information we need from the market. Therefore we can represent the information on the market equivalently in terms of the following parameters:

$$i_T \equiv \left\{\widehat{\mathbf{B}}, \widehat{\boldsymbol{\Sigma}}; T\right\}. \tag{7.57}$$

Prior knowledge

We model the investor's prior as a *normal-inverse-Wishart (NIW) distribution*. In other words, it is convenient to factor the joint distribution of $\mathbf{B}$ and $\boldsymbol{\Sigma}$ into the conditional distribution of $\mathbf{B}$ given $\boldsymbol{\Sigma}$ and the marginal distribution of $\boldsymbol{\Sigma}$.

We model the conditional prior on $\mathbf{B}$ given $\boldsymbol{\Sigma}$ as a matrix-valued normal distribution (2.181) with the following parameters:

$$\mathbf{B}|\boldsymbol{\Sigma} \sim \mathrm{N}\left(\mathbf{B}_0, \frac{\boldsymbol{\Sigma}}{T_0}, \boldsymbol{\Sigma}_{F,0}^{-1}\right), \tag{7.58}$$

where $\mathbf{B}_0$ is an $N \times K$ matrix, $\boldsymbol{\Sigma}_{F,0}$ is a $K \times K$ symmetric and positive matrix and T_0 is a positive scalar.

We model the marginal prior on $\boldsymbol{\Sigma}$ as an inverse-Wishart distribution. In other words, it is easier to model the distribution of the inverse of $\boldsymbol{\Sigma}$, which we assume Wishart-distributed with the following parameters:

$$\boldsymbol{\Sigma}^{-1} \sim \mathrm{W}\left(\nu_0, \frac{\boldsymbol{\Sigma}_0^{-1}}{\nu_0}\right), \tag{7.59}$$

where $\boldsymbol{\Sigma}_0$ is an $N \times N$ positive definite matrix and ν_0 is a positive scalar.

To analyze the role played by the parameters that appear in the above expressions, we compute the unconditional (marginal) prior on $\mathbf{B}$. We show in Appendix www.7.8 that this distribution is a matrix-valued Student t distribution (2.198) with the following parameters:

$$\mathbf{B} \sim \mathrm{St}\left(\nu_0 + K - N, \mathbf{B}_0, \frac{\nu_0}{\nu_0 + K - N}\boldsymbol{\Sigma}_0, \frac{\boldsymbol{\Sigma}_{F,0}^{-1}}{T_0}\right). \tag{7.60}$$

From this expression we see that the parameter $\mathbf{B}_0$ in (7.58) reflects the investor's view on the parameter $\mathbf{B}$. Indeed, from (2.203) we obtain:

$$\mathrm{E}\{\mathbf{B}\} = \mathbf{B}_0. \tag{7.61}$$

On the other hand from (2.206) the parameter $\boldsymbol{\Sigma}_{F,0}^{-1}$ in (7.58) yields the covariance structure between the m-th and n-th row of $\mathbf{B}$, i.e. the sensitivities of the m-th and n-th market invariant to the factors:

$$\mathrm{Cov}\left\{\mathbf{B}_{(m)}, \mathbf{B}_{(n)}\right\} = \frac{1}{T_0} \frac{\nu_0}{\nu_0 - N + K - 2} \left[\boldsymbol{\Sigma}_0\right]_{mn} \boldsymbol{\Sigma}_{F,0}^{-1}. \tag{7.62}$$

This also shows that the parameter T_0 in (7.58) reflects the investor's confidence on his view on $\mathbf{B}$, as a large T_0 corresponds to small variances and covariances in the prior on the factor loadings.

The parameter $\boldsymbol{\Sigma}_0$ in (7.59) reflects the investor's view on the dispersion parameter $\boldsymbol{\Sigma}$. Indeed, from (2.227) the prior expectation reads:

$$\mathrm{E}\left\{\boldsymbol{\Sigma}^{-1}\right\} = \boldsymbol{\Sigma}_0^{-1}. \tag{7.63}$$

On the other hand, the parameter ν_0 in (7.59) describes the investor's confidence in this view. Indeed, from (2.229) we obtain:

$$\mathrm{Cov}\left\{\mathrm{vec}\left[\boldsymbol{\Sigma}^{-1}\right]\right\} = \frac{1}{\nu_0}\left(\mathbf{I}_{N^2} + \mathbf{K}_{NN}\right)\left(\boldsymbol{\Sigma}_0^{-1} \otimes \boldsymbol{\Sigma}_0^{-1}\right), \tag{7.64}$$

where vec is the operator (A.104) that stacks the columns of $\boldsymbol{\Sigma}^{-1}$ into a vector, $\mathbf{I}$ is the identity matrix, $\mathbf{K}$ is the commutation matrix (A.108) and $\otimes$ is the Kronecker product (A.96). Therefore a large value ν_0 corresponds to little uncertainty about the view on $\boldsymbol{\Sigma}^{-1}$ and thus about the view on $\boldsymbol{\Sigma}$.

To summarize, the investor's experience and his confidence are described by the following prior parameters:

$$e_C \equiv \left\{\mathbf{B}_0, \boldsymbol{\Sigma}_0, \boldsymbol{\Sigma}_{F,0}; T_0, \nu_0\right\}. \tag{7.65}$$

To determine the values of these parameters in financial applications we can use the techniques discussed in Section 7.4.

Posterior distribution

Given the above assumptions it is possible to carry out the integration in (7.15) explicitly.

As we show in Appendix www.7.6, the posterior distribution of $\mathbf{B}$ and $\boldsymbol{\Sigma}$ is, like the prior, a normal-inverse-Wishart (NIW) distribution. Indeed, recall (7.54)-(7.56) and define the following additional parameters:

$$T_1\left[i_T, e_C\right] \equiv T_0 + T \tag{7.66}$$

$$\boldsymbol{\Sigma}_{F,1}\left[i_T, e_C\right] \equiv \frac{T_0 \boldsymbol{\Sigma}_{F,0} + T\widehat{\boldsymbol{\Sigma}}_F}{T_0 + T} \tag{7.67}$$

$$\mathbf{B}_1\left[i_T, e_C\right] \equiv \left(T_0 \mathbf{B}_0 \boldsymbol{\Sigma}_{F,0} + T\widehat{\mathbf{B}}\widehat{\boldsymbol{\Sigma}}_F\right)\left(T_0 \boldsymbol{\Sigma}_{F,0} + T\widehat{\boldsymbol{\Sigma}}_F\right)^{-1} \tag{7.68}$$

$$\nu_1\left[i_T, e_C\right] \equiv T + \nu_0 \tag{7.69}$$

$$\begin{aligned}\boldsymbol{\Sigma}_1\left[i_T, e_C\right] \equiv \frac{1}{\nu_1}\Big[&T\widehat{\boldsymbol{\Sigma}} + \nu_0 \boldsymbol{\Sigma}_0 + T_0 \mathbf{B}_0 \boldsymbol{\Sigma}_{F,0} \mathbf{B}_0' \\ &+ T\widehat{\mathbf{B}}\widehat{\boldsymbol{\Sigma}}_F \widehat{\mathbf{B}}' - T_1 \mathbf{B}_1 \boldsymbol{\Sigma}_{F,1} \mathbf{B}_1'\Big].\end{aligned} \tag{7.70}$$

Then the dispersion parameter is inverse-Wishart-distributed:

$$\boldsymbol{\Sigma}^{-1} \sim \mathrm{W}\left(\nu_1, \frac{\boldsymbol{\Sigma}_1^{-1}}{\nu_1}\right). \tag{7.71}$$

On the other hand the distribution of the factor loadings conditioned on the dispersion parameter is a matrix-valued normal distribution (2.181) with the following parameters:

$$\mathbf{B}|\boldsymbol{\Sigma} \sim \mathrm{N}\left(\mathbf{B}_1, \frac{\boldsymbol{\Sigma}}{T_1}, \boldsymbol{\Sigma}_{F,1}^{-1}\right). \tag{7.72}$$

Also, since prior and posterior are both normal-inverse-Wishart distributions, from (7.60) we immediately derive the unconditional distribution of the factor loadings, which is a matrix-valued Student t distribution:

$$\mathbf{B} \sim \mathrm{St}\left(\nu_1 + K - N, \mathbf{B}_1, \frac{\nu_1}{\nu_1 + K - N}\boldsymbol{\Sigma}_1, \frac{\boldsymbol{\Sigma}_{F,1}^{-1}}{T_1}\right). \tag{7.73}$$

7.3.2 Summarizing the posterior distribution

We can summarize the main features of the posterior distribution of $\mathbf{B}$ and $\boldsymbol{\Sigma}$ by means of its location-dispersion ellipsoid, as discussed in Section 7.1.2.

We have two options: we can consider the two separate location-dispersion ellipsoids of the marginal posterior distributions of $\mathbf{B}$ and $\boldsymbol{\Sigma}$ respectively, or we can consider the single location-dispersion ellipsoid of the joint distribution of $\mathbf{B}$ and $\boldsymbol{\Sigma}$. Since both approaches find applications in allocation problems, we present both cases.

Marginal posterior distribution of the factor loadings B

As far as $\mathbf{B}$ is concerned, its marginal posterior distribution is the matrix-valued Student t distribution (7.73).

First we compute the classical-equivalent estimator of $\mathbf{B}$, i.e. a parameter of location of the marginal posterior distribution of $\mathbf{B}$. Choosing the expected value (7.5) as location parameter, we obtain from (2.203) the following classical-equivalent estimator of the factor loadings:

$$\widehat{\mathbf{B}}_{\text{ce}}\left[i_T, e_C\right] = \left(T_0 \mathbf{B}_0 \boldsymbol{\Sigma}_{F,0} + T \widehat{\mathbf{B}} \widehat{\boldsymbol{\Sigma}}_F\right)\left(T_0 \boldsymbol{\Sigma}_{F,0} + T \widehat{\boldsymbol{\Sigma}}_F\right)^{-1}. \tag{7.74}$$

It is easy to check that, as the number of observations T increases, this classical-equivalent estimator shrinks towards the OLS estimator $\widehat{\mathbf{B}}$. On the other hand, as the investor's confidence T_0 in his experience regarding $\mathbf{B}$ increases, the classical-equivalent estimator (7.74) shrinks toward the investor's view $\mathbf{B}_0$. Notice the symmetric role that the confidence level T_0 and the number of observations T play in (7.74): the confidence level T_0 can be interpreted as the number of "pseudo-observations" that would be necessary in a classical setting to support the investor's confidence about his view $\mathbf{B}_0$.

Now we turn to the dispersion parameter for $\mathbf{B}$. Choosing the covariance (7.7) as scatter parameter we obtain from (2.204) the following result:

$$\mathbf{S}_{\mathbf{B}}\left[i_T, e_C\right] = \frac{1}{T_1} \frac{\nu_1}{\nu_1 + K - N - 2} \boldsymbol{\Sigma}_{F,1}^{-1} \otimes \boldsymbol{\Sigma}_1, \tag{7.75}$$

where $\otimes$ is the Kronecker product (A.95) and where the explicit dependence on information and experience is given in (7.66)-(7.70).

The location and dispersion parameters (7.74) and (7.75) respectively define the location-dispersion uncertainty ellipsoid (7.10) for $\mathbf{B}$ with radius proportional to q:

$$\mathcal{E}^q_{\widehat{\mathbf{B}}_{\text{ce}}, \mathbf{S}_{\mathbf{B}}} \equiv \left\{\mathbf{B} : \operatorname{vec}\left[\mathbf{B} - \widehat{\mathbf{B}}_{\text{ce}}\right]' \mathbf{S}_{\mathbf{B}}^{-1} \operatorname{vec}\left[\mathbf{B} - \widehat{\mathbf{B}}_{\text{ce}}\right] \leq q^2\right\}, \tag{7.76}$$

where vec is the operator (A.104) that stacks the columns of a matrix into a vector.

From (7.75) and the definitions (7.66)-(7.70) we observe that when either the number of observations T or the confidence in the views T_0 tends to infinity, the Bayesian setting becomes the classical setting. Indeed, in this case the uncertainty ellipsoid (7.76) shrinks to the single point $\widehat{\mathbf{B}}_{\text{ce}}$, no matter the radius factor q. In other words, the marginal posterior distribution of $\mathbf{B}$ becomes infinitely peaked around its classical-equivalent estimator.

Marginal posterior distribution of the perturbation covariance Σ

As far as $\boldsymbol{\Sigma}$ is concerned, its marginal posterior distribution is the inverse-Wishart distribution (7.71).

First we compute the classical-equivalent estimator of $\mathbf{\Sigma}$, i.e. a parameter of location of the marginal posterior distribution of $\mathbf{\Sigma}$. Choosing the mode (7.6) as location parameter, we show in Appendix www.7.4 that the ensuing classical-equivalent estimator reads:

$$\widehat{\mathbf{\Sigma}}_{\text{ce}}\left[i_T, e_C\right] = \frac{1}{\nu_0 + T + N + 1}\Big[T\widehat{\mathbf{\Sigma}} + \nu_0\mathbf{\Sigma}_0 + T_0\mathbf{B}_0\mathbf{\Sigma}_{F,0}\mathbf{B}_0' \\ + T\widehat{\mathbf{B}}\widehat{\mathbf{\Sigma}}_F\widehat{\mathbf{B}}' - T_1\mathbf{B}_1\mathbf{\Sigma}_{F,1}\mathbf{B}_1'\Big] . \tag{7.77}$$

It is easy to check that, as the number of observations T increases, the classical-equivalent estimator (7.77) shrinks towards the sample covariance $\widehat{\mathbf{\Sigma}}$. On the other hand, as the investor's confidence ν_0 in his experience regarding $\mathbf{\Sigma}$ increases, the classical-equivalent estimator (7.77) shrinks toward the investor's view $\mathbf{\Sigma}_0$. Notice the symmetric role that the confidence level ν_0 and the number of observations T play in (7.77): the confidence level ν_0 can be interpreted as the number of "pseudo-observations" that would be necessary in a classical setting to support the investor's confidence about his view $\mathbf{\Sigma}_0$.

Now we turn to the dispersion parameter for $\mathbf{\Sigma}$. Since $\mathbf{\Sigma}$ is symmetric, we disregard the redundant elements above the diagonal. In other words we only consider the vector vech $[\mathbf{\Sigma}]$, where vech is the operator that stacks the columns of a matrix skipping the redundant entries above the diagonal. Choosing the modal dispersion (7.8) as scatter parameter, we show in Appendix www.7.4 that the dispersion of vech $[\mathbf{\Sigma}]$ reads:

$$\mathbf{S}_{\mathbf{\Sigma}}\left[i_T, e_C\right] = \frac{2\nu_1^2}{\left(\nu_1 + N + 1\right)^3}\left(\mathbf{D}_N'\left(\mathbf{\Sigma}_1^{-1} \otimes \mathbf{\Sigma}_1^{-1}\right)\mathbf{D}_N\right)^{-1}, \tag{7.78}$$

where $\mathbf{D}_N$ is the duplication matrix (A.113); $\otimes$ is the Kronecker product (A.95); and the explicit dependence on information and experience is given in (7.66)-(7.70).

The location and dispersion parameters (7.77) and (7.78) define the location-dispersion uncertainty ellipsoid (7.10) for $\mathbf{\Sigma}$ of radius proportional to q:

$$\mathcal{E}^q_{\widehat{\mathbf{\Sigma}}_{\text{ce}}, \mathbf{S}_{\mathbf{\Sigma}}} \equiv \left\{\mathbf{\Sigma} : \text{ vech}\left[\mathbf{\Sigma} - \widehat{\mathbf{\Sigma}}_{\text{ce}}\right]' \mathbf{S}_{\mathbf{\Sigma}}^{-1} \text{vech}\left[\mathbf{\Sigma} - \widehat{\mathbf{\Sigma}}_{\text{ce}}\right] \leq q^2\right\}. \tag{7.79}$$

Notice that the matrices $\mathbf{\Sigma}$ in this ellipsoid are always symmetric, because the vech operator only spans the non-redundant elements of a matrix. When the radius factor q is small enough, the matrices $\mathbf{\Sigma}$ in this ellipsoid are also positive, because positivity is a continuous property and $\widehat{\mathbf{\Sigma}}_{\text{ce}}$ is positive, see Figure 7.3.

From (7.78) and the definitions (7.66)-(7.70) we observe that when either the number of observations T or the confidence in the views ν_0 tends to infinity, the Bayesian setting becomes the classical setting. Indeed in this case the uncertainty ellipsoid (7.79) shrinks to the single point $\widehat{\mathbf{\Sigma}}_{\text{ce}}$, no matter the radius factor q. In other words, the marginal posterior distribution of $\mathbf{\Sigma}$ becomes infinitely peaked around its classical-equivalent estimator.

Joint posterior distribution of B and Σ

We now turn to the analysis of the joint posterior distribution of

$$\boldsymbol{\theta} \equiv \begin{pmatrix} \operatorname{vec}[\mathbf{B}] \\ \operatorname{vech}[\mathbf{\Omega}] \end{pmatrix}, \tag{7.80}$$

where $\mathbf{\Omega} \equiv \mathbf{\Sigma}^{-1}$. Indeed, it is much easier to parameterize the joint distribution of $\mathbf{B}$ and $\mathbf{\Sigma}$ in terms of the inverse of $\mathbf{\Sigma}$.

In Appendix www.7.7 we compute the mode (7.6) of the posterior distribution of $\boldsymbol{\theta}$, which reads:

$$\widehat{\boldsymbol{\theta}}_{\text{ce}}[i_T, e_C] \equiv \begin{pmatrix} \mathbf{B}_1 \\ \frac{\nu_1+K-N-1}{\nu_1} \operatorname{vech}\left[\mathbf{\Sigma}_1^{-1}\right] \end{pmatrix}, \tag{7.81}$$

where the explicit dependence on information and experience is given in (7.66)-(7.70).

In Appendix www.7.7 we also compute the modal dispersion (7.8) of the posterior distribution of $\boldsymbol{\theta}$, which reads:

$$\mathbf{S}_{\boldsymbol{\theta}}[i_T, e_C] = \begin{pmatrix} \mathbf{S}_{\mathbf{B}} & \mathbf{0}_{(NK)^2 \times (N(N+1)/2)^2} \\ \mathbf{0}_{(N(N+1)/2)^2 \times (NK)^2} & \mathbf{S}_{\mathbf{\Omega}} \end{pmatrix}. \tag{7.82}$$

In this expression:

$$\mathbf{S}_{\mathbf{B}}[i_T, e_C] \equiv \frac{1}{T_1} \frac{\nu_1}{\nu_1 + K - N - 1} \mathbf{K}_{NK} \left(\mathbf{\Sigma}_1 \otimes \mathbf{\Sigma}_{F,1}^{-1}\right) \mathbf{K}_{KN} \tag{7.83}$$

$$\mathbf{S}_{\mathbf{\Omega}}[i_T, e_C] \equiv \frac{2}{\nu_1} \frac{\nu_1 + K - N - 1}{\nu_1} \left[\mathbf{D}_N' \left(\mathbf{\Sigma}_1 \otimes \mathbf{\Sigma}_1\right) \mathbf{D}_N\right]^{-1}, \tag{7.84}$$

where $\mathbf{K}_{NK}$ is the commutation matrix ($A.108$); $\mathbf{D}_N$ is the duplication matrix ($A.113$); $\otimes$ is the Kronecker product ($A.95$); and the explicit dependence on information and experience is given in (7.66)-(7.70).

The location and dispersion parameters (7.81) and (7.82) respectively define the joint location-dispersion uncertainty ellipsoid (7.10) with radius factor q. It is straightforward to check that all the comments regarding the self-adjusting nature of the location-dispersion ellipsoids (7.76) and (7.79) also apply to the joint location-dispersion ellipsoid.

7.4 Determining the prior

In Section 7.1 we discussed how the Bayesian approach to parameter estimation relies on the investor's prior knowledge of the market parameters $\boldsymbol{\theta}$, which is modeled in terms of the prior probability density function $f_{\text{pr}}(\boldsymbol{\theta})$.

The parametric expression of the prior density is typically determined by a location parameter $\boldsymbol{\theta}_0$, which corresponds to the "peak" of the prior beliefs, and a set of scalars that define the level of dispersion of the prior density, i.e. the confidence in the prior beliefs.

The confidence in the investor's beliefs is usually left as a free parameter that can be tweaked on a case-by-base basis. Therefore specifying the prior corresponds to determining the value of the location parameter $\boldsymbol{\theta}_0$.

For example, assume that the market consists of equity-like securities. Therefore, the linear returns are market invariants:

$$\mathbf{L}_t \equiv \operatorname{diag}\left(\mathbf{P}_{t-\tau}\right)^{-1} \mathbf{P}_t - \mathbf{1}, \tag{7.85}$$

see Section 3.1.1. Assume that the linear returns are normally distributed:

$$\mathbf{L}_t | \boldsymbol{\mu}, \boldsymbol{\Sigma} \sim \mathrm{N}\left(\boldsymbol{\mu}, \boldsymbol{\Sigma}\right). \tag{7.86}$$

This is the multivariate normal Bayesian model (7.16), where the prior is determined by the following parameters:

$$\boldsymbol{\theta}_0 \equiv \left(\boldsymbol{\mu}_0, \boldsymbol{\Sigma}_0\right), \tag{7.87}$$

see (7.23) and (7.25).

In this section we present some techniques to quantify the investor's experience, i.e. to define the prior parameters $\boldsymbol{\theta}_0$ that determine the prior and thus the whole Bayesian estimation process.

These techniques rely on the unconstrained *allocation function*, which is the unconstrained optimal allocation (6.33) considered as a function of the parameters $\boldsymbol{\theta}$ that determine the distribution of the underlying market invariants:

$$\boldsymbol{\theta} \mapsto \boldsymbol{\alpha}\left(\boldsymbol{\theta}\right) \equiv \underset{\boldsymbol{\alpha}}{\operatorname{argmax}} \left\{\mathcal{S}_{\boldsymbol{\theta}}\left(\boldsymbol{\alpha}\right)\right\}. \tag{7.88}$$

To illustrate, we consider the leading example in Section 6.1.

From (7.86) the prices at the investment horizon are normally distributed:

$$\mathbf{P}_{T+\tau}^{\boldsymbol{\mu},\boldsymbol{\Sigma}} \sim \mathrm{N}\left(\boldsymbol{\xi}, \boldsymbol{\Phi}\right), \tag{7.89}$$

where the parameters $\boldsymbol{\xi}$ and $\boldsymbol{\Phi}$ follow from (7.85) and read:

$$\boldsymbol{\xi} \equiv \operatorname{diag}\left(\mathbf{p}_T\right)\left(\mathbf{1} + \boldsymbol{\mu}\right), \quad \boldsymbol{\Phi} \equiv \operatorname{diag}\left(\mathbf{p}_T\right) \boldsymbol{\Sigma} \operatorname{diag}\left(\mathbf{p}_T\right). \tag{7.90}$$

The lower-case notation $\mathbf{p}_T$ in the above expressions stresses that the current prices are realized random variables, i.e. they are known.

The index of satisfaction is the certainty-equivalent (6.21), which after substituting (7.90) reads:

$$\mathrm{CE}_{\boldsymbol{\mu},\boldsymbol{\Sigma}}\left(\boldsymbol{\alpha}\right)=\boldsymbol{\alpha}'\operatorname{diag}\left(\mathbf{p}_T\right)\left(\mathbf{1}+\boldsymbol{\mu}\right)-\frac{1}{2\zeta}\boldsymbol{\alpha}'\operatorname{diag}\left(\mathbf{p}_T\right)\boldsymbol{\Sigma}\operatorname{diag}\left(\mathbf{p}_T\right)\boldsymbol{\alpha}. \tag{7.91}$$

Maximizing this expression, from the first-order conditions we obtain the allocation function:

$$\left(\boldsymbol{\mu},\boldsymbol{\Sigma}\right)\mapsto\boldsymbol{\alpha}\left(\boldsymbol{\mu},\boldsymbol{\Sigma}\right)=\zeta\operatorname{diag}\left(\mathbf{p}_T\right)^{-1}\boldsymbol{\Sigma}^{-1}\left(\mathbf{1}+\boldsymbol{\mu}\right). \tag{7.92}$$

7.4.1 Allocation-implied parameters

Here we present in a more general context the approach proposed by Sharpe (1974) and Black and Litterman (1990), see also Grinold (1996) and He and Litterman (2002).

Typically, investors have a vague, qualitative idea of a suitable value for the prior parameters $\boldsymbol{\theta}_0$. Nonetheless, they usually have a very precise idea of what should be considered a suitable portfolio composition $\boldsymbol{\alpha}_0$, which we call the *prior allocation*.

By inverting the allocation function (7.88), we can set the prior parameters $\boldsymbol{\theta}_0$ as the parameters implied by the prior allocation $\boldsymbol{\alpha}_0$:

$$\boldsymbol{\theta}_0\equiv\boldsymbol{\theta}\left(\boldsymbol{\alpha}_0\right). \tag{7.93}$$

In other words, if the market parameters were $\boldsymbol{\theta}_0$, the optimal allocation would be $\boldsymbol{\alpha}_0$: therefore $\boldsymbol{\theta}_0$ is a prior parameter specification consistent with the prior allocation $\boldsymbol{\alpha}_0$.

In general, the dimension of the market parameters, namely the number S of entries in the vector $\boldsymbol{\theta}$, is larger than the dimension of the market, namely the number N of entries in the vector $\boldsymbol{\alpha}$: therefore the function (7.88) cannot be inverted. This problem can be overcome by pinning down some of the parameters by means of some alternative technique.

In our leading example the N-variate allocation function (7.92) is determined by the $S\equiv N\left(N+3\right)/2$ free parameters in $\left(\boldsymbol{\mu},\boldsymbol{\Sigma}\right)$. Fixing a value $\overline{\boldsymbol{\Sigma}}$ for the covariance, for instance by means of a shrinkage estimate (4.160), we obtain the following inverse function:

$$\boldsymbol{\mu}\left(\boldsymbol{\alpha}\right)=\frac{1}{\zeta}\overline{\boldsymbol{\Sigma}}\operatorname{diag}\left(\mathbf{p}_T\right)\boldsymbol{\alpha}-\mathbf{1}. \tag{7.94}$$

This function yields the *implied expected returns* of an allocation. Thus we can set the prior (7.87) as follows:

$$\boldsymbol{\mu}_0\equiv\boldsymbol{\mu}\left(\boldsymbol{\alpha}_0\right),\quad\boldsymbol{\Sigma}_0\equiv\overline{\boldsymbol{\Sigma}}. \tag{7.95}$$

More in general, we can impose a set of constraints $\mathcal{C}$ on the allocation function (7.88). Indeed, imposing constraints on portfolios leads to better out-of-sample results, see Frost and Savarino (1988). This way the allocation function results defined as follows:

$$\boldsymbol{\theta} \mapsto \boldsymbol{\alpha}(\boldsymbol{\theta}) \equiv \underset{\boldsymbol{\alpha} \in \mathcal{C}}{\operatorname{argmax}} \{\mathcal{S}_{\boldsymbol{\theta}}(\boldsymbol{\alpha})\}. \tag{7.96}$$

As in (7.93), the implied prior parameters $\boldsymbol{\theta}_0$ are obtained by first inverting this function, possibly fixing some of the parameters with different techniques, and then evaluating the inverse function in the prior allocation $\boldsymbol{\alpha}_0$.

For instance we can assume a budget constraint:

$$\mathcal{C}_1 : \boldsymbol{\alpha}'\mathbf{p}_T = w_T. \tag{7.97}$$

Also, we can impose that specific portfolios, i.e. linear combinations of securities, should not exceed given thresholds:

$$\mathcal{C}_2 : \underline{\mathbf{g}} \leq \mathbf{G}\boldsymbol{\alpha} \leq \overline{\mathbf{g}}, \tag{7.98}$$

where the $K \times N$ matrix $\mathbf{G}$ determines the specific portfolios and the K-dimensional vectors $\underline{\mathbf{g}}$ and $\overline{\mathbf{g}}$ determine the upper and lower thresholds respectively.

In Appendix www.7.9 we show that by adding the constraints $\mathcal{C}_1$ and $\mathcal{C}_2$ in our leading example the inverse function (7.94) is replaced by the following expression:

$$\boldsymbol{\alpha} \mapsto \boldsymbol{\mu}(\boldsymbol{\alpha}) + [\operatorname{diag}(\mathbf{p}_T)]^{-1} \mathbf{G}' (\overline{\boldsymbol{\gamma}} - \underline{\boldsymbol{\gamma}}). \tag{7.99}$$

In this expression $\boldsymbol{\mu}(\boldsymbol{\alpha})$ are the expected returns implied by the constraint (7.97), defined implicitly as follows:

$$\boldsymbol{\mu}(\boldsymbol{\alpha}) - \frac{\mathbf{1}'\overline{\boldsymbol{\Sigma}}^{-1}\boldsymbol{\mu}(\boldsymbol{\alpha})}{\mathbf{1}'\overline{\boldsymbol{\Sigma}}^{-1}\mathbf{1}}\mathbf{1} = \frac{1}{\zeta}\left(\overline{\boldsymbol{\Sigma}} \operatorname{diag}(\mathbf{p}_T)\boldsymbol{\alpha} - \frac{w_T}{\mathbf{1}'\overline{\boldsymbol{\Sigma}}^{-1}\mathbf{1}}\mathbf{1}\right); \tag{7.100}$$

and $(\overline{\boldsymbol{\gamma}}, \underline{\boldsymbol{\gamma}})$ are the Lagrange multipliers relative to the inequality constraints (7.98) which satisfy the Kuhn-Tucker conditions:

$$\overline{\boldsymbol{\gamma}}, \underline{\boldsymbol{\gamma}} \geq \mathbf{0} \tag{7.101}$$

$$\sum_{n=1}^{N} \underline{\gamma}_k G_{kn} \underline{g}_n = \sum_{n=1}^{N} \overline{\gamma}_k G_{kn} \overline{g}_n = 0, \quad k = 1, \ldots, K. \tag{7.102}$$

This is the result of Grinold and Easton (1998), see also Grinold and Kahn (1999).

7.4.2 Likelihood maximization

A different approach to quantify the investor's experience consists in defining the prior parameters $\boldsymbol{\theta}_0$ as a constrained classical maximum likelihood estimate, where the constraint is imposed in terms of the allocation function, see Jagannathan and Ma (2003) for the specific case which we outline in the example below.

Consider the standard maximum likelihood estimator of the market invariants (4.66), which in the Bayesian notation (7.13) of this chapter reads:

$$\begin{aligned}\widehat{\boldsymbol{\theta}} &\equiv \underset{\boldsymbol{\theta}\in\boldsymbol{\Theta}}{\operatorname{argmax}}\, f_{I_T|\boldsymbol{\theta}}\left(i_T|\boldsymbol{\theta}\right) \\ &= \underset{\boldsymbol{\theta}\in\boldsymbol{\Theta}}{\operatorname{argmax}} \left\{\sum_{t=1}^{T} \ln f\left(\mathbf{x}_t|\boldsymbol{\theta}\right)\right\},\end{aligned} \tag{7.103}$$

where the terms $\mathbf{x}_t$ represent the observed time series of the market invariants.

Now consider a set $\mathcal{C}$ of investment constraints, see Frost and Savarino (1988). By means of the allocation function $\boldsymbol{\alpha}(\boldsymbol{\theta})$ defined in (7.88) we select a subset in the domain $\boldsymbol{\Theta}$ of possible values for the parameter market parameters:

$$\widetilde{\boldsymbol{\Theta}} \equiv \{\boldsymbol{\theta} \in \boldsymbol{\Theta} \text{ such that } \boldsymbol{\alpha}(\boldsymbol{\theta}) \in \mathcal{C}\}. \tag{7.104}$$

In our example (7.91), consider an investor who has no risk propensity, i.e. such that $\zeta \rightarrow 0$ in his exponential utility function. Assume there exists a budget constraint and a no-short-sale constraint:

$$\mathcal{C}_1 : \boldsymbol{\alpha}'\mathbf{p}_T = w_T, \tag{7.105}$$

$$\mathcal{C}_2 : \boldsymbol{\alpha} \geq \mathbf{0}. \tag{7.106}$$

In Appendix www.7.9 we show that the constrained allocation function gives rise to the following constraints for the covariance matrix:

$$\widetilde{\boldsymbol{\Theta}} \equiv \left\{\boldsymbol{\Sigma} \text{ such that } \boldsymbol{\Sigma} \succeq \mathbf{0}, \boldsymbol{\Sigma}^{-1}\mathbf{1} \geq \mathbf{0}\right\}, \tag{7.107}$$

where the notation "$\succeq \mathbf{0}$" stands for "symmetric and positive".

The prior parameters $\boldsymbol{\theta}_0$ can be defined as the maximum likelihood estimate (7.103) of the market parameters constrained to the subset (7.104). In other words, the prior parameters are defined as follows:

$$\begin{aligned}\boldsymbol{\theta}_0 &\equiv \underset{\boldsymbol{\theta}\in\widetilde{\boldsymbol{\Theta}}}{\operatorname{argmax}}\, f_{I_T|\boldsymbol{\theta}}\left(i_T|\boldsymbol{\theta}\right) \\ &= \underset{\boldsymbol{\theta}\in\widetilde{\boldsymbol{\Theta}}}{\operatorname{argmax}} \left\{\sum_{t=1}^{T} \ln f\left(\mathbf{x}_t|\boldsymbol{\theta}\right)\right\}.\end{aligned} \tag{7.108}$$

From the log-likelihood under the normal hypothesis (7.86) in terms of the inverse of the covariance $\mathbf{\Omega} \equiv \mathbf{\Sigma}^{-1}$ and the constraints (7.107) we obtain:

$$\mathbf{\Omega}_0 = \underset{\substack{\mathbf{\Omega} \succeq \mathbf{0} \\ \mathbf{\Omega 1} \geq \mathbf{0}}}{\operatorname{argmax}} \left\{ \frac{T}{2} \sum_{t=1}^{T} \ln |\mathbf{\Omega}| - \frac{T}{2} \sum_{t=1}^{T} \operatorname{tr} \left[\widehat{\mathbf{\Sigma}} \mathbf{\Omega} \right] \right\}, \tag{7.109}$$

where $\widehat{\mathbf{\Sigma}}$ is the sample covariance (7.18). In turn, this expression defines the prior $\mathbf{\Sigma}_0 \equiv \mathbf{\Omega}_0^{-1}$ in (7.87).

8

Evaluating allocations

The classical approach to allocation evaluation discussed in the second part of the book assumes known the distribution of the market. In reality, the distribution of the market is not known and can only be estimated with some error. Therefore we need to update the evaluation criteria of a generic allocation in such a way that they account for estimation risk: this is the subject of the present chapter.

In Section 8.1 we realize that, since the distribution of the market is not known, an allocation cannot be a simple number. Instead, it is the outcome of a decision, contingent on the specific realization of the available information: the same allocation decision would have outputted different portfolios if the time series of market invariants had assumed different values. In order to evaluate an allocation decision it is important to track its dependence on the available information and stress test its performance in a set of different information scenarios. This is the same approach used to assess the performance of an estimator: the natural equivalent of the estimator's loss in this context is the opportunity cost, a positive quantity that the investor should try to minimize.

In Section 8.2 we apply the above evaluation process to the simplest allocation strategy: the prior allocation decision. This is a decision that completely disregards any historical information from the market, as it only relies on the investor's prior beliefs. Such an extreme approach is doomed to yield suboptimal results. Indeed, in the language of estimators the prior allocation is an extremely biased strategy. Nonetheless, the investor's experience is a key ingredient in allocation problems: a milder version of the prior allocation should somehow enter an optimal allocation decision.

In Section 8.3 we evaluate the most intuitive allocation strategy: the sample-based allocation decision. This decision is obtained by substituting the unknown market parameters with their estimated values in the maximization problem that defines the classical optimal allocation. Intuitively, when the estimates are backed by plenty of reliable data the final allocation is close to the truly optimal, yet unattainable, allocation. Nevertheless, if the amount of information is limited and the estimation process is naive, this approach is heavily sub-optimal. In the language of estimators, the sample-based strategy is an extremely inefficient allocation. We discuss in detail all the causes of this inefficiency, which include the leverage effect of estimation risk due to ill-conditioned estimates.

8.1 Allocations as decisions

A generic allocation $\boldsymbol{\alpha}$ is more than just a vector that represents the number of units of the securities in a given portfolio. An allocation is the outcome of a decision process that filters the available information. Had the available information been different, the same decision process would have yielded a different allocation vector.

In order to evaluate an allocation we need to evaluate the decision process behind it. This can be accomplished with the same approach used to evaluate an estimator. The recipe goes as follows: first, we introduce a natural measure of sub-optimality for a generic allocation, namely the opportunity cost; then we track the dependence of the opportunity cost on the unknown market parameters; then we compute the distribution of the opportunity cost of the given allocation decision under different information scenarios; finally we evaluate the distribution of the opportunity cost of the given allocation decision as the market parameters vary in a suitable stress test range.

8.1.1 Opportunity cost of a sub-optimal allocation

The optimal allocation $\boldsymbol{\alpha}^*$ was defined in (6.33) as the one that maximizes the investor's satisfaction, given his constraints:

$$\boldsymbol{\alpha}^* \equiv \underset{\boldsymbol{\alpha} \in \mathcal{C}}{\operatorname{argmax}} \left\{ \mathcal{S}(\boldsymbol{\alpha}) \right\}. \tag{8.1}$$

For instance, in the leading example discussed Section 6.1, the constraints are the budget constraint (6.24):

$$\mathcal{C}_1 : \ \mathbf{p}_T' \boldsymbol{\alpha} = w_T; \tag{8.2}$$

and the value at risk constraint (6.26):

$$\mathcal{C}_2 : \ \operatorname{Var}_c(\boldsymbol{\alpha}) \leq \gamma w_T. \tag{8.3}$$

The investor's satisfaction is modeled by the certainty-equivalent of final wealth (6.21), which reads:

$$\operatorname{CE}(\boldsymbol{\alpha}) = \boldsymbol{\xi}' \boldsymbol{\alpha} - \frac{1}{2\zeta} \boldsymbol{\alpha}' \boldsymbol{\Phi} \boldsymbol{\alpha}. \tag{8.4}$$

The optimal allocation (6.39) reads:

$$\boldsymbol{\alpha}^* \equiv \zeta \boldsymbol{\Phi}^{-1} \boldsymbol{\xi} + \frac{w_T - \zeta \mathbf{p}_T' \boldsymbol{\Phi}^{-1} \boldsymbol{\xi}}{\mathbf{p}_T' \boldsymbol{\Phi}^{-1} \mathbf{p}_T} \boldsymbol{\Phi}^{-1} \mathbf{p}_T. \tag{8.5}$$

This allocation maximizes the certainty equivalent (8.4). Geometrically, this allocation corresponds to the higher iso-satisfaction line compatible with the investment constraints in the risk/reward plane of the allocations, see Figure 8.1 and refer to Figure 6.2 for a more detailed description.

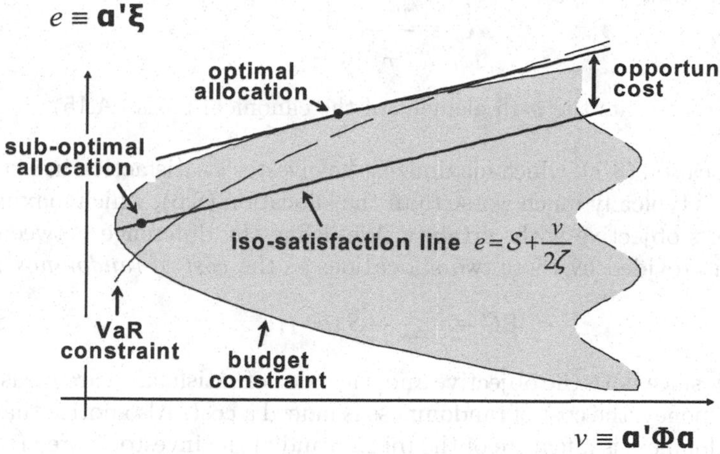

Fig. 8.1. Leading allocation example: opportunity cost of a sub-optimal allocation

In a hypothetical deterministic world where the investor has complete foresight of the market, the investor's main objective Ψ, whether it is final wealth as in (5.3), or relative wealth, as in (5.4), or net profits, as in (5.8), or possibly other specifications, becomes a deterministic function of the allocation, instead of being a random variable. As discussed on p. 241, in this hypothetical deterministic environment the investor does not need to evaluate an allocation based on an index of satisfaction $\mathcal{S}$. Instead, he considers directly his main objective and determines the optimal allocation as the one that maximizes his objective, given his constraints:

$$\boldsymbol{\alpha}_d \equiv \underset{\boldsymbol{\alpha} \in \mathcal{C}}{\operatorname{argmax}} \{\psi_{\boldsymbol{\alpha}}\}, \tag{8.6}$$

where "d" stands for "deterministic" and the lower-case notation stresses that the objective ψ is a deterministic value.

In our example, from (6.13) the markets are normally distributed:

$$\mathbf{P}_{T+\tau} \sim \mathrm{N}(\boldsymbol{\xi}, \boldsymbol{\Phi}). \tag{8.7}$$

The investor's main objective is (6.4), namely final wealth:

$$\Psi_{\boldsymbol{\alpha}} \equiv \boldsymbol{\alpha}' \mathbf{P}_{T+\tau}. \tag{8.8}$$

Assume that the investor knows that the first security will display the largest return over the investment horizon. Then he will invest all his budget in the first security:

$$\boldsymbol{\alpha}_d \equiv \frac{w_T}{p_T^{(1)}} \boldsymbol{\delta}^{(1)}, \tag{8.9}$$

where $\boldsymbol{\delta}^{(n)}$ represents the n-th element of the canonical basis (A.15).

The allocation (8.1), which maximizes the investor's satisfaction in a statistical sense, is typically much worse than the allocation (8.6), which maximizes the investor's objective with certainty. We define the difference between the satisfaction provided by these two allocations as the *cost of randomness*:

$$\mathrm{RC} \equiv \psi_{\boldsymbol{\alpha}_d} - \mathcal{S}(\boldsymbol{\alpha}^*). \tag{8.10}$$

Notice that, since both the objective and the index of satisfaction are measured in terms of money, the cost of randomness is indeed a cost. Also notice that the cost of randomness is a feature of the market and of the investor's preferences: it is not a feature of a specific allocation.

In our example it is immediate to understand that in hindsight the cash pocketed for having picked the winner as in (8.9) exceeds the certainty-equivalent of the suitably diversified portfolio (8.5).

Although the cost of randomness can be large, this cost is inevitable. Therefore what we defined as the optimal solution (8.1) is indeed optimal. As a result, the optimal allocation is the benchmark against which to evaluate any allocation.

Indeed, consider a generic allocation $\boldsymbol{\alpha}$ that satisfies the investment constraints. The difference between the satisfaction provided by the optimal allocation $\boldsymbol{\alpha}^*$ and the satisfaction provided by the generic allocation $\boldsymbol{\alpha}$ is the *opportunity cost* of the generic allocation:

$$\mathrm{OC}(\boldsymbol{\alpha}) \equiv \mathcal{S}(\boldsymbol{\alpha}^*) - \mathcal{S}(\boldsymbol{\alpha}). \tag{8.11}$$

Notice that the opportunity cost is always non-negative, since by definition the optimal solution $\boldsymbol{\alpha}^*$ provides the maximum amount of satisfaction given the constraints.

In our example, consider the deterministic allocation (8.9). This allocation, which turns out to be ideal ex-post, is actually sub-optimal ex-ante, when the investment decision is made, because it is not diversified.

The deterministic allocation satisfies the budget constraint (8.2) and, for suitable choices of the confidence level c and the budget at risk γ, it also satisfies the VaR constraint (8.3), see Figure 8.1.

From (6.38) the equation in the risk/reward plane of Figure 8.1 of the iso-satisfaction line corresponding to a generic allocation $\boldsymbol{\alpha}$ reads:

$$e = \mathrm{CE}(\boldsymbol{\alpha}) + \frac{v}{2\zeta}. \tag{8.12}$$

Therefore the opportunity cost of the deterministic allocation $\boldsymbol{\alpha}_d$ is the vertical distance between the iso-satisfaction line that corresponds to $\boldsymbol{\alpha}^*$ and the iso-satisfaction line that corresponds to $\boldsymbol{\alpha}_d$.

More in general we can evaluate any allocation, not necessarily an allocation that respects the investment constraints, by defining a cost, measured in terms of money, whenever an allocation $\boldsymbol{\alpha}$ violates the investment constraints. We denote this cost as $\mathcal{C}^+(\boldsymbol{\alpha})$.

For instance, if the indices of satisfaction $\widetilde{\mathcal{S}}$ associated with the investor's multiple objectives (6.9) are translation invariant, i.e. they satisfy (5.72), the cost of violating the respective constraints (6.25) reads:

$$\mathcal{C}^+(\boldsymbol{\alpha}) = \max\left\{0, \widetilde{s} - \widetilde{\mathcal{S}}(\boldsymbol{\alpha})\right\}. \tag{8.13}$$

In our example, the investor evaluates his profits in terms of the value at risk, which from (5.165) is translation invariant. Therefore the cost of violating the VaR constraint (8.3) reads:

$$\mathcal{C}_2^+(\boldsymbol{\alpha}) = \max\left\{0, \mathrm{Var}_c(\boldsymbol{\alpha}) - \gamma w_T\right\}. \tag{8.14}$$

In general, it is always possible to associate a cost with the violation of a given constraint, although possibly in a more ad-hoc way.

For instance, a possible constraint is the requirement that among the N securities in the market only a smaller number M appear in the optimal allocation. It is possible to model the cost for violating this constraint as follows:

$$\mathcal{C}^+(\boldsymbol{\alpha}) \equiv g(\#(\boldsymbol{\alpha}) - M), \tag{8.15}$$

where the function # counts the non-null entries of a given allocation vector $\boldsymbol{\alpha}$ and the function g is null when its argument is negative or null and it is otherwise increasing.

The opportunity cost of a generic allocation $\boldsymbol{\alpha}$ that does not necessarily satisfy the constraints reads:

$$\mathrm{OC}(\boldsymbol{\alpha}) \equiv \mathcal{S}(\boldsymbol{\alpha}^*) - \mathcal{S}(\boldsymbol{\alpha}) + \mathcal{C}^+(\boldsymbol{\alpha}). \tag{8.16}$$

Again, notice that the opportunity cost is always non-negative, given that by definition the optimal solution $\boldsymbol{\alpha}^*$ provides the maximum amount of satisfaction given the constraints. Also notice that the opportunity cost has the dimensions of money, since the investor's satisfaction is measured in terms of money: thus the opportunity cost indeed represents a cost.

8.1.2 Opportunity cost as function of the market parameters

The distribution of the market invariants, and thus the distribution of the market at the investment horizon, is fully determined by a set of unknown market parameters $\boldsymbol{\theta}^{\mathrm{t}}$. Consequently, the optimal allocation (8.1), i.e. the allocation that maximizes the investor's index of satisfaction given his constraints, depends on these market parameters. Similarly, the opportunity cost (8.16) of a generic allocation also depends on the on the market unknown parameters $\boldsymbol{\theta}^{\mathrm{t}}$. In view of evaluating an allocation, in this section we track the dependence on the underlying market parameters of the optimal allocation and of the opportunity cost of a generic suboptimal allocation.

The distribution of the market prices at the investment horizon $\mathbf{P}_{T+\tau}$ is determined by the distribution of the market invariants relative to the investment horizon $\mathbf{X}_{T+\tau}$. This distribution in turn is the projection to the investment horizon of the distribution of the market invariants relative to the estimation interval, which is fully determined by a set of parameters $\boldsymbol{\theta}$:

$$\boldsymbol{\theta} \overset{(3.64)}{\mapsto} \mathbf{X}^{\boldsymbol{\theta}}_{T+\tau} \overset{(3.79)}{\mapsto} \mathbf{P}^{\boldsymbol{\theta}}_{T+\tau}. \tag{8.17}$$

In our leading example we assume that the market consists of equity-like securities. Therefore from Section 3.1.1 the linear returns are market invariants:

$$\mathbf{L}_t \equiv \mathrm{diag}\left(\mathbf{P}_{t-\tau}\right)^{-1} \mathbf{P}_t - \mathbf{1}. \tag{8.18}$$

The simple projection formula (3.64) actually applies to the compounded returns. Nevertheless, by assuming that the estimation interval $\widetilde{\tau}$ and the investment horizon τ coincide, the more complex projection formula for the linear returns (3.78) becomes trivial. Also, we assume that the investment interval is fixed and we drop it from the notation.

In order to be consistent with (8.7), the linear returns are normally distributed:

$$\mathbf{L}^{\boldsymbol{\mu},\boldsymbol{\Sigma}}_t \sim \mathrm{N}(\boldsymbol{\mu}, \boldsymbol{\Sigma}), \tag{8.19}$$

where the parameters of the market invariants $\boldsymbol{\mu}$ and $\boldsymbol{\Sigma}$ are the N-dimensional vector of expected returns and the $N \times N$ covariance matrix respectively. Then the prices at the investment horizon are normally distributed:

$$\mathbf{P}_{T+\tau}^{\boldsymbol{\mu},\boldsymbol{\Sigma}} \sim \mathrm{N}\left(\boldsymbol{\xi}\left(\boldsymbol{\mu}\right), \boldsymbol{\Phi}\left(\boldsymbol{\Sigma}\right)\right), \tag{8.20}$$

where from (8.18) we obtain:

$$\boldsymbol{\xi}\left(\boldsymbol{\mu}\right) \equiv \operatorname{diag}\left(\mathbf{p}_T\right)\left(\mathbf{1}+\boldsymbol{\mu}\right), \quad \boldsymbol{\Phi}\left(\boldsymbol{\Sigma}\right) \equiv \operatorname{diag}\left(\mathbf{p}_T\right)\boldsymbol{\Sigma}\operatorname{diag}\left(\mathbf{p}_T\right). \tag{8.21}$$

The lower-case notation $\mathbf{p}_T$ stresses that the current prices are realized random variables, i.e. they are known.

In this context (8.17) reads:

$$\left(\boldsymbol{\mu}, \boldsymbol{\Sigma}\right) \overset{(3.78)}{\mapsto} \mathbf{L}_t^{\boldsymbol{\mu},\boldsymbol{\Sigma}} \overset{(3.79)}{\mapsto} \mathbf{P}_{T+\tau}^{\boldsymbol{\mu},\boldsymbol{\Sigma}}. \tag{8.22}$$

Consider an allocation $\boldsymbol{\alpha}$. The market prices $\mathbf{P}_{T+\tau}^{\boldsymbol{\theta}}$ and the allocation $\boldsymbol{\alpha}$ determine the investor's objective Ψ, which in turn determines the investor's satisfaction $\mathcal{S}$:

$$\left(\boldsymbol{\alpha}, \mathbf{P}_{T+\tau}^{\boldsymbol{\theta}}\right) \overset{(5.10)\text{-}(5.15)}{\mapsto} \Psi_{\boldsymbol{\alpha}}^{\boldsymbol{\theta}} \overset{(5.52)}{\mapsto} \mathcal{S}_{\boldsymbol{\theta}}\left(\boldsymbol{\alpha}\right). \tag{8.23}$$

In our example the investor's primary objective is his final wealth (8.8):

$$\Psi_{\boldsymbol{\alpha}}^{\boldsymbol{\mu},\boldsymbol{\Sigma}} \equiv \boldsymbol{\alpha}'\mathbf{P}_{T+\tau}^{\boldsymbol{\mu},\boldsymbol{\Sigma}}. \tag{8.24}$$

His satisfaction from the generic allocation $\boldsymbol{\alpha}$, modeled as the certainty-equivalent of an exponential utility function, follows from (8.4) and (8.21) and reads:

$$\begin{aligned} \mathrm{CE}_{\boldsymbol{\mu},\boldsymbol{\Sigma}}\left(\boldsymbol{\alpha}\right) = {} & \boldsymbol{\alpha}'\operatorname{diag}\left(\mathbf{p}_T\right)\left(\mathbf{1}+\boldsymbol{\mu}\right) \\ & -\frac{1}{2\zeta}\boldsymbol{\alpha}'\operatorname{diag}\left(\mathbf{p}_T\right)\boldsymbol{\Sigma}\operatorname{diag}\left(\mathbf{p}_T\right)\boldsymbol{\alpha}. \end{aligned} \tag{8.25}$$

A chain similar to (8.23) holds for the investor's constraints ensuing from the investor's multiple secondary objectives:

$$\left(\boldsymbol{\alpha}, \mathbf{P}_{T+\tau}^{\boldsymbol{\theta}}\right) \overset{(5.10)\text{-}(5.15)}{\mapsto} \widetilde{\Psi}_{\boldsymbol{\alpha}}^{\boldsymbol{\theta}} \overset{(5.52)}{\mapsto} \widetilde{\mathcal{S}}_{\boldsymbol{\theta}}\left(\boldsymbol{\alpha}\right) \overset{(6.25)}{\mapsto} \mathcal{C}_{\boldsymbol{\theta}}. \tag{8.26}$$

In our example the investor's secondary objective are the profits since inception (6.11):

$$\widetilde{\Psi}_{\boldsymbol{\alpha}}^{\boldsymbol{\mu},\boldsymbol{\Sigma}} \equiv \boldsymbol{\alpha}'\left(\mathbf{P}_{T+\tau}^{\boldsymbol{\mu},\boldsymbol{\Sigma}} - \mathbf{p}_T\right). \tag{8.27}$$

The investor monitors his profits by means of the value at risk. From (6.22) and (8.21) the dependence of the VaR on the market parameters reads:

$$\begin{aligned} \mathrm{Var}_{\boldsymbol{\mu},\boldsymbol{\Sigma}}\left(\boldsymbol{\alpha}\right) = {} & -\boldsymbol{\mu}'\operatorname{diag}\left(\mathbf{p}_T\right)\boldsymbol{\alpha} \\ & +\sqrt{2\boldsymbol{\alpha}'\operatorname{diag}\left(\mathbf{p}_T\right)\boldsymbol{\Sigma}\operatorname{diag}\left(\mathbf{p}_T\right)\boldsymbol{\alpha}}\operatorname{erf}^{-1}\left(2c-1\right). \end{aligned} \tag{8.28}$$

Therefore the investor's VaR constraint (8.3) reads:

$$\mathcal{C}_{\boldsymbol{\mu},\boldsymbol{\Sigma}}: 0 \geq -\gamma w_T - \boldsymbol{\mu}' \operatorname{diag}(\mathbf{p}_T)\boldsymbol{\alpha} + \sqrt{2\boldsymbol{\alpha}' \operatorname{diag}(\mathbf{p}_T)\boldsymbol{\Sigma}\operatorname{diag}(\mathbf{p}_T)\boldsymbol{\alpha}}\operatorname{erf}^{-1}(2c-1). \tag{8.29}$$

The optimal allocation (8.1) is the one that maximizes the investor's satisfaction (8.23) given the investor's constraints (8.26). As such, the optimal allocation depends on the underlying market parameters:

$$\boldsymbol{\alpha}(\boldsymbol{\theta}) \equiv \underset{\boldsymbol{\alpha}\in\mathcal{C}_{\boldsymbol{\theta}}}{\operatorname{argmax}}\{\mathcal{S}_{\boldsymbol{\theta}}(\boldsymbol{\alpha})\}. \tag{8.30}$$

This is the *optimal allocation function*. The optimal allocation gives rise to the maximum possible level of satisfaction, which also depends on the market parameters:

$$\overline{\mathcal{S}}(\boldsymbol{\theta}) \equiv \mathcal{S}_{\boldsymbol{\theta}}(\boldsymbol{\alpha}(\boldsymbol{\theta})) \equiv \max_{\boldsymbol{\alpha}\in\mathcal{C}_{\boldsymbol{\theta}}}\{\mathcal{S}_{\boldsymbol{\theta}}(\boldsymbol{\alpha})\}. \tag{8.31}$$

In our example, substituting (8.21) in (8.5) we obtain the functional dependence of the optimal allocation on the parameters $\boldsymbol{\mu}$ and $\boldsymbol{\Sigma}$ that determine the distribution of the market invariants:

$$\boldsymbol{\alpha}(\boldsymbol{\mu},\boldsymbol{\Sigma}) = [\operatorname{diag}(\mathbf{p}_T)]^{-1}\boldsymbol{\Sigma}^{-1}\left(\zeta\boldsymbol{\mu} + \frac{w_T - \zeta\mathbf{1}'\boldsymbol{\Sigma}^{-1}\boldsymbol{\mu}}{\mathbf{1}'\boldsymbol{\Sigma}^{-1}\mathbf{1}}\mathbf{1}\right). \tag{8.32}$$

As we prove in Appendix www.8.1 the maximum satisfaction reads:

$$\overline{\mathrm{CE}}(\boldsymbol{\mu},\boldsymbol{\Sigma}) = \frac{\zeta}{2}\left(C - \frac{B^2}{A}\right) + w_T\left(1 + \frac{B}{A} - \frac{w_T}{\zeta}\frac{1}{2A}\right), \tag{8.33}$$

where

$$A \equiv \mathbf{1}'\boldsymbol{\Sigma}^{-1}\mathbf{1}, \quad B \equiv \mathbf{1}'\boldsymbol{\Sigma}^{-1}\boldsymbol{\mu}, \quad C \equiv \boldsymbol{\mu}'\boldsymbol{\Sigma}^{-1}\boldsymbol{\mu}. \tag{8.34}$$

A generic allocation $\boldsymbol{\alpha}$ is suboptimal because the satisfaction that the investor draws from it is less than the maximum possible level (8.31). Furthermore, the generic allocation $\boldsymbol{\alpha}$ might violate the investment constraints. From the constraint specification $\mathcal{C}_{\boldsymbol{\theta}}$ as a function of the market parameters that follows from (8.26) we also derive the cost $\mathcal{C}_{\boldsymbol{\theta}}^{+}(\boldsymbol{\alpha})$ of the generic allocation violating the constraints.

For instance, if the indices of satisfaction $\widetilde{\mathcal{S}}$ associated with the investor's multiple objectives are translation invariant the cost of violating the respective constraints follows from (8.13) and reads:

$$\mathcal{C}_{\boldsymbol{\theta}}^{+}(\boldsymbol{\alpha}) = \max\left\{0, \widetilde{s} - \widetilde{\mathcal{S}}_{\boldsymbol{\theta}}(\boldsymbol{\alpha})\right\}. \tag{8.35}$$

In our example the cost of violating the VaR constraint is given by (8.14). From the expression of the VaR (8.28) as a function of the market parameters, the cost of violating the VaR constraint reads:

$$\begin{aligned} \mathcal{C}^{+}_{\boldsymbol{\mu},\boldsymbol{\Sigma}}(\boldsymbol{\alpha}) = \max\{&0, -\gamma w_T - \boldsymbol{\mu}' \operatorname{diag}(\mathbf{p}_T)\boldsymbol{\alpha} \\ &+ \sqrt{2\boldsymbol{\alpha}' \operatorname{diag}(\mathbf{p}_T)\boldsymbol{\Sigma}\operatorname{diag}(\mathbf{p}_T)\boldsymbol{\alpha}} \operatorname{erf}^{-1}(2c-1)\Big\}. \end{aligned} \tag{8.36}$$

From the maximum level of satisfaction (8.31), the satisfaction provided by a generic allocation (8.23) and the cost of violating the constraints (8.35) we obtain the expression of the opportunity cost (8.16) of a generic allocation $\boldsymbol{\alpha}$ as a function of the underlying parameters of the market invariants:

$$\mathrm{OC}_{\boldsymbol{\theta}}(\boldsymbol{\alpha}) \equiv \overline{\mathcal{S}}(\boldsymbol{\theta}) - \mathcal{S}_{\boldsymbol{\theta}}(\boldsymbol{\alpha}) + \mathcal{C}^{+}_{\boldsymbol{\theta}}(\boldsymbol{\alpha}). \tag{8.37}$$

In our example the opportunity cost of a generic allocation $\boldsymbol{\alpha}$ that satisfies the budget constraint is the difference between the optimal level of satisfaction (8.33) and the satisfaction provided by the generic allocation (8.25), plus the cost of violating the VaR constraint (8.36).

8.1.3 Opportunity cost as loss of an estimator

A generic allocation, not necessarily the optimal allocation, is a *decision*. As discussed in (6.15), this decision processes the information i_T available in the market and based on the investor's profile, which we consider fixed in this chapter, outputs a vector that represents the amount to invest in each security in a given market:

$$\boldsymbol{\alpha}[\cdot] : i_T \mapsto \mathbb{R}^N, \tag{8.38}$$

If the true parameters $\boldsymbol{\theta}^{\mathrm{t}}$ that determine the distribution of the market were known, i.e. $\boldsymbol{\theta}^{\mathrm{t}} \in i_T$, then these would represent all the information required to compute the optimal allocation: no additional information on the market could lead to a better allocation. As a consequence, there would be no need to consider any alternative allocation decision, as the only sensible decision would be the optimal allocation function (8.30) evaluated in the true value of the market parameters:

$$\boldsymbol{\alpha}[i_T] \equiv \boldsymbol{\alpha}(\boldsymbol{\theta}^{\mathrm{t}}). \tag{8.39}$$

Nevertheless, the true value of the market parameters $\boldsymbol{\theta}^{\mathrm{t}}$ is not known, i.e. $\boldsymbol{\theta}^{\mathrm{t}}$ is *not* part of the information i_T available at the time the investment is made: $\boldsymbol{\theta}^{\mathrm{t}} \notin i_T$. At best, the parameters $\boldsymbol{\theta}^{\mathrm{t}}$ can be estimated with some error. In other words, the truly optimal allocation (8.39) cannot be implemented.

Therefore the investor needs to *decide* how to process the information i_T available in the market in order to determine a suitable vector of securities.

For instance, but not necessarily, an investor might rely on estimates $\widehat{\boldsymbol{\theta}}[i_T]$ of the market parameters in (8.39).

Consider a generic allocation decision $\boldsymbol{\alpha}[i_T]$ as in (8.38). The information on the market is typically summarized in the time series of the past observations of a set of market invariants:

$$i_T \equiv \{\mathbf{x}_1, \ldots, \mathbf{x}_T\}, \tag{8.40}$$

where the lower-case notation stresses that these are realizations of random variables.

In our leading example, the market invariants are the linear returns (8.19) and the information on the market is contained in the time series of the past non-overlapping observations of these returns:

$$i_T \equiv \{\mathbf{l}_1, \ldots, \mathbf{l}_T\}. \tag{8.41}$$

Consider for instance a very simplistic allocation decision, according to which all the initial budget w_T is invested in the best performer over the last period. This strategy only processes part of the available information, namely the last observation in the time series (8.41). Indeed, this allocation decision is defined as follows:

$$\boldsymbol{\alpha}[i_T] \equiv w_T \frac{\boldsymbol{\delta}^{(b)}}{p_T^{(b)}}. \tag{8.42}$$

In this expression $\boldsymbol{\delta}^{(n)}$ denotes the n-th element of the canonical basis (A.15) and b is the index of the best among the realized returns:

$$b \equiv \underset{n \in \{1,\ldots,N\}}{\operatorname{argmax}} \{l_{T,n}\}, \tag{8.43}$$

where $l_{T,n}$ denotes the last-period return of the n-th security.

The generic allocation decision $\boldsymbol{\alpha}[i_T]$ gives rise to an opportunity cost (8.37), which depends on the underlying market parameters:

$$\mathrm{OC}_{\boldsymbol{\theta}}(\boldsymbol{\alpha}[i_T]) \equiv \overline{\mathcal{S}}(\boldsymbol{\theta}) - \mathcal{S}_{\boldsymbol{\theta}}(\boldsymbol{\alpha}[i_T]) + \mathcal{C}_{\boldsymbol{\theta}}^{+}(\boldsymbol{\alpha}[i_T]). \tag{8.44}$$

The satisfaction ensuing from the best-performer decision (8.42) follows from (8.25) and reads:

$$\mathrm{CE}_{\boldsymbol{\mu},\boldsymbol{\Sigma}}(\boldsymbol{\alpha}[i_T]) = w_T(1+\mu_b) - \frac{w_T^2}{2\zeta}\Sigma_{bb}. \tag{8.45}$$

The cost of the best-performer strategy violating the VaR constraint follows from substituting (8.42) in (8.36) and reads:

$$\mathcal{C}_{\boldsymbol{\mu},\boldsymbol{\Sigma}}^{+}(\boldsymbol{\alpha}[i_T]) = w_T \max\left\{0, \sqrt{2\Sigma_{bb}}\operatorname{erf}^{-1}(2c-1) - \mu_b - \gamma\right\}. \tag{8.46}$$

Recalling the expression (8.33) of the maximum possible satisfaction, the opportunity cost of the best-performer strategy reads:

$$\begin{aligned}\mathrm{OC}_{\boldsymbol{\mu},\boldsymbol{\Sigma}}\left(\boldsymbol{\alpha}\left[i_T\right]\right) = & \frac{\zeta}{2}\left(C-\frac{B^2}{A}\right)+w_T\left(1+\frac{B}{A}-\frac{w_T}{\zeta}\frac{1}{2A}\right) \\ & -w_T\left(1+\mu_b\right)+\frac{w_T^2}{2\zeta}\Sigma_{bb} \\ & +w_T \max\left\{0, \sqrt{2\Sigma_{bb}}\operatorname{erf}^{-1}(2c-1)-\mu_b-\gamma\right\},\end{aligned} \tag{8.47}$$

where A, B and C are the constants defined in (8.34). Notice that since ζ and w have the dimension of money and all the other quantities are a-dimensional, the opportunity cost is measured in terms of money.

Nevertheless, the opportunity cost (8.44) is not deterministic. Indeed, the times series (8.40) that feeds the generic allocation decision $\boldsymbol{\alpha}\left[i_T\right]$ is the specific realization of a set of T random variables, namely the market invariants:

$$I_T^{\boldsymbol{\theta}} \equiv \left\{\mathbf{X}_1^{\boldsymbol{\theta}}, \ldots, \mathbf{X}_T^{\boldsymbol{\theta}}\right\}. \tag{8.48}$$

The distribution of the invariants depends on the underlying unknown market parameters $\boldsymbol{\theta}$. In different markets, or even in the same market $\boldsymbol{\theta}$ but in different scenarios, the realization of the time series would have assumed a different value i_T' and thus the given allocation decision would have outputted a different set of values $\boldsymbol{\alpha}\left[i_T'\right]$.

This is the same situation encountered in the evaluation of an estimator, see (4.15). Therefore, in order to evaluate a generic allocation we have to proceed as in Figure 4.2. In other words, we replace the specific outcome of the market information i_T with the random variable (8.48). This way the given generic allocation decision (8.38) yields a random variable:

$$\boldsymbol{\alpha}\left[\cdot\right]: I_T^{\boldsymbol{\theta}} \mapsto \mathbb{R}^N. \tag{8.49}$$

We stress that the distribution of the random variable $\boldsymbol{\alpha}\left[I_T^{\boldsymbol{\theta}}\right]$ depends on the underlying assumption $\boldsymbol{\theta}$ on the distribution of the market invariants.

In our leading example, the time series of the past non-overlapping linear returns (8.41) is a specific realization of a set of T random variables identically distributed as in (8.19) and independent across time:

$$I_T^{\boldsymbol{\mu},\boldsymbol{\Sigma}} \equiv \left\{\mathbf{L}_1^{\boldsymbol{\mu},\boldsymbol{\Sigma}}, \mathbf{L}_2^{\boldsymbol{\mu},\boldsymbol{\Sigma}}, \ldots, \mathbf{L}_T^{\boldsymbol{\mu},\boldsymbol{\Sigma}}\right\}. \tag{8.50}$$

By substituting in (8.43) the last observation in the time series (8.41) with the last of the set of random variables (8.50) we obtain a discrete random variable B that takes values among the first N integers:

$$B(\boldsymbol{\mu}, \boldsymbol{\Sigma}) \equiv \underset{n \in \{1,\ldots,N\}}{\operatorname{argmax}} \left\{ L_{T,n}^{\boldsymbol{\mu},\boldsymbol{\Sigma}} \right\}. \tag{8.51}$$

In turn, the scenario-dependent version of the best-performer strategy (8.42) is defined in terms of the random variable B as follows:

$$\boldsymbol{\alpha}\left[I_T^{\boldsymbol{\mu},\boldsymbol{\Sigma}}\right] \equiv w_T \frac{\boldsymbol{\delta}^{(B)}}{P_T^{(B)}}. \tag{8.52}$$

This is a discrete random variable that depends on the assumptions on the underlying market parameters $(\boldsymbol{\mu}, \boldsymbol{\Sigma})$ through (8.51).

The random variable $\boldsymbol{\alpha}\left[I_T^{\boldsymbol{\theta}}\right]$ in turn gives rise to an opportunity cost (8.44) which also becomes a random variable that depends on the underlying assumption on the market parameters:

$$\begin{aligned} \text{Loss}\left(\boldsymbol{\alpha}\left[I_T^{\boldsymbol{\theta}}\right], \boldsymbol{\alpha}(\boldsymbol{\theta})\right) &\equiv \text{OC}_{\boldsymbol{\theta}}\left(\boldsymbol{\alpha}\left[I_T^{\boldsymbol{\theta}}\right]\right) \\ &\equiv \overline{\mathcal{S}}(\boldsymbol{\theta}) - \mathcal{S}_{\boldsymbol{\theta}}\left(\boldsymbol{\alpha}\left[I_T^{\boldsymbol{\theta}}\right]\right) + \mathcal{C}_{\boldsymbol{\theta}}^{+}\left(\boldsymbol{\alpha}\left[I_T^{\boldsymbol{\theta}}\right]\right). \end{aligned} \tag{8.53}$$

In the context of estimators, the opportunity cost is the (non-quadratic) loss (4.19) of the generic allocation decision with respect to the optimal allocation: indeed this random variable is never negative and is zero only in those scenarios where the outcome of the allocation decision happens to coincide with the optimal strategy.

The satisfaction ensuing from the stochastic version of the best-performer strategy (8.52) replaces the satisfaction (8.45) ensuing from the specific realization of the last-period returns:

$$\text{CE}_{\boldsymbol{\mu},\boldsymbol{\Sigma}}\left(\boldsymbol{\alpha}\left[I_T^{\boldsymbol{\mu},\boldsymbol{\Sigma}}\right]\right) = w_T\left(1 + \mu_B\right) - \frac{w_T^2}{2\zeta}\Sigma_{BB}. \tag{8.54}$$

This is a random variable, defined in terms of the random variable (8.51). More precisely, this is a discrete random variable, since its realizations can only take on a number of values equal to the number N of securities in the market, see Figure 8.2.

Similarly, the cost of violating the VaR constraint ensuing from the stochastic version of the best-performer strategy (8.52) replaces the cost (8.46) ensuing from the specific realization of the last-period returns:

$$\mathcal{C}_{\boldsymbol{\mu},\boldsymbol{\Sigma}}^{+}\left(\boldsymbol{\alpha}\left[I_T^{\boldsymbol{\mu},\boldsymbol{\Sigma}}\right]\right) = w_T \max\left\{0, \sqrt{2\Sigma_{BB}}\,\text{erf}^{-1}(2c-1) - \mu_B - \gamma\right\}. \tag{8.55}$$

This is also a discrete random variable, defined in terms of the random variable (8.51), see Figure 8.2.

The difference between the optimal satisfaction (8.33) and the actual satisfaction (8.54) plus the cost of violating the VaR constraint (8.55) represents the opportunity cost of the best-performer strategy (8.52). This opportunity cost is a discrete random variable which replaces the opportunity cost (8.47) ensuing from the specific realization of the last-period returns, see Figure 8.2:

$$\begin{aligned} \mathrm{OC}_{\boldsymbol{\mu},\boldsymbol{\Sigma}}\left(\boldsymbol{\alpha}\left[I_T^{\boldsymbol{\mu},\boldsymbol{\Sigma}}\right]\right) = {} & \frac{\zeta}{2}\left(C-\frac{B^2}{A}\right)+w_T\left(1+\frac{B}{A}-\frac{w_T}{\zeta}\frac{1}{2A}\right) \\ & -w_T\left(1+\mu_B\right)+\frac{w_T^2}{2\zeta}\Sigma_{BB} \\ & +w_T \max\left\{0, \sqrt{2\Sigma_{BB}}\operatorname{erf}^{-1}(2c-1)-\mu_B-\gamma\right\}, \end{aligned} \tag{8.56}$$

where A, B and C are the constants defined in (8.34).

8.1.4 Evaluation of a generic allocation decision

With the expression of the opportunity cost (8.53) we can evaluate an allocation decision for any value of the parameters $\boldsymbol{\theta}$ that determine the underlying distribution of the market invariants. Quite obviously, we only care about the performance of the allocation decision for the true value $\boldsymbol{\theta}^{\mathrm{t}}$ of the market parameters. Nevertheless, even more obviously, we do not know the true value $\boldsymbol{\theta}^{\mathrm{t}}$, otherwise we would simply implement the optimal allocation (8.39).

Therefore, in order to evaluate the given allocation decision, we consider the opportunity cost (8.53) of that strategy as a function of the underlying market parameters as we let the market parameters $\boldsymbol{\theta}$ vary in a suitable range $\boldsymbol{\Theta}$ that is broad enough to most likely include the true, unknown parameter $\boldsymbol{\theta}^{\mathrm{t}}$:

$$\boldsymbol{\theta} \mapsto \mathrm{OC}_{\boldsymbol{\theta}}\left(\boldsymbol{\alpha}\left[I_T^{\boldsymbol{\theta}}\right]\right), \quad \boldsymbol{\theta} \in \boldsymbol{\Theta}. \tag{8.57}$$

If the distribution of the opportunity cost is tightly peaked around a positive value very close to zero for all the markets $\boldsymbol{\theta}$ in the given range $\boldsymbol{\Theta}$, in particular it is close to zero in all the scenarios in correspondence of the true, yet unknown, value $\boldsymbol{\theta}^{\mathrm{t}}$. In this case the given allocation strategy is guaranteed to perform well and is close to optimal. This is the definition of optimality for an allocation decision in the presence of estimation risk: it is the same approach used to evaluate an estimator, see Figure 8.2 and compare with Figure 4.4.

In order to reduce the dimension of the market parameters and display the results of our evaluation, we assume in our example (8.19) that the correlation matrix of the linear returns has the following structure:

$$\boldsymbol{\Xi}(\rho) \equiv \begin{pmatrix} 1 & \rho & \cdots & \rho \\ \rho & \ddots & & \vdots \\ \vdots & & \ddots & \rho \\ \rho & \cdots & \rho & 1 \end{pmatrix}, \quad \rho \in \boldsymbol{\Theta} \equiv [0,1). \tag{8.58}$$

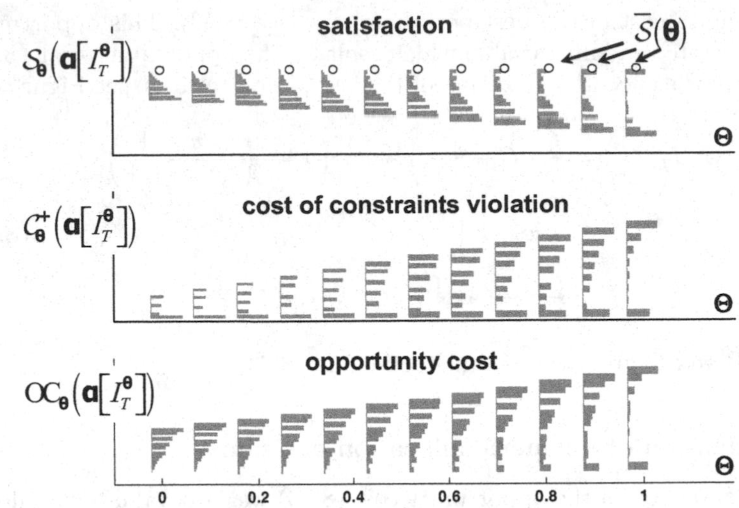

Fig. 8.2. Evaluation of allocation decisions as estimators

For the standard deviations and the expected values we assume the following structure:

$$\sqrt{\operatorname{diag}\left(\boldsymbol{\Sigma}\left(\rho\right)\right)} \equiv \left(1+\xi\rho\right)\mathbf{v}, \quad \boldsymbol{\mu} \equiv p\sqrt{\operatorname{diag}\left(\boldsymbol{\Sigma}\left(\rho\right)\right)}, \tag{8.59}$$

where $\mathbf{v}$ is a fixed vector of volatilities and ξ and p are fixed positive scalars. In other words, we assume that more correlated markets are more volatile, see Loretan and English (2000) and Forbes and Rigobon (2002) for comments regarding this assumption; furthermore, we assume that there exists a fixed risk premium for volatility. This way we obtain a one-parameter family of markets steered by the overall level of correlation among the securities.

In the top plot in Figure 8.2 we display the maximum satisfaction (8.33), which is not attainable:

$$\rho \mapsto \overline{\mathrm{CE}}\left(\boldsymbol{\mu}\left(\rho\right), \boldsymbol{\Sigma}\left(\rho\right)\right), \quad \rho \in \boldsymbol{\Theta} \equiv [0, 1). \tag{8.60}$$

In the same plot we display the distribution of the satisfaction (8.54) ensuing from the best-performer strategy:

$$\rho \mapsto \mathrm{CE}_{\boldsymbol{\mu}(\rho),\boldsymbol{\Sigma}(\rho)}\left(\boldsymbol{\alpha}\left[I_T^{\boldsymbol{\mu}(\rho),\boldsymbol{\Sigma}(\rho)}\right]\right), \quad \rho \in \boldsymbol{\Theta} \equiv [0, 1). \tag{8.61}$$

In the middle plot in Figure 8.2 we display the distribution of the cost (8.55) of the best-performer strategy violating the value at risk constraint:

$$\rho \mapsto \mathcal{C}^{+}_{\boldsymbol{\mu}(\rho),\boldsymbol{\Sigma}(\rho)}\left(\boldsymbol{\alpha}\left[I_T^{\boldsymbol{\mu}(\rho),\boldsymbol{\Sigma}(\rho)}\right]\right), \quad \rho \in \boldsymbol{\Theta} \equiv [0, 1). \tag{8.62}$$

In the bottom plot in Figure 8.2 we display the distribution of the opportunity cost (8.56) of the best-performer strategy:

$$\rho \mapsto \mathrm{OC}_{\boldsymbol{\mu}(\rho),\boldsymbol{\Sigma}(\rho)}\left(\boldsymbol{\alpha}\left[I_T^{\boldsymbol{\mu}(\rho),\boldsymbol{\Sigma}(\rho)}\right]\right), \quad \rho \in \boldsymbol{\Theta} \equiv [0,1). \tag{8.63}$$

Refer to symmys.com for more details on these plots.

We remark that since the opportunity cost (8.57) of an allocation decision is a random variable, the evaluation of its distribution is rather subjective. In principle, we should develop a theory to evaluate the distribution of the opportunity cost that parallels the discussion in Chapter 5. Nonetheless, aside from the additional computational burden, modeling the investor's attitude toward estimation risk is an even harder task than modeling his attitude toward risk. Given the scope of the book, we do not dwell on this topic, leaving the evaluation of the distribution of the opportunity cost on the more qualitative level provided by a graphical inspection, see Figure 8.2.

8.2 Prior allocation

The simplest allocation strategy consists in investing in a pre-defined portfolio that reflects the investor's experience, models, or prior beliefs and disregards any historical information from the market. In this section we analyze this strategy along the guidelines discussed in Section 8.1.

8.2.1 Definition

The *prior allocation decision* is a strategy that neglects the information i_T contained in the time series of the market invariants:

$$\boldsymbol{\alpha}_{\mathrm{p}}[i_T] \equiv \boldsymbol{\alpha}, \tag{8.64}$$

where "p" stand for "prior" and $\boldsymbol{\alpha}$ is a vector that satisfies all the constraints that do not depend on the unknown market parameters.

We remark that the prior allocation is a viable decision of the form (8.38), i.e. it is a decision that processes (by disregarding) only the information available on the market at the time the investment is made.

An example of such an allocation decision is the *equally-weighted portfolio* (6.16), which we report here:

$$\boldsymbol{\alpha}_{\mathrm{p}} \equiv \frac{w_T}{N} \operatorname{diag}(\mathbf{p}_T)^{-1} \mathbf{1}, \tag{8.65}$$

where w_T is the initial budget, $\mathbf{p}_T$ are the current market prices and $\mathbf{1}$ is an N-dimensional vector of ones.

8.2.2 Evaluation

In order to evaluate the prior allocation we proceed as in Section 8.1.

First we consider a set $\boldsymbol{\Theta}$ of market parameters that is broad enough to most likely include the true, unknown value $\boldsymbol{\theta}^{\mathrm{t}}$.

For each value $\boldsymbol{\theta}$ of the market parameters in the stress test set $\boldsymbol{\Theta}$ we compute as in (8.30) the optimal allocation function:

$$\boldsymbol{\alpha}(\boldsymbol{\theta}) \equiv \underset{\boldsymbol{\alpha} \in \mathcal{C}_{\boldsymbol{\theta}}}{\operatorname{argmax}} \{\mathcal{S}_{\boldsymbol{\theta}}(\boldsymbol{\alpha})\}; \tag{8.66}$$

Then we compute as in (8.31) the optimal level of satisfaction if $\boldsymbol{\theta}$ are the underlying market parameters, namely $\overline{\mathcal{S}}(\boldsymbol{\theta})$.

In our leading example the optimal allocation is (8.32), which provides the optimal level of satisfaction (8.33).

Next, we should randomize as in (8.48) the information from the market i_T, generating a distribution of information scenarios $I_T^{\boldsymbol{\theta}}$ that depends on the assumption $\boldsymbol{\theta}$ on the market parameters and then we should compute the outcome of the prior allocation decision (8.64) applied to the information scenarios, obtaining the random variable $\boldsymbol{\alpha}_{\mathrm{p}}\left[I_T^{\boldsymbol{\theta}}\right]$. Nevertheless, since by definition the prior allocation does not depend on the information on the market, we do not need to perform this step.

Therefore we move on to the next step and compute from (8.23) the satisfaction $\mathcal{S}_{\boldsymbol{\theta}}(\boldsymbol{\alpha}_{\mathrm{p}})$ ensuing from the prior allocation decision under the assumption $\boldsymbol{\theta}$ for the market parameters. Similarly, from (8.26) and expressions such as (8.35) we compute the cost of the prior allocation decision violating the constraints $\mathcal{C}_{\boldsymbol{\theta}}^{+}(\boldsymbol{\alpha}_{\mathrm{p}})$ under the assumption $\boldsymbol{\theta}$ for the market parameters. We stress that, unlike in the general case, in the case of the prior allocation decision both satisfaction and cost of constraint violation are deterministic.

Then we compute the opportunity cost (8.53) of the prior allocation, which is the difference between the satisfaction from the unattainable optimal allocation and the satisfaction from the prior allocation, plus the cost of the prior allocation violating the constraints:

$$\mathrm{OC}_{\boldsymbol{\theta}}(\boldsymbol{\alpha}_{\mathrm{p}}) \equiv \overline{\mathcal{S}}(\boldsymbol{\theta}) - \mathcal{S}_{\boldsymbol{\theta}}(\boldsymbol{\alpha}_{\mathrm{p}}) + \mathcal{C}_{\boldsymbol{\theta}}^{+}(\boldsymbol{\alpha}_{\mathrm{p}}). \tag{8.67}$$

Again, unlike in the general case, in the case of the prior allocation decision the opportunity cost is not a random variable.

The satisfaction provided by the equally weighted portfolio (8.65) follows from (8.25) and reads:

$$\mathrm{CE}_{\boldsymbol{\mu},\boldsymbol{\Sigma}}(\boldsymbol{\alpha}_{\mathrm{p}}) = w_T\left(1 + \frac{(\boldsymbol{\mu}'\mathbf{1})}{N}\right) - \frac{w_T^2}{2\zeta}\frac{\mathbf{1}'\boldsymbol{\Sigma}\mathbf{1}}{N^2}. \tag{8.68}$$

The cost of the equally weighted portfolio (8.65) violating the VaR constraint follows from (8.36) and reads:

$$\mathcal{C}^{+}_{\boldsymbol{\mu},\boldsymbol{\Sigma}}(\boldsymbol{\alpha}_{\mathrm{P}}) = w_T \max\left\{0, -\gamma - \frac{\mathbf{1}'\boldsymbol{\mu}}{N} + \frac{\sqrt{2\mathbf{1}'\boldsymbol{\Sigma}\mathbf{1}}}{N}\operatorname{erf}^{-1}(2c-1)\right\}. \tag{8.69}$$

Therefore the opportunity cost of the equally weighted portfolio under the assumption $\boldsymbol{\mu}$ and $\boldsymbol{\Sigma}$ for the market parameters reads:

$$\mathrm{OC}_{\boldsymbol{\mu},\boldsymbol{\Sigma}}(\boldsymbol{\alpha}_{\mathrm{P}}) \equiv \overline{\mathrm{CE}}(\boldsymbol{\mu},\boldsymbol{\Sigma}) - \mathrm{CE}_{\boldsymbol{\mu},\boldsymbol{\Sigma}}(\boldsymbol{\alpha}_{\mathrm{P}}) + \mathcal{C}^{+}_{\boldsymbol{\mu},\boldsymbol{\Sigma}}(\boldsymbol{\alpha}_{\mathrm{P}}), \tag{8.70}$$

where the first term on the right hand side is given in (8.33).

Finally we consider as in (8.57) the opportunity cost of the prior allocation as a function of the underlying assumptions $\boldsymbol{\theta}$ on the market, as $\boldsymbol{\theta}$ varies in the stress test range:

$$\boldsymbol{\theta} \mapsto \mathrm{OC}_{\boldsymbol{\theta}}(\boldsymbol{\alpha}_{\mathrm{P}}), \quad \boldsymbol{\theta} \in \boldsymbol{\Theta}, \tag{8.71}$$

see Figure 8.3. If this function is close to zero for each value $\boldsymbol{\theta}$ of the market parameters in the stress test set $\boldsymbol{\Theta}$ then the prior allocation is close to optimal.

In order to display the results in Figure 8.3 we let the underlying market parameters vary according to (8.58)-(8.59), obtaining a one-parameter family of markets, parameterized by the overall level of correlation ρ. Refer to `symmys.com` for more details on these plots.

In the top plot in Figure 8.3 we display the maximum satisfaction (8.33), which is not attainable:

$$\rho \mapsto \overline{\mathrm{CE}}(\boldsymbol{\mu}(\rho), \boldsymbol{\Sigma}(\rho)), \quad \rho \in \boldsymbol{\Theta} \equiv [0,1). \tag{8.72}$$

In the same plot we display the satisfaction (8.68) ensuing from the equally weighted portfolio (8.65):

$$\rho \mapsto \mathrm{CE}_{\boldsymbol{\mu}(\rho),\boldsymbol{\Sigma}(\rho)}(\boldsymbol{\alpha}_{\mathrm{P}}), \quad \rho \in \boldsymbol{\Theta} \equiv [0,1). \tag{8.73}$$

In the plot in the middle of Figure 8.3 we display the cost (8.69) of the equally weighted portfolio violating the VaR constraint:

$$\rho \mapsto \mathcal{C}^{+}_{\boldsymbol{\mu}(\rho),\boldsymbol{\Sigma}(\rho)}(\boldsymbol{\alpha}_{\mathrm{P}}), \quad \rho \in \boldsymbol{\Theta} \equiv [0,1). \tag{8.74}$$

Notice that for large enough values of the overall market correlation the value at risk constraint is not satisfied: therefore the investor pays a price that affects his total satisfaction.

In the bottom plot in Figure 8.3 we display the opportunity cost (8.70) of the equally weighted portfolio:

$$\rho \mapsto \mathrm{OC}_{\boldsymbol{\mu}(\rho),\boldsymbol{\Sigma}(\rho)}\left(\boldsymbol{\alpha}_{\mathrm{P}}\right), \quad \rho \in \boldsymbol{\Theta} \equiv [0,1) . \tag{8.75}$$

It appears that for our investor the equally weighted portfolio is only suitable if the market is sufficiently diversified.

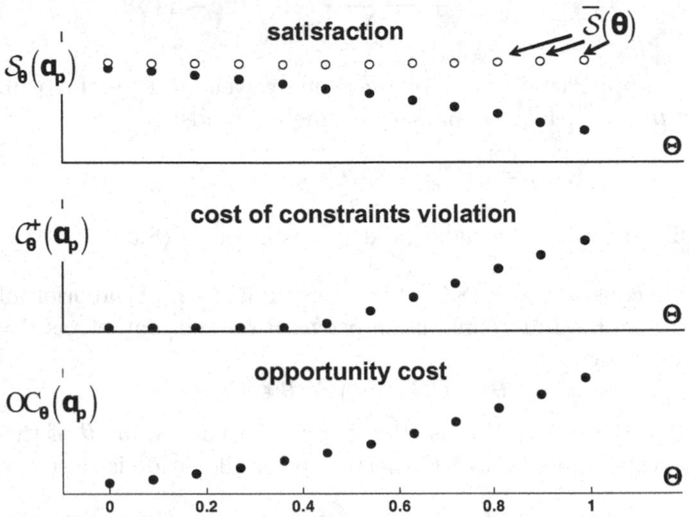

Fig. 8.3. Prior allocation: evaluation

8.2.3 Discussion

In general the opportunity cost of a prior allocation is large. The reason why the prior allocation decision is sub-optimal is quite obvious: just like the hands of a broken watch, which happen to correctly indicate the time only twice a day, the prior allocation is only good in those markets, if any, where the optimal allocation happens to be close to the prior allocation.

Notice the resemblance of this situation with the failure of the "fixed" estimator (4.32). Indeed, like in the case of the fixed estimator, the prior allocation is extremely efficient, meaning that the loss, namely the opportunity cost (8.67), is a deterministic variable, instead of a random variable like in the general case (8.53). Nevertheless, since the information on the market is disregarded, the prior allocation does not track the market parameters $\boldsymbol{\theta}$ as these vary in the stress test set $\boldsymbol{\Theta}$, see Figure 8.3. As a result, in the language of estimators the prior allocation is extremely biased.

8.3 Sample-based allocation

In this section we discuss the most intuitive approach to allocation, namely the sample-based allocation decision. This decision consists in replacing the true unknown value of the market parameters with estimates in the optimal allocation function.

We evaluate the sample-based allocation decision by computing its opportunity cost along the guidelines discussed in Section 8.1. Since the opportunity cost is caused by the error in the estimation of the market parameters, in this context the opportunity cost is called *estimation risk*.

As it turns out, the large estimation risk of sample-based allocation decisions is due to the extreme sensitivity of the optimal allocation function to the input market parameters: in other words, the optimization process leverages the estimation error already present in the estimates of the market parameters, see also Jobson and Korkie (1980), Best and Grauer (1991), Green and Hollifield (1992), Chopra and Ziemba (1993) and Britten-Jones (1999).

8.3.1 Definition

Consider the optimal allocation function (8.30):

$$\boldsymbol{\alpha}(\boldsymbol{\theta}) \equiv \underset{\boldsymbol{\alpha} \in \mathcal{C}_{\boldsymbol{\theta}}}{\operatorname{argmax}} \left\{ \mathcal{S}_{\boldsymbol{\theta}}(\boldsymbol{\alpha}) \right\}. \tag{8.76}$$

The truly optimal allocation (8.39) cannot be implemented because it relies on knowledge of the true market parameters $\boldsymbol{\theta}^{\mathrm{t}}$, which are unknown.

In our leading example the optimal allocation function is (8.32):

$$\boldsymbol{\alpha}(\boldsymbol{\mu}, \boldsymbol{\Sigma}) = \left[\operatorname{diag}(\mathbf{p}_T)\right]^{-1} \boldsymbol{\Sigma}^{-1} \left(\zeta \boldsymbol{\mu} + \frac{w_T - \zeta \mathbf{1}' \boldsymbol{\Sigma}^{-1} \boldsymbol{\mu}}{\mathbf{1}' \boldsymbol{\Sigma}^{-1} \mathbf{1}} \mathbf{1} \right). \tag{8.77}$$

The market parameters $\boldsymbol{\mu}$ and $\boldsymbol{\Sigma}$ are unknown.

Nevertheless, these parameters can be estimated by means of an estimator $\widehat{\boldsymbol{\theta}}$ that processes the information available in the market i_T as described in Chapter 4:

$$\widehat{\boldsymbol{\theta}}[i_T] \approx \boldsymbol{\theta}^{\mathrm{t}}. \tag{8.78}$$

In our leading example, from the time series of the past observations of the non-overlapping linear returns (8.41) we can estimate the parameters $\boldsymbol{\mu}$ and $\boldsymbol{\Sigma}$ that determine the distribution of the market (8.19). For instance, we can estimate these parameters by means of the sample mean (4.98), which in this context reads:

$$\widehat{\boldsymbol{\mu}}[i_T] \equiv \frac{1}{T} \sum_{t=1}^{T} \mathbf{l}_t; \tag{8.79}$$

and sample covariance (4.99), which in this context reads:

$$\widehat{\Sigma}[i_T] \equiv \frac{1}{T}\sum_{t=1}^{T}(\mathbf{l}_t - \widehat{\boldsymbol{\mu}})(\mathbf{l}_t - \widehat{\boldsymbol{\mu}})'. \tag{8.80}$$

It is intuitive to replace the unknown market parameters $\boldsymbol{\theta}^{\mathrm{t}}$ that should ideally feed the optimal allocation function (8.76) with their estimates (8.78). This way we obtain the *sample-based allocation decision*:

$$\begin{aligned}\boldsymbol{\alpha}_{\mathrm{s}}[i_T] &\equiv \boldsymbol{\alpha}\left(\widehat{\boldsymbol{\theta}}[i_T]\right) \qquad (8.81)\\ &\equiv \underset{\boldsymbol{\alpha}\in\mathcal{C}_{\widehat{\boldsymbol{\theta}}[i_T]}}{\operatorname{argmax}}\left\{\mathcal{S}_{\widehat{\boldsymbol{\theta}}[i_T]}(\boldsymbol{\alpha})\right\}.\end{aligned}$$

We stress that, unlike the truly optimal allocation (8.39) which cannot be implemented, the sample-based allocation decision is indeed a decision and thus it can be implemented. In other words, the sample-based allocation decision processes the information available on the market at the time the investment decision is made, i.e. it is of the general form (8.38).

In our leading example, the sample-based allocation follows from replacing (8.79) and (8.80) in (8.77) and reads:

$$\boldsymbol{\alpha}_{\mathrm{s}} = [\operatorname{diag}(\mathbf{p}_T)]^{-1}\,\widehat{\Sigma}^{-1}\left(\zeta\widehat{\boldsymbol{\mu}} + \frac{w_T - \zeta\mathbf{1}'\widehat{\Sigma}^{-1}\widehat{\boldsymbol{\mu}}}{\mathbf{1}'\widehat{\Sigma}^{-1}\mathbf{1}}\mathbf{1}\right). \tag{8.82}$$

8.3.2 Evaluation

In order to evaluate the sample-based allocation we proceed as in Section 8.1.

First we consider a set $\boldsymbol{\Theta}$ of market parameters that is broad enough to most likely include the true, unknown value $\boldsymbol{\theta}^{\mathrm{t}}$.

For each value $\boldsymbol{\theta}$ of the market parameters in the stress test set $\boldsymbol{\Theta}$ we compute the optimal allocation function $\boldsymbol{\alpha}(\boldsymbol{\theta})$, see (8.76).

Then we compute as in (8.31) the optimal level of satisfaction if $\boldsymbol{\theta}$ are the underlying market parameters, namely $\overline{\mathcal{S}}(\boldsymbol{\theta})$.

In our leading example the optimal allocation (8.77) provides the optimal level of satisfaction (8.33).

Then, as in (8.48), for each value $\boldsymbol{\theta}$ of the market parameters in the stress test set $\boldsymbol{\Theta}$ we randomize the information from the market i_T, generating a distribution of information scenarios $I_T^{\boldsymbol{\theta}}$ that depends on the assumption $\boldsymbol{\theta}$ on the market parameters:

$$I_T^{\boldsymbol{\theta}} \equiv \left\{ \mathbf{X}_1^{\boldsymbol{\theta}}, \ldots, \mathbf{X}_T^{\boldsymbol{\theta}} \right\}. \tag{8.83}$$

By applying the estimator $\widehat{\boldsymbol{\theta}}$ to the different information scenarios (8.83) instead of the specific realization i_T as in (8.78) we obtain a random variable:

$$\widehat{\boldsymbol{\theta}}\left[i_T\right] \mapsto \widehat{\boldsymbol{\theta}}\left[I_T^{\boldsymbol{\theta}}\right]. \tag{8.84}$$

We stress that the distribution of this random variable is determined by the underlying assumption $\boldsymbol{\theta}$ on market parameters.

In our leading example, we replace i_T, i.e. the specific observations of the past linear returns (8.41), with the set $I_T^{\boldsymbol{\mu},\boldsymbol{\Sigma}}$ of independent and identically distributed variables (8.50). This way the estimators (8.79) and (8.80) become random variables, whose distribution follows from (4.102) and (4.103) respectively:

$$\widehat{\boldsymbol{\mu}}\left[I_T^{\boldsymbol{\mu},\boldsymbol{\Sigma}}\right] \sim \mathrm{N}\left(\boldsymbol{\mu}, \frac{\boldsymbol{\Sigma}}{T}\right) \tag{8.85}$$

$$T\widehat{\boldsymbol{\Sigma}}\left[I_T^{\boldsymbol{\mu},\boldsymbol{\Sigma}}\right] \sim \mathrm{W}\left(T-1, \boldsymbol{\Sigma}\right), \tag{8.86}$$

where the two random variables $\widehat{\boldsymbol{\mu}}$ and $\widehat{\boldsymbol{\Sigma}}$ are independent.

In turn, the sample-based allocation decision (8.81) in the different information scenarios yields a random variable whose distribution depends on the underlying market parameters:

$$\begin{aligned} \boldsymbol{\alpha}_{\mathrm{s}}\left[I_T^{\boldsymbol{\theta}}\right] &\equiv \boldsymbol{\alpha}\left(\widehat{\boldsymbol{\theta}}\left[I_T^{\boldsymbol{\theta}}\right]\right) \\ &\equiv \underset{\boldsymbol{\alpha} \in \mathcal{C}_{\widehat{\boldsymbol{\theta}}\left[I_T^{\boldsymbol{\theta}}\right]}}{\operatorname{argmax}} \left\{ \mathcal{S}_{\widehat{\boldsymbol{\theta}}\left[I_T^{\boldsymbol{\theta}}\right]}\left(\boldsymbol{\alpha}\right) \right\}. \end{aligned} \tag{8.87}$$

This step corresponds to (8.49).

In our example, the distribution of the sample-based allocation (8.82) under the assumptions (8.85) and (8.86) is not known analytically but we can easily compute it numerically. We generate a large number J of Monte Carlo scenarios from (8.85) and (8.86), which are independent of each other:

$${}_j\widehat{\boldsymbol{\mu}}^{\boldsymbol{\mu},\boldsymbol{\Sigma}}, \quad {}_j\widehat{\boldsymbol{\Sigma}}^{\boldsymbol{\mu},\boldsymbol{\Sigma}}, \qquad j = 1, \ldots, J. \tag{8.88}$$

Then we compute the respective sample-based allocation (8.82) in each of these scenarios:

$$\begin{aligned} {}_j\boldsymbol{\alpha}_{\mathrm{s}}^{\boldsymbol{\mu},\boldsymbol{\Sigma}} &\equiv \zeta \left[\operatorname{diag}\left(\mathbf{p}_T\right)\right]^{-1} {}_j\widehat{\boldsymbol{\Sigma}}^{-1} {}_j\widehat{\boldsymbol{\mu}}. \\ &+ \frac{w_T - \zeta \mathbf{1}' {}_j\widehat{\boldsymbol{\Sigma}}^{-1} {}_j\widehat{\boldsymbol{\mu}}}{\mathbf{1}' {}_j\widehat{\boldsymbol{\Sigma}}^{-1}\mathbf{1}} \left[\operatorname{diag}\left(\mathbf{p}_T\right)\right]^{-1} {}_j\widehat{\boldsymbol{\Sigma}}^{-1}\mathbf{1}. \end{aligned} \tag{8.89}$$

Notice that the allocations generated this way depend on the underlying parameters $\boldsymbol{\mu}$ and $\boldsymbol{\Sigma}$ through the sample estimators (8.88).

Next we compute as in (8.23) the satisfaction $\mathcal{S}_{\boldsymbol{\theta}}\left(\boldsymbol{\alpha}_{\mathrm{s}}\left[I_T^{\boldsymbol{\theta}}\right]\right)$ ensuing from each scenario of the sample-based allocation decision (8.87) under the assumption $\boldsymbol{\theta}$ for the market parameters, which, we recall, is a random variable.

Similarly, from (8.26) and expressions such as (8.35) we compute the cost of the sample-based allocation decision violating the constraints $\mathcal{C}_{\boldsymbol{\theta}}^{+}\left(\boldsymbol{\alpha}_{\mathrm{s}}\left[I_T^{\boldsymbol{\theta}}\right]\right)$ in each scenario under the assumption $\boldsymbol{\theta}$ for the market parameters, which is also a random variable.

In our example we compute according to (8.25) the satisfaction ensuing from each Monte Carlo scenario (8.89) of the sample-based allocation:

$$\mathrm{CE}_{\boldsymbol{\mu},\boldsymbol{\Sigma}}\left(\boldsymbol{\alpha}_{\mathrm{s}}\left[I_T^{\boldsymbol{\mu},\boldsymbol{\Sigma}}\right]\right) \approx \mathrm{CE}_{\boldsymbol{\mu},\boldsymbol{\Sigma}}\left({}_j\boldsymbol{\alpha}_{\mathrm{s}}^{\boldsymbol{\mu},\boldsymbol{\Sigma}}\right), \quad j = 1 \ldots J. \tag{8.90}$$

The respective histogram represents the numerical probability density function of the satisfaction from the sample-based allocation.

Similarly we compute according to (8.36) the cost of violating the value at risk constraint ensuing from each Monte Carlo scenario (8.89) of the sample-based allocation:

$$\mathcal{C}_{\boldsymbol{\mu},\boldsymbol{\Sigma}}^{+}\left(\boldsymbol{\alpha}_{\mathrm{s}}\left[I_T^{\boldsymbol{\mu},\boldsymbol{\Sigma}}\right]\right) \approx \mathcal{C}_{\boldsymbol{\mu},\boldsymbol{\Sigma}}^{+}\left({}_j\boldsymbol{\alpha}_{\mathrm{s}}^{\boldsymbol{\mu},\boldsymbol{\Sigma}}\right), \quad j = 1 \ldots J. \tag{8.91}$$

The respective histogram represents the numerical probability density function of the cost of the sample-based allocation violating the VaR constraint.

Then we compute the opportunity cost (8.53) of the sample-based allocation under the assumption $\boldsymbol{\theta}$ for the market parameters, which is the difference between the satisfaction from the unattainable optimal allocation and the satisfaction from the sample-based allocation, plus the cost of the sample-based allocation violating the constraints:

$$\mathrm{OC}_{\boldsymbol{\theta}}\left(\boldsymbol{\alpha}_{\mathrm{s}}\left[I_T^{\boldsymbol{\theta}}\right]\right) \equiv \overline{\mathcal{S}}(\boldsymbol{\theta}) - \mathcal{S}_{\boldsymbol{\theta}}\left(\boldsymbol{\alpha}_{\mathrm{s}}\left[I_T^{\boldsymbol{\theta}}\right]\right) + \mathcal{C}_{\boldsymbol{\theta}}^{+}\left(\boldsymbol{\alpha}_{\mathrm{s}}\left[I_T^{\boldsymbol{\theta}}\right]\right). \tag{8.92}$$

We stress that the opportunity cost is a general concept: whenever the investor misses the optimal, unattainable allocation he is exposed to a loss. When the sub-optimality of his allocation decision is due to the error in the estimates of the underlying market parameters, like in the case of the sample-based allocation, the loss, or the opportunity cost, is called *estimation risk*.

Finally, as in (8.57) we let the market parameters $\boldsymbol{\theta}$ vary in the stress test range $\boldsymbol{\Theta}$, analyzing the opportunity cost of the sample-based strategy as a function of the underlying market parameters:

$$\boldsymbol{\theta} \mapsto \mathrm{OC}_{\boldsymbol{\theta}}\left(\boldsymbol{\alpha}_{\mathrm{s}}\left[I_T^{\boldsymbol{\theta}}\right]\right), \quad \boldsymbol{\theta} \in \boldsymbol{\Theta}, \tag{8.93}$$

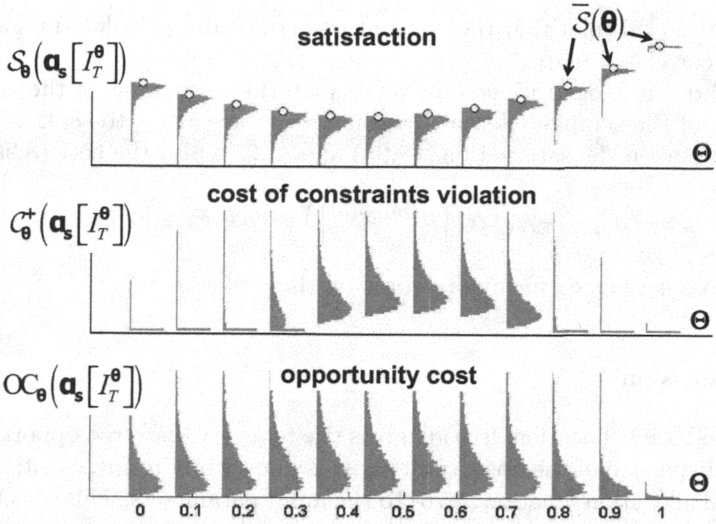

Fig. 8.4. Sample-based allocation: evaluation

see Figure 8.4.

If the distribution of the opportunity cost (8.93) is tightly peaked around a positive value very close to zero for all the markets $\boldsymbol{\theta}$ in the stress test range $\boldsymbol{\Theta}$, in particular it is close to zero in all the scenarios in correspondence of the true, yet unknown, value $\boldsymbol{\theta}^{\mathrm{t}}$. In this case the sample-based allocation decision is guaranteed to perform well and is close to optimal.

In order to display the results in our leading example we let the underlying market parameters vary according to (8.58)-(8.59), obtaining a one-parameter family of markets, parameterized by the overall level of correlation ρ.

In the top plot in Figure 8.4 we display the unattainable maximum satisfaction (8.33) as a function of the overall correlation:

$$\rho \mapsto \overline{\mathrm{CE}}\left(\boldsymbol{\mu}\left(\rho\right), \boldsymbol{\Sigma}\left(\rho\right)\right), \quad \rho \in \boldsymbol{\Theta} \equiv [0, 1) . \tag{8.94}$$

In the same plot we display the histograms of the satisfaction (8.90) from the sample-based allocation:

$$\rho \mapsto \mathrm{CE}_{\boldsymbol{\mu}(\rho), \boldsymbol{\Sigma}(\rho)}\left(\boldsymbol{\alpha}_{\mathrm{s}}\left[I_T^{\boldsymbol{\mu}(\rho), \boldsymbol{\Sigma}(\rho)}\right]\right), \quad \rho \in \boldsymbol{\Theta} \equiv [0, 1) . \tag{8.95}$$

In the plot in middle of Figure 8.4 we display the histograms of the cost (8.91) of violating the value at risk constraint:

$$\rho \mapsto \mathcal{C}^{+}_{\boldsymbol{\mu}(\rho), \boldsymbol{\Sigma}(\rho)}\left(\boldsymbol{\alpha}_{\mathrm{s}}\left[I_T^{\boldsymbol{\mu}(\rho), \boldsymbol{\Sigma}(\rho)}\right]\right), \quad \rho \in \boldsymbol{\Theta} \equiv [0, 1) . \tag{8.96}$$

We notice from this plot that the value at risk constraint is violated regularly in slightly correlated markets.

In the bottom plot in Figure 8.4 we display the histograms of the opportunity cost of the sample-based allocation, which, according to (8.92), is the difference between the satisfactions (8.94) and (8.95), plus the cost (8.96):

$$\rho \mapsto \mathrm{OC}_{\boldsymbol{\mu}(\rho),\boldsymbol{\Sigma}(\rho)}\left(\boldsymbol{\alpha}_{\mathrm{s}}\left[I_T^{\boldsymbol{\mu}(\rho),\boldsymbol{\Sigma}(\rho)}\right]\right), \quad \rho \in \boldsymbol{\Theta} \equiv [0,1). \tag{8.97}$$

Refer to `symmys.com` for more details on these plots.

8.3.3 Discussion

The sample-based allocation decision gives rise to a very scattered opportunity cost. The dispersion of the opportunity cost is due mainly to the sensitivity of the optimal allocation function (8.76) to the input parameters. This sensitivity gives rise to a leveraged propagation of the estimation error, as we proceed to discuss.

In the first place, the scenario-dependent estimates $\widehat{\boldsymbol{\theta}}\left[I_T^{\boldsymbol{\theta}}\right]$ provided by sample-based estimators are in general quite dispersed around the underlying market parameter $\boldsymbol{\theta}$. In other words, sample-based estimators are quite inefficient.

In our example the distribution of the estimator is given in (8.85) and (8.86). These estimates are very disperse when the number of observations T in the sample is low, see (4.109) and (4.119).

In the second place, the inefficiency of the estimators propagates into the estimates of the investor's satisfaction $\mathcal{S}_{\widehat{\boldsymbol{\theta}}}$ and of the constraints $\mathcal{C}_{\widehat{\boldsymbol{\theta}}}$ that appear in the definition of the sample-based allocation (8.87).

In our example, two variables fully determine the investor's satisfaction (8.25) and the cost of constraint violation (8.36), namely:

$$v \equiv \boldsymbol{\alpha}' \operatorname{diag}(\mathbf{p}_T)\, \boldsymbol{\Sigma} \operatorname{diag}(\mathbf{p}_T)\, \boldsymbol{\alpha} \tag{8.98}$$

$$e \equiv \boldsymbol{\alpha}' \operatorname{diag}(\mathbf{p}_T)\,(\mathbf{1}+\boldsymbol{\mu}). \tag{8.99}$$

The natural estimators of these variables in terms of the estimators (8.85) and (8.86) read:

$$\widehat{v} \equiv \boldsymbol{\alpha}' \operatorname{diag}(\mathbf{p}_T)\, \widehat{\boldsymbol{\Sigma}} \operatorname{diag}(\mathbf{p}_T)\, \boldsymbol{\alpha} \tag{8.100}$$

$$\widehat{e} \equiv \boldsymbol{\alpha}' \operatorname{diag}(\mathbf{p}_T)\,(\mathbf{1}+\widehat{\boldsymbol{\mu}}). \tag{8.101}$$

In Appendix www.8.2 we show that the distributions of the estimators (8.100) and (8.101) read respectively:

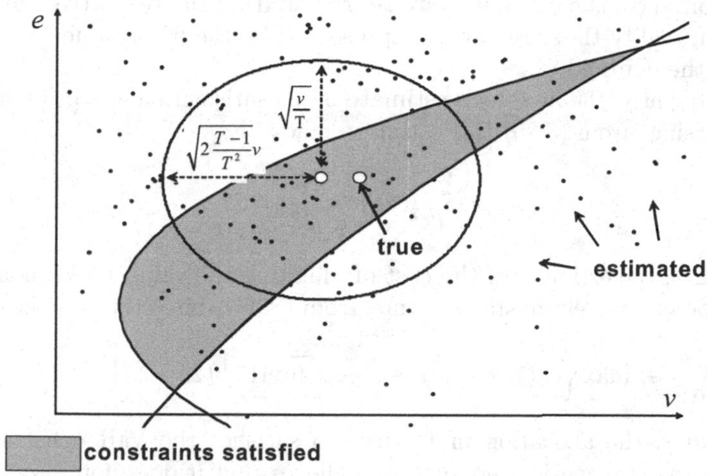

Fig. 8.5. Sample-based allocation: error in satisfaction and constraints assessment

$$\widehat{e} \sim \mathrm{N}\left(e, \frac{v}{T}\right), \quad T\widehat{v} \sim \mathrm{Ga}(T-1, v). \tag{8.102}$$

To gain insight into the main joint properties of $\widehat{v}$ and $\widehat{e}$, which fully determine the quantities of interest to the investor, we consider the location-dispersion ellipsoid of $(\widehat{v}, \widehat{e})$ in the plane of coordinates (v, e), see Figure 8.5 and refer to Figure 8.1. Also refer to Section 2.4.3 for a thorough discussion of the location-dispersion ellipsoid in a general context and to Appendix www.8.2 for a proof of the results that follow.

The center of the location-dispersion ellipsoid of (8.100)-(8.101) reads:

$$\mathrm{E}\{\widehat{v}\} = \frac{T-1}{T}v, \quad \mathrm{E}\{\widehat{e}\} = e. \tag{8.103}$$

In other words, there exists a bias that disappears as the number of observations grows. Since $\widehat{v}$ and $\widehat{e}$ are independent, the principal axes of their location-dispersion ellipsoid are aligned with the reference axes.

The semi-lengths of the two principal axes of the location-dispersion ellipsoid of (8.100)-(8.101), which represent the standard deviations of each estimator respectively, read:

$$\mathrm{Sd}\{\widehat{v}\} = \sqrt{2\frac{T-1}{T^2}v}, \quad \mathrm{Sd}\{\widehat{e}\} = \sqrt{\frac{v}{T}}. \tag{8.104}$$

In Figure 8.5 we plot the location-dispersion ellipsoid along with several possible outcomes (small dots) of the estimation process. In each scenario the investor estimates that the variables v and e, which fully determine his satisfaction and his constraints, are represented by the respective small dot, whereas in reality they are always represented by the fixed value close to the center of the ellipsoid.

Consequently, the investor's estimate of his satisfaction can be completely mistaken, since from (8.25) this estimate reads:

$$\mathcal{S}_{\widehat{\mu},\widehat{\Sigma}} \equiv \widehat{e} - \frac{\widehat{v}}{2\zeta}. \tag{8.105}$$

Similarly, the estimate of the cost of violating the value at risk constraint can also be completely mistaken, since from (8.36) this estimate reads:

$$\mathcal{C}^{+}_{\widehat{\mu},\widehat{\Sigma}} \equiv \max\left\{0, (1-\gamma)\, w_T - \widehat{e} + \sqrt{2\widehat{v}}\, \mathrm{erf}^{-1}(2c-1)\right\}. \tag{8.106}$$

In particular, the allocation in Figure 8.5 satisfies the VaR constraint, although in many scenarios the investor believes that it does not.

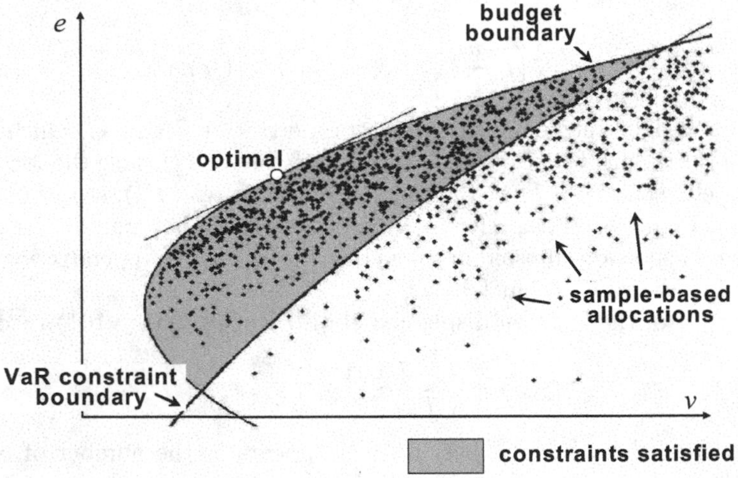

Fig. 8.6. Sample-based allocation: leverage of estimation error

Finally, the optimal allocation function is extremely sensitive to the value of the market parameters. In other words, the maximization in (8.87) leverages the dispersion of the estimates of satisfaction and constraints.

In our example the solution ${}_j\boldsymbol{\alpha}_{\mathrm{s}}$ defined in (8.89) of the allocation optimization problem in the j-th Monte Carlo scenario involves the inverse of the sample covariance matrix ${}_j\widehat{\boldsymbol{\Sigma}}$ of the linear returns.

Consider as in (4.148) the PCA decomposition of the true covariance matrix and of its sample estimator in each of the J Monte Carlo scenarios:

$$\boldsymbol{\Sigma} \equiv \mathbf{E}\boldsymbol{\Lambda}\mathbf{E}', \quad {}_j\widehat{\boldsymbol{\Sigma}} \equiv {}_j\widehat{\mathbf{E}}_j\widehat{\boldsymbol{\Lambda}}_j\widehat{\mathbf{E}}'. \tag{8.107}$$

In this expression $\boldsymbol{\Lambda}$ is the diagonal matrix of the eigenvalues sorted in decreasing order:

$$\boldsymbol{\Lambda} \equiv \operatorname{diag}(\lambda_1, \ldots, \lambda_N); \tag{8.108}$$

the matrix $\mathbf{E}$ is the juxtaposition of the respective normalized eigenvectors; and the same notation holds for all the sample ("hat") counterparts.

The sample estimator of the covariance matrix tends to push the lowest eigenvalues of the sample covariance matrix toward zero, see Figure 4.15. Therefore the inverse of the sample covariance matrix displays a small-denominator effect:

$${}_j\widehat{\boldsymbol{\Sigma}}^{-1} = {}_j\widehat{\mathbf{E}} \operatorname{diag}\left(\frac{1}{{}_j\widehat{\lambda}_1}, \ldots, \frac{1}{{}_j\widehat{\lambda}_N}\right) {}_j\widehat{\mathbf{E}}'. \tag{8.109}$$

These small denominators push the inverse matrix (8.109) toward infinity. As a consequence, the ensuing allocations ${}_j\boldsymbol{\alpha}_{\mathrm{s}}$ become both very extreme and very sensitive.

In turn, the above extreme allocations ${}_j\boldsymbol{\alpha}_{\mathrm{s}}$ give rise to very poor levels of satisfaction and badly violate the constraints. Indeed, consider the true coordinates (8.98) and (8.99) (*not* the estimated coordinates (8.100) and (8.101)) of the sample-based allocations in the j-th Monte-Carlo scenario:

$$\begin{aligned} {}_jv &\equiv {}_j\boldsymbol{\alpha}_{\mathrm{s}}' \operatorname{diag}(\mathbf{p}_T)\, \boldsymbol{\Sigma} \operatorname{diag}(\mathbf{p}_T)\, {}_j\boldsymbol{\alpha}_{\mathrm{s}} & (8.110)\\ {}_je &\equiv {}_j\boldsymbol{\alpha}_{\mathrm{s}}' \operatorname{diag}(\mathbf{p}_T)(\mathbf{1} + \boldsymbol{\mu}). & (8.111) \end{aligned}$$

In Figure 8.6 we plot the coordinates (8.110) and (8.111) obtained in the Monte Carlo scenarios, also refer to Figure 8.1.

From (8.25) the investor's satisfaction from the generic allocation ${}_j\boldsymbol{\alpha}_{\mathrm{s}}$ in the j-th scenario is completely determined by the coordinates (8.110) and (8.111):

$$\mathrm{CE}({}_j\boldsymbol{\alpha}_{\mathrm{s}}) = {}_je - \frac{{}_jv}{2\zeta}. \tag{8.112}$$

Similarly, we see from (8.36) that these coordinates also determine the cost of violating the value at risk constraint:

$$\mathcal{C}^+({}_j\boldsymbol{\alpha}_{\mathrm{s}}) = \max\left\{0, (1-\gamma)\, w_T - {}_je + \sqrt{2\,{}_jv}\, \operatorname{erf}^{-1}(2c-1)\right\}. \tag{8.113}$$

The sample-based allocation satisfies the budget constraint: therefore all the allocations lie in suboptimal positions within the budget-constraint boundary. Nevertheless, the value at risk constraint is not satisfied in many scenarios. We see from Figure 8.4 that the situation is not exceptional, as the VaR constraint is violated regularly for a wide range of market parameters.

For the allocations that satisfy the VaR constraint the opportunity cost, or estimation risk, is the vertical distance between the allocation's iso satisfaction line and the optimal iso-satisfaction line as in Figure 8.1.

For the allocations that do not satisfy the VaR constraint, the cost of violating the VaR constraint kicks in, and the opportunity cost becomes the vertical distance between the allocation's iso-satisfaction line and the optimal iso-satisfaction line, plus the term (8.113).

The opportunity cost associated with a generic allocation decision can be interpreted as a loss in the context of estimators, see (8.53).

Unlike the prior allocation, which disregards the information available on the market, the sample-based allocation processes that information. In particular, the sample-based allocation tracks the market parameters $\boldsymbol{\theta}$ through the estimator $\widehat{\boldsymbol{\theta}}$ as these vary in the stress test range. Therefore the center of the distribution of the opportunity cost of the sample-based allocation is quite close to zero for all the values of the market parameters in the stress test range, see Figure 8.4 and compare with Figure 8.3: in the language of estimators, the sample-based allocation decision is not too biased.

On the other hand, the extreme sensitivity of the allocation optimization process to the market parameters leverages the estimation error of the estimator $\widehat{\boldsymbol{\theta}}$, making the distribution of the opportunity cost very disperse: in the language of estimators, the sample-based allocation decision is very inefficient.

We stress that the above remarks depend on the choice of the estimator $\widehat{\boldsymbol{\theta}}$ chosen in (8.78) to estimate the market parameters. For instance, we can lower the inefficiency of the sample-based allocation decision by using shrinkage estimators, refer to Section 4.4. Indeed, in the extreme case where the estimator is fully shrunk toward the shrinkage target, the ensuing sample-based allocation degenerates into a prior allocation: as discussed in Section 8.2, the prior allocation is extremely efficient.

We revisit "shrinkage" allocation decisions in a more general Bayesian context in Chapter 9.

9

Optimizing allocations

The classical approach to allocation optimization discussed in the second part of the book assumes that the distribution of the market is known. The sample-based allocation, discussed in the previous chapter, is a two-step process: first the market distribution is estimated and then the estimate is inputted in the classical allocation optimization problem. Since this process leverages the estimation error, portfolio managers, traders and professional investors in a broader sense mistrust these two-step "optimal" approaches and prefer to resort to ad-hoc recipes, or trust their prior knowledge/experience.

In this chapter we discuss allocation strategies that account for estimation risk within the allocation decision process. These strategies must be optimal according to the evaluation criteria introduced in the previous chapter: in other words, the overall opportunity cost of these strategies must be as low as possible.

The main reasons why estimation risk plays such an important role in financial applications is the extreme sensitivity of the optimal allocation function to the unknown parameters that determine the distribution of the market. In Section 9.1 we use the Bayesian approach to estimation to limit this sensitivity. We present Bayesian allocations in terms of the predictive distribution of the market, as well as the classical-equivalent Bayesian allocation, which relies on Bayes-Stein shrinkage estimators of the market parameters. The Bayesian approach provides a mechanism that mixes the positive features of the prior allocation and the sample-based allocation: the estimate of the market is shrunk towards the investor's prior in a self-adjusting way and the overall opportunity cost is reduced.

In Section 9.2 we present the Black-Litterman approach to control the extreme sensitivity of the optimal allocation function to the unknown market parameters. Like the Bayesian approach, the Black-Litterman methodology makes use of Bayes' rule. In this case the market is directly shrunk towards the investor's prior views, rather than indirectly through the market parameters. We present the theory in a general context, performing the computations explicitly in the case of normally distributed markets. Then we apply those

A. Meucci, *Risk and Asset Allocation,* Springer Finance,

results to the mean-variance framework. Finally we propose a methodology to assess and tweak the investor's prior views.

In Section 9.3 we present Michaud's resampling technique. The rationale behind this approach consists in limiting the extreme sensitivity of the optimal allocation function to the market parameters by averaging several sample-based allocations in different scenarios. After presenting the resampled allocation in both the mean-variance and in a more general setting, we discuss the advantages and the limitations of this technique.

In Section 9.4 we discuss robust allocation decisions. Rather than trying to limit the sensitivity of the optimal allocation function, the robust approach aims at determining the "best" allocation in the presence of estimation risk, according to the evaluation criteria discussed in Chapter 8. In other words, robust allocations minimize the opportunity cost over a reasonable set of potential markets. The conceptually intuitive robust approach is hard to implement in the general case. Therefore, we resort to the two-step mean-variance framework: under suitable assumptions for the investment constraints the optimal allocations solve a second-order cone programming problem: as a result, the optimal allocations can be efficiently determined numerically.

In Section 9.5 we blend the optimality properties of the robust approach with the smoothness and self-adjusting nature of the Bayesian approach. Indeed, the robust approach presents only two disadvantages: the possible markets considered in the robust optimization are defined quite arbitrarily and the investor's prior views are not taken into account. By means of the Bayesian posterior we can select naturally a notable set of markets and smoothly blend the investor's experience with the information from the market. We present first the robust Bayesian method in a general context, showing how this approach includes the previous allocation strategies as limit cases. Then we apply the general theory to the two-step mean-variance framework, discussing the self-adjusting mechanism of robust Bayesian allocations strategies.

9.1 Bayesian allocation

Consider the optimal allocation function (8.30), which for each value of the market parameters $\boldsymbol{\theta}$ maximizes the investor's satisfaction given his investment constraints:

$$\boldsymbol{\alpha}(\boldsymbol{\theta}) \equiv \underset{\boldsymbol{\alpha} \in \mathcal{C}_{\boldsymbol{\theta}}}{\operatorname{argmax}} \left\{ \mathcal{S}_{\boldsymbol{\theta}}(\boldsymbol{\alpha}) \right\}. \tag{9.1}$$

Since the true value $\boldsymbol{\theta}^{\mathrm{t}}$ of the market parameters is not known, the truly optimal allocation cannot be implemented. Furthermore, as discussed in Chapter 8, the allocation function (9.1) is extremely sensitive to the input parameters $\boldsymbol{\theta}$: a slightly wrong input can give rise to a very large opportunity cost.

In this section we use the Bayesian approach to parameter estimation to define allocation decisions whose opportunity cost is not as large.

9.1.1 Utility maximization

Expected utility has been historically the first and most prominent approach to model the investor's preferences. Therefore Bayesian theory was first applied to allocation problems in the context of expected utility maximization, see Zellner and Chetty (1965), and Bawa, Brown, and Klein (1979).

We recall from Section 5.4 that in the expected utility framework the investor's index of satisfaction is modeled by the certainty-equivalent ensuing from an increasing utility function u:

$$\mathcal{S}(\boldsymbol{\alpha}) \equiv u^{-1}\left(\mathrm{E}\left\{u\left(\Psi_{\boldsymbol{\alpha}}\right)\right\}\right). \tag{9.2}$$

In this expression the investor's objective Ψ, namely absolute wealth, relative wealth, net profits, or other specifications, is a linear function of the allocation and the market vector: $\Psi \equiv \boldsymbol{\alpha}'\mathbf{M}$. The market vector $\mathbf{M}$ is a simple affine function of the market prices at the investment horizon: its distribution can be represented in terms of a probability density function $f_{\boldsymbol{\theta}}(\mathbf{m})$ which is fully determined by a set of market parameters $\boldsymbol{\theta}$.

Due to (5.99), in this context the optimal allocation function (9.1) can be expressed equivalently as follows:

$$\begin{aligned} \boldsymbol{\alpha}(\boldsymbol{\theta}) &\equiv \underset{\boldsymbol{\alpha}\in\mathcal{C}_{\boldsymbol{\theta}}}{\operatorname{argmax}}\left\{\mathrm{E}\left\{u\left(\Psi_{\boldsymbol{\alpha}}^{\boldsymbol{\theta}}\right)\right\}\right\} \\ &= \underset{\boldsymbol{\alpha}\in\mathcal{C}_{\boldsymbol{\theta}}}{\operatorname{argmax}}\left\{\int u\left(\boldsymbol{\alpha}'\mathbf{m}\right) f_{\boldsymbol{\theta}}(\mathbf{m})\, d\mathbf{m}\right\}. \end{aligned} \tag{9.3}$$

Consider an investor with exponential utility function. His expected utility reads:

$$\begin{aligned} \mathrm{E}\left\{u\left(\Psi_{\boldsymbol{\alpha}}^{\boldsymbol{\theta}}\right)\right\} &= -\mathrm{E}\left\{e^{-\frac{1}{\zeta}\boldsymbol{\alpha}'\mathbf{M}}\right\} \\ &= -\int e^{-\frac{1}{\zeta}\boldsymbol{\alpha}'\mathbf{m}} f_{\boldsymbol{\theta}}(\mathbf{m})\, d\mathbf{m} \equiv -\phi_{\boldsymbol{\theta}}\left(\frac{i}{\zeta}\boldsymbol{\alpha}\right), \end{aligned} \tag{9.4}$$

where $\phi_{\boldsymbol{\theta}}$ denotes the characteristic function of the market vector. Assume that the market is normally distributed. From (2.157) the characteristic function reads:

$$\phi_{\boldsymbol{\xi},\boldsymbol{\Phi}}(\mathbf{x}) = e^{i\boldsymbol{\xi}'\mathbf{x}-\frac{1}{2}\mathbf{x}'\boldsymbol{\Phi}\mathbf{x}}. \tag{9.5}$$

Then the allocation optimization (9.3) becomes:

$$\boldsymbol{\alpha}(\boldsymbol{\xi},\boldsymbol{\Phi}) \equiv \underset{\boldsymbol{\alpha}\in\mathcal{C}_{\boldsymbol{\xi},\boldsymbol{\Phi}}}{\operatorname{argmax}}\left\{-e^{-\frac{1}{\zeta}\left(\boldsymbol{\xi}'\boldsymbol{\alpha}-\frac{1}{2\zeta}\boldsymbol{\alpha}'\boldsymbol{\Phi}\boldsymbol{\alpha}\right)}\right\}. \tag{9.6}$$

This problem is clearly equivalent to the maximization of the certainty equivalent (8.4).

The optimal allocation function (9.3) is extremely sensitive to the unknown market parameters $\boldsymbol{\theta}$.

On the other hand, in the Bayesian framework the unknown parameters $\boldsymbol{\theta}$ are a random variable whose possible outcomes are described by the posterior probability density function $f_{\mathrm{po}}(\boldsymbol{\theta})$. Assume that the investment constraints in the allocation function (9.3) do not depend on the unknown parameters $\boldsymbol{\theta}$. In order to smoothen the sensitivity of the allocation function to the parameters it is quite natural to consider the weighted average of the argument of the optimization (9.3) over all the possible outcomes of the market parameters:

$$\overline{\boldsymbol{\alpha}} \equiv \underset{\boldsymbol{\alpha} \in \mathcal{C}}{\operatorname{argmax}} \left\{ \int \mathrm{E}\left\{ u\left(\Psi_{\boldsymbol{\alpha}}^{\boldsymbol{\theta}} \right) \right\} f_{\mathrm{po}}(\boldsymbol{\theta}) \, d\boldsymbol{\theta} \right\}. \tag{9.7}$$

The posterior distribution of the parameters depends on both the information on the market i_T and the investor's experience e_C, see (7.15). Consider the *predictive distribution* of the market, which is defined in terms of the posterior distribution of the parameters as follows:

$$f_{\mathrm{prd}}(\mathbf{m}; i_T, e_C) \equiv \int f_{\boldsymbol{\theta}}(\mathbf{m}) \, f_{\mathrm{po}}(\boldsymbol{\theta}; i_T, e_C) \, d\boldsymbol{\theta}. \tag{9.8}$$

This expression is indeed a probability density function, i.e. it satisfies (2.5) and (2.6). Like the posterior distribution of the parameters, also the predictive distribution of the market depends on both information and experience: it describes the statistical features of the market vector $\mathbf{M}$, keeping into account that the value of $\boldsymbol{\theta}$ is not known with certainty, i.e., accounting for estimation risk.

Using the definition of the predictive density in the average allocation (9.7) and exchanging the order of integration it is immediate to check that the average allocation can be written as follows:

$$\begin{aligned} \boldsymbol{\alpha}_{\mathrm{B}}[i_T, e_C] &= \underset{\boldsymbol{\alpha} \in \mathcal{C}}{\operatorname{argmax}} \left\{ \int u(\boldsymbol{\alpha}'\mathbf{m}) \, f_{\mathrm{prd}}(\mathbf{m}; i_T, e_C) \, d\mathbf{m} \right\} \\ &\equiv \underset{\boldsymbol{\alpha} \in \mathcal{C}}{\operatorname{argmax}} \left\{ \mathrm{E}\left\{ u\left(\Psi_{\boldsymbol{\alpha}}^{i_T, e_C} \right) \right\} \right\}. \end{aligned} \tag{9.9}$$

This is the *Bayesian allocation decision,* which maximizes the expected utility of the investor' objective, where the expectation is computed according to the predictive distribution of the market. In other words, the Bayesian allocation decision is the standard Von Neumann-Morgenstern optimal allocation where instead of the unknown market distribution we use its predictive distribution.

Since the predictive distribution accounts for estimation risk and includes the investor's experience, so does the Bayesian allocation decision.

Assume that in our example (9.5) the covariance $\boldsymbol{\Phi}$ is known, and that the posterior distribution of the expected value is normal:

$$\boldsymbol{\xi} \sim \mathrm{N}\left(\boldsymbol{\xi}_1[i_T, e_C], \frac{\boldsymbol{\Phi}}{T_1} \right). \tag{9.10}$$

When $\boldsymbol{\Phi}$ is known, this specification is consistent with the posterior (7.32).

We show in Appendix www.9.7 that the predictive distribution of the normal market (9.5) with the normal posterior for the parameters (9.10) is also normal:

$$\phi_{\text{prd}}(\mathbf{x}; i_T, e_C) = e^{i\mathbf{x}'\boldsymbol{\xi}_1[i_T,e_C] - \frac{1}{2}\mathbf{x}'\frac{1+T_1}{T_1}\boldsymbol{\Phi}\mathbf{x}}. \tag{9.11}$$

Therefore, from (9.4) the Bayesian allocation decision reads:

$$\boldsymbol{\alpha}_{\text{B}} \equiv \underset{\boldsymbol{\alpha}\in\mathcal{C}}{\operatorname{argmax}} \left\{ -e^{-\frac{1}{\zeta}\left(\boldsymbol{\alpha}'\boldsymbol{\xi}_1 - \frac{1+T_1}{2\zeta T_1}\boldsymbol{\alpha}'\boldsymbol{\Phi}\boldsymbol{\alpha}\right)} \right\}. \tag{9.12}$$

Allocation decisions based on the predictive distribution continue to find applications in finance, see for instance Jorion (1986). See also Pastor (2000) and Pastor and Stambaugh (2002) for applications based on explicit factor models.

9.1.2 Classical-equivalent maximization

Consider the more general case where the investment constraints in the optimal allocation function (9.1) depend on the unknown parameters $\boldsymbol{\theta}$, or the investor's satisfaction cannot be modeled by the certainty-equivalent. Then the Bayesian allocation (9.9) is not a viable option.

To generalize the Bayesian approach to this context, instead of averaging the distribution of the market by means of the predictive distribution (9.8) we average the distribution of the market parameters that feed the optimal allocation function. In other words, we replace the true unknown market parameters in (9.1) with a classical-equivalent estimator $\widehat{\boldsymbol{\theta}}_{\text{ce}}$, such as the expected value of the posterior distribution (7.5) or the mode of the posterior distribution (7.6). This way we obtain the *classical-equivalent Bayesian allocation decision*:

$$\begin{aligned} \boldsymbol{\alpha}_{\text{ce}}[i_T, e_C] &\equiv \boldsymbol{\alpha}\left(\widehat{\boldsymbol{\theta}}_{\text{ce}}[i_T, e_C]\right) \qquad (9.13) \\ &\equiv \underset{\boldsymbol{\alpha}\in\mathcal{C}_{\widehat{\boldsymbol{\theta}}_{\text{ce}}[i_T,e_C]}}{\operatorname{argmax}} \left\{ \mathcal{S}_{\widehat{\boldsymbol{\theta}}_{\text{ce}}[i_T,e_C]}(\boldsymbol{\alpha}) \right\}. \end{aligned}$$

This allocation decision depends through the classical-equivalent estimate on both the market information available i_T and the investor's experience e_C.

Consider the leading example (8.18), where we assumed that the market consists of equity-like securities for which the linear returns are market invariants:

$$\mathbf{L}_t \equiv \operatorname{diag}(\mathbf{P}_{t-\tau})^{-1}\mathbf{P}_t - \mathbf{1}. \tag{9.14}$$

We assume as in (8.19) that the linear returns are normally distributed:

$$\mathbf{L}_t|\boldsymbol{\mu},\boldsymbol{\Sigma} \sim \mathrm{N}(\boldsymbol{\mu},\boldsymbol{\Sigma}). \tag{9.15}$$

This is the multivariate normal Bayesian model (7.16).

The available information on the market is represented by the time series of the past linear returns, see (8.41). As in (7.19) this information can be summarized by the sample mean of the observed linear returns (8.79), their sample covariance (8.80) and the length of the time series:

$$i_T \equiv \left\{\widehat{\boldsymbol{\mu}}, \widehat{\boldsymbol{\Sigma}}; T\right\}. \tag{9.16}$$

As in (7.27) the investor's experience is summarized by the following parameters:

$$e_C \equiv \{\boldsymbol{\mu}_0, \boldsymbol{\Sigma}_0; T_0, \nu_0\}. \tag{9.17}$$

As in (7.20)-(7.21) the investor's experience is modeled as a normal-inverse-Wishart distribution:

$$\boldsymbol{\mu}|\boldsymbol{\Sigma} \sim \mathrm{N}\left(\boldsymbol{\mu}_0, \frac{\boldsymbol{\Sigma}}{T_0}\right), \quad \boldsymbol{\Sigma}^{-1} \sim \mathrm{W}\left(\nu_0, \frac{\boldsymbol{\Sigma}_0^{-1}}{\nu_0}\right). \tag{9.18}$$

The classical-equivalent estimators of $\boldsymbol{\mu}$ and $\boldsymbol{\Sigma}$ are (7.35) and (7.38), which we report here:

$$\widehat{\boldsymbol{\mu}}_{\text{ce}}(i_T, e_C) = \frac{T_0\boldsymbol{\mu}_0 + T\widehat{\boldsymbol{\mu}}}{T_0 + T}, \tag{9.19}$$

$$\widehat{\boldsymbol{\Sigma}}_{\text{ce}}(i_T, e_C) = \frac{1}{\nu_0 + T + N + 1}\left[\nu_0\boldsymbol{\Sigma}_0 + T\widehat{\boldsymbol{\Sigma}}\right. \tag{9.20}$$
$$\left. + \frac{(\boldsymbol{\mu}_0 - \widehat{\boldsymbol{\mu}})(\boldsymbol{\mu}_0 - \widehat{\boldsymbol{\mu}})'}{\frac{1}{T} + \frac{1}{T_0}}\right].$$

In our leading example the optimal allocation function is (8.32). Substituting the classical-equivalent estimators into the functional expression of the optimal allocation function we obtain the classical-equivalent Bayesian allocation:

$$\boldsymbol{\alpha}_{\text{ce}} \equiv [\mathrm{diag}(\mathbf{p}_T)]^{-1}\widehat{\boldsymbol{\Sigma}}_{\text{ce}}^{-1}\left(\zeta\widehat{\boldsymbol{\mu}}_{\text{ce}} + \frac{w_T - \zeta\mathbf{1}'\widehat{\boldsymbol{\Sigma}}_{\text{ce}}^{-1}\widehat{\boldsymbol{\mu}}_{\text{ce}}}{\mathbf{1}'\widehat{\boldsymbol{\Sigma}}_{\text{ce}}^{-1}\mathbf{1}}\mathbf{1}\right). \tag{9.21}$$

9.1.3 Evaluation

To evaluate the classical-equivalent Bayesian allocation we proceed as in Chapter 8, computing the distribution of the opportunity cost as the underlying market parameters $\boldsymbol{\theta}$ vary in a suitable stress test range $\boldsymbol{\Theta}$, which in this case is naturally defined as the domain of the posterior distribution.

Therefore, for each value $\boldsymbol{\theta}$ of the market parameters in the domain $\boldsymbol{\Theta}$ of the posterior distribution we compute the optimal allocation function $\boldsymbol{\alpha}(\boldsymbol{\theta})$ as defined in (9.1). Then we compute as in (8.31) the optimal level of satisfaction if $\boldsymbol{\theta}$ are the underlying market parameters, namely $\overline{\mathcal{S}}(\boldsymbol{\theta})$.

In our leading example the optimal allocation function is (8.32) and the respective optimal level of satisfaction is (8.33).

Next, for each value $\boldsymbol{\theta}$ of the market parameters in the stress test set $\boldsymbol{\Theta}$ we randomize as in (8.48) the information from the market i_T, generating a distribution of information scenarios $I_T^{\boldsymbol{\theta}}$ that depends on the assumption $\boldsymbol{\theta}$ on the market parameters. This way the classical-equivalent estimator becomes a random variable:

$$\widehat{\boldsymbol{\theta}}_{\text{ce}}\left[i_T, e_C\right] \mapsto \widehat{\boldsymbol{\theta}}_{\text{ce}}\left[I_T^{\boldsymbol{\theta}}, e_C\right]. \tag{9.22}$$

We stress that the distribution of this random variable is determined by the underlying assumption $\boldsymbol{\theta}$ on market parameters.

In our example, we replace i_T, i.e. the specific observations of the past linear returns, with a set $I_T^{\boldsymbol{\mu},\boldsymbol{\Sigma}}$ of T independent and identically distributed variables (9.15). This way the sample mean and the sample covariance become random variables distributed according to (8.85) and (8.86) respectively. As a result, the classical-equivalent estimators (9.19) and (9.20) become random variables, whose distribution can be simulated by a large number J of Monte Carlo scenarios as in (8.88):

$${}_j\widehat{\boldsymbol{\mu}}_{\text{ce}}^{\boldsymbol{\mu},\boldsymbol{\Sigma}}, \quad {}_j\widehat{\boldsymbol{\Sigma}}_{\text{ce}}^{\boldsymbol{\mu},\boldsymbol{\Sigma}}, \qquad j = 1, \ldots, J. \tag{9.23}$$

Notice that this distribution depends on the assumption $(\boldsymbol{\mu}, \boldsymbol{\Sigma})$ on the market parameters.

In turn, the classical-equivalent Bayesian allocation decision (9.13) yields a random variable whose distribution depends on the underlying market parameters:

$$\boldsymbol{\alpha}_{\text{ce}}\left[I_T^{\boldsymbol{\theta}}, e_C\right] \equiv \boldsymbol{\alpha}\left(\widehat{\boldsymbol{\theta}}_{\text{ce}}\left[I_T^{\boldsymbol{\theta}}, e_C\right]\right). \tag{9.24}$$

In our example we substitute (9.23) in (9.21), obtaining J allocations ${}_j\boldsymbol{\alpha}_{\text{ce}}^{\boldsymbol{\mu},\boldsymbol{\Sigma}}$.

Next we compute as in (8.23) the satisfaction $\mathcal{S}_{\boldsymbol{\theta}}\left(\boldsymbol{\alpha}_{\text{ce}}\left[I_T^{\boldsymbol{\theta}}, e_C\right]\right)$ ensuing from each scenario of the classical-equivalent Bayesian allocation decision (9.24) under the assumption $\boldsymbol{\theta}$ for the market parameters, which, we recall, is a random variable. Similarly, from (8.26) and expressions such as (8.35) we compute the cost of the classical-equivalent Bayesian allocation decision violating the constraints $\mathcal{C}_{\boldsymbol{\theta}}^{+}\left(\boldsymbol{\alpha}_{\text{ce}}\left[I_T^{\boldsymbol{\theta}}, e_C\right]\right)$ in each scenario, which is also a random variable.

In our example we proceed as in (8.90)-(8.91).

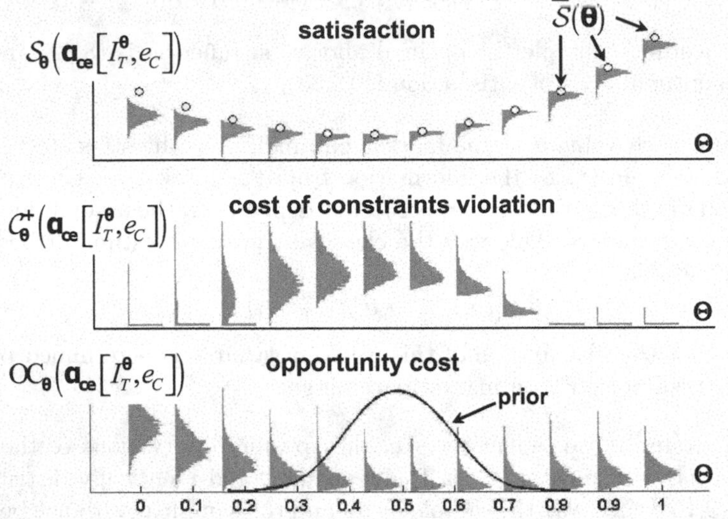

Fig. 9.1. Bayesian classical-equivalent allocation: evaluation

Then we compute the opportunity cost (8.53) of the classical-equivalent Bayesian allocation under the assumption $\boldsymbol{\theta}$ for the market parameters, which is the difference between the satisfaction from the unattainable optimal allocation and the satisfaction from the classical-equivalent Bayesian allocation, plus the cost of the classical-equivalent Bayesian allocation violating the constraints:

$$\begin{aligned} \mathrm{OC}_{\boldsymbol{\theta}}\left(\boldsymbol{\alpha}_{\mathrm{ce}}\left[I_T^{\boldsymbol{\theta}}, e_C\right]\right) \equiv \overline{\mathcal{S}}(\boldsymbol{\theta}) - \mathcal{S}_{\boldsymbol{\theta}}\left(\boldsymbol{\alpha}_{\mathrm{ce}}\left[I_T^{\boldsymbol{\theta}}, e_C\right]\right) \\ + \mathcal{C}_{\boldsymbol{\theta}}^{+}\left(\boldsymbol{\alpha}_{\mathrm{ce}}\left[I_T^{\boldsymbol{\theta}}, e_C\right]\right). \end{aligned} \tag{9.25}$$

Finally, as in (8.57) we let the market parameters $\boldsymbol{\theta}$ vary in the stress test range $\boldsymbol{\Theta}$, analyzing the opportunity cost of the classical-equivalent Bayesian allocation as a function of the underlying market parameters:

$$\boldsymbol{\theta} \mapsto \mathrm{OC}_{\boldsymbol{\theta}}\left(\boldsymbol{\alpha}_{\mathrm{ce}}\left[I_T^{\boldsymbol{\theta}}, e_C\right]\right). \tag{9.26}$$

If the distribution of the opportunity cost (9.26) is tightly peaked around a positive value very close to zero for all the markets $\boldsymbol{\theta}$ in the stress test range $\boldsymbol{\Theta}$, in particular it is close to zero in all the scenarios in correspondence of the true, yet unknown, value $\boldsymbol{\theta}^{\mathrm{t}}$. In this case the classical-equivalent Bayesian allocation decision is guaranteed to perform well and is close to optimal.

In our example we proceed as in (8.94)-(8.97), see Figure 9.1 and compare with Figure 8.4. Refer to symmys.com for more details on these plots.

9.1.4 Discussion

As discussed in Section 7.1.2, due to (7.4) the classical-equivalent estimator is a shrinkage estimator of the market parameters. Indeed it is a Bayes-Stein shrinkage estimator, where the shrinkage target is represented by the investor's prior experience $\boldsymbol{\theta}_0$. When the information available in the market is much larger than the investor's confidence in his experience, i.e. $T \gg C$, the classical-equivalent estimator converges to the sample estimate $\widehat{\boldsymbol{\theta}}$. On the other hand, when the investor's confidence in his experience is much larger than the information from the market, i.e. $C \gg T$, the classical-equivalent estimator shrinks to the prior $\boldsymbol{\theta}_0$.

Therefore, when $T \gg C$, the classical-equivalent Bayesian allocation (9.13) tends to the sample-based allocation (8.81). On the other hand, when $C \gg T$, the classical-equivalent Bayesian allocation tends to the prior allocation (8.64) which is fully determined by the prior parameters inputted by the investor and completely disregards the information from the market.

In the general case, the classical-equivalent Bayesian allocation is a blend of the sample-based allocation and the allocation determined by the prior. In other words, the classical-equivalent Bayesian allocation strategy can be interpreted as a "shrinkage" of the sample-based allocation towards the investor's prior/experience, where the amount of shrinkage is adjusted naturally by the relation between the amount information T and the confidence level C.

We recall from (8.53) that the opportunity cost of an allocation decision can be interpreted as the loss of an estimator. The same way as shrinkage estimators are a little more biased but less inefficient than sample estimators and thus display a lower error, so classical-equivalent Bayesian allocations generate opportunity costs that are less scattered than in the case of the sample-based strategy, at least for those values of the market parameters close to the prior assumption.

We see this in Figure 9.1, which refers to the classical-equivalent Bayesian allocation (9.21). Compare this figure with the evaluation of the prior allocation in Figure 8.3 and with the evaluation of the sample-based allocation in Figure 8.4.

The market parameters vary as in (8.58)-(8.59), i.e. the market is determined by the overall level of correlation. We plot the distribution of the prior overall correlation as implied by (9.18), which we compute by means of simulations.

Since the Bayesian estimate includes the investor's experience, the classical-equivalent Bayesian allocation automatically yields better results when the stress test (9.26) is run in the neighborhood of the prior assumptions on the market parameters, although coincidentally the cost of constraints violation is larger in the same region.

9.2 Black-Litterman allocation

Consider the optimal allocation function (8.30), which for each value of the market parameters $\boldsymbol{\theta}$ maximizes the investor's satisfaction given his investment constraints:

$$\boldsymbol{\alpha}(\boldsymbol{\theta}) \equiv \underset{\boldsymbol{\alpha} \in \mathcal{C}_{\boldsymbol{\theta}}}{\operatorname{argmax}} \left\{ \mathcal{S}_{\boldsymbol{\theta}}(\boldsymbol{\alpha}) \right\}. \tag{9.27}$$

Since the true value $\boldsymbol{\theta}^{\mathrm{t}}$ of the market parameters is not known, the truly optimal allocation cannot be implemented. Furthermore, as discussed in Chapter 8, the allocation function (9.66) is extremely sensitive to the input parameters $\boldsymbol{\theta}$: a slightly wrong input can give rise to a very large opportunity cost.

Like the Bayesian approach, the approach to asset allocation of Black and Litterman (1990) applies Bayes' rule to limit the sensitivity of the optimal allocation function to the input parameters. Nevertheless, the Black-Litterman framework differs from the classical-equivalent approach in that in the classical-equivalent approach the estimates of the market parameters are shrunk toward the investor's prior, whereas in the Black-Litterman approach it is the market distribution that is shrunk toward the investor's prior[1].

We present first the theory for the general case, where the market is described by a generic distribution and the investor can express views on any function of the market. Then we detail the computations that lead to the Black-Litterman allocation decision for the case where the investor expresses views on linear combinations of a normally distributed market.

9.2.1 General definition

Consider a market represented by the multivariate random variable $\mathbf{X}$. This could be the set of market invariants, or directly the set of market prices at the investment horizon, or any other variable that directly or indirectly fully determines the market.

Assume that it is possible to determine the distribution of this random variable, as represented for instance by the probability density function $f_{\mathbf{X}}$, by means of a reliable model/estimation technique. We call this the "official" distribution of the market. For instance, we could estimate this distribution by one of the techniques discussed in Chapter 4, or by means of general equilibrium arguments.

[1] The interpretation in terms of shrinkage of market parameters is also possible, see He and Litterman (2002).

Consider for example the case where the market X is represented by the daily return on the S&P 500 index, and suppose that X is normally distributed:

$$X \sim \mathrm{N}\left(\mu, \sigma^2\right). \tag{9.28}$$

We represent this distribution on the horizontal axis in Figure 9.2.

The distribution $f_{\mathbf{X}}$ is affected by estimation risk. To smoothen the effect of estimation risk, the statistician asks the investor's opinion on the market. The opinion is the investor's view on the outcome of the market $\mathbf{X}$. The investor's opinion is not a one-shot statement: the investor must be an expert, must have built a track-record and will be asked an opinion on a regular basis.

When asked by the statistician, the investor assesses that the outcome of the market is $\mathbf{V}$, a random variable that, possibly depending on the market scenario, is larger or smaller than the value $\mathbf{X}$ predicted by the "official" model. In other words, when the variable $\mathbf{X}$ assumes a specific value $\mathbf{x}$, the investor believes that the real outcome differs from $\mathbf{x}$ by a random amount. Therefore, the view $\mathbf{V}$ is a perturbation of the "official" outcome, and as such it is expressed as a conditional distribution $\mathbf{V}|\mathbf{x}$. The choice of the model for this conditional distribution, as represented for instance by the probability density function $f_{\mathbf{V}|\mathbf{x}}$, reflects the statistician's confidence in the investor.

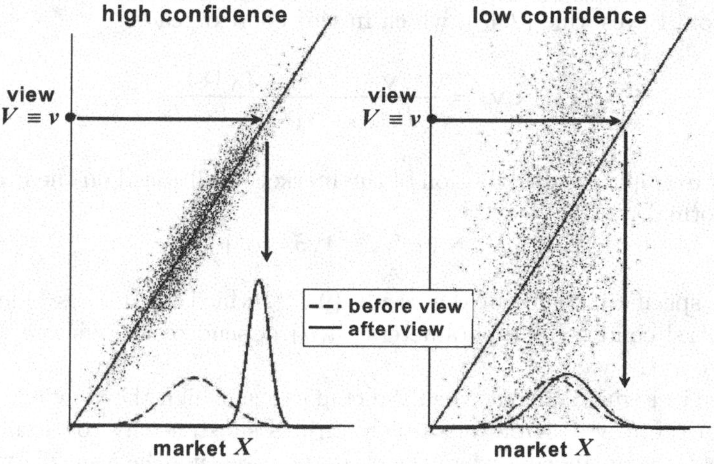

Fig. 9.2. Black-Litterman approach to market estimation

For example, the investor's opinion on the return of the S&P 500 index could be modeled as a normal perturbation to the "official" distribution:

$$V|x \sim \mathrm{N}\left(x, \phi^2\right). \tag{9.29}$$

If the statistician considers the investor unreliable, i.e. if he assumes that the investor's view will significantly depart from the "official" distribution (9.28) on a regular basis, he will choose a large value for the conditional standard deviation ϕ of the view. Viceversa, if the statistician trusts the investor he will model the view with a low value of ϕ.

In Figure 9.2 we see that when the confidence is high, the investor's statement is very close to the "official" distribution (a tight clouds of points). Viceversa, when the confidence is low, the cloud is very scattered.

More in general, the investor's opinion might regard a specific area of expertise of the market. In other words, instead of regarding directly the market $\mathbf{X}$, the view refers to a generic multivariate function $\mathbf{g}(\mathbf{X})$ on the market. Therefore the conditional model for the view becomes of the form $\mathbf{V}|\mathbf{x} \equiv \mathbf{V}|\mathbf{g}(\mathbf{x})$ and is represented for instance by the respective conditional probability density function $f_{\mathbf{V}|\mathbf{g}(\mathbf{x})}$.

Once the model has been set up, the statistician will ask the investor's opinion. The investor will produce a specific number $\mathbf{v}$, namely his prediction on $\mathbf{V}$.

At this point the statistician processes the above inputs and computes the distribution of the market conditioned on the investor's opinion $\mathbf{X}|\mathbf{v}$. The representation of this distribution in terms of its probability density function follows from Bayes' rule (2.43), which in this context reads:

$$f_{\mathbf{X}|\mathbf{v}}(\mathbf{x}|\mathbf{v}) = \frac{f_{\mathbf{V}|\mathbf{g}(\mathbf{x})}(\mathbf{v}|\mathbf{x}) f_{\mathbf{X}}(\mathbf{x})}{\int f_{\mathbf{V}|\mathbf{g}(\mathbf{x})}(\mathbf{v}|\mathbf{x}) f_{\mathbf{X}}(\mathbf{x}) d\mathbf{x}}. \tag{9.30}$$

In our example the distribution of the market conditioned on the investor's view is normal:

$$X|v \sim \mathrm{N}\left(\widetilde{\mu}\left(v, \phi^2\right), \widetilde{\sigma}^2\left(\phi^2\right)\right). \tag{9.31}$$

This is a specific instance of the result (9.44), which we discuss below in a more general context. The parameters $(\widetilde{\mu}, \widetilde{\sigma})$ depend on the view v and the confidence in the view ϕ^2.

We see in Figure 9.2 that when the confidence is high the view has a large impact on the new distribution, which shrinks substantially towards the investor's statement. Indeed, when the cloud representing the joint distribution is tight, knowledge of one coordinate (the view) almost completely determines the other (the market). When the confidence is low, the market distribution is almost unaffected by the investor's statement.

To summarize, in order to include the investor's view in the "official" market model, we proceed as follows: we start from the "official" distribution of the market $f_{\mathbf{X}}$; then we determine the investor's area of expertise, i.e. a function $\mathbf{g}$ of the market; then we specify a model $f_{\mathbf{V}|\mathbf{g}(\mathbf{x})}$ for the conditional distribution of the investor's view given the market; then we record the investor's input, i.e. the specific value $\mathbf{v}$ of his view; finally we compute the conditional distribution (9.30) of the market given the investor's view.

At this point we can define the *Black-Litterman allocation decision* as the optimal allocation function (9.27) computed using the market (9.30) determined by the view:

$$\boldsymbol{\alpha}_{\mathrm{BL}}[\mathbf{v}] \equiv \underset{\boldsymbol{\alpha}\in\mathcal{C}_{\mathbf{v}}}{\operatorname{argmax}}\left\{\mathcal{S}_{\mathbf{v}}(\boldsymbol{\alpha})\right\}. \tag{9.32}$$

Unlike in the other allocation strategies discussed in this chapter, the dependence of the Black-Litterman allocation on the contingent realization of the information i_T is not explicit.

Suppose that the market consists of the S&P500, whose return is X, and a risk-free security with null return. Assume that the investor has a budget w_T. Then an allocation is fully determined by the relative weight $\omega \equiv \alpha / w_T$ of the investment α in the risky security.

Assume that the investor's objective is final wealth, that his index of satisfaction is the expected value, and that he is bound by the no-short sale constraint. Then the Black-Litterman allocation reads:

$$\omega_{\mathrm{BL}}[v] \equiv \underset{0\leq\omega\leq 1}{\operatorname{argmax}}\left\{\omega\widetilde{\mu}\right\}, \tag{9.33}$$

where $\widetilde{\mu}$ is the expected value in (9.31).

9.2.2 Practicable definition: linear expertise on normal markets

Black and Litterman (1990) compute and discuss the analytical solution to (9.30) in a specific, yet quite general, case, see also Black and Litterman (1992).

First of all, the "official" model for the N-dimensional market vector $\mathbf{X}$ is assumed normal[2]:

$$\mathbf{X} \sim \mathrm{N}(\boldsymbol{\mu}, \boldsymbol{\Sigma}). \tag{9.34}$$

To illustrate, we consider an institution that adopts the RiskMetrics model to optimize the allocation of an international fund that invests in the following six stock indices: Italy, Spain, Switzerland, Canada, US and Germany. In this case the market are the daily compounded returns:

$$\mathbf{C} \sim \mathrm{N}(\boldsymbol{\mu}, \boldsymbol{\Sigma}). \tag{9.35}$$

Notice that this corresponds to the standard distributional assumption in Black and Scholes (1973).

[2] In the original paper the market is represented by the linear returns on a set of securities and the parameters $(\boldsymbol{\mu}, \boldsymbol{\Sigma})$ satisfy a general equilibrium model.

The expected value of the daily returns is assumed zero:

$$\boldsymbol{\mu} \equiv (0,0,0,0,0,0)' . \tag{9.36}$$

The covariance matrix of the daily returns on the above asset classes is estimated by exponential smoothing of the observed daily returns and is made publicly available by RiskMetrics. The matrix in our example was estimated in August 1999. Its decomposition in terms of standard deviations and correlations reads respectively:

$$\sqrt{\operatorname{diag}(\boldsymbol{\Sigma})} \equiv 0.01 \times (1.34, 1.52, 1.53, 1.55, 1.82, 1.97)' \tag{9.37}$$

and (we report only the non-trivial elements)

$$\operatorname{Cor}\{\mathbf{C}\} = \begin{pmatrix} \cdot & 54\% & 62\% & \mathbf{25\%} & 41\% & 59\% \\ \cdot & \cdot & 69\% & \mathbf{29\%} & 36\% & \mathbf{83\%} \\ \cdot & \cdot & \cdot & \mathbf{15\%} & 46\% & 65\% \\ \cdot & \cdot & \cdot & \cdot & \mathbf{47\%} & \mathbf{39\%} \\ \cdot & \cdot & \cdot & \cdot & \cdot & 38\% \\ \cdot & \cdot & \cdot & \cdot & \cdot & \cdot \end{pmatrix} . \tag{9.38}$$

Second, the investor's area of expertise is a linear function of the market:

$$\mathbf{g}(\mathbf{x}) \equiv \mathbf{P}\mathbf{x}, \tag{9.39}$$

where $\mathbf{P}$ is the "pick" matrix: each of its K rows is an N-dimensional vector that corresponds to one view and selects the linear combination of the market involved in that view.

The specification (9.39) is very flexible, in that the investor does not necessarily need to express views on all the market variables. Furthermore, views do not necessarily need to be expressed in absolute terms for each market variable considered, as any linear combination of the market constitutes a potential view.

A fund manager might assess absolute views on three markets: the Spanish, the Canadian and the German index. Therefore, the "pick" matrix reads:

$$\mathbf{P} \equiv \begin{pmatrix} 0 & 1 & 0 & 0 & 0 & 0 \\ 0 & 0 & 0 & 1 & 0 & 0 \\ 0 & 0 & 0 & 0 & 0 & 1 \end{pmatrix} . \tag{9.40}$$

Notice from (9.38) that the Spanish and the German markets are highly correlated (83%) and that the Canadian index is relatively independent of the other markets.

Third, the conditional distribution of the investor's views given the outcome of the market is assumed normal:

$$\mathbf{V}|\mathbf{Px} \sim \mathrm{N}(\mathbf{Px}, \boldsymbol{\Omega}), \tag{9.41}$$

where the symmetric and positive matrix $\boldsymbol{\Omega}$ denotes the statistician's confidence in the investor's opinion.

A particularly convenient choice for the uncertainty matrix is

$$\boldsymbol{\Omega} \equiv \left(\frac{1}{c} - 1\right) \mathbf{P\Sigma P}', \tag{9.42}$$

where c is a positive scalar. This corresponds to an "empirical Bayesian" approach: the statistician gives relatively speaking more leeway to the investor's assessment on those combinations that are more volatile according to the official market model (9.34). The scalar c tweaks the absolute confidence in the investor's skills, see Figure 9.2. The case $c \to 0$ gives rise to an infinitely disperse distribution of the views: this means that the investor's views have no impact, i.e. the investor is not trusted. The case $c \to 1$ gives rise to an infinitely peaked distribution of the views: this means that the investor is trusted completely over the official market model. The case $c \equiv 1/2$ corresponds to the situation where the investor is trusted as much as the official market model.

In our example we define $\boldsymbol{\Omega}$ as in (9.42), where we set $c \equiv 1/2$.

Fourth, the investor is asked his opinion on his area of expertise. This will turn into a specific value $\mathbf{v}$ of the views $\mathbf{V}$.

The fund manager assesses that the Spanish index will remain unvaried, the Canadian stock index will score a negative return of 2% and the German index will experience a positive change of 2%. Therefore the views read:

$$\mathbf{v} \equiv 0.01 \times (0, -2, 2)'. \tag{9.43}$$

By means of Bayes' rule (9.30) it is possible to compute the distribution of the market conditioned on the investor's views. We show in Appendix www.9.3 that the *Black-Litterman distribution* is normal:

$$\mathbf{X}|\mathbf{v} \sim \mathrm{N}(\boldsymbol{\mu}_{\mathrm{BL}}, \boldsymbol{\Sigma}_{\mathrm{BL}}), \tag{9.44}$$

where the expected values read:

$$\boldsymbol{\mu}_{\mathrm{BL}}(\mathbf{v}, \boldsymbol{\Omega}) \equiv \boldsymbol{\mu} + \boldsymbol{\Sigma}\mathbf{P}' \left(\mathbf{P\Sigma P}' + \boldsymbol{\Omega}\right)^{-1} (\mathbf{v} - \mathbf{P}\boldsymbol{\mu}); \tag{9.45}$$

and the covariance matrix reads:

$$\boldsymbol{\Sigma}_{\mathrm{BL}}(\boldsymbol{\Omega}) \equiv \boldsymbol{\Sigma} - \boldsymbol{\Sigma}\mathbf{P}' \left(\mathbf{P\Sigma P}' + \boldsymbol{\Omega}\right)^{-1} \mathbf{P\Sigma}. \tag{9.46}$$

Notice that the expression of the covariance is not affected by the value of the views $\mathbf{v}$. This is a peculiarity of the normal setting.

The expression of the Black-Litterman market distribution can be used to determine the optimal asset allocation that includes the investor's views.

In our example we consider an investor who has an initial budget w_T, and who is subject to the full-investment and the no-short-sale constraints:

$$\mathcal{C}: \boldsymbol{\alpha}'\mathbf{p}_T = w_T, \quad \boldsymbol{\alpha} \geq \mathbf{0}. \tag{9.47}$$

Furthermore, we assume that the investor's objective is final wealth:

$$\Psi_{\boldsymbol{\alpha}} \equiv \boldsymbol{\alpha}'\mathbf{P}_{T+\tau}. \tag{9.48}$$

In order to determine the optimal allocation we consider the two-step mean-variance framework. First we compute the efficient frontier(6.74), which in this context reads:

$$\boldsymbol{\alpha}(v) \equiv \underset{\boldsymbol{\alpha}}{\operatorname{argmax}}\, \boldsymbol{\alpha}' \operatorname{E}\{\mathbf{P}_{T+\tau}\} \tag{9.49}$$
$$\text{subject to} \begin{cases} \boldsymbol{\alpha}'\mathbf{p}_T = w_T \\ \boldsymbol{\alpha} \geq \mathbf{0} \\ \boldsymbol{\alpha}' \operatorname{Cov}\{\mathbf{P}_{T+\tau}\}\boldsymbol{\alpha} = v. \end{cases}$$

To compute the market inputs, namely $\operatorname{E}\{\mathbf{P}_{T+\tau}\}$ and $\operatorname{Cov}\{\mathbf{P}_{T+\tau}\}$, we need the characteristic function (2.157) of the Black-Litterman distribution (9.44) of the compounded returns:

$$\phi_{\mathbf{C}}(\boldsymbol{\omega}) = e^{i\boldsymbol{\mu}_{\mathrm{BL}}'\boldsymbol{\omega} - \frac{1}{2}\boldsymbol{\omega}'\boldsymbol{\Sigma}_{\mathrm{BL}}\boldsymbol{\omega}}. \tag{9.50}$$

Dropping "BL" from the notation, from (3.95) the expected values of the prices read:

$$\begin{aligned} \operatorname{E}\left\{P_{T+\tau}^{(n)}\right\} &= P_T^{(n)}\phi_{\mathbf{C}}\left(-i\boldsymbol{\delta}^{(n)}\right) \\ &= P_T^{(n)} e^{\left(\mu_n + \frac{\Sigma_{nn}}{2}\right)}. \end{aligned} \tag{9.51}$$

Similarly, from (3.96) we obtain the covariance matrix of the market:

$$\begin{aligned} \operatorname{Cov}\left\{P_{T+\tau}^{(m)}, P_{T+\tau}^{(n)}\right\} &= P_T^{(m)}P_T^{(n)}\phi_{\mathbf{C}}\left(-i\boldsymbol{\delta}^{(m)} - i\boldsymbol{\delta}^{(n)}\right) \\ &\quad -\operatorname{E}\left\{P_{T+\tau}^{(m)}\right\}\operatorname{E}\left\{P_{T+\tau}^{(n)}\right\} \\ &= P_T^{(m)}P_T^{(n)} e^{(\mu_m+\mu_n)} e^{\frac{1}{2}(\Sigma_{mm}+\Sigma_{nn})}\left(e^{\Sigma_{mn}} - 1\right). \end{aligned} \tag{9.52}$$

Formulas (9.51) and (9.52) yield the inputs of the mean-variance optimization as functions of the Black-Litterman parameters (9.45) and (9.46). Substituting these expressions in (9.49) we obtain for any level of variance v the respective efficient allocation that includes the investor's views, see Figure 9.4. In a second stage the investor chooses the efficient portfolio that best suits his profile, as in Figure 6.23.

9.2.3 Evaluation

The Black-Litterman approach can, but does not need to, rely on the contingent historical information i_T available when the investment decision is made. Indeed, this approach blends two models for the market, namely the investor's and the official models: these models can be based on historical information, or they can rely on prior information, or other rationales, such as general equilibrium arguments, etc. Therefore we cannot apply the approach discussed in Section 8.1 to the evaluation of the Black-Litterman allocation.

On the other hand, the expected value $\boldsymbol{\mu}_{\mathrm{BL}}$ tilted by the views $\mathbf{v}$ according to the Black-Litterman formula (9.45) might be in strong contrast with the value $\boldsymbol{\mu}$ that appears in the official market model (9.34). In this section we discuss a technique to measure this difference and tweak the most extreme views accordingly, see also Fusai and Meucci (2003). Notice that we only need to consider the tilted expected values, since the explicit value of the views $\mathbf{v}$ does not enter the expression for the covariance matrix (9.46).

First we recall the definition (1.35) of z-score, widely used by practitioners: the distance of a suspicious value x of the random variable X from the accepted expected value μ divided by the standard deviation σ of X. In a multivariate environment the z-score becomes the Mahalanobis distance (2.61).

Under the normal hypothesis (9.34) for the official market model, the square Mahalanobis distance of the market $\mathbf{X}$ from its expected value $\boldsymbol{\mu}$ through the metric induced by its covariance $\boldsymbol{\Sigma}$ is distributed as a chi-square with N degrees of freedom:

$$M^2 \equiv (\mathbf{X} - \boldsymbol{\mu})' \boldsymbol{\Sigma}^{-1} (\mathbf{X} - \boldsymbol{\mu}) \sim \chi^2_N, \tag{9.53}$$

see Appendix www.7.1.

In our context the "suspicious" value is the Black-Litterman vector of expected values $\boldsymbol{\mu}_{\mathrm{BL}}$. If we consider $\boldsymbol{\mu}_{\mathrm{BL}}$ as a realization of the random variable $\mathbf{X}$, we can compute the respective realization of the square Mahalanobis distance accordingly:

$$m^2_{\mathbf{v}} \equiv (\boldsymbol{\mu}_{\mathrm{BL}}(\mathbf{v}) - \boldsymbol{\mu})' \boldsymbol{\Sigma}^{-1} (\boldsymbol{\mu}_{\mathrm{BL}}(\mathbf{v}) - \boldsymbol{\mu}). \tag{9.54}$$

Intuitively, if the square distance $m^2_{\mathbf{v}}$ is small, the views are not too far from the market model and the consistence of the Black-Litterman expectations with the market model is high. In turn, the realization $m^2_{\mathbf{v}}$ of the random variable M^2 can be considered small if M^2 is likely to be larger than $m^2_{\mathbf{v}}$.

Therefore, we define the index of consistence $C(\mathbf{v})$ of the Black-Litterman expectations with the market model as the probability that the random variable M^2 is larger than the realization $m^2_{\mathbf{v}}$:

$$C(\mathbf{v}) \equiv \mathbb{P}\left(M^2 \geq m^2_{\mathbf{v}}\right) = 1 - F^{\mathrm{Ga}}_{N,1}\left(m^2_{\mathbf{v}}\right). \tag{9.55}$$

In this expression $F^{\mathrm{Ga}}_{N,1}$ represents the cumulative density function of the chi-square distribution with N degrees of freedom, which is a special case of the gamma cumulative density function (1.111).

In the extreme case where the realization $m_{\mathbf{v}}^2$ is zero, i.e. when $\boldsymbol{\mu}_{\mathrm{BL}}$ coincide with the model value $\boldsymbol{\mu}$, the random variable M^2 is certainly larger than the realized value and thus the consistence of the Black-Litterman expectations with the market model is total, i.e. one. As the realized value $m_{\mathbf{v}}^2$ increases, i.e. as $\boldsymbol{\mu}_{\mathrm{BL}}$ drifts apart from the model value $\boldsymbol{\mu}$, the random variable M^2 becomes less and less likely to be larger than the observed value and the consistence of the Black-Litterman expectations with the market model decreases.

We remark that the consistence C of the Black-Litterman expectations with the market model plays a dual role with the statistician's confidence c in the investor that appears in (9.42). Indeed, when the confidence c in the investor is zero, the views are ignored and the Black-Litterman distribution becomes the market distribution. Therefore the Mahalanobis distance of the Black-Litterman model from the official market model becomes null and the consistence $C(\mathbf{v})$ of the Black-Litterman expectations with the market model is total. As the confidence c in the investor increases, so does the Mahalanobis distance of the Black-Litterman model from the official market model and thus the consistence C of the Black-Litterman expectations with the market decreases.

When the overall consistence (9.55) is below an agreed threshold, often a slight shift in only one of the views suffices to boost the consistence level. Therefore, another natural problem is how to detect the "boldest" views, and how to fix them accordingly. To solve this problem, we compute the sensitivity of the consistence index to the views. From the chain rule of calculus, this sensitivity reads:

$$\begin{aligned}\frac{\partial C(\mathbf{v})}{\partial \mathbf{v}} &= \frac{dC}{dm^2}\frac{\partial m^2}{\partial \boldsymbol{\mu}_{\mathrm{BL}}}\frac{\partial \boldsymbol{\mu}_{\mathrm{BL}}}{\partial \mathbf{v}} \\ &= -2 f_{N;1}^{\mathrm{Ga}}\left(m_{\mathbf{v}}^2\right)\left(\mathbf{P}\boldsymbol{\Sigma}\mathbf{P}' + \boldsymbol{\Omega}\right)^{-1}\mathbf{P}\left(\boldsymbol{\mu}_{\mathrm{BL}} - \boldsymbol{\mu}\right).\end{aligned} \tag{9.56}$$

In this expression $f_{N;1}^{\mathrm{Ga}}$ is the probability density function of the chi-square distribution with N degrees of freedom, which is a special case of the gamma probability density function (1.110).

In order to tweak the views, the investor simply needs to compute (9.56) and find the entry with the largest absolute value. If that entry is positive (negative), the respective view must be increased (decreased) slightly.

To illustrate, we apply this recipe to our example. We start with the views (9.43), which we report here:

$$\mathbf{v} \equiv 0.01 \times (0, -2, 2)'. \tag{9.57}$$

The consistence index (9.55) and the consistence sensitivities (9.56) read respectively:

$$C = 93.8\%, \quad \frac{\partial C}{\partial \mathbf{v}} = (8.1, 5.6, -9.0)'. \tag{9.58}$$

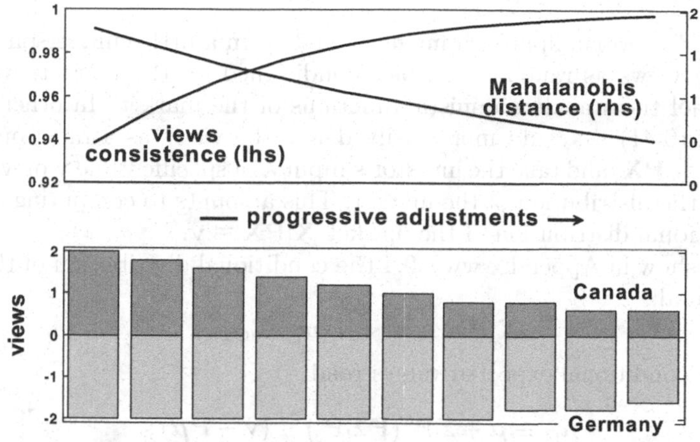

Fig. 9.3. Black-Litterman approach: views assessment

The consistence index is relatively insensitive to the second view on Canada, although it is of the same magnitude as the third view on Germany, namely 200 basis points. On the other hand the first view on Spain, which is apparently innocuous, has a larger effect on the consistence index: this is not unexpected, since the second view refers to a relatively independent market, whereas the first and third views state contrasting opinions on highly correlated markets, see (9.38).

Suppose that a consistence of at least 95% is required. To reach this level one should fine-tune, and actually decrease, the third view on the German index. It turns out that a 20 basis point shift, that changes (9.57) as follows

$$\mathbf{v} = 0.01 \times (0, -2, 1.8)', \tag{9.59}$$

brings the overall consistence above the desired level:

$$C = 95.4\%. \tag{9.60}$$

In Figure 9.3 we see the effect on the consistence index of progressively reducing the boldness of the views: in the lower plot we display different views on the performance of Canada and Germany starting from the initial views +2% and -2% respectively; in the upper plot of the figure we report the progressively increasing consistence index (9.55) corresponding to less and less extreme views, along with the respective progressively decreasing square Mahalanobis distance (9.54) between the Black-Litterman expectations and the market expectations.

9.2.4 Discussion

The Black-Litterman approach might at first seem a little cumbersome. Why model the views as random variables conditioned on the market, when we could model them as deterministic functions of the market? In other words, instead of (9.41) we could more easily define the views as a function of the market $\mathbf{V} \equiv \mathbf{PX}$, and take the investor's input as a specific value $\mathbf{v}$ on which to condition the distribution of the market. This amounts to computing directly the conditional distribution of the market $\mathbf{X}|\mathbf{PX} \equiv \mathbf{v}$.

As we show in Appendix www.9.4 the conditional distribution of the market is normal:

$$\mathbf{X}|\mathbf{PX} \equiv \mathbf{v} \sim \mathrm{N}\left(\boldsymbol{\mu}_C, \boldsymbol{\Sigma}_C\right), \tag{9.61}$$

where the conditional expected values read:

$$\boldsymbol{\mu}_C \equiv \boldsymbol{\mu} + \boldsymbol{\Sigma}\mathbf{P}'\left(\mathbf{P}\boldsymbol{\Sigma}\mathbf{P}'\right)^{-1}\left(\mathbf{v} - \mathbf{P}\boldsymbol{\mu}\right); \tag{9.62}$$

and the conditional covariance matrix reads:

$$\boldsymbol{\Sigma}_C \equiv \boldsymbol{\Sigma} - \boldsymbol{\Sigma}\mathbf{P}'\left(\mathbf{P}\boldsymbol{\Sigma}\mathbf{P}'\right)^{-1}\mathbf{P}\boldsymbol{\Sigma}. \tag{9.63}$$

It is immediate to check that, as expected, this distribution is degenerate on the views:

$$\mathbf{PX}|\mathbf{PX} \equiv \mathbf{v} \sim \mathrm{N}\left(\mathbf{v}, \mathbf{0}\right). \tag{9.64}$$

Indeed, by definition of conditional distribution, the views $\mathbf{PX} = \mathbf{v}$ are supposed to take place with certainty. This is the reason why the direct conditional approach to modeling the views is not appropriate: the conditional approach yields a too "spiky" distribution. Therefore, since the allocation optimization process is very sensitive to the input parameters, when the optimal allocations are computed directly according to the conditional model, the resulting portfolios are extremely different from those computed according to the "official" market model and often give rise to corner solutions, see Figure 9.4.

Instead, the Black-Litterman distribution (9.44) blends smoothly the "official" market model (9.34) with the investor's blunt opinion, represented by the conditional distribution (9.61).

Indeed, the conditional distribution represents an extreme case of the Black-Litterman distribution, namely the case where the scatter matrix $\boldsymbol{\Omega}$ is null, i.e. the statistician's confidence in the investor's views is total. On the other hand, the "official" market model represents the opposite extreme case of the the Black-Litterman distribution, namely the case where the scatter matrix $\boldsymbol{\Omega}$ is infinite, i.e. the statistician's confidence in the investor is null:

$$\begin{array}{ccl} & & \mathbf{X} \sim \mathrm{N}\left(\boldsymbol{\mu}, \boldsymbol{\Sigma}\right) \qquad (\boldsymbol{\Omega} \to \infty) \\ & \nearrow & \\ \mathbf{X} \sim \mathrm{N}\left(\boldsymbol{\mu}_{\mathrm{BL}}, \boldsymbol{\Sigma}_{\mathrm{BL}}\right) & & \\ & \searrow & \\ & & \mathbf{X} \sim \mathrm{N}\left(\boldsymbol{\mu}_C, \boldsymbol{\Sigma}_C\right) \qquad (\boldsymbol{\Omega} \to \mathbf{0}). \end{array} \tag{9.65}$$

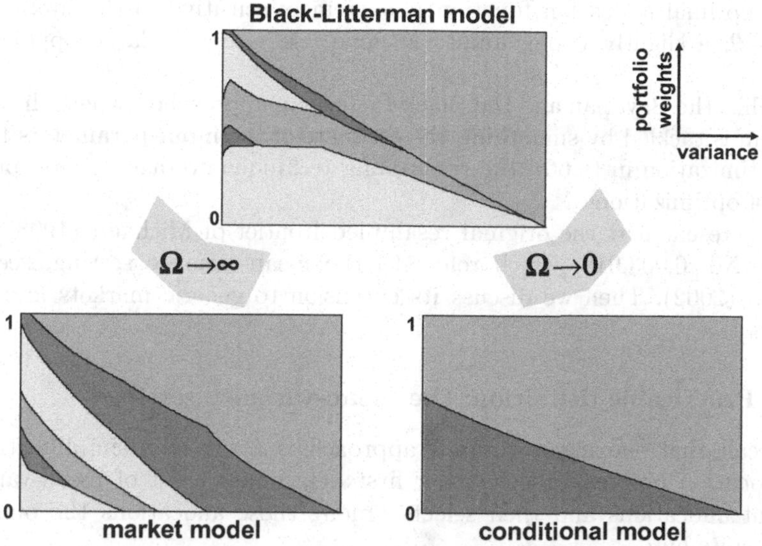

Fig. 9.4. Black-Litterman approach: sensitivity to the input parameters

For the intermediate cases, as the confidence in the investor's views decreases, the Black-Litterman distribution smoothly shifts away from the conditional model towards the "official" market model. This mechanism lessens the effect of the input parameters on the final allocations.

In Figure 9.4 we plot the efficient portfolios in terms of their relative weights computed according to the Black-Litterman distribution as in (9.49). We consider the general Black-Litterman distribution, as well as its limit cases, namely the "official" market model (9.34) and the distribution conditioned on the investor's views (9.61). Notice that the conditional distribution gives rise to corner solutions, i.e. highly concentrated portfolios.

9.3 Resampled allocation

Consider the optimal allocation function (8.30), which for each value of the market parameters $\boldsymbol{\theta}$ maximizes the investor's satisfaction given his investment constraints:

$$\boldsymbol{\alpha}(\boldsymbol{\theta}) \equiv \underset{\boldsymbol{\alpha} \in \mathcal{C}_{\boldsymbol{\theta}}}{\operatorname{argmax}} \left\{ \mathcal{S}_{\boldsymbol{\theta}}(\boldsymbol{\alpha}) \right\}. \tag{9.66}$$

Since the true value $\boldsymbol{\theta}^{\mathrm{t}}$ of the market parameters is not known, the truly optimal allocation cannot be implemented. Furthermore, as discussed in Chapter

8, the optimal allocation function is extremely sensitive to the input parameters $\boldsymbol{\theta}$: a slightly wrong input can give rise to a very large opportunity cost.

Unlike the Bayesian and the Black-Litterman approaches, where the above problem is tackled by smoothing the estimate of the input parameters before the optimization in (9.66), the resampling technique averages the outputs of a set of optimizations.

We present first the original resampled frontier of Michaud (1998), U.S. Patent No. 6,003,018, which refers to the mean-variance setting, see also Scherer (2002). Then we discuss its extension to generic markets and preferences.

9.3.1 Practicable definition: the mean-variance setting

We recall that the mean-variance approach is a two-step simplification of an allocation problem: the investor first determines a set of mean-variance efficient allocations and then selects among those allocations the one that better suits him.

The assumptions of the original resampling recipe are the following: first, the investor's objective admits the mean-variance formulations in terms of linear returns and relative weights, see Section 6.3.4; second, the market consists of equity-like securities for which the linear returns are market invariants, see Section 3.1.1; third, the investment horizon and the estimation interval coincide, see Section 6.5.4; fourth, the investment constraints are such that the dual formulation is correct, see Section 6.5.3; fifth, the constraints do not depend on unknown market parameters.

Under the above assumptions the mean-variance problem can be written as in (6.147), which in the dual formulation (6.146) reads:

$$\mathbf{w}^{(i)} = \underset{\substack{\mathbf{w}\in\mathcal{C} \\ \mathbf{w}'\boldsymbol{\mu}\geq e^{(i)}}}{\operatorname{argmin}} \mathbf{w}'\boldsymbol{\Sigma}\mathbf{w}, \quad i = 1, \ldots, I. \tag{9.67}$$

In this expression $\boldsymbol{\mu}$ and $\boldsymbol{\Sigma}$ are the expected values and the covariances of the linear returns of the securities relative to the investment horizon; the set $\{e^{(1)}, \ldots, e^{(I)}\}$ is a significative grid of target expected values; and $\mathcal{C}$ is the set of investment constraints.

To determine the efficient portfolio weights (9.67) the resampling recipe follows these steps.

Step 1. Estimate the inputs ${}_0\widehat{\boldsymbol{\mu}}$ and ${}_0\widehat{\boldsymbol{\Sigma}}$ of the mean-variance framework from the analysis of the observed time series i_T of the past linear returns:

$$i_T \equiv \{\mathbf{l}_1, \ldots, \mathbf{l}_T\}. \tag{9.68}$$

This can be done for instance, but not necessarily, by means of the sample estimators (8.79) and (8.80).

Step 2a. Consider the time series i_T as the realization of a set of market invariants, i.e. independent and identically distributed returns:

$$I_T \equiv \{\mathbf{L}_1, \mathbf{L}_2, \ldots, \mathbf{L}_T\}. \tag{9.69}$$

Step 2b. Make assumptions on the distribution generating the returns (9.69), for instance assuming normality, and set the estimated parameters as the true parameters that determine the distribution of the returns:

$$\mathbf{L}_t \sim \mathrm{N}\left({}_0\widehat{\boldsymbol{\mu}}, {}_0\widehat{\boldsymbol{\Sigma}}\right). \tag{9.70}$$

Step 2c. Resample a large number Q of Monte Carlo scenarios of realizations of (9.69) from the distribution (9.70):

$${}_q i_T \equiv \{{}_q\mathbf{l}_1, \ldots, {}_q\mathbf{l}_T\}, \quad q = 1, \ldots, Q. \tag{9.71}$$

Step 3. Estimate the inputs ${}_q\widehat{\boldsymbol{\mu}}$ and ${}_q\widehat{\boldsymbol{\Sigma}}$ of the mean-variance framework from the resampled time series (9.71) as in Step 1.

Step 4a. Compute the global minimum-variance portfolio from each of the resampled inputs:

$${}_q\mathbf{w}_{MV} = \underset{\mathbf{w}\in\mathcal{C}}{\operatorname{argmin}}\, \mathbf{w}'{}_q\widehat{\boldsymbol{\Sigma}}\mathbf{w}, \quad q = 1, \ldots, Q. \tag{9.72}$$

Step 4b. Compute the respective estimated expected value in each scenario:

$${}_q\underline{e} \equiv {}_q\mathbf{w}'_{MV}{}_q\widehat{\boldsymbol{\mu}}, \quad q = 1, \ldots, Q. \tag{9.73}$$

Step 4c. Compute the maximum estimated expected value in each scenario:

$${}_q\overline{e} \equiv \max\left\{{}_q\widehat{\boldsymbol{\mu}}'\boldsymbol{\delta}^{(1)}, \ldots, {}_q\widehat{\boldsymbol{\mu}}'\boldsymbol{\delta}^{(N)}\right\}, \quad q = 1, \ldots, Q, \tag{9.74}$$

where $\boldsymbol{\delta}$ is the canonical basis ($A.15$).

Step 4d. For each scenario q determine a grid $\left\{{}_q e^{(1)}, \ldots, {}_q e^{(I)}\right\}$ of equally-spaced target expected values as follows:

$$\begin{aligned}
{}_q e^{(1)} &\equiv {}_q\underline{e} \\
&\vdots \\
{}_q e^{(i)} &\equiv {}_q\underline{e} + \frac{{}_q\overline{e} - {}_q\underline{e}}{I-1}(i-1) \\
&\vdots \\
{}_q e^{(I)} &\equiv {}_q\overline{e}.
\end{aligned} \tag{9.75}$$

Step 4e. Solve the mean-variance dual problem (9.67) for all the Monte Carlo scenarios $q = 1, \ldots, Q$ and all the target expected values $i = 1, \ldots, I$:

$${}_q\mathbf{w}^{(i)} = \underset{\substack{\mathbf{w}\in\mathcal{C} \\ \mathbf{w}'{}_q\widehat{\boldsymbol{\mu}}\geq {}_q e^{(i)}}}{\operatorname{argmin}}\, \mathbf{w}'{}_q\widehat{\boldsymbol{\Sigma}}\mathbf{w}. \tag{9.76}$$

Step 5. Define the *resampled efficient frontier* as the average of the above allocations, possibly rejecting some outliers:

$$\mathbf{w}_{\mathrm{rs}}^{(i)} \equiv \frac{1}{Q} \sum_{q=1}^{Q} {}_q\mathbf{w}^{(i)}, \quad i = 1, \ldots, I, \tag{9.77}$$

where "rs" stands for "resampled".

Step 6. Compute the efficient allocations from the respective relative weights:

$$\boldsymbol{\alpha}_{\mathrm{rs}}^{(i)} \equiv w_T \operatorname{diag}(\mathbf{p}_T)^{-1} \mathbf{w}_{\mathrm{rs}}^{(i)}, \quad i = 1, \ldots, I, \tag{9.78}$$

where w_T is the initial budget.

Following the steps 1-6 we obtain a set of allocations, namely (9.78), from which the investor can choose according to his preferences.

9.3.2 General definition

It is not difficult to generalize the rationale behind the resampled frontier to a more general setting, which does not necessarily rely on the two-step mean-variance approach. We modify the steps 1-6 that led to the resampled frontier respectively as follows.

Step 0. Instead of the expected values and the covariances of the linear returns, in general the market at the investment horizon is determined by a set of parameters $\boldsymbol{\theta}$, which steer the parametric distribution of the market invariants $\mathbf{X}_t^{\boldsymbol{\theta}}$, see (8.17).

Step 1. Using one of the techniques discussed in Chapter 4, estimate the parameters ${}_0\widehat{\boldsymbol{\theta}} \equiv \widehat{\boldsymbol{\theta}}[i_T]$ from the available time series of the market invariants:

$$i_T \equiv \{\mathbf{x}_1, \ldots, \mathbf{x}_T\}. \tag{9.79}$$

We stress that the market invariants are not necessarily the linear returns: depending on the market, they could be for instance changes in yield to maturity, or other quantities, see Section 3.1.

Step 2. Generate a large number of Monte Carlo realizations of the time series ${}_q i_T$ of the market invariants, assuming that the distribution underlying the market invariants in Step 0 is determined by the estimated values. In other words, generate a large number Q of Monte Carlo realizations:

$${}_q i_T \equiv \{{}_q\mathbf{x}_1, \ldots, {}_q\mathbf{x}_T\}, \quad q = 1, \ldots, Q, \tag{9.80}$$

from the following set of random variables:

$$I_T^{{}_0\widehat{\boldsymbol{\theta}}} \equiv \left\{\mathbf{X}_1^{{}_0\widehat{\boldsymbol{\theta}}}, \ldots, \mathbf{X}_T^{{}_0\widehat{\boldsymbol{\theta}}}\right\}. \tag{9.81}$$

Step 3. In each scenario q estimate as in Step 1 the parameters ${}_q\widehat{\boldsymbol{\theta}} \equiv \widehat{\boldsymbol{\theta}}[{}_q i_T]$ from the resampled time series (9.80).

Step 4. Instead of determining the efficient frontier, in general the investor maximizes his primary index of satisfaction given his constraints, which depend on the market parameters, see (9.66). Therefore replace the optimization (9.76) with the following expression:

$$ {}_q\boldsymbol{\alpha} \equiv \underset{\boldsymbol{\alpha} \in \mathcal{C}_{{}_q\widehat{\boldsymbol{\theta}}}}{\operatorname{argmax}} \left\{ \mathcal{S}_{{}_q\widehat{\boldsymbol{\theta}}} (\boldsymbol{\alpha}) \right\}, \tag{9.82} $$

for all the Monte Carlo scenarios $q = 1, \ldots, Q$.

Step 5. Determine the resampled allocation by averaging the Monte Carlo optimal allocations:

$$ \boldsymbol{\alpha}_{\mathrm{rs}} \equiv \frac{1}{Q} \sum_{q=1}^{Q} {}_q\boldsymbol{\alpha}. \tag{9.83} $$

We stress that the resampled allocation is a decision $\boldsymbol{\alpha}_{\mathrm{rs}} [i_T]$, which depends on the available information (9.79) through the following chain, which summarizes the whole resampling technique:

$$ i_T \overset{\text{estimate}}{\longmapsto} {}_0\widehat{\boldsymbol{\theta}} \overset{\text{resample}}{\longmapsto} {}_q i_T \overset{\text{estimate}}{\longmapsto} {}_q\widehat{\boldsymbol{\theta}} \overset{\text{optimize}}{\longmapsto} {}_q\boldsymbol{\alpha} \overset{\text{average}}{\longmapsto} \boldsymbol{\alpha}_{\mathrm{rs}}. \tag{9.84} $$

We can further simplify the generic definition of the resampled allocation by avoiding the above sequential steps 1-5. Indeed the q-th scenario of the resampled allocation (9.82) is the optimal allocation function $\boldsymbol{\alpha}(\boldsymbol{\theta})$ defined in (9.66) applied to the estimate from the q-th scenario of the Monte-Carlo-generated time series:

$$ {}_q\boldsymbol{\alpha} = \boldsymbol{\alpha} \left(\widehat{\boldsymbol{\theta}} \left[{}_q i_T \right] \right). \tag{9.85} $$

Furthermore, the q-th scenario of the time series ${}_q i_T$ is a realization of the random variable $I_T^{\widehat{\boldsymbol{\theta}}[i_T]}$, see (9.81). Therefore the average of the Monte Carlo scenarios (9.83) is the expectation of the allocations induced by the random variable $I_T^{\widehat{\boldsymbol{\theta}}[i_T]}$. In other words, the general definition of the *resampled allocation* can be summarized as follows:

$$ \boldsymbol{\alpha}_{\mathrm{rs}} [i_T] \equiv \mathrm{E} \left\{ \boldsymbol{\alpha} \left(\widehat{\boldsymbol{\theta}} \left[I_T^{\widehat{\boldsymbol{\theta}}[i_T]} \right] \right) \right\}, \tag{9.86} $$

where "rs" stands for "resampled". This is indeed an allocation decision, which processes the currently available information, see (8.38).

In all the cases of practical interest, the resampled allocation cannot be computed in analytical closed form from the definition (9.86). Therefore, to implement the resampling technique we need to follow all the steps in (9.84).

Consider a random vector $\mathbf{u}$ distributed as follows:

$$ \mathbf{u} \sim \mathrm{N} \left(\widehat{\boldsymbol{\mu}} [i_T], \frac{1}{T} \widehat{\boldsymbol{\Sigma}} [i_T] \right), \tag{9.87} $$

where $\widehat{\boldsymbol{\mu}}[i_T]$ and $\widehat{\boldsymbol{\Sigma}}[i_T]$ are the sample mean and covariance of the linear returns (8.79) and (8.80) respectively. Now consider the positive and symmetric random matrix $\mathbf{V}$ distributed as follows:

$$T\mathbf{V}^{-1} \sim \mathrm{W}\left(T-1, \widehat{\boldsymbol{\Sigma}}[i_T]\right). \tag{9.88}$$

Furthermore, assume that $\mathbf{w}$ and $\mathbf{V}$ are independent. From (8.85) and (8.86) we obtain the distribution of the sample estimators applied to the time series distributed according to the estimated parameters:

$$\widehat{\boldsymbol{\mu}}\left[I_T^{\widehat{\boldsymbol{\mu}}[i_T],\widehat{\boldsymbol{\Sigma}}[i_T]}\right] \stackrel{d}{=} \mathbf{u}, \quad \widehat{\boldsymbol{\Sigma}}\left[I_T^{\widehat{\boldsymbol{\mu}}[i_T],\widehat{\boldsymbol{\Sigma}}[i_T]}\right] \stackrel{d}{=} \mathbf{V}^{-1}. \tag{9.89}$$

In our leading example the optimal allocation function is (8.32):

$$\boldsymbol{\alpha}(\boldsymbol{\mu}, \boldsymbol{\Sigma}) = [\mathrm{diag}(\mathbf{p}_T)]^{-1} \boldsymbol{\Sigma}^{-1}\left(\zeta\boldsymbol{\mu} + \frac{w_T - \zeta\mathbf{1}'\boldsymbol{\Sigma}^{-1}\boldsymbol{\mu}}{\mathbf{1}'\boldsymbol{\Sigma}^{-1}\mathbf{1}}\mathbf{1}\right). \tag{9.90}$$

Therefore the allocations induced by the random variable $I_T^{\widehat{\boldsymbol{\mu}}[i_T],\widehat{\boldsymbol{\Sigma}}[i_T]}$ read:

$$\begin{aligned}\boldsymbol{\alpha}\left(\widehat{\boldsymbol{\mu}}\left[I_T^{\widehat{\boldsymbol{\mu}}[i_T],\widehat{\boldsymbol{\Sigma}}[i_T]}\right], \widehat{\boldsymbol{\Sigma}}\left[I_T^{\widehat{\boldsymbol{\mu}}[i_T],\widehat{\boldsymbol{\Sigma}}[i_T]}\right]\right) &\stackrel{d}{=} \zeta[\mathrm{diag}(\mathbf{p}_T)]^{-1}\mathbf{V}\mathbf{u} \\ &+ \frac{w_T - \zeta\mathbf{1}'\mathbf{V}\mathbf{u}}{\mathbf{1}'\mathbf{V}\mathbf{1}}[\mathrm{diag}(\mathbf{p}_T)]^{-1}\mathbf{V}\mathbf{1}.\end{aligned} \tag{9.91}$$

In turn the resampled allocation, which is the expected value of the above allocations, reads:

$$\begin{aligned}\boldsymbol{\alpha}_{\mathrm{rs}}[i_T] &\equiv \mathrm{E}\left\{\boldsymbol{\alpha}\left(\widehat{\boldsymbol{\mu}}\left[I_T^{\widehat{\boldsymbol{\mu}}[i_T],\widehat{\boldsymbol{\Sigma}}[i_T]}\right], \widehat{\boldsymbol{\Sigma}}\left[I_T^{\widehat{\boldsymbol{\mu}}[i_T],\widehat{\boldsymbol{\Sigma}}[i_T]}\right]\right)\right\} \\ &= [\mathrm{diag}(\mathbf{p}_T)]^{-1}\left(\zeta\,\mathrm{E}\left\{\mathbf{V}\widehat{\boldsymbol{\mu}} - \frac{\mathbf{1}'\mathbf{V}\widehat{\boldsymbol{\mu}}}{\mathbf{1}'\mathbf{V}\mathbf{1}}\mathbf{V}\mathbf{1}\right\} + w_T\,\mathrm{E}\left\{\frac{\mathbf{V}\mathbf{1}}{\mathbf{1}'\mathbf{V}\mathbf{1}}\right\}\right),\end{aligned} \tag{9.92}$$

see Appendix www.9.1.

The expectations in (9.92) are not known in analytical form. Therefore we generate a large number Q of Monte Carlo scenarios from (9.88):

$${}_q\mathbf{V}^{i_T}, \quad q = 1, \ldots, Q, \tag{9.93}$$

where we emphasized that the distribution that generates the Monte Carlo scenarios is determined by the available time series of market invariants i_T. Then we compute the resampled allocation (9.92) as follows:

$$\begin{aligned}\boldsymbol{\alpha}_{\mathrm{rs}}[i_T] \equiv [\mathrm{diag}(\mathbf{p}_T)]^{-1}\Bigg(&\frac{\zeta}{Q}\sum_{q=1}^{Q} {}_q\mathbf{V}^{i_T}\widehat{\boldsymbol{\mu}} \\ &-\frac{\zeta}{Q}\sum_{q=1}^{Q}\frac{\mathbf{1}'{}_q\mathbf{V}^{i_T}\widehat{\boldsymbol{\mu}}}{\mathbf{1}'{}_q\mathbf{V}^{i_T}\mathbf{1}}{}_q\mathbf{V}\mathbf{1} + \frac{w_T}{Q}\sum_{q=1}^{Q}\frac{{}_q\mathbf{V}^{i_T}\mathbf{1}}{\mathbf{1}'{}_q\mathbf{V}^{i_T}\mathbf{1}}\Bigg).\end{aligned} \tag{9.94}$$

Notice that the resampled allocation depends on the available time series of market invariants i_T, because this determines the Monte Carlo simulations (9.93) through (9.87) and (9.88).

In Figure 9.5 we display the resampled allocation $\boldsymbol{\alpha}_{\text{rs}}[i_T]$ along with the sample-based allocation that $\boldsymbol{\alpha}_{\text{s}}[i_T]$ in the plane of the coordinates that determine the investor's satisfaction and constraints, see (8.25) and (8.36):

$$v \equiv \boldsymbol{\alpha}' \operatorname{diag}(\mathbf{p}_T) \boldsymbol{\Sigma} \operatorname{diag}(\mathbf{p}_T) \boldsymbol{\alpha} \tag{9.95}$$

$$e \equiv \boldsymbol{\alpha}' \operatorname{diag}(\mathbf{p}_T)(\mathbf{1} + \boldsymbol{\mu}). \tag{9.96}$$

In the specific case plotted in the figure the resampling process generates an allocation with less opportunity cost than the sample-based allocation. Furthermore the resampled allocation satisfies the constraints, as opposed to the sample-based allocation, compare with Figure 8.6 and the respective discussion. Nonetheless, we remark that there is no guarantee that this will always be the case, see the discussion below in Section 9.3.4.

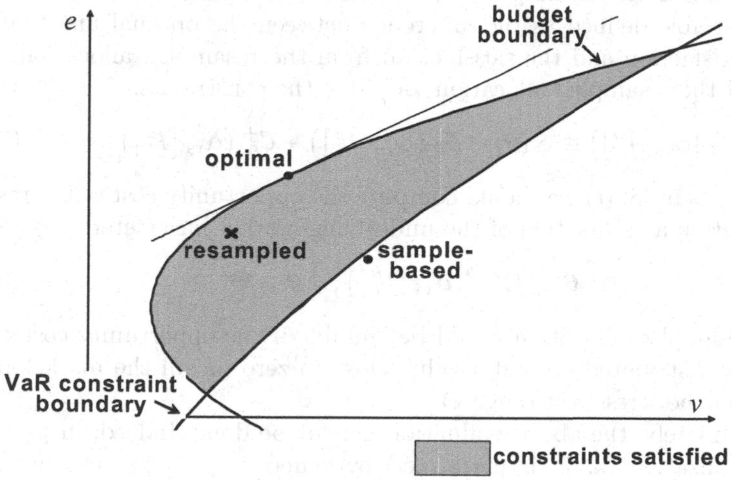

Fig. 9.5. Resampled allocation: comparison with sample-based allocation

9.3.3 Evaluation

To evaluate the sample-based allocation decision we should proceed in principle as in Chapter 8.

First we should consider a set $\boldsymbol{\Theta}$ of market parameters that is broad enough to most likely include the true, unknown value $\boldsymbol{\theta}^{\text{t}}$.

For each value $\boldsymbol{\theta}$ of the market parameters in the stress test set $\boldsymbol{\Theta}$ we should compute the optimal allocation function, see $\boldsymbol{\alpha}(\boldsymbol{\theta})$ (9.66). Then we should compute as in (8.31) the optimal level of satisfaction if $\boldsymbol{\theta}$ are the underlying market parameters, namely $\overline{\mathcal{S}}(\boldsymbol{\theta})$.

Next, we should randomize as in (8.48) the information from the market i_T, generating a distribution of information scenarios that depends on the assumption $\boldsymbol{\theta}$ on the market parameters:

$$I_T^{\boldsymbol{\theta}} \equiv \left\{ \mathbf{X}_1^{\boldsymbol{\theta}}, \ldots, \mathbf{X}_T^{\boldsymbol{\theta}} \right\}. \tag{9.97}$$

Then we should compute the resampled allocation (9.86) from the randomized information, obtaining the random variable $\boldsymbol{\alpha}_{\text{rs}}\left[I_T^{\boldsymbol{\theta}}\right]$.

Next we should compute as in (8.23) the satisfaction $\mathcal{S}_{\boldsymbol{\theta}}\left(\boldsymbol{\alpha}_{\text{rs}}\left[I_T^{\boldsymbol{\theta}}\right]\right)$ ensuing from each scenario of the resampled allocation decision under the assumption $\boldsymbol{\theta}$ for the market parameters, which, we recall, is a random variable. Similarly, from (8.26) and expressions such as (8.35) we should compute the cost of the resampled allocation decision violating the constraints $\mathcal{C}_{\boldsymbol{\theta}}^{+}\left(\boldsymbol{\alpha}_{\text{rs}}\left[I_T^{\boldsymbol{\theta}}\right]\right)$ in each scenario, which is also a random variable.

Then we should compute the opportunity cost (8.53) of the resampled allocation under the assumption $\boldsymbol{\theta}$ for the market parameters, namely the random variable defined as the difference between the optimal unattainable level of satisfaction and the satisfaction from the resampled allocation, plus the cost of the resampled allocation violating the constraints:

$$\text{OC}_{\boldsymbol{\theta}}\left(\boldsymbol{\alpha}_{\text{rs}}\left[I_T^{\boldsymbol{\theta}}\right]\right) \equiv \overline{\mathcal{S}}(\boldsymbol{\theta}) - \mathcal{S}_{\boldsymbol{\theta}}\left(\boldsymbol{\alpha}_{\text{rs}}\left[I_T^{\boldsymbol{\theta}}\right]\right) + \mathcal{C}_{\boldsymbol{\theta}}^{+}\left(\boldsymbol{\alpha}_{\text{rs}}\left[I_T^{\boldsymbol{\theta}}\right]\right). \tag{9.98}$$

Finally, as in (8.57) we should compute the opportunity cost of the resampled allocation as a function of the underlying market parameters:

$$\boldsymbol{\theta} \mapsto \text{OC}_{\boldsymbol{\theta}}\left(\boldsymbol{\alpha}_{\text{rs}}\left[I_T^{\boldsymbol{\theta}}\right]\right), \quad \boldsymbol{\theta} \in \boldsymbol{\Theta}. \tag{9.99}$$

The resampled allocation would be suitable if the opportunity cost turns out tightly distributed around a value close to zero for all the market parameters $\boldsymbol{\theta}$ in the stress test range $\boldsymbol{\Theta}$.

Unfortunately, the above evaluation cannot be done. Indeed, in practice, the randomization (9.97) is performed by generating a large number J of Monte Carlo realizations of the time series of the market invariants:

$$\boldsymbol{\theta} \overset{\text{stress test}}{\mapsto} {}^j i_T \equiv \left\{ {}^j\mathbf{x}_1, \ldots, {}^j\mathbf{x}_T \right\}, \quad j = 1, \ldots, J. \tag{9.100}$$

In turn, in each scenario j the resampled allocation is obtained by implementing a second Monte Carlo simulation as in (9.84). In other words, the distribution of the opportunity cost as a function of the assumptions on the underlying parameters (9.99) is obtained trough the following chain of steps:

$$\begin{aligned} \boldsymbol{\theta} &\overset{\text{stress test}}{\mapsto} {}^j i_T \overset{\text{estimate}}{\mapsto} {}^j_0\widehat{\boldsymbol{\theta}} \overset{\text{resample}}{\mapsto} {}^j_q i_T \overset{\text{estimate}}{\mapsto} {}^j_q\widehat{\boldsymbol{\theta}} \\ &\overset{\text{optimize}}{\mapsto} {}^j_q\boldsymbol{\alpha} \overset{\text{average}}{\mapsto} {}^j\boldsymbol{\alpha}_{\text{rs}} \overset{\text{evaluate}}{\mapsto} \text{OC}_{\boldsymbol{\theta}}\left({}^j\boldsymbol{\alpha}_{\text{rs}}\right). \end{aligned} \tag{9.101}$$

To implement this chain we need to solve an optimization problem for each Monte Carlo scenario q stemming from another Monte Carlo scenario j: the computational burden of this operation is prohibitive.

9.3.4 Discussion

The resampling technique is very innovative. It displays several advantages but also a few drawbacks, see also Markowitz and Usmen (2003) and Ceria and Stubbs (2004).

In the first place, intuitively the expectation in the definition (9.86) of the resampled allocation decision reduces the sensitivity to the market parameters, and thus it gives rise to a less disperse opportunity cost than the sample-based allocation decision. Nonetheless, the proof of this statement for generic markets and preferences is not obvious.

Furthermore, the expectation in the definition (9.86) of the resampled allocation can give rise to resampled allocations that violate the investment constraints, not only in the case where the constraints depend on the unknown market parameters. For instance, consider the constraint (8.15) of not investing in more than M of the N securities in the market: each allocation ${}_q\boldsymbol{\alpha}$ in the average (9.83) satisfies this constraint, but the ensuing resampled allocation does not.

Finally, it is very hard to stress test the performance of this technique due to the excessive computational burden, see (9.101) and comments thereafter.

9.4 Robust allocation

So far the pursuit of optimal allocation strategies has focused on fixing the excessive sensitivity to the input parameters of the optimal allocation function. The robust approach aims directly at determining the "best" allocation, according to the evaluation criteria discussed in Chapter 8.

First we formalize the intuitive definition of robust allocation decisions for general markets and preferences. Then, in order to compute the solution of a robust allocation problem in practice, we resort to the two-step mean-variance framework.

9.4.1 General definition

Consider the opportunity cost of a generic allocation $\boldsymbol{\alpha}$ that satisfies the investment constraints, which is defined in (8.37) as the difference between the maximum possible satisfaction and the actual satisfaction provided by the given allocation:

$$\mathrm{OC}_{\boldsymbol{\theta}}(\boldsymbol{\alpha}) \equiv \overline{\mathcal{S}}(\boldsymbol{\theta}) - \mathcal{S}_{\boldsymbol{\theta}}(\boldsymbol{\alpha}). \tag{9.102}$$

According to the discussion in Section 8.1, since the true value of the market parameters $\boldsymbol{\theta}$ is not known, an allocation is optimal if it gives rise to a minimal opportunity cost for all the values of the market parameters in an uncertainty range $\boldsymbol{\Theta}$ that is broad enough to most likely include the true, unknown value $\boldsymbol{\theta}^{\mathrm{t}}$ of the market parameters. This way in particular the opportunity cost is guaranteed to be low in correspondence of the unknown value $\boldsymbol{\theta}^{\mathrm{t}}$.

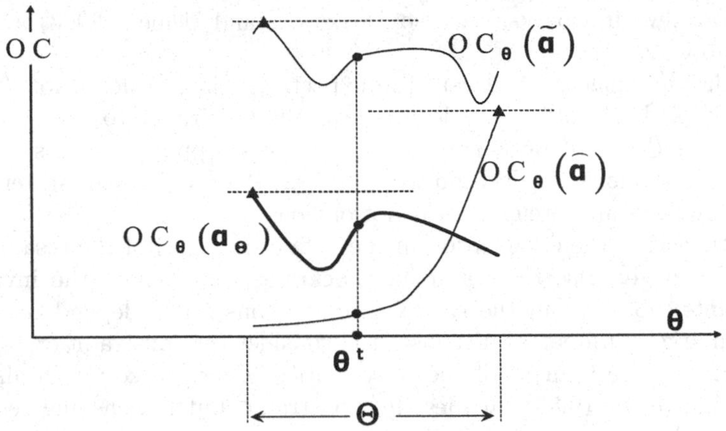

Fig. 9.6. Opportunity cost as function of the market parameters

The robust approach aims precisely at determining an allocation $\boldsymbol{\alpha}$ such that the opportunity cost is uniformly minimal for all the values $\boldsymbol{\theta}$ in the uncertainty range $\boldsymbol{\Theta}$. To make sure that the opportunity cost is uniformly low for all the values $\boldsymbol{\theta}$ in $\boldsymbol{\Theta}$ we take a conservative approach and monitor its maximum over the range $\boldsymbol{\Theta}$, see Figure 9.6. Furthermore, we require that the allocation $\boldsymbol{\alpha}$ satisfies the constraints for all the values $\boldsymbol{\theta}$ in the given range $\boldsymbol{\Theta}$, a condition which we denote as follows:

$$\boldsymbol{\alpha} \in \mathcal{C}_{\boldsymbol{\Theta}} \equiv \{\boldsymbol{\alpha} \in \mathcal{C}_{\boldsymbol{\theta}} \text{ for all } \boldsymbol{\theta} \in \boldsymbol{\Theta}\}. \tag{9.103}$$

In other words, we consider the allocation such that the maximum opportunity cost (9.102) on the given range is the lowest possible:

$$\boldsymbol{\alpha}_{\boldsymbol{\Theta}} \equiv \underset{\boldsymbol{\alpha} \in \mathcal{C}_{\boldsymbol{\Theta}}}{\operatorname{argmin}} \left\{ \max_{\boldsymbol{\theta} \in \boldsymbol{\Theta}} \left\{ \overline{\mathcal{S}}(\boldsymbol{\theta}) - \mathcal{S}_{\boldsymbol{\theta}}(\boldsymbol{\alpha}) \right\} \right\}. \tag{9.104}$$

Notice that this allocation in general does not give rise the least possible opportunity cost in correspondence of the true parameters $\boldsymbol{\theta}^{\mathrm{t}}$, although the damage is guaranteed to be contained, see Figure 9.6.

The allocation (9.104) and its quality depend on the choice for the uncertainty range $\boldsymbol{\Theta}$ of the market parameters, see Figure 9.7. The smaller the

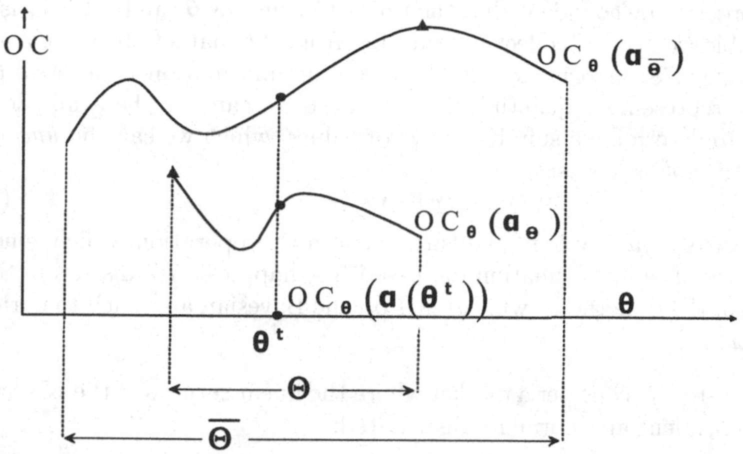

Fig. 9.7. Quality of robust allocation as function of the uncertainty range

range Θ, the lower the maximum value of the opportunity cost generated by α_{Θ} and thus the higher the quality of α_{Θ}. Indeed, in the limit case where the evaluation set is the single true value θ^{t} the ensuing allocation (9.104) becomes the truly optimal solution (8.39), which gives rise to a null opportunity cost. As we expand the evaluation set Θ, the opportunity cost of the best allocation (9.104), although it is uniformly the least among all the possible allocations, increases.

To summarize, we built a recipe to pursue the best allocation by accounting for estimation risk: first, determine an uncertainty range Θ of market parameters that contains the true parameter θ^{t}, and yet it is as small as possible; then solve the optimization (9.104).

Consider our leading example where satisfaction is determined by the certainty-equivalent of an exponential utility function and the investor has a full-investment budget constraint and a value at risk constraint. Assume that we determined a suitable range Θ for μ and Σ. The allocation recipe (9.104) reads in this context:

$$\alpha_{\Theta} \equiv \underset{\alpha}{\operatorname{argmin}} \left\{ \max_{\mu,\Sigma \in \Theta} \left\{ \overline{\mathrm{CE}}(\mu,\Sigma) - \mathrm{CE}_{\mu,\Sigma}(\alpha) \right\} \right\} \tag{9.105}$$

$$\text{subject to} \begin{cases} \alpha' \mathbf{p}_T = w_T \\ \mathrm{Var}_{\mu,\Sigma}(\alpha) \leq \gamma w_T, \text{ for all } \mu,\Sigma \in \Theta, \end{cases}$$

where the explicit expression of the certainty equivalent and the VaR are provided in (8.25), (8.28) and (8.33).

In order to be confident that the range $\boldsymbol{\Theta}$ contains $\boldsymbol{\theta}^{\mathrm{t}}$ and yet it is as small as possible we need to collect information from the market. Just like a generic estimator (8.78) associates with the available information i_T a *value* $\widehat{\boldsymbol{\theta}}$ that suitably represents a quantity of interest, so we can use the available information to determine a suitable *range* of values, which we call the *uncertainty set*, or the *robustness set*:

$$i_T \mapsto \widehat{\boldsymbol{\Theta}}\left[i_T\right]. \tag{9.106}$$

There exists a variety of methods to perform this operation, which generalize the theory of point estimation discussed in Chapter 4. We discuss in Section 9.5 one of these methods, which relies on the Bayesian approach to parameter estimation.

For instance, consider a market where the linear returns of the N securities are independent and normally distributed:

$$\mathbf{L}_t \sim \mathrm{N}\left(\boldsymbol{\mu}^{\mathrm{t}}, \boldsymbol{\Sigma}^{\mathrm{t}}\right), \tag{9.107}$$

where $\boldsymbol{\mu}^{\mathrm{t}}$ and $\boldsymbol{\Sigma}^{\mathrm{t}}$ are the true expected values and covariance matrix respectively.

Assume that the covariance $\boldsymbol{\Sigma}^{\mathrm{t}}$ is known. We have to determine a suitable uncertainty set $\widehat{\boldsymbol{\Theta}}_{\boldsymbol{\mu}}$ for $\boldsymbol{\mu}$ such that we can be confident that the true parameter $\boldsymbol{\mu}^{\mathrm{t}}$ lies within its boundaries. Consider the sample estimator $\widehat{\boldsymbol{\mu}}\left[i_T\right]$ defined in (8.79), and define the uncertainty set $\widehat{\boldsymbol{\Theta}}$ as follows:

$$\widehat{\boldsymbol{\Theta}}_{\boldsymbol{\mu}}\left[i_T\right] \equiv \left\{\boldsymbol{\mu} \text{ such that } \mathrm{Ma}^2\left(\boldsymbol{\mu}, \widehat{\boldsymbol{\mu}}\left[i_T\right], \boldsymbol{\Sigma}^{\mathrm{t}}\right) \leq \frac{Q_{\chi^2_N}(p)}{T}\right\}, \tag{9.108}$$

where Ma is the Mahalanobis distance (2.61) of $\boldsymbol{\mu}$ from $\widehat{\boldsymbol{\mu}}$ induced by the metric $\boldsymbol{\Sigma}^{\mathrm{t}}$; $Q_{\chi^2_N}(p)$ is the quantile of the chi-square distribution with N degrees of freedom (1.109) for a confidence level $p \in (0,1)$; and T is the number of observations in the time series of the returns that we use to estimate $\widehat{\boldsymbol{\mu}}$.

The set (9.108) is an ellipsoid centered in $\widehat{\boldsymbol{\mu}}$, with shape determined by $\boldsymbol{\Sigma}^{\mathrm{t}}$ and with radius proportional to $1/\sqrt{T}$, see (A.73) and comments thereafter. As we show in Appendix www.9.2 the following result holds for the probability that the range (9.108) captures the true expected values:

$$\mathbb{P}\left\{\boldsymbol{\mu}^{\mathrm{t}} \in \widehat{\boldsymbol{\Theta}}_{\boldsymbol{\mu}}\left[i_T\right]\right\} = p. \tag{9.109}$$

In other words, with a confidence p that we can set arbitrarily, the true parameter $\boldsymbol{\mu}^{\mathrm{t}}$ lies within the set (9.108): as we require a higher confidence, the quantile in (9.108) increases, and so does the size of the ellipsoid. As intuition suggests, for a given confidence p, the more information is available, i.e. the larger the number of observations T in the time series of the returns, the smaller the uncertainty ellipsoid.

By letting the evaluation range in the optimization problem (9.104) be determined by currently available information as in (9.106), we obtain the definition of the *robust allocation decision*:

$$\boldsymbol{\alpha}_{\mathrm{r}}\left[i_T\right] \equiv \underset{\boldsymbol{\alpha}\in\mathcal{C}_{\widehat{\Theta}[i_T]}}{\operatorname{argmin}} \left\{ \max_{\boldsymbol{\theta}\in\widehat{\Theta}[i_T]} \left\{ \overline{\mathcal{S}}\left(\boldsymbol{\theta}\right) - \mathcal{S}_{\boldsymbol{\theta}}\left(\boldsymbol{\alpha}\right)\right\} \right\}. \tag{9.110}$$

This is indeed a decision, which processes the currently available information, see (8.38).

The smaller the uncertainty set $\widehat{\Theta}$ in (9.110), the less conservative the investor from the point of view of estimation risk. Indeed, in the limit where the robustness set consists of only one point, namely the point estimate $\widehat{\boldsymbol{\theta}}$, the robust allocation decision becomes the sample-based allocation decision (8.81). Nevertheless, we stress that if the uncertainty set is very likely to include the true unknown parameters, the smaller the uncertainty set, the better the quality of the robust allocation, see Figure 9.7.

In Appendix www.9.5 we show that using the uncertainty set (9.108) in (9.105) the ensuing robust allocation decision solves the following problem:

$$\boldsymbol{\alpha}_{\mathrm{r}} \equiv \underset{\boldsymbol{\alpha}}{\operatorname{argmin}} \left\{ \max_{\boldsymbol{\mu}\in\widehat{\Theta}_{\boldsymbol{\mu}}} \left\{ \begin{array}{l} \boldsymbol{\mu}'\mathbf{T}\boldsymbol{\mu} + \frac{w_T}{A}\mathbf{1}'\left(\boldsymbol{\Sigma}^{\mathrm{t}}\right)^{-1}\boldsymbol{\mu} \\ -\boldsymbol{\alpha}'\operatorname{diag}\left(\mathbf{p}_T\right)\boldsymbol{\mu} + \frac{1}{2\zeta}\left\|\boldsymbol{\Lambda}^{1/2}\mathbf{E}'\boldsymbol{\alpha}\right\|^2 \end{array} \right\} \right\}, \tag{9.111}$$

subject to:

$$\left\{ \begin{array}{l} \boldsymbol{\alpha}'\mathbf{p}_T = w_T \\ \sqrt{2}\tau\left\|\boldsymbol{\Lambda}^{1/2}\mathbf{E}'\boldsymbol{\alpha}\right\| \leq \widehat{\boldsymbol{\mu}}'\operatorname{diag}\left(\mathbf{p}_T\right)\boldsymbol{\alpha} + \gamma w_T \\ \qquad\qquad + \left\|\boldsymbol{\Lambda}^{1/2}\mathbf{E}'\boldsymbol{\alpha}\right\|^2 - \frac{\sqrt{Q_N(p)/T}}{\left\|\boldsymbol{\Lambda}^{1/2}\mathbf{E}'\boldsymbol{\alpha}\right\|}. \end{array} \right. \tag{9.112}$$

In this expression

$$\begin{aligned} A &\equiv \mathbf{1}'\left(\boldsymbol{\Sigma}^{\mathrm{t}}\right)^{-1}\mathbf{1} \\ \tau &\equiv \operatorname{erf}^{-1}\left(2c-1\right) \\ \mathbf{T} &\equiv \frac{\zeta}{2}\left(\boldsymbol{\Sigma}^{\mathrm{t}}\right)^{-1}\left(\mathbf{I} - \frac{1}{A}\mathbf{1}\mathbf{1}'\left(\boldsymbol{\Sigma}^{\mathrm{t}}\right)^{-1}\right); \end{aligned} \tag{9.113}$$

and $\boldsymbol{\Lambda}$ and $\mathbf{E}$ are the eigenvalues and the eigenvectors respectively of the following spectral decomposition:

$$\operatorname{diag}\left(\mathbf{p}_T\right)\boldsymbol{\Sigma}^{\mathrm{t}}\operatorname{diag}\left(\mathbf{p}_T\right) \equiv \mathbf{E}\boldsymbol{\Lambda}^{1/2}\boldsymbol{\Lambda}^{1/2}\mathbf{E}'. \tag{9.114}$$

The maximization for a given $\boldsymbol{\alpha}$ in (9.111) is satisfied by the tangency condition of ellipsoidal contours in the variable $\boldsymbol{\mu}$ with a fixed ellipsoid: this problem does not admit analytical solutions, as it is a modification of the spectral equation, see (A.68). Therefore, the second optimization, namely the minimization in (9.111) cannot be performed. Furthermore, the VaR constraint in (9.112) is not a conic constraint. Therefore the solution of the robust allocation is not numerically tractable, see Section 6.2.

9.4.2 Practicable definition: the mean-variance setting

Although the rationale behind the robust allocation decision is conceptually simple, solving the min-max optimization (9.110) is close to impossible even under simple assumptions on preferences, markets and constraints, as we have seen in the example (9.111)-(9.112).

Therefore robust allocation is tackled in practice within the two-step mean-variance framework. This is not surprising, since we resorted to the mean-variance approximation even in the classical setting that disregards estimation risk. When the robust allocation problem is set in the mean-variance framework we can apply recent results on robust optimization, see El Ghaoui and Lebret (1997) and Ben-Tal and Nemirovski (1995), see also Ben-Tal and Nemirovski (2001).

We recall from Section 6.3 that the mean-variance approach is a two-step simplification of a generic allocation problem: the investor first determines a set of mean-variance efficient allocations and then he selects among those allocations the one that better suits him.

We assume that the investment constraints $\mathcal{C}$ do not depend on the unknown market parameters and are such that the inequality version (6.144) of the mean-variance problem applies, see Section 6.5.3. In this setting the mean-variance problem can be written as follows:

$$\begin{aligned} \boldsymbol{\alpha}^{(i)} &= \underset{\boldsymbol{\alpha}}{\operatorname{argmax}}\, \boldsymbol{\alpha}'\boldsymbol{\mu} \\ &\text{subject to } \begin{cases} \boldsymbol{\alpha} \in \mathcal{C} \\ \boldsymbol{\alpha}'\boldsymbol{\Sigma}\boldsymbol{\alpha} \leq v^{(i)}. \end{cases} \end{aligned} \tag{9.115}$$

In this expression $\boldsymbol{\mu}$ and $\boldsymbol{\Sigma}$ are the expected value and the covariance matrix respectively of the market vector $\mathbf{M}$:

$$\boldsymbol{\mu} \equiv \mathrm{E}\{\mathbf{M}\}, \quad \boldsymbol{\Sigma} \equiv \mathrm{Cov}\{\mathbf{M}\}; \tag{9.116}$$

the market vector $\mathbf{M}$ in turn is the affine transformation of the prices at the investment horizon $\mathbf{P}_{T+\tau}$ which together with the allocation vector $\boldsymbol{\alpha}$ determines the investor's objective $\Psi \equiv \boldsymbol{\alpha}'\mathbf{M}$, see (5.10); the set $\{v^{(1)}, \ldots, v^{(I)}\}$ is a significative grid of target variances of the investor's objective.

According to (9.110), the robust version of the mean-variance problem (9.115) reads:

$$\boldsymbol{\alpha}_{\mathrm{r}}^{(i)} = \underset{\boldsymbol{\alpha}}{\operatorname{argmax}} \left\{ \min_{\boldsymbol{\mu} \in \widehat{\boldsymbol{\Theta}}_{\boldsymbol{\mu}}} \{\boldsymbol{\alpha}'\boldsymbol{\mu}\} \right\} \tag{9.117}$$

$$\text{subject to} \begin{cases} \boldsymbol{\alpha} \in \mathcal{C} \\ \max_{\boldsymbol{\Sigma} \in \widehat{\boldsymbol{\Theta}}_{\boldsymbol{\Sigma}}} \{\boldsymbol{\alpha}'\boldsymbol{\Sigma}\boldsymbol{\alpha}\} \leq v^{(i)}, \end{cases}$$

where $\widehat{\boldsymbol{\Theta}}_{\boldsymbol{\mu}}$ and $\widehat{\boldsymbol{\Theta}}_{\boldsymbol{\Sigma}}$ are uncertainty sets for the market parameters (9.116) that are estimated from the available information i_T. Depending on the specification of these uncertainty sets, the resulting robust problem assumes different forms.

- **Known covariances, elliptical set for expected values**

A possible specification for the uncertainty sets assumes an ellipsoidal shape for the uncertainty on the parameter $\boldsymbol{\mu}$ and no uncertainty for $\boldsymbol{\Sigma}$:

$$\widehat{\boldsymbol{\Theta}}_{\boldsymbol{\mu}} \equiv \left\{\boldsymbol{\mu} \text{ such that } \operatorname{Ma}^2(\boldsymbol{\mu}, \mathbf{m}, \mathbf{T}) \leq q^2\right\} \tag{9.118}$$

$$\widehat{\boldsymbol{\Theta}}_{\boldsymbol{\Sigma}} \equiv \widehat{\boldsymbol{\Sigma}}. \tag{9.119}$$

In this expression $\widehat{\boldsymbol{\Sigma}}$ is a point estimate of $\boldsymbol{\Sigma}$; $\mathbf{m}$ is an N-dimensional vector; $\mathbf{T}$ is an $N \times N$ symmetric and positive matrix; Ma is the Mahalanobis distance (2.61) of $\boldsymbol{\mu}$ from $\mathbf{m}$ induced by the metric $\mathbf{T}$; and

$$q^2 \equiv Q_{\chi^2_N}(p) \tag{9.120}$$

is the quantile of the chi-square distribution with N degrees of freedom (1.109) for a confidence level $p \in (0,1)$.

Ceria and Stubbs (2004) consider the following specification in (9.118):

$$\mathbf{m} \equiv \widehat{\boldsymbol{\mu}}[i_T], \ \mathbf{T} \text{ exogenous}, \tag{9.121}$$

where $\widehat{\boldsymbol{\mu}}$ is a sample-based estimator of the true parameter.

De Santis and Foresi (2002) blend a market model with the investor's views by specifying the parameters in (9.118) in terms of the Black-Litterman posterior distribution (9.44):

$$\mathbf{m} \equiv \boldsymbol{\mu}_{\mathrm{BL}}, \quad \mathbf{T} \equiv \boldsymbol{\Sigma}_{\mathrm{BL}}. \tag{9.122}$$

The uncertainty set (9.118) is an ellipsoid centered in $\mathbf{m}$ whose shape is determined by $\mathbf{T}$, see (A.73) and comments thereafter. The rationale behind the assumption (9.118) is that the uncertainty about $\boldsymbol{\mu}$ is approximately normally distributed:

$$\boldsymbol{\mu} \sim \mathrm{N}(\mathbf{m}, \mathbf{T}), \tag{9.123}$$

see also (9.108). In Appendix www.7.1 we show that in this case the following result holds for the probability that the range captures the true expected values:

$$\mathbb{P}\left\{\boldsymbol{\mu} \in \widehat{\boldsymbol{\Theta}}_{\boldsymbol{\mu}}\right\} = p. \tag{9.124}$$

If the investor considers small ellipsoids by setting p close to zero, he is little worried about missing the true expected values in the optimization (9.117). In other words, he is very aggressive as far as estimation risk is concerned. On the other hand, if the the investor sets p close to one, he is very cautious from the point of view of estimation risk.

As we discuss in Section 9.4.3, if the investment constraints $\mathcal{C}$ are sufficiently regular, the optimization (9.117) simplifies to a second-order cone programming problem and thus the robust frontier can be computed numerically.

- **Box set for expected values, elliptical set for covariances**

An alternative specification of the uncertainty sets in the robust optimization (9.117) is adopted by Goldfarb and Iyengar (2003). The uncertainty set for the expected values is of the box-form:

$$\widehat{\boldsymbol{\Theta}}_{\boldsymbol{\mu}} \equiv \left\{\boldsymbol{\mu} \text{ such that } \underline{\boldsymbol{\mu}} \leq \boldsymbol{\mu} \leq \overline{\boldsymbol{\mu}}\right\}. \tag{9.125}$$

The uncertainty set for the covariance matrix follows from a K-factor model such as (3.119), where factors and perturbations are uncorrelated. In other words, the uncertainty set for the covariance matrix is specified as follows:

$$\widehat{\boldsymbol{\Theta}}_{\boldsymbol{\Sigma}} \equiv \left\{\mathbf{BGB}' + \operatorname{diag}(\mathbf{d})\right\}. \tag{9.126}$$

In this expression $\underline{\mathbf{d}} \leq \mathbf{d} \leq \overline{\mathbf{d}}$; the covariance $\mathbf{G}$ of the factors is assumed known, and each row $\mathbf{b}_{(n)}$ of the $N \times K$ matrix of the factor loadings $\mathbf{B}$ belongs to an ellipsoid such as $(A.73)$:

$$\mathbf{b}_{(n)} \in \mathcal{E}_n, \quad n = 1, \ldots, N. \tag{9.127}$$

As it turns out, when the investment constraints $\mathcal{C}$ are sufficiently regular, this specification also gives rise to a second-order cone programming problem. Therefore the robust frontier can computed numerically, see Section 6.2.

- **Box set for expected values, box set for covariances**

A third possible specification of the uncertainty sets in (9.117) is provided by Halldorsson and Tutuncu (2003), who assume box-sets for all the parameters:

$$\widehat{\boldsymbol{\Theta}}_{\boldsymbol{\mu}} \equiv \left\{\boldsymbol{\mu} \text{ such that } \underline{\boldsymbol{\mu}} \leq \boldsymbol{\mu} \leq \overline{\boldsymbol{\mu}}\right\} \tag{9.128}$$

$$\widehat{\boldsymbol{\Theta}}_{\boldsymbol{\Sigma}} \equiv \left\{\boldsymbol{\Sigma} \succeq \mathbf{0} \text{ such that } \underline{\boldsymbol{\Sigma}} \leq \boldsymbol{\Sigma} \leq \overline{\boldsymbol{\Sigma}}\right\}, \tag{9.129}$$

where the notation $\boldsymbol{\Sigma} \succeq \mathbf{0}$ stands for symmetric and positive matrices. Under further assumptions on the investment constraints $\mathcal{C}$, the ensuing robust mean-variance problem can be cast in the form of a saddle-point search and solved numerically with an interior-point algorithm.

9.4.3 Discussion

As we show in Appendix www.9.6, the robust mean-variance problem (9.117) under the specifications (9.118)-(9.119) for the robustness sets can be written equivalently as follows:

$$\boldsymbol{\alpha}_{\mathrm{r}}^{(i)} = \underset{\boldsymbol{\alpha}}{\operatorname{argmax}} \left\{ \boldsymbol{\alpha}'\mathbf{m} - q\sqrt{\boldsymbol{\alpha}'\mathbf{T}\boldsymbol{\alpha}} \right\} \tag{9.130}$$
$$\text{subject to } \begin{cases} \boldsymbol{\alpha} \in \mathcal{C} \\ \boldsymbol{\alpha}'\widehat{\boldsymbol{\Sigma}}\boldsymbol{\alpha} \leq v^{(i)}. \end{cases}$$

If the investment constraints $\mathcal{C}$ are regular enough, this problem can be cast in the form of a second-order cone programming problem (6.55), see Appendix www.9.6. Therefore the robust frontier can computed numerically.

The robust efficient frontier (9.130) represents a two-parameter family of allocations, i.e. a surface, determined by the target variance v, which represents market risk, and the size of the uncertainty ellipsoid, which is directly related to q and represents aversion to estimation risk, see (9.124) and comments thereafter.

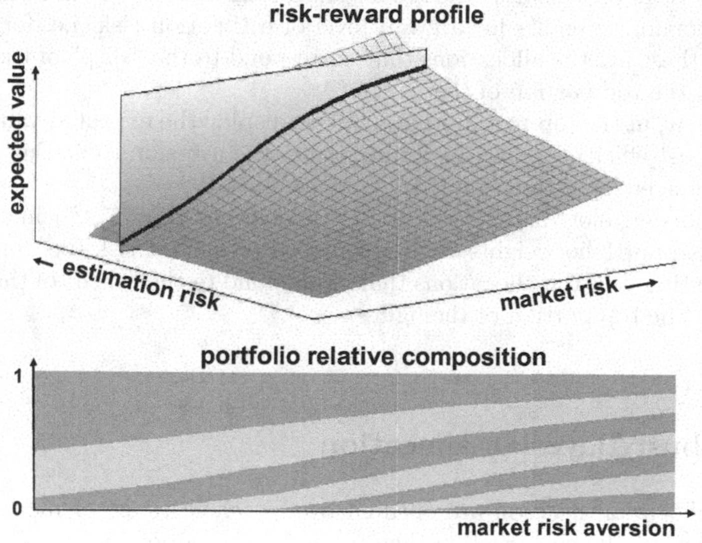

Fig. 9.8. Robust efficient allocations: fixed aversion to estimation risk

We can parameterize the robust surface (9.130) equivalently in terms of a market risk Lagrange multiplier $\gamma_m \geq 0$ and an estimation risk multiplier $\gamma_e \equiv q$ as follows:

$$\boldsymbol{\alpha}_r(\gamma_m, \gamma_e) = \underset{\boldsymbol{\alpha} \in \mathcal{C}}{\operatorname{argmax}} \left\{ \boldsymbol{\alpha}'\mathbf{m} - \gamma_m \sqrt{\boldsymbol{\alpha}'\widehat{\boldsymbol{\Sigma}}\boldsymbol{\alpha}} - \gamma_e \sqrt{\boldsymbol{\alpha}'\mathbf{T}\boldsymbol{\alpha}} \right\}. \tag{9.131}$$

This way we obtain the following interpretation of the two-parameter robust frontier: investors balance the trade off between the expected value of their objective, represented by the term $\boldsymbol{\alpha}'\mathbf{m}$, and risk. Risk appears in two forms: market risk, represented by the market volatility $\sqrt{\boldsymbol{\alpha}'\widehat{\boldsymbol{\Sigma}}\boldsymbol{\alpha}}$, and estimation risk, represented by the estimation uncertainty $\sqrt{\boldsymbol{\alpha}'\mathbf{T}\boldsymbol{\alpha}}$.

Larger values of the multiplier γ_m give rise to allocations that suit investors who are more averse to market risk: therefore we can interpret γ_m as a market risk aversion parameter. Similarly, larger values of the multiplier γ_e give rise to allocations that suit investors who are more averse to estimation risk: therefore we can interpret γ_e as an estimation risk aversion parameter.

In Figure 9.8 we compute the robust efficient frontier for a market of $N \equiv 7$ securities, under the standard constraints of no short-selling, namely $\boldsymbol{\alpha} \geq \mathbf{0}$, and of full investment of the initial budget w_T, namely $\boldsymbol{\alpha}'\mathbf{p}_T = w_T$

The top plot displays the expected value $\mathbf{m}'\boldsymbol{\alpha}_\mathrm{r}$ of the robust efficient surface (9.131) as a function of the aversion to market risk γ_m and of the aversion to estimation risk γ_e.

The bottom plot displays the robust allocations $\boldsymbol{\alpha}_\mathrm{r}(\gamma_m, \overline{\gamma_e})$ in terms of the relative portfolio weights for a given level of estimation risk, i.e. for a fixed value $\overline{\gamma_e}$: these are the allocations that correspond to the "slice" of the robust surface in the top portion of the figure.

Similarly, in the top plot in Figure 9.9 we display the expected value $\mathbf{m}'\boldsymbol{\alpha}_\mathrm{r}$ of the robust efficient surface as a function of the aversion to market risk γ_m and of the aversion estimation risk γ_e.

The bottom plot displays the robust allocations $\boldsymbol{\alpha}_\mathrm{r}(\overline{\gamma_m}, \gamma_e)$ in terms of the relative portfolio weights for a given level of market risk, i.e. for a fixed value $\overline{\gamma_m}$: these are the allocations that correspond to the "slice" of the robust surface in the top portion of the figure.

9.5 Robust Bayesian allocation

Robust allocation decision are optimal over a whole range of market parameters, because by construction they minimize the opportunity cost over the given range. Nevertheless, in the classical approach, the choice of the robustness range is quite arbitrary.

Using the Bayesian approach to estimation we can naturally identify a suitable robustness range for the market parameters: robust Bayesian allocation decisions account for estimation risk over a range of market parameters that includes both the available information and the investor's experience according to a self-adjusting mechanism.

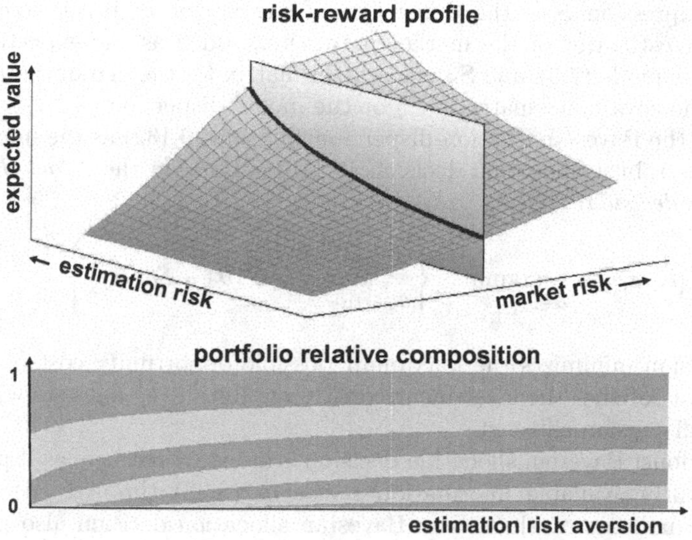

Fig. 9.9. Robust efficient allocations: fixed aversion to market risk

9.5.1 General definition

The robust allocation decision (9.110) minimizes the opportunity cost due to estimation risk uniformly over the uncertainty set $\widehat{\boldsymbol{\Theta}}$ for market parameters. The choice of the uncertainty set is crucial for the success of the respective allocation strategy: on the one hand $\widehat{\boldsymbol{\Theta}}$ should be as small as possible, in order to keep the maximum possible opportunity cost low; on the other hand $\widehat{\boldsymbol{\Theta}}$ should be as large as possible, in order to most likely include the true unknown parameters.

The Bayesian framework defines uncertainty sets in a natural way. Indeed, in the Bayesian framework the unknown market parameters $\boldsymbol{\theta}$ are random variables. The likelihood that the parameters assume given values is described by the posterior probability density function $f_{\text{po}}(\boldsymbol{\theta})$, which is determined by the available information i_T and by the investor's experience e_C, see Figure 7.1. The region where the posterior distribution displays a higher concentration deserves more attention than the tails of the distribution: this region is a natural choice for the uncertainty set $\widehat{\boldsymbol{\Theta}}$.

From the discussion in Section 7.1.2, the region where the posterior distribution displays a higher concentration is represented by the location-dispersion ellipsoid of the market parameters (7.10), see Figure 7.2:

$$\widehat{\boldsymbol{\Theta}}^q [i_T, e_C] \equiv \left\{ \boldsymbol{\theta} : \left(\boldsymbol{\theta} - \widehat{\boldsymbol{\theta}}_{\text{ce}} \right)' \mathbf{S}_{\boldsymbol{\theta}}^{-1} \left(\boldsymbol{\theta} - \widehat{\boldsymbol{\theta}}_{\text{ce}} \right) \leq q^2 \right\}. \quad (9.132)$$

In this expression S is the dimension of the vector $\boldsymbol{\theta}$; $\widehat{\boldsymbol{\theta}}_{\text{ce}}$ is a classical-equivalent estimator of the market parameters, such as the expected value (7.5) or the mode (7.6); and $\mathbf{S}_{\boldsymbol{\theta}}$ is a scatter matrix for the market parameters, such as the covariance matrix (7.7) or the modal dispersion (7.8).

Using the Bayesian location-dispersion ellipsoid (9.132) as the uncertainty set for the robust allocation decision (9.110) we obtain the *robust Bayesian allocation decision*:

$$\boldsymbol{\alpha}_{\text{rB}}\left[i_T, e_C\right] \equiv \underset{\boldsymbol{\alpha} \in \mathcal{C}_{\widehat{\Theta}^q[i_T, e_C]}}{\operatorname{argmin}} \left\{ \max_{\boldsymbol{\theta} \in \widehat{\Theta}^q[i_T, e_C]} \left\{ \overline{\mathcal{S}}\left(\boldsymbol{\theta}\right) - \mathcal{S}_{\boldsymbol{\theta}}\left(\boldsymbol{\alpha}\right) \right\} \right\}. \tag{9.133}$$

This decision minimizes the maximum possible opportunity cost of an allocation that satisfies the investment constraints for all the markets within the location-dispersion ellipsoid.

The robust Bayesian allocation decision is indeed a decision, as it processes the currently available information i_T as in (8.38) through the ellipsoid (9.132). Furthermore, the robust Bayesian allocation decision also processes the investor's experience e_C within a sound statistical framework. Finally, the robust Bayesian allocation decision also depends on the radius factor q. From (7.11) and (7.12) we can interpret q as the investor's aversion to estimation risk: the smaller q, the smaller the ellipsoid, the higher the chances that the true value of the market parameters are not included within the boundaries of the uncertainty set.

The interplay among the available information i_T, the investor's experience e_C and the investor's aversion to estimation risk q shapes the uncertainty set (9.132) and thus the robust Bayesian allocation decision (9.133) in a self-adjusting way.

Due to (7.4), when the confidence C in the investor's experience e_C is very large compared to the amount of information T from the market, the posterior distribution becomes extremely peaked around the prior $\boldsymbol{\theta}_0$. Therefore, no matter the aversion to estimation risk q, the robustness set (9.132) shrinks to the point $\boldsymbol{\theta}_0$, see the discussion in Section 7.1.2. In other words, the robust Bayesian allocation decision (9.133) becomes:

$$\boldsymbol{\alpha}_{\text{p}} \equiv \underset{\boldsymbol{\alpha} \in \mathcal{C}_{\boldsymbol{\theta}_0}}{\operatorname{argmax}} \left\{ \mathcal{S}_{\boldsymbol{\theta}_0}\left(\boldsymbol{\alpha}\right) \right\}. \tag{9.134}$$

This is a prior allocation decision, see (8.64).

Similarly, due to (7.4), when the amount T of information on the market i_T is very large compared to the confidence C in the investor's experience e_C, the posterior distribution becomes extremely peaked around its classical-equivalent estimator, which is determined by the sample i_T. Therefore, no matter the aversion to estimation risk q, the robustness set (9.132) shrinks to a point, namely the sample estimate $\widehat{\boldsymbol{\theta}}\left[i_T\right]$, see the discussion in Section 7.1.2. In other words the robust Bayesian allocation decision (9.133) becomes:

$$\boldsymbol{\alpha}_{\mathrm{s}}[i_T] \equiv \underset{\boldsymbol{\alpha} \in \mathcal{C}_{\widehat{\boldsymbol{\theta}}[i_T]}}{\operatorname{argmax}} \left\{ \mathcal{S}_{\widehat{\boldsymbol{\theta}}[i_T]}(\boldsymbol{\alpha}) \right\}. \tag{9.135}$$

This is the sample-based allocation decision, see (8.81).

When the aversion to estimation risk q in the definition of the robustness set (9.132) tends to zero, the radius of the ellipsoid shrinks to zero and thus the ellipsoid degenerates to a point, its center, which is the classical-equivalent estimator $\widehat{\boldsymbol{\theta}}_{\mathrm{ce}}$. Therefore the robust Bayesian allocation decision (9.133) becomes:

$$\boldsymbol{\alpha}_{\mathrm{ce}}[i_T, e_C] \equiv \underset{\boldsymbol{\alpha} \in \mathcal{C}_{\widehat{\boldsymbol{\theta}}_{\mathrm{ce}}[i_T, e_C]}}{\operatorname{argmax}} \left\{ \mathcal{S}_{\widehat{\boldsymbol{\theta}}_{\mathrm{ce}}[i_T, e_C]}(\boldsymbol{\alpha}) \right\}. \tag{9.136}$$

This is the classical-equivalent Bayesian allocation decision, see (9.13).

For all the intermediate cases, the robust Bayesian allocation decision smoothly blends the information from the market with the investor's experience, at the same time accounting for estimation risk, within a sound, self-adjusting statistical framework.

9.5.2 Practicable definition: the mean-variance setting

The conceptually simple robust Bayesian allocation decision (9.133) cannot be computed in practice even under simple assumptions on preferences, markets and constraints. Therefore, it must be implemented within the two-step mean-variance framework, where the investor first determines a set of efficient allocations and then selects among those allocations the one that best suits him.

We assume that the investment constraints $\mathcal{C}$ do not depend on the unknown market parameters and are such that the inequality version (6.144) of the mean-variance problem applies. Furthermore, it is convenient to set up the mean-variance problem in terms of relative weights and linear returns, see Section 6.3.4.

With these settings the mean-variance problem can be written as follows:

$$\mathbf{w}^{(i)} = \underset{\mathbf{w}}{\operatorname{argmax}}\, \mathbf{w}'\boldsymbol{\mu} \tag{9.137}$$
$$\text{subject to } \begin{cases} \mathbf{w} \in \mathcal{C} \\ \mathbf{w}'\boldsymbol{\Sigma}\mathbf{w} \leq v^{(i)}, \end{cases}$$

where $\boldsymbol{\mu}$ and $\boldsymbol{\Sigma}$ represent the expected values and the covariances of the linear returns on the securities relative to the investment horizon:

$$\boldsymbol{\mu} \equiv \mathrm{E}\{\mathbf{L}_{T+\tau,\tau}\}, \quad \boldsymbol{\Sigma} \equiv \mathrm{Cov}\{\mathbf{L}_{T+\tau,\tau}\}; \tag{9.138}$$

and the set $\{v^{(1)}, \ldots, v^{(I)}\}$ is a significative grid of target variances of the return on the portfolio.

According to (9.133), the robust Bayesian version of the mean-variance problem (9.137) reads:

$$\mathbf{w}_{\text{rB}}^{(i)} = \underset{\mathbf{w}}{\operatorname{argmax}} \left\{ \min_{\boldsymbol{\mu} \in \widehat{\boldsymbol{\Theta}}_{\boldsymbol{\mu}}} \{\mathbf{w}'\boldsymbol{\mu}\} \right\} \tag{9.139}$$

$$\text{subject to} \begin{cases} \mathbf{w} \in \mathcal{C} \\ \max_{\boldsymbol{\Sigma} \in \widehat{\boldsymbol{\Theta}}_{\boldsymbol{\Sigma}}} \{\mathbf{w}'\boldsymbol{\Sigma}\mathbf{w}\} \leq v^{(i)}, \end{cases}$$

where $\widehat{\boldsymbol{\Theta}}_{\boldsymbol{\mu}}$ and $\widehat{\boldsymbol{\Theta}}_{\boldsymbol{\Sigma}}$ are location-dispersion ellipsoids for $\boldsymbol{\mu}$ and $\boldsymbol{\Sigma}$ respectively, defined in terms of the Bayesian posterior distribution of these parameters.

In order to specify the posterior distribution of $\boldsymbol{\mu}$ and $\boldsymbol{\Sigma}$ we make a few further assumptions, see also Meucci (2005): first, the market consists of equity-like securities for which the linear returns are market invariants, see Section 3.1.1; second, the investment horizon and the estimation interval coincide, see Section 6.5.4; third, the linear returns are normally distributed:

$$\mathbf{L}_{t,\tau}|\boldsymbol{\mu}, \boldsymbol{\Sigma} \sim \mathrm{N}(\boldsymbol{\mu}, \boldsymbol{\Sigma}). \tag{9.140}$$

Furthermore, we model the investor's prior experience as a normal-inverse-Wishart distribution:

$$\boldsymbol{\mu}|\boldsymbol{\Sigma} \sim \mathrm{N}\left(\boldsymbol{\mu}_0, \frac{\boldsymbol{\Sigma}}{T_0}\right), \quad \boldsymbol{\Sigma}^{-1} \sim \mathrm{W}\left(\nu_0, \frac{\boldsymbol{\Sigma}_0^{-1}}{\nu_0}\right). \tag{9.141}$$

We recall from Section 7.2 that $(\boldsymbol{\mu}_0, \boldsymbol{\Sigma}_0)$ represents the investor's experience on the parameters. On the other hand, (T_0, ν_0) represents the respective confidence. Therefore the investor's experience is summarized in:

$$e_C \equiv \{\boldsymbol{\mu}_0, \boldsymbol{\Sigma}_0; T_0, \nu_0\}. \tag{9.142}$$

Under the above hypotheses it is possible to compute the posterior distribution of $\boldsymbol{\mu}$ and $\boldsymbol{\Sigma}$ analytically, see Section 7.2. The information from the market is summarized by the sample mean and the sample covariance of the past realizations of the linear returns, namely

$$\widehat{\boldsymbol{\mu}} \equiv \frac{1}{T}\sum_{t=1}^{T} \mathbf{l}_{t,\tau}, \quad \widehat{\boldsymbol{\Sigma}} \equiv \frac{1}{T}\sum_{t=1}^{T} (\mathbf{l}_{t,\tau} - \widehat{\boldsymbol{\mu}})(\mathbf{l}_{t,\tau} - \widehat{\boldsymbol{\mu}}), \tag{9.143}$$

plus the length of the time-series:

$$i_T \equiv \left\{\widehat{\boldsymbol{\mu}}, \widehat{\boldsymbol{\Sigma}}; T\right\}. \tag{9.144}$$

The posterior distribution, like the prior distribution (9.141), is also normal-inverse-Wishart, where the respective parameters read:

$$T_1[i_T, e_C] \equiv T_0 + T \tag{9.145}$$

$$\boldsymbol{\mu}_1[i_T, e_C] \equiv \frac{1}{T_1}[T_0\boldsymbol{\mu}_0 + T\widehat{\boldsymbol{\mu}}] \tag{9.146}$$

$$\nu_1[i_T, e_C] \equiv \nu_0 + T \tag{9.147}$$

$$\boldsymbol{\Sigma}_1[i_T, e_C] \equiv \frac{1}{\nu_1}\left[\nu_0\boldsymbol{\Sigma}_0 + T\widehat{\boldsymbol{\Sigma}} + \frac{(\boldsymbol{\mu}_0 - \widehat{\boldsymbol{\mu}})(\boldsymbol{\mu}_0 - \widehat{\boldsymbol{\mu}})'}{\frac{1}{T} + \frac{1}{T_0}}\right]. \tag{9.148}$$

The uncertainty set for $\boldsymbol{\mu}$ is the location-dispersion ellipsoid (7.37) of the marginal posterior distribution of $\boldsymbol{\mu}$:

$$\widehat{\boldsymbol{\Theta}}_{\mu} \equiv \left\{\boldsymbol{\mu} : \; (\boldsymbol{\mu} - \widehat{\boldsymbol{\mu}}_{\text{ce}})' \mathbf{S}_{\mu}^{-1} (\boldsymbol{\mu} - \widehat{\boldsymbol{\mu}}_{\text{ce}}) \leq q_{\mu}^2\right\}. \tag{9.149}$$

In this expression q_{μ} is the radius factor that represents aversion to estimation risk for $\boldsymbol{\mu}$; $\widehat{\boldsymbol{\mu}}_{\text{ce}}$ is the classical-equivalent estimator of $\boldsymbol{\mu}$, which from (7.35) reads explicitly:

$$\widehat{\boldsymbol{\mu}}_{\text{ce}} \left[i_T, e_C\right] = \boldsymbol{\mu}_1; \tag{9.150}$$

and $\mathbf{S}_{\mu}$ is the scatter matrix for $\boldsymbol{\mu}$, which from (7.36) reads explicitly:

$$\mathbf{S}_{\mu} \left[i_T, e_C\right] = \frac{1}{T_1} \frac{\nu_1}{\nu_1 - 2} \boldsymbol{\Sigma}_1. \tag{9.151}$$

The uncertainty set for $\boldsymbol{\Sigma}$ is the location-dispersion ellipsoid (7.40) of the marginal posterior distribution of $\boldsymbol{\Sigma}$:

$$\widehat{\boldsymbol{\Theta}}_{\Sigma} \equiv \left\{\boldsymbol{\Sigma} : \; \operatorname{vech}\left[\boldsymbol{\Sigma} - \widehat{\boldsymbol{\Sigma}}_{\text{ce}}\right]' \mathbf{S}_{\Sigma}^{-1} \operatorname{vech}\left[\boldsymbol{\Sigma} - \widehat{\boldsymbol{\Sigma}}_{\text{ce}}\right] \leq q_{\Sigma}^2\right\}. \tag{9.152}$$

In this expression vech is the operator that stacks the columns of a matrix skipping the redundant entries above the diagonal; q_{Σ} is the radius factor that represents aversion to estimation risk for $\boldsymbol{\Sigma}$; $\widehat{\boldsymbol{\Sigma}}_{\text{ce}}$ is the classical-equivalent estimator of $\boldsymbol{\Sigma}$, which from (7.38) reads explicitly:

$$\widehat{\boldsymbol{\Sigma}}_{\text{ce}} \left[i_T, e_C\right] = \frac{\nu_1}{\nu_1 + N + 1} \boldsymbol{\Sigma}_1; \tag{9.153}$$

and $\mathbf{S}_{\Sigma}$ is the scatter matrix for vech $[\boldsymbol{\Sigma}]$. From (7.39) the scatter matrix reads explicitly as follows:

$$\mathbf{S}_{\Sigma} \left[i_T, e_C\right] = \frac{2\nu_1^2}{(\nu_1 + N + 1)^3} \left(\mathbf{D}_N' \left(\boldsymbol{\Sigma}_1^{-1} \otimes \boldsymbol{\Sigma}_1^{-1}\right) \mathbf{D}_N\right)^{-1}, \tag{9.154}$$

where $\mathbf{D}_N$ is the duplication matrix (A.113) and $\otimes$ is the Kronecker product (A.95).

9.5.3 Discussion

In Appendix www.9.8 we show that the robust Bayesian mean-variance problem (9.139) with the robustness uncertainty sets specified as in (9.149) and (9.152) simplifies as follows:

$$\mathbf{w}_{\text{rB}}^{(i)} = \underset{\mathbf{w}}{\operatorname{argmax}} \left\{\mathbf{w}'\boldsymbol{\mu}_1 - \gamma_{\mu}\sqrt{\mathbf{w}'\boldsymbol{\Sigma}_1\mathbf{w}}\right\} \tag{9.155}$$

$$\text{subject to } \begin{cases} \mathbf{w} \in \mathcal{C} \\ \mathbf{w}'\boldsymbol{\Sigma}_1\mathbf{w} \leq \gamma_{\Sigma}^{(i)}, \end{cases}$$

where:

$$\gamma_\mu \equiv \sqrt{\frac{q_\mu^2}{T_1} \frac{\nu_1}{\nu_1 - 2}} \tag{9.156}$$

$$\gamma_\Sigma^{(i)} \equiv \frac{v^{(i)}}{\frac{\nu_1}{\nu_1+N+1} + \sqrt{\frac{2\nu_1^2 q_\Sigma^2}{(\nu_1+N+1)^3}}}. \tag{9.157}$$

This maximization is in the same form as the robust allocation decision (9.130). Like that problem, under regularity assumption for the constraints $\mathcal{C}$ also this maximization can be can be cast in the form of a second-order cone programming problem (6.55). Therefore the robust Bayesian frontier (9.155) can computed numerically.

The original robust Bayesian mean-variance problem (9.139) with the robustness uncertainty sets (9.149) and (9.152) is parametrized by the aversion to estimation risk for the expected values, represented by q_μ, the aversion to estimation risk for the covariances, represented by q_Σ, and the exposure to market risk, represented by $v^{(i)}$. Therefore, in principle, the robust Bayesian mean-variance efficient frontier should constitute a three-dimensional surface in the N-dimensional space of the allocations.

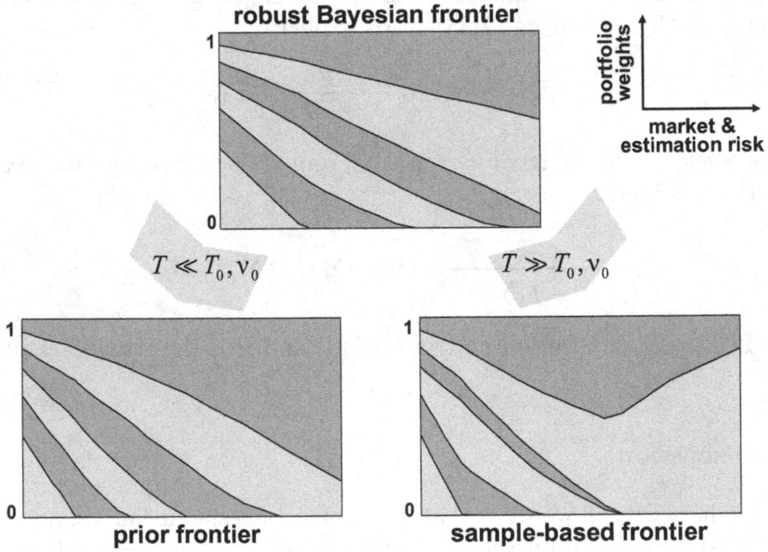

Fig. 9.10. Robust Bayesian mean-variance efficient allocations

Nevertheless, the efficient allocations (9.155) can be parametrized equivalently in terms of one single positive multiplier λ as follows:

$$\mathbf{w}_{\mathrm{rB}}(\lambda) = \underset{\mathbf{w}\in\mathcal{C}}{\operatorname{argmax}} \left\{ \mathbf{w}'\boldsymbol{\mu}_1 - \lambda\sqrt{\mathbf{w}'\boldsymbol{\Sigma}_1\mathbf{w}} \right\}. \tag{9.158}$$

The multiplier λ is determined by the scalars (9.156) and (9.157). It is easy to check that the value of λ is directly related to the aversion to estimation risk $(q_{\boldsymbol{\mu}}, q_{\boldsymbol{\Sigma}})$ and inversely related to the exposure to market risk $v^{(i)}$. Accordingly, the term under the square root in (9.158) represents both estimation and market risk and the coefficient λ represents aversion to both types of risk.

In other words, the a-priori three-dimensional robust Bayesian efficient frontier collapses to a line. Hence the robust Bayesian mean-variance efficient frontier is conceptually similar to, and just as parsimonious as, the classical mean-variance efficient frontier (9.137). Nevertheless, in the classical setting the coefficient of risk aversion only refers to market risk, whereas in the robust Bayesian setting the coefficient of risk aversion blends aversion to both market risk and estimation risk.

From (9.145)-(9.148) the expected values $\boldsymbol{\mu}_1$ and the covariance matrix $\boldsymbol{\Sigma}_1$ in (9.158) are self-adjusting mixtures of the classical estimators $\left(\widehat{\boldsymbol{\mu}}, \widehat{\boldsymbol{\Sigma}}\right)$ and of the prior parameters $(\boldsymbol{\mu}_0, \boldsymbol{\Sigma}_0)$. In particular, when the number of observations T is large with respect to the confidence levels T_0 and ν_0 in the investor's prior, the expected values $\boldsymbol{\mu}_1$ tend to the sample mean $\widehat{\boldsymbol{\mu}}$ and the covariance matrix $\boldsymbol{\Sigma}_1$ tends to the sample covariance $\widehat{\boldsymbol{\Sigma}}$. Therefore we obtain a sample-based efficient frontier:

$$\mathbf{w}_{\mathrm{s}}(\lambda) = \underset{\mathbf{w}\in\mathcal{C}}{\operatorname{argmax}} \left\{ \mathbf{w}'\widehat{\boldsymbol{\mu}} - \lambda\sqrt{\mathbf{w}'\widehat{\boldsymbol{\Sigma}}\mathbf{w}} \right\}. \tag{9.159}$$

Similarly, when the confidence levels T_0 and ν_0 in the investor's prior are large with respect to the number of observations T, the expected values $\boldsymbol{\mu}_1$ tend to the prior $\boldsymbol{\mu}_0$ and the covariance matrix $\boldsymbol{\Sigma}_1$ tends to the prior $\boldsymbol{\Sigma}_0$. Therefore we obtain a prior efficient frontier that disregards any information from the market:

$$\mathbf{w}_{\mathrm{p}}(\lambda) = \underset{\mathbf{w}\in\mathcal{C}}{\operatorname{argmax}} \left\{ \mathbf{w}'\boldsymbol{\mu}_0 - \lambda\sqrt{\mathbf{w}'\boldsymbol{\Sigma}_0\mathbf{w}} \right\}. \tag{9.160}$$

Consider a market of $N \equiv 6$ stocks from the utilities sector of the S&P 500. We estimate the sample mean and covariance from a database of weekly returns. We specify the prior with an equilibrium argument, as in (8.58)-(8.59), where we assume a correlation of 0.5.

Suppose that the investor is bound by the standard budget constraint $\mathbf{w}'\mathbf{1} = 1$ and the standard no-short-sale constraint $\mathbf{w} \geq \mathbf{0}$.

In Figure 9.10 we plot the general robust Bayesian efficient frontier (9.158) and the limit cases (9.159) and (9.160), refer to `symmys.com` for more details.

Part IV

Appendices

A

Linear algebra

In this appendix we review the main concepts of linear algebra. We stress the geometrical interpretation wherever possible and we do not shun loose expressions in order to appeal to intuition. For a thorough introduction to linear algebra the reader is referred to references such as Lang (1997).

A.1 Vector space

The natural environment of linear algebra are finite-dimensional vector spaces. A *vector space* is a set on whose elements we can perform certain operations. In practice, we focus our attention on the Euclidean space $\mathbb{R}^N$. We can represent geometrically the Euclidean space $\mathbb{R}^N$ as the space generated by N axes, as in the left portion of Figure A.1

A vector in $\mathbb{R}^N$ can be represented as a column of N real numbers

$$\mathbf{v} \equiv (v_1, \ldots, v_N)', \tag{A.1}$$

where the symbol $'$ denotes transposition. Geometrically, it is natural to represent a vector as an arrow whose tail sits on the origin of the N axes that generate the space and whose tip is the N-tuple $(A.1)$.

Alternatively, it is useful to think of an analytical representation of a vector as a function that with each of the first N integers associates a real number, the "entry" on the respective axis:

$$\mathbf{v} : \; n \in \{1, \ldots, N\} \rightarrow v_n \in \mathbb{R}. \tag{A.2}$$

Refer again to Figure A.1 for an interpretation.

The set of such vectors is a vector space, since the following operations are properly defined on its elements.

The *sum* of two vectors is defined component-wise as follows:

$$[\mathbf{u} + \mathbf{v}]_n \equiv u_n + v_n. \tag{A.3}$$

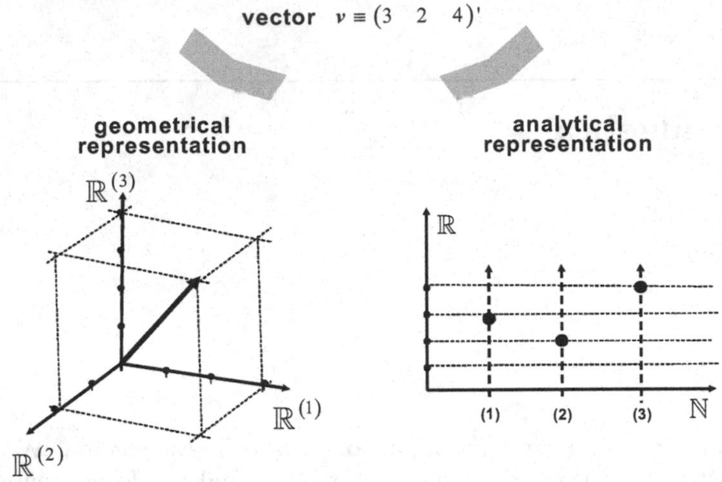

Fig. A.1. Representations of a vector

This is the *parallelogram rule*: the sum of two arrows stemming from the origin is the diagonal of the parallelogram spanned by the arrows.

The *multiplication by a scalar* is defined component-wise as follows:

$$[\alpha \mathbf{v}]_n \equiv \alpha v_n. \tag{A.4}$$

This is a stretch by a factor α in the direction of $\mathbf{v}$.

Combining sums and multiplications by a scalar we obtain *linear combinations* of vectors.

All possible linear combinations of an arbitrary set of vectors $\{\mathbf{v}_1, \ldots, \mathbf{v}_K\}$ in a vector space generates a *vector subspace* of that vector space. To visualize a subspace, consider the parallelotope described by the vertices of a set of vectors $\{\mathbf{v}_1, \ldots, \mathbf{v}_K\}$. The subspace generated by these vectors is the parallelotope obtained by stretching all the vertices to plus and minus infinity.

Vectors are *linearly independent* if the parallelotope they generate is non-degenerate. We see in Figure A.2 the case of three vectors, respectively linearly independent and linearly dependent.

The last important feature of the Euclidean space $\mathbb{R}^N$ is the existence of an *inner product*, an operation that allows to define useful concepts such as orthogonality and length. The inner product is defined as the sum of the entry-by-entry multiplication of two vectors:

$$\langle \mathbf{u}, \mathbf{v} \rangle \equiv \sum_{n=1}^{N} u_n v_n. \tag{A.5}$$

By means of the inner product we can define the length of a vector in $\mathbb{R}^N$, also called the *norm*:

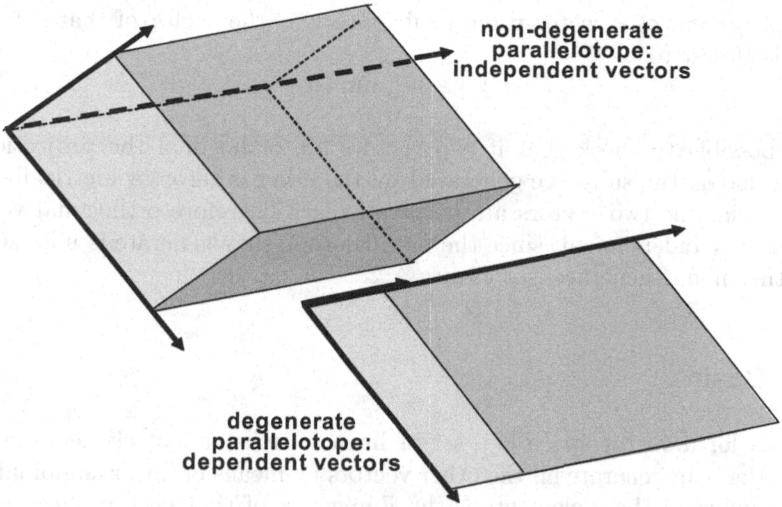

Fig. A.2. Linear (in)dependence among vectors

$$\|\mathbf{v}\| \equiv \sqrt{\langle \mathbf{v}, \mathbf{v} \rangle}. \tag{A.6}$$

The reader will recognize that the norm is indeed the length, as its definition can be interpreted in geometric terms as the *Pythagorean theorem*. Furthermore, the norm displays the following intuitive properties of a length:

$$\begin{aligned} \|\mathbf{v}\| &\geq 0 \\ \|\mathbf{v}\| &= 0 \Leftrightarrow \mathbf{v} = 0 \\ \|\alpha \mathbf{v}\| &= |\alpha| \, \|\mathbf{v}\| \\ \|\mathbf{u} + \mathbf{v}\| &\leq \|\mathbf{u}\| + \|\mathbf{v}\| \, . \end{aligned} \tag{A.7}$$

The last property is called *triangular inequality* and follows from the *Cauchy-Schwartz inequality*:

$$|\langle \mathbf{u}, \mathbf{v} \rangle| \leq \|\mathbf{u}\| \, \|\mathbf{v}\| \, , \tag{A.8}$$

in which the equality holds if and only if $\mathbf{u} \equiv \alpha \mathbf{v}$ for some scalar α. If the scalar α is positive:

$$\langle \mathbf{u}, \mathbf{v} \rangle = \|\mathbf{u}\| \, \|\mathbf{v}\| \, ; \tag{A.9}$$

if the scalar α is negative:

$$\langle \mathbf{u}, \mathbf{v} \rangle = - \|\mathbf{u}\| \, \|\mathbf{v}\| \, . \tag{A.10}$$

We omit the (easy) proof.

Two vectors $\mathbf{u}$ and $\mathbf{v}$ are *orthogonal* if their inner product is null:

$$\langle \mathbf{u}, \mathbf{v} \rangle = 0. \tag{A.11}$$

The *projection* of a vector $\mathbf{u}$ on a subspace V is the vector of that subspace that is closest to $\mathbf{u}$:

$$P(\mathbf{u}, V) \equiv \underset{\mathbf{v} \in V}{\operatorname{argmin}} \|\mathbf{u} - \mathbf{v}\| . \tag{A.12}$$

It is possible to check that if two vectors are orthogonal the projection of either one on the subspace generated by the other is zero: geometrically, this means that the two vectors are perpendicular. Therefore orthogonal vectors are linearly independent, since the parallelotope they generate is not skewed, and thus non-degenerate, see Figure A.2.

A.2 Basis

A *basis* for a vector space is a set of linearly independent elements of that space that can generate all the other vectors by means of linear combinations. The number of these elements is the *dimension* of that vector space. In the case of the Euclidean space $\mathbb{R}^N$, this number is N. Therefore, a basis is a set of vectors

$$\mathbf{e}^{(n)}, \quad n = 1, \ldots, N, \tag{A.13}$$

such that, for suitable scalars $\alpha_1, \ldots, \alpha_N$, any vector $\mathbf{v}$ of $\mathbb{R}^N$ can be expressed as a linear combination:

$$\mathbf{v} = \sum_{n=1}^{N} \alpha_n \mathbf{e}^{(n)}. \tag{A.14}$$

The *canonical basis* is the following set of vectors:

$$\begin{aligned} \boldsymbol{\delta}^{(1)} &\equiv (1, 0, \ldots, 0)' \\ &\vdots \\ \boldsymbol{\delta}^{(N)} &\equiv (0, 0, \ldots, 1)' . \end{aligned} \tag{A.15}$$

It is possible to check that the canonical basis is the only set of vectors such that the inner product of one of them, say $\boldsymbol{\delta}^{(n)}$, with a generic vector $\mathbf{v}$ in $\mathbb{R}^N$ yields the n-th entry of that vector:

$$\left\langle \mathbf{v}, \boldsymbol{\delta}^{(n)} \right\rangle = v_n . \tag{A.16}$$

The generic element $\boldsymbol{\delta}^{(n)}$ of this basis is called the *Kronecker delta centered in n*. This name stems from the analytical representation of the vector $\boldsymbol{\delta}^{(n)}$ as in the right portion of Figure A.1, which is a function peaked on the integer n.

A.3 Linear transformations

Consider a function A that maps vectors $\mathbf{v}$ of the Euclidean space $\mathbb{R}^N$ into vectors that belong to the same Euclidean space $\mathbb{R}^N$, or to another Euclidean space $\mathbb{R}^M$:

$$A: \mathbf{v} \in \mathbb{R}^N \mapsto \mathbf{u} \equiv A[\mathbf{v}] \in \mathbb{R}^M. \tag{A.17}$$

The function A is a *linear transformation*, or a *linear application*, if it preserves the sum and the multiplication by a scalar:

$$\begin{aligned} A[\mathbf{u}+\mathbf{v}] &= A[\mathbf{u}] + A[\mathbf{v}] \\ A[\alpha\mathbf{v}] &= \alpha A[\mathbf{v}]. \end{aligned} \tag{A.18}$$

In Figure A.3 we sketch the graphical meaning of a linear application. Consider the parallelotope $\mathfrak{P}$ described by the vertices of a set of K vectors $\{\mathbf{v}_1, \ldots, \mathbf{v}_K\}$. Now consider the parallelotope $\mathfrak{P}'$ described by the vertices of the set of vectors $\{A[\mathbf{v}_1], \ldots, A[\mathbf{v}_K]\}$. A transformation A is linear if $A(\mathfrak{P}) = \mathfrak{P}'$, i.e. if parallelotopes are mapped into parallelotopes: it is called a linear application because it does not bend straight lines. This interpretation

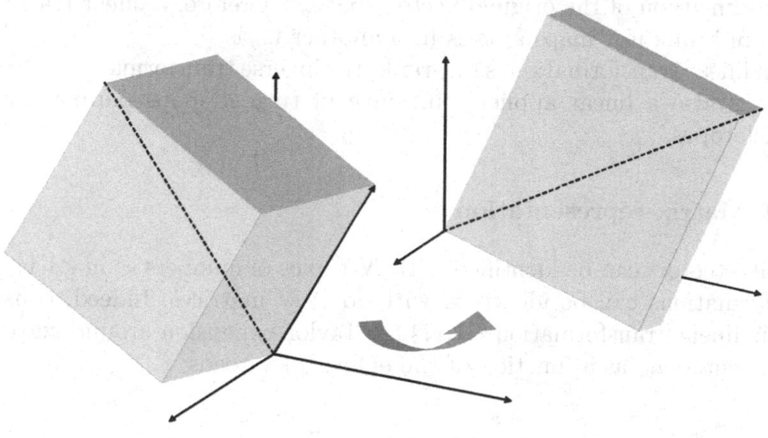

Fig. A.3. Geometrical representation of a linear transformation

makes it immediate to see that a sequence of two linear applications

$$(B \circ A)[\mathbf{v}] \equiv B[A[\mathbf{v}]] \tag{A.19}$$

is a linear application.

The inverse A^{-1} of a linear transformation A is the transformation that applied either before or after the linear transformation A cancels the effect of the transformation A. In other words, for all vectors $\mathbf{v}$ the inverse transformation A^{-1} satisfies:

$$\left(A^{-1} \circ A\right)[\mathbf{v}] = \mathbf{v} = \left(A \circ A^{-1}\right)[\mathbf{v}] . \tag{A.20}$$

The inverse of a linear application is not always defined: if a linear transformation A "squeezes" a parallelotope into a degenerate parallelotope it is not possible to recover univocally the vectors that generated the original parallelotope. In this case the dimension of the image space $A\left[\mathbb{R}^N\right]$ is less than the dimension N of the original space.

The dimension of the image space is called the *rank* of the application A:

$$\operatorname{rank}(A) \equiv \dim\left(A\left[\mathbb{R}^N\right]\right) . \tag{A.21}$$

Since a linear application can either squeeze a vector space or preserve its dimension, it follows from the definition (A.21) of rank that:

$$\operatorname{rank}(B \circ A) \leq \min\left(\operatorname{rank}(A), \operatorname{rank}(B)\right) . \tag{A.22}$$

A linear transformation is invertible if it is full-rank, i.e. if its rank is equal to the dimension of the original vector space. Therefore, a linear transformation is full-rank if it maps a basis into another basis.

If a linear transformation is full-rank, the inverse transformation A^{-1} exists and it is also a linear application, since in turn it maps parallelotopes in parallelotopes.

A.3.1 Matrix representation

Just like vectors can be identified with N-tuples of numbers as in (A.1), linear transformations can be identified with $M \times N$ matrices. Indeed, consider a generic linear transformation (A.17). A Taylor expansion around zero of the generic entry u_m as a function of the entries of $\mathbf{v}$ reads:

$$u_m = A_m + \sum_{n=1}^{N} A_{mn} v_n + \sum_{n,l=1}^{N} A_{mnl} v_n v_l + \cdots , \tag{A.23}$$

where $A_{\ldots}$ are suitable constant coefficients. In order for (A.18) to hold only the coefficients A_{mn} in the second term can contain non-zero elements. Collecting these terms in a matrix $\mathbf{A}$ we can represent the linear transformation (A.17) by means of its *matrix representation* as follows:

$$\mathbf{u} \equiv A[\mathbf{v}] \equiv \mathbf{A}\mathbf{v}, \tag{A.24}$$

where the product of a matrix by a vector is defined as:

$$[\mathbf{Av}]_m \equiv \sum_{n=1}^{N} A_{mn} v_n. \tag{A.25}$$

For example, consider the *identity transformation* defined as follows:

$$I[\mathbf{v}] \equiv \mathbf{v}. \tag{A.26}$$

It is immediate to check that the identity transformation is represented by the *identity matrix*, defined as follows:

$$\mathbf{I}_N \equiv \begin{pmatrix} 1 & 0 & \cdots & 0 \\ 0 & 1 & \ddots & \vdots \\ \vdots & \ddots & \ddots & 0 \\ 0 & \cdots & 0 & 1 \end{pmatrix}. \tag{A.27}$$

From (A.25) we also derive the "row-by-column" multiplication rule for matrices. Indeed, it is easy to check that the matrix representation $\mathbf{C}$ of the transformation $C \equiv B \circ A$ defined in (A.19) reads:

$$C_{mn} = \sum_l B_{ml} A_{ln}. \tag{A.28}$$

Notice that a matrix can be seen as a function from the two-dimensional grid of integer coordinates to the real numbers:

$$\mathbf{A} : (m,n) \in \{1,\ldots,M\} \times \{1,\ldots,N\} \to A_{mn} \in \mathbb{R}. \tag{A.29}$$

This definition parallels the analytical definition (A.2) of a vector.

A.3.2 Rotations

Rotations are special kinds of linear transformations. As intuition suggests, a linear transformation R is a *rotation* in the Euclidean space $\mathbb{R}^N$ if it does not alter the length[1], i.e. the norm (A.6), of any vector in $\mathbb{R}^N$:

$$\|R[\mathbf{v}]\| = \|\mathbf{v}\|. \tag{A.30}$$

A rotation is always invertible, since it does not "squeeze" parallelotopes and therefore it does not make them degenerate. Moreover, the inverse of a rotation is a rotation.

From the definition of rotation (A.30), the definition of norm (A.6), the rule for the representation of the composition of two linear applications (A.28)

[1] More precisely, this is the definition of *isometries*, which include rotations, reflections and inversions.

and the representation of the identity ($A.27$), it is easy to derive the following result for the matrix representation $\mathbf{R}$ of the rotation R:

$$\mathbf{R}^{-1} = \mathbf{R}'. \tag{A.31}$$

In words, a linear transformation R is a rotation if and only if the representation of its inverse is the transpose of its representation.

For example, for any θ the matrix

$$\mathbf{R}_\theta \equiv \begin{pmatrix} \cos\theta & -\sin\theta \\ \sin\theta & \cos\theta \end{pmatrix} \tag{A.32}$$

satisfies ($A.31$), and thus it represents a rotation in $\mathbb{R}^2$. Indeed, it represents a counterclockwise rotation of an angle θ: this can be easily verified by checking the result of applying ($A.32$) to the two vectors of the canonical basis ($A.15$).

Furthermore, it can be proved that any rotation in $\mathbb{R}^2$ can be represented by a matrix of the form ($A.32$) for a suitable angle θ.

A.4 Invariants

Consider a generic linear transformation A from $\mathbb{R}^N$ to itself. Consider now another transformation $\widetilde{A}$ obtained in terms of the composition ($A.19$) with an invertible transformation B as follows:

$$\widetilde{A} \equiv B^{-1} \circ A \circ B. \tag{A.33}$$

We call the transformations A and $\widetilde{A}$ *equivalent linear transformations*. Indeed, $\widetilde{A}$ brings the original reference frame into an equivalent one by means of the invertible transformation B, then performs the same operation as A and finally brings the result back to the original reference frame by means of the inverse transformation B^{-1}.

Two equivalent transformations A and $\widetilde{A}$ must share many properties. Nevertheless, their matrix representations $\mathbf{A}$ and $\widetilde{\mathbf{A}}$ might be very different. Therefore, it can be hard to detect equivalent transformations from their representations. In this section we describe some features that are common to any representation of equivalent transformations.

A.4.1 Determinant

Consider the parallelotope $\mathfrak{P}$ described by the vertices of a set of independent vectors. We recall that the linear transformation A by definition maps this parallelotope into another parallelotope $\mathfrak{P}'$, see Figure A.3. In so doing, A stretches and turns $\mathfrak{P}$ and therefore modifies its volume by some factor. This factor does not depend on the particular choice of $\mathfrak{P}$: the linearity of A implies that the volume of any parallelotope is modified by the same factor. We call this factor, modulo a sign, the determinant. In other words, the *determinant* of the transformation A is the number $\det(A)$ such that

$$\mathrm{Vol}(\mathfrak{P}') = \pm \det(A) \,\mathrm{Vol}(\mathfrak{P}), \tag{A.34}$$

where "Vol" denotes the volume and the sign is positive (negative) if the transformation includes an even (odd) number of reflections.

In particular, the transformation A is not invertible if and only if $\mathfrak{P}'$ is degenerate, i.e. if its volume is zero. Therefore, a transformation A is not invertible if and only if

$$\det(A) = 0. \tag{A.35}$$

Furthermore, we see that for the composite transformation (A.19) the following rule holds:

$$\det(B \circ A) = \det(B) \det(A). \tag{A.36}$$

In particular, since the identity transformation (A.26) does not alter the volumes:

$$1 = \det\left(B \circ B^{-1}\right) = \det(B) \det\left(B^{-1}\right). \tag{A.37}$$

Now we can prove that the determinant is indeed an invariant. If a linear transformation $\widetilde{A}$ is equivalent to a linear transformation A as in (A.33), then:

$$\det\left(\widetilde{A}\right) = \det\left(B^{-1}\right) \det(A) \det(B) = \det(A). \tag{A.38}$$

It can be proved that the formula to compute explicitly the determinant of a linear transformation A in terms of its matrix representation $\mathbf{A}$ reads:

$$\det(A) \equiv |\mathbf{A}| = \sum_{\{i_1,\ldots,i_N\} \in \mathcal{P}} \pm A_{i_1 1} \cdots A_{i_N N}, \tag{A.39}$$

where the sum is taken over all the permutations $\mathcal{P}$ of the first N integers and the sign is positive for even permutations (i.e. obtained by a sequence of an even number of switches) and negative for odd permutations.

For example, the formula for the determinant of a generic 2×2 matrix

$$\mathbf{A} \equiv \begin{pmatrix} A_{11} & A_{12} \\ A_{21} & A_{22} \end{pmatrix} \tag{A.40}$$

is

$$|\mathbf{A}| = A_{11} A_{22} - A_{21} A_{12}. \tag{A.41}$$

In one situation the determinant is particularly easy to compute. Consider a diagonal matrix $\mathbf{D}$, i.e. a matrix where all elements D_{mn} for $m \neq n$ are zero. Geometrically, a diagonal matrix $\mathbf{D}$ represents a stretch by a factor D_{nn}

along the generic n-th axis. In this situation a parallelotope is stretched into a new parallelotope whose volume is multiplied by $D_{11} \cdots D_{NN}$. Therefore the determinant in this case reads

$$|\mathbf{D}| = \prod_{n=1}^{N} D_{nn}, \tag{A.42}$$

i.e., the determinant is the product of the diagonal elements. Notice that $(A.42)$ automatically accounts for the change in sign due to reflections, since a reflection is associated with a negative entry on the diagonal.

Since the determinant is an invariant, the result is the same for any equivalent representation $\widetilde{\mathbf{A}}$. Therefore, it is particularly convenient to find, if possible, equivalent representations $\widetilde{\mathbf{A}}$ of a generic linear transformation A that are diagonal.

A.4.2 Trace

The *trace* of a generic linear transformation A from $\mathbb{R}^N$ to itself is defined in terms of its matrix representation $\mathbf{A}$ as the sum of the diagonal entries:

$$\operatorname{tr}(A) \equiv \operatorname{tr}(\mathbf{A}) \equiv \sum_{n=1}^{N} A_{nn}. \tag{A.43}$$

From this definition and the multiplication rule $(A.28)$ we obtain the *circular property* of the trace:

$$\operatorname{tr}(A \circ B \circ C) = \operatorname{tr}(B \circ C \circ A). \tag{A.44}$$

Consider now two equivalent linear transformations A and $\widetilde{A}$ as in $(A.33)$. Then the following result holds:

$$\operatorname{tr}\left(\widetilde{A}\right) = \operatorname{tr}\left(B^{-1} \circ A \circ B\right) = \operatorname{tr}\left(B \circ B^{-1} \circ A\right) = \operatorname{tr}(A). \tag{A.45}$$

This proves that the trace is indeed an invariant.

A.4.3 Eigenvalues

An *eigenvector* of a linear transformation A from $\mathbb{R}^N$ to $\mathbb{R}^N$ is a vector $\mathbf{v}$ that is not rotated by the transformation, i.e. such that for a suitable scalar λ the following holds:

$$A[\mathbf{v}] = \lambda \mathbf{v}. \tag{A.46}$$

The number λ is called the *eigenvalue* relative to the eigenvector $\mathbf{v}$. Notice that if $\mathbf{v}$ is an eigenvector of A, so is any multiple $\alpha\mathbf{v}$. In general, a linear transformation A does not admit eigenvalues. Nevertheless, if some eigenvalues exist, it becomes much easier to analyze the properties of A.

If they exist, eigenvalues are invariants (beware: eigenvectors are *not* invariants). Indeed, if there exists a pair $(\lambda, \mathbf{v})$ that satisfies $(A.46)$ then, for any equivalent transformation $\widetilde{A}$ as in $(A.33)$ we can see that $\mathbf{w} \equiv \mathbf{B}^{-1}\mathbf{v}$ is an eigenvector for the same eigenvalue:

$$\widetilde{A}[\mathbf{w}] = \mathbf{B}^{-1}\mathbf{A}\mathbf{B}\mathbf{w} = \mathbf{B}^{-1}\mathbf{A}\mathbf{v} = \lambda\mathbf{B}^{-1}\mathbf{v} = \lambda\mathbf{w}. \tag{A.47}$$

In order to compute the eigenvalues of A, or to realize that they do not exist, notice from the definition $(A.46)$ that λ is an eigenvalue if and only if the linear application $A - \lambda I$, where I is the identity $(A.26)$, "squeezes" a specific direction, i.e. the direction spanned by the eigenvector $\mathbf{v}$, into the zero vector. This can happen only if $A - \lambda I$ is not invertible. Therefore, from $(A.35)$ an eigenvalue λ solves the equation

$$\det(A - \lambda I) = 0. \tag{A.48}$$

In general, this equation does not necessarily admit real solutions.

For example, consider a generic 2×2 matrix $(A.40)$. Making use of $(A.41)$ it is easy to check that $(A.48)$ becomes:

$$0 = \lambda^2 - \lambda \operatorname{tr}(A) + \det(A). \tag{A.49}$$

The possible solutions read:

$$\lambda = \frac{1}{2}\left(\operatorname{tr}(A) \pm \sqrt{\operatorname{tr}(A)^2 - 4\det(A)}\right). \tag{A.50}$$

This shows that if $\operatorname{tr}(A)^2 < 4\det(A)$ there is no solution. Otherwise, the two solutions are invariants, as they only depend on trace and determinant, which are invariants.

A.5 Spectral theorem

The spectral theorem is an extremely useful result whose interpretation and application involve all the invariants described in Section A.4.

A.5.1 Analytical result

In general a linear transformation does not admit eigenvectors and eigenvalues. Nevertheless, in a special, yet very important, case it is possible to find a whole basis of orthogonal eigenvectors.

First we need two definitions. A linear application S is *symmetric* if its matrix representation is symmetric with respect to the diagonal, i.e. it is equal to its transpose:

$$\mathbf{S} = \mathbf{S}'. \tag{A.51}$$

A linear application S is *positive* if for any $\mathbf{v} \in \mathbb{R}^N$ its matrix representation satisfies the following inequality:[2]

$$\langle \mathbf{v}, \mathbf{S}\mathbf{v} \rangle \geq 0. \tag{A.52}$$

We stress that a positive matrix can have negative entries.

The *spectral theorem* states that a symmetric matrix admits an orthogonal basis of eigenvectors. In other words, if a square matrix $\mathbf{S}$ satisfies (A.51), then there exist N numbers $(\lambda_1, \ldots, \lambda_N)$ and N vectors $\left(\mathbf{e}^{(1)}, \ldots, \mathbf{e}^{(N)}\right)$ such that

$$\mathbf{S}\mathbf{e}^{(n)} = \lambda_n \mathbf{e}^{(n)}, \tag{A.53}$$

and, if $m \neq n$,

$$\left\langle \mathbf{e}^{(m)}, \mathbf{e}^{(n)} \right\rangle = 0. \tag{A.54}$$

If in addition the matrix $\mathbf{S}$ is positive, due to (A.52) all the eigenvalues must be positive. Furthermore, we can always rearrange the eigenvalues, and their respective eigenvectors, in such a way that:

$$\lambda_1 \geq \ldots \geq \lambda_N \geq 0. \tag{A.55}$$

Finally, we can always normalize the eigenvectors in such a way that their length is unitary:

$$\left\| \mathbf{e}^{(n)} \right\| = 1, \quad n = 1, \ldots, N. \tag{A.56}$$

Under the restrictions (A.55) and (A.56), and modulo a reflection of the eigenvectors, there exists only one such set of eigenvalue-eigenvector pairs $\left\{\lambda_n, \mathbf{e}^{(n)}\right\}$.

For example the matrix

$$\mathbf{S} \equiv \begin{pmatrix} \frac{9}{4} & \frac{\sqrt{3}}{4} \\ \frac{\sqrt{3}}{4} & \frac{11}{4} \end{pmatrix} \tag{A.57}$$

is symmetric and positive definite. Indeed, the eigenvalues can be computed as in (A.50) and read:

$$\lambda_1 = 3, \quad \lambda_2 = 2. \tag{A.58}$$

Solving (A.46) for the eigenvectors

$$\begin{pmatrix} \frac{9}{4} - 3 & \frac{\sqrt{3}}{4} \\ \frac{\sqrt{3}}{4} & \frac{11}{4} - 3 \end{pmatrix} \mathbf{e}^{(1)} = \mathbf{0}, \quad \begin{pmatrix} \frac{9}{4} - 2 & \frac{\sqrt{3}}{4} \\ \frac{\sqrt{3}}{4} & \frac{11}{4} - 2 \end{pmatrix} \mathbf{e}^{(2)} = \mathbf{0}, \tag{A.59}$$

[2] It is customary to define a matrix as *positive definite* if the inequality in (A.52) is strict and *positive semi-definite* if that inequality is slack.

we obtain:

$$\mathbf{e}^{(1)} = \rho \begin{pmatrix} 1 \\ \sqrt{3} \end{pmatrix} = \widetilde{\rho} \begin{pmatrix} \cos\frac{\pi}{3} \\ \sin\frac{\pi}{3} \end{pmatrix} \tag{A.60}$$

$$\mathbf{e}^{(2)} = \kappa \begin{pmatrix} -\sqrt{3} \\ 1 \end{pmatrix} = \widetilde{\kappa} \begin{pmatrix} -\sin\frac{\pi}{3} \\ \cos\frac{\pi}{3} \end{pmatrix}. \tag{A.61}$$

In this expression $\rho, \widetilde{\rho}, \kappa$ and $\widetilde{\kappa}$ are arbitrary constants: imposing ($A.56$) we obtain $\widetilde{\rho} \equiv \widetilde{\kappa} \equiv 1$.

Notice that ($A.54$) and ($A.56$) imply that the following matrix, defined as the juxtaposition of the eigenvectors:

$$\mathbf{E} \equiv \left(\mathbf{e}^{(1)}, \ldots, \mathbf{e}^{(N)}\right), \tag{A.62}$$

satisfies:

$$\mathbf{E}\mathbf{E}' = \mathbf{I}_N. \tag{A.63}$$

Comparing ($A.63$) with ($A.31$) we see that $\mathbf{E}$ represents a rotation in $\mathbb{R}^N$ and thus does not alter the norm of a vector:

$$\|\mathbf{E}\mathbf{v}\| = \|\mathbf{v}\| = \|\mathbf{E}'\mathbf{v}\|. \tag{A.64}$$

Defining:

$$\boldsymbol{\Lambda} \equiv \operatorname{diag}(\lambda_1, \ldots, \lambda_N), \tag{A.65}$$

we can restate the spectral theorem ($A.53$) as follows:

$$\mathbf{S} = \mathbf{E}\boldsymbol{\Lambda}\mathbf{E}'. \tag{A.66}$$

From the invariance of the trace ($A.45$).we obtain the following relation between the diagonal elements of $\mathbf{S}$ and the sum of its eigenvalues:

$$\sum_{n=1}^{N} S_{nn} \equiv \operatorname{tr}(\mathbf{S}) = \operatorname{tr}(\boldsymbol{\Lambda}) \equiv \sum_{n=1}^{N} \lambda_n, \tag{A.67}$$

Notice also that the first, largest eigenvalue of the symmetric and positive matrix $\mathbf{S}$ satisfies the following identity:

$$\begin{aligned} \lambda_1 &= \max_{\|\mathbf{u}\|=1} \{\mathbf{u}'\boldsymbol{\Lambda}\mathbf{u}\} = \max_{\|\mathbf{E}'\mathbf{z}\|=1} \left\{(\mathbf{E}'\mathbf{z})'\boldsymbol{\Lambda}(\mathbf{E}'\mathbf{z})\right\} \\ &= \max_{\|\mathbf{z}\|=1} \{\mathbf{z}'\mathbf{S}\mathbf{z}\} = \max_{\mathbf{z}} \left\{\frac{\mathbf{z}'\mathbf{S}\mathbf{z}}{\mathbf{z}'\mathbf{z}}\right\}. \end{aligned} \tag{A.68}$$

Similarly, the last, smallest eigenvalue of $\mathbf{S}$ satisfies:

$$\lambda_N = \min_{\mathbf{z}} \frac{\mathbf{z}\mathbf{S}\mathbf{z}}{\mathbf{z}'\mathbf{z}}. \tag{A.69}$$

We conclude mentioning that if all the entries of a symmetric and positive matrix $\mathbf{S}$ are positive, the *Perron-Frobenius theorem* implies that the entries of the eigenvector relative to the largest eigenvalue are all positive, see Smirnov (1970). In other words, the first eigenvector points in the direction of the first orthant in the geometrical representation on the left of Figure A.1.

A.5.2 Geometrical interpretation

By means of the spectral theorem we can provide an intuitive geometrical representation of a symmetric and positive matrix. First of all, we write the spectral theorem (*A*.66) as follows:

$$\mathbf{S} = \mathbf{E}\sqrt{\mathbf{\Lambda}}\sqrt{\mathbf{\Lambda}}\mathbf{E}', \tag{A.70}$$

where $\mathbf{\Lambda}$ is the diagonal matrix (*A*.65) of the positive eigenvalues of $\mathbf{S}$ and $\mathbf{E}$ is the juxtaposition of the eigenvectors of $\mathbf{S}$ as defined in (*A*.62).

In our example (*A*.57) we have

$$\sqrt{\mathbf{\Lambda}} \equiv \operatorname{diag}\left(\sqrt{3}, \sqrt{2}\right), \tag{A.71}$$

and

$$\mathbf{E} \equiv \begin{pmatrix} \cos\frac{\pi}{3} & -\sin\frac{\pi}{3} \\ \sin\frac{\pi}{3} & \cos\frac{\pi}{3} \end{pmatrix}. \tag{A.72}$$

Consider the following locus:

$$\mathcal{E}_{\mathbf{m},\mathbf{S}} \equiv \left\{\mathbf{x} \in \mathbb{R}^N \text{ such that } (\mathbf{x} - \mathbf{m})' \mathbf{S}^{-1} (\mathbf{x} - \mathbf{m}) \leq 1\right\}, \tag{A.73}$$

where $\mathbf{m}$ is any fixed vector in $\mathbb{R}^N$. This equation represents an ellipsoid. Indeed, consider a new set of coordinates $\mathbf{y}$ in $\mathbb{R}^N$, obtained by the following affine transformation:

$$\mathbf{y} \equiv \mathbf{\Lambda}^{-\frac{1}{2}}\mathbf{E}' (\mathbf{x} - \mathbf{m}). \tag{A.74}$$

Using (*A*.63) we invert this relation as follows:

$$\mathbf{x} = \mathbf{m} + \mathbf{E}\mathbf{\Lambda}^{\frac{1}{2}}\mathbf{y}. \tag{A.75}$$

Substituting this expression in (*A*.73) we see that $\mathcal{E}_{\mathbf{m},\mathbf{S}}$ is the equation of the unit sphere in the new coordinates:

$$\mathcal{E}_{\mathbf{m},\mathbf{S}} \equiv \left\{\mathbf{y} \in \mathbb{R}^N \text{ such that } y_1^2 + \cdots + y_N^2 \leq 1\right\}. \tag{A.76}$$

On the other hand, from (*A*.75) it follows that the locus (*A*.73) is obtained by first left-multiplying each point $\mathbf{y}$ on the unit sphere by the matrix $\mathbf{\Lambda}^{\frac{1}{2}}$; then by left-multiplying the outcome by the matrix $\mathbf{E}$; and finally by adding the vector $\mathbf{m}$.

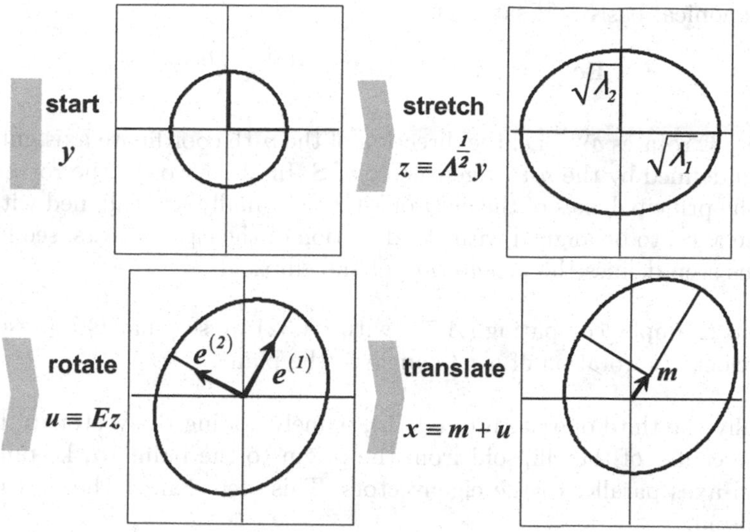

Fig. A.4. Representation of symmetric positive matrices as ellipsoids

Since the matrix $\mathbf{\Lambda}^{\frac{1}{2}}$ is diagonal, the first operation in (A.75), namely the multiplication by $\mathbf{\Lambda}^{\frac{1}{2}}$, corresponds to stretching the unit sphere along each coordinate axis by an amount equal to the square root of the respective eigenvalue, see Figure A.4. Therefore the sphere becomes an ellipsoid whose principal axes are aligned with the reference axes and where, for each $n = 1, \ldots, N$, the length of the n-th principal axis is the square root of the n-th eigenvalue of $\mathbf{S}$. This step defines the *shape* of the ellipsoid. In particular, the volume of the ellipsoid is proportional to the product of the lenghts of the principal axes:

$$\text{Vol}\{\mathcal{E}_{\mathbf{m},\mathbf{S}}\} = \gamma_N \sqrt{\lambda_1} \cdots \sqrt{\lambda_N} = \gamma_N \sqrt{|\mathbf{\Lambda}|} = \gamma_N \sqrt{|\mathbf{S}|}, \tag{A.77}$$

In this expression the constant γ_N is the volume of the unit sphere in N dimensions:

$$\gamma_N \equiv \frac{\pi^{\frac{N}{2}}}{\Gamma\left(\frac{N}{2}+1\right)}, \tag{A.78}$$

where Γ is the gamma function (B.80), see Fang, Kotz, and Ng (1990), p. 74.

In our example from (A.71) the first reference axis is stretched by a factor $\sqrt{3}$ and the second reference axis is stretched by a factor $\sqrt{2}$. Thus the area of the ellipsoid is $\pi\sqrt{6}$.

As for the second operation in (A.75), namely the multiplication by the rotation $\mathbf{E}$, from (A.15) and (A.62) the rotation $\mathbf{E}$ applied to the n-th element of the canonical basis $\boldsymbol{\delta}^{(n)}$ satisfies:

$$\mathbf{E}\boldsymbol{\delta}^{(n)} = \left(\mathbf{e}^{(1)}, \cdots, \mathbf{e}^{(N)}\right)\boldsymbol{\delta}^{(n)} = \mathbf{e}^{(n)}. \tag{A.79}$$

Therefore $\mathbf{E}$ rotates $\boldsymbol{\delta}^{(n)}$, i.e. the direction of the n-th coordinate axis, into the direction defined by the n-th eigenvector of $\mathbf{S}$. In other words, the rotation $\mathbf{E}$ brings the principal axes of the ellipsoid, that originally were aligned with the reference axes, to be aligned with the direction of the eigenvectors, see Figure A.4. This step defines the *orientation* of the ellipsoid.

In our example, comparing $(A.72)$ with $(A.32)$ we see that $\mathbf{E}$ represents a counterclockwise rotation of a $\pi/3$ angle in the plane.

Finally the third operation in $(A.75)$, namely adding the vector $\mathbf{m}$, translates the center of the ellipsoid from the origin to the point $\mathbf{m}$, keeping the principal axes parallel to the eigenvectors. This step defines the *location* of the ellipsoid.

In our example we assumed:

$$\mathbf{m} \equiv (0.3, 0.4)'. \tag{A.80}$$

Therefore the ellipsoid is translated in such a way that $(A.80)$ becomes its center.

To summarize, the locus $\mathcal{E}_{\mathbf{m},\mathbf{S}}$ defined in $(A.73)$ is an ellipsoid. The principal axes of this ellipsoid are parallel to the eigenvectors of $\mathbf{S}$ and the lenghts of the principal axes are the square roots of the eigenvalues of $\mathbf{S}$. Hence, the orientation and the shape of the ellipsoid $\mathcal{E}_{\mathbf{m},\mathbf{S}}$ contain all the information about $\mathbf{S}$, namely the information about eigenvalues and eigenvectors: therefore the orientation and the shape of $\mathcal{E}_{\mathbf{m},\mathbf{S}}$ are a representation of $\mathbf{S}$. Similarly, the ellipsoid $\mathcal{E}_{\mathbf{m},\mathbf{S}}$ is centered in $\mathbf{m}$. Hence, the location of the ellipsoid $\mathcal{E}_{\mathbf{m},\mathbf{S}}$ contains all the information about $\mathbf{m}$ and thus the location of $\mathcal{E}_{\mathbf{m},\mathbf{S}}$ is a representation of $\mathbf{m}$.

A.6 Matrix operations

We present here some matrix operations that we apply in the main text to tackle financial problems. See Searle (1982), Magnus and Neudecker (1999), and references therein for more on this subject.

A.6.1 Useful identities

From $\mathbf{I} = \mathbf{A}\mathbf{A}^{-1}$ and $\mathbf{I} = \mathbf{I}'$ we obtain the following identity:

$$(\mathbf{A}')^{-1} = (\mathbf{A}^{-1})'. \tag{A.81}$$

From (*A*.36) we derive:

$$|\mathbf{BA}| = |\mathbf{B}|\,|\mathbf{A}|. \tag{A.82}$$

In particular, from (*A*.37) we obtain:

$$|\mathbf{A}^{-1}| = \frac{1}{|\mathbf{A}|}. \tag{A.83}$$

Changing the matrix $\mathbf{A}$ into its transpose $\mathbf{A}'$ in the computation of the determinant (*A*.39) does not affect the result, therefore:

$$|\mathbf{A}'| = |\mathbf{A}|. \tag{A.84}$$

From (*A*.44) we obtain:

$$\operatorname{tr}(\mathbf{ABC}) = \operatorname{tr}(\mathbf{BCA}). \tag{A.85}$$

Finally, partition a generic $N \times N$ invertible matrix $\mathbf{M}$ as follows:

$$\mathbf{M} \equiv \begin{pmatrix} \mathbf{A} & \mathbf{B} \\ \mathbf{C} & \mathbf{D} \end{pmatrix}, \tag{A.86}$$

where the $K \times K$ matrix $\mathbf{A}$ is invertible and so is the $(N-K) \times (N-K)$ matrix $\mathbf{D}$, the size of the remaining matrices being determined accordingly. Define the *Schur complements* of $\mathbf{A}$ and $\mathbf{D}$ respectively:

$$(\mathbf{M}|\mathbf{A}) \equiv \mathbf{D} - \mathbf{C}\mathbf{A}^{-1}\mathbf{B}, \quad (\mathbf{M}|\mathbf{D}) \equiv \mathbf{A} - \mathbf{B}\mathbf{D}^{-1}\mathbf{C}; \tag{A.87}$$

and define:

$$(\mathbf{B}|\mathbf{M}) \equiv (\mathbf{M}|\mathbf{D})^{-1}\,\mathbf{B}\mathbf{D}^{-1}, \quad (\mathbf{C}|\mathbf{M}) \equiv \mathbf{D}^{-1}\mathbf{C}\,(\mathbf{M}|\mathbf{D})^{-1}. \tag{A.88}$$

Then

$$\mathbf{M}^{-1} = \begin{pmatrix} (\mathbf{M}|\mathbf{D})^{-1} & -(\mathbf{B}|\mathbf{M}) \\ -(\mathbf{C}|\mathbf{M}) & (\mathbf{M}|\mathbf{A})^{-1} \end{pmatrix}. \tag{A.89}$$

In particular, some algebra shows that the following identity holds for any conformable matrices:

$$\left(\mathbf{A} - \mathbf{B}\mathbf{D}^{-1}\mathbf{C}\right)^{-1} = \mathbf{A}^{-1} - \mathbf{A}^{-1}\mathbf{B}\left(\mathbf{C}\mathbf{A}^{-1}\mathbf{B} - \mathbf{D}\right)^{-1}\mathbf{C}\mathbf{A}^{-1}. \tag{A.90}$$

Also, the relation below follows:

$$|\mathbf{I}_J + \mathbf{CB}| = |\mathbf{I}_K + \mathbf{BC}|, \tag{A.91}$$

where $J \equiv N - K$ is the number of rows in $\mathbf{C}$, which is arbitrary, since N and K are arbitrary.

A.6.2 Tensors and Kronecker product

Loosely speaking, vectors can be considered as matrices with only one side. Matrices have two sides. Tensors are matrices with three or more sides. Tensors are the subject of multilinear analysis. A *tensor* of order p is a function from the p-dimensional grid of coordinates to $\mathbb{R}$:

$$T: \{1,\ldots,N_1\} \times \cdots \times \{1,\ldots,N_p\} \mapsto T_{n_1\cdots n_p} \in \mathbb{R}. \tag{A.92}$$

For example, from (*A*.2) a vector is a tensor of order 1:

$$\mathbf{v}: \{1,\ldots,N\} \mapsto v_n \in \mathbb{R}. \tag{A.93}$$

Similarly, from (*A*.29) a matrix is a tensor of order 2:

$$\mathbf{A}: \{1,\ldots,M\} \times \{1,\ldots,N\} \mapsto A_{mn} \in \mathbb{R}. \tag{A.94}$$

The set of tensors of a given order is a vector space whose elements enjoy remarkable transformation properties. A less superficial discussion of this subject is beyond the scope of this book.

The *Kronecker product* is an operation defined between two generic matrices $\mathbf{A}$ and $\mathbf{B}$ of dimensions $M \times N$ and $P \times Q$ respectively. The result is a tensor of order four:

$$[\mathbf{A} \otimes \mathbf{B}]_{mnpq} \equiv A_{mn} B_{pq}. \tag{A.95}$$

Given the special structure of the tensor (*A*.95), we can represent the Kronecker product equivalently as the following $MP \times NQ$ matrix:

$$\mathbf{A} \otimes \mathbf{B} \equiv \begin{pmatrix} A_{11}\mathbf{B} & \cdots & A_{1N}\mathbf{B} \\ \vdots & \ddots & \vdots \\ A_{M1}\mathbf{B} & \cdots & A_{MN}\mathbf{B} \end{pmatrix}. \tag{A.96}$$

We can check from the definition (*A*.96) that the Kronecker product is distributive with respect to the sum and associative:

$$\begin{aligned} \mathbf{A} \otimes (\mathbf{B} + \mathbf{C}) &= \mathbf{A} \otimes \mathbf{B} + \mathbf{A} \otimes \mathbf{C} \\ (\mathbf{B} + \mathbf{C}) \otimes \mathbf{A} &= \mathbf{B} \otimes \mathbf{A} + \mathbf{C} \otimes \mathbf{A} \\ \mathbf{A} \otimes (\mathbf{B} \otimes \mathbf{C}) &= (\mathbf{A} \otimes \mathbf{B}) \otimes \mathbf{C}. \end{aligned} \tag{A.97}$$

Nevertheless, it is not commutative:

$$\mathbf{A} \otimes \mathbf{B} \neq \mathbf{B} \otimes \mathbf{A}. \tag{A.98}$$

Also, the Kronecker product satisfies:

$$(\mathbf{A} \otimes \mathbf{B})' = \mathbf{A}' \otimes \mathbf{B}' \tag{A.99}$$

and

$$(\mathbf{A} \otimes \mathbf{B})(\mathbf{C} \otimes \mathbf{D}) = \mathbf{AC} \otimes \mathbf{BD}. \tag{A.100}$$

If $\mathbf{A}$ is an $N \times N$ invertible matrix and $\mathbf{B}$ is a $K \times K$ invertible matrix, from $(A.100)$ it follows immediately:

$$(\mathbf{A} \otimes \mathbf{B})^{-1} = \mathbf{A}^{-1} \otimes \mathbf{B}^{-1}. \tag{A.101}$$

Also, for the determinant of the Kronecker product it follows:

$$|\mathbf{A} \otimes \mathbf{B}| = |\mathbf{A}|^{K} |\mathbf{B}|^{N}, \tag{A.102}$$

and for the trace of the Kronecker product:

$$\operatorname{tr}(\mathbf{A} \otimes \mathbf{B}) = \operatorname{tr}(\mathbf{A}) \operatorname{tr}(\mathbf{B}). \tag{A.103}$$

A.6.3 The "vec" and "vech" operators

The vec *operator* stacks the K columns of a generic $N \times K$ matrix $\mathbf{A} \equiv (\mathbf{a}^{(1)}, \ldots, \mathbf{a}^{(K)})$ into an NK-dimensional column vector:

$$\operatorname{vec}[\mathbf{A}] \equiv \begin{pmatrix} \mathbf{a}^{(1)} \\ \vdots \\ \mathbf{a}^{(K)} \end{pmatrix}. \tag{A.104}$$

For instance, in the case $N \equiv 2$:

$$\operatorname{vec}\left[\begin{pmatrix} a_{11} & a_{12} \\ a_{21} & a_{22} \end{pmatrix}\right] \equiv \begin{pmatrix} a_{11} \\ a_{21} \\ a_{12} \\ a_{22} \end{pmatrix}. \tag{A.105}$$

A notable link between the vec operator and the Kronecker product is the following relation, that holds for any conformable matrices:

$$\operatorname{vec}[\mathbf{ABC}] = (\mathbf{C}' \otimes \mathbf{A}) \operatorname{vec}[\mathbf{B}]. \tag{A.106}$$

Also notice the simple relation between the vec operator and the trace:

$$\operatorname{tr}(\mathbf{AB}) = \operatorname{vec}[\mathbf{A}']' \operatorname{vec}[\mathbf{B}]. \tag{A.107}$$

If instead of stacking the columns of $\mathbf{A}$ we stacked the columns of its transpose $\mathbf{A}'$ we would obtain an NK-dimensional vector with the same entries, but in different order. The matrix $\mathbf{K}$ that transforms one vector into the other is called the *commutation matrix* and is thus defined by the following identity:

$$\mathrm{vec}\left[\mathbf{A}\right] \equiv \mathbf{K}_{NK}\,\mathrm{vec}\left[\mathbf{A}'\right]. \tag{A.108}$$

The commutation matrix satisfies:

$$\mathbf{K}'_{NK} = \mathbf{K}^{-1}_{NK} = \mathbf{K}_{KN}. \tag{A.109}$$

The explicit expression of the commutation matrix is given in terms of the canonical basis $(A.15)$ as follows:

$$\mathbf{K}_{NK} \equiv \sum_{n=1}^{N}\sum_{k=1}^{K}\left(\left[\boldsymbol{\delta}^{(n)}\right]\left[\boldsymbol{\delta}^{(k)}\right]' \otimes \left[\boldsymbol{\delta}^{(k)}\right]\left[\boldsymbol{\delta}^{(n)}\right]'\right). \tag{A.110}$$

For instance, in the case $N \equiv K \equiv 2$:

$$\mathbf{K}_{22} \equiv \begin{pmatrix} 1\,0\,0\,0 \\ 0\,0\,1\,0 \\ 0\,1\,0\,0 \\ 0\,0\,0\,1 \end{pmatrix}. \tag{A.111}$$

Consider now a symmetric $N \times N$ square matrix $\boldsymbol{\Omega}$. To deal only with the non-redundant entries of $\boldsymbol{\Omega}$ we introduce the vech operator, which stacks the columns of a $\boldsymbol{\Omega}$ skipping the entries above the diagonal. The result is an $N\left(N+1\right)/2$-dimensional column vector.

For instance, in the case $N \equiv 2$:

$$\mathrm{vech}\left[\begin{pmatrix} \omega_{11}\ \omega_{21} \\ \omega_{21}\ \omega_{22} \end{pmatrix}\right] \equiv \begin{pmatrix} \omega_{11} \\ \omega_{21} \\ \omega_{22} \end{pmatrix}. \tag{A.112}$$

Since $\mathrm{vec}\left[\boldsymbol{\Omega}\right]$ contains the redundant entries of $\boldsymbol{\Omega}$, it can be obtained from $\mathrm{vech}\left[\boldsymbol{\Omega}\right]$ by means of a suitable constant matrix $\mathbf{D}$, called the *duplication matrix*, which is defined by the following identity:

$$\mathrm{vec}\left[\boldsymbol{\Omega}\right] \equiv \mathbf{D}_{N}\,\mathrm{vech}\left[\boldsymbol{\Omega}\right]. \tag{A.113}$$

For instance in the case $N \equiv 2$:

$$\mathbf{D}_{2} \equiv \begin{pmatrix} 1\,0\,0 \\ 0\,1\,0 \\ 0\,1\,0 \\ 0\,0\,1 \end{pmatrix}. \tag{A.114}$$

A.6.4 Matrix calculus

We assume known the rules of calculus for smooth real-valued functions $f(\mathbf{x})$, where $\mathbf{x}$ is a vector in $\mathbb{R}^N$. Consider an $N \times K$ matrix of variables $\mathbf{X}$ and a smooth real-valued function $f(\mathbf{X})$. By means of the vec operator we can extend the rules of calculus to this new environment. Indeed, the function f can be seen equivalently as feeding on NK-dimensional vectors:

$$f(\mathbf{X}) \equiv f(\text{vec}[\mathbf{X}]), \tag{A.115}$$

where vec is the operator (A.104). In view of optimization problems, we are mainly interested in computing the gradient $\mathbf{g}$, which is an NK-dimensional vector

$$\mathbf{g} \equiv \frac{\partial f}{\partial \, \text{vec}[\mathbf{X}]}, \tag{A.116}$$

and the Hessian $\mathbf{H}$, which is an $NK \times NK$ symmetric matrix:

$$\mathbf{H} \equiv \frac{\partial^2 f}{\partial \, \text{vec}[\mathbf{X}] \, \partial \, \text{vec}[\mathbf{X}]'}. \tag{A.117}$$

Since the direct computation of these quantities from the definition might be hard, we propose alternative routes to obtain the desired results, based on a Taylor expansion. Indeed, if we manage to express the first variation of the function f due to an infinitesimal change $d\mathbf{X}$ as follows:

$$df = \mathbf{g}' \, \text{vec}[d\mathbf{X}], \tag{A.118}$$

then $\mathbf{g}$ is the gradient (A.116). For instance the following result holds:

$$df = \text{tr}(\mathbf{G} d\mathbf{X}) \Rightarrow \frac{\partial f}{\partial \, \text{vec}[\mathbf{X}]} = \text{vec}[\mathbf{G}'], \tag{A.119}$$

which follows from (A.118) and the set of equalities:

$$\text{tr}(\mathbf{G} d\mathbf{X}) = \sum_{m,n=1}^{N} [\mathbf{G}']_{nm} \, d\mathbf{X}_{nm} = \text{vec}[\mathbf{G}']' \, \text{vec}[d\mathbf{X}]. \tag{A.120}$$

Similarly, if we manage to express the second variation of the function f due to an infinitesimal change $d\mathbf{X}$ as follows:

$$d(df) = \text{vec}[d\mathbf{X}]' \, \mathbf{H} \, \text{vec}[d\mathbf{X}], \tag{A.121}$$

where $\mathbf{H}$ is symmetric, then $\mathbf{H}$ is the Hessian (A.117).

As an application we derive the gradient and the Hessian of $\ln|\mathbf{X}|$, where $\mathbf{X}$ is a square $N \times N$ matrix. Consider first a matrix $\boldsymbol{\epsilon}$ of small elements. A direct computation of the determinant (A.39) shows that:

$$|\mathbf{I} + \boldsymbol{\epsilon}| \approx 1 + \operatorname{tr}(\boldsymbol{\epsilon}) + \cdots, \tag{A.122}$$

where the dots contain products of two or more small terms ϵ_{mn} which are second-order with respect to the leading terms. Then:

$$\begin{aligned} d|\mathbf{X}| &\equiv |\mathbf{X} + d\mathbf{X}| - |\mathbf{X}| = |\mathbf{X}| \left(\left| \mathbf{I} + \mathbf{X}^{-1} d\mathbf{X} \right| - 1 \right) \\ &= |\mathbf{X}| \operatorname{tr}\left(\mathbf{X}^{-1} d\mathbf{X} \right), \end{aligned} \tag{A.123}$$

and thus:

$$d \ln |\mathbf{X}| = \operatorname{tr}\left(\mathbf{X}^{-1} d\mathbf{X} \right). \tag{A.124}$$

Applying the general rule ($A.119$) to this specific case we obtain:

$$\frac{\partial \ln |\mathbf{X}|}{\partial \operatorname{vec}[\mathbf{X}]} = \operatorname{vec}\left[(\mathbf{X}')^{-1} \right]. \tag{A.125}$$

To compute the Hessian of $\ln |\mathbf{X}|$ first of all we differentiate $\mathbf{I} = \mathbf{X}\mathbf{X}^{-1}$ to obtain:

$$d\left(\mathbf{X}^{-1} \right) = -\mathbf{X}^{-1} (d\mathbf{X}) \mathbf{X}^{-1}. \tag{A.126}$$

Computing the second differential from ($A.124$) we obtain:

$$d(d \ln |\mathbf{X}|) = \operatorname{tr}\left(d\left(\mathbf{X}^{-1} \right) d\mathbf{X} \right) = -\operatorname{tr}\left(\mathbf{X}^{-1} (d\mathbf{X}) \mathbf{X}^{-1} d\mathbf{X} \right). \tag{A.127}$$

Using ($A.106$), ($A.107$) and ($A.108$) we arrive at the following expression:

$$\begin{aligned} d(d \ln |\mathbf{X}|) &= -\operatorname{vec}[d\mathbf{X}']' \operatorname{vec}\left[\mathbf{X}^{-1} (d\mathbf{X}) \mathbf{X}^{-1} \right] \\ &= -\operatorname{vec}[d\mathbf{X}]' \mathbf{K}_{NN} \left((\mathbf{X}')^{-1} \otimes \mathbf{X}^{-1} \right) \operatorname{vec}[d\mathbf{X}]. \end{aligned} \tag{A.128}$$

Therefore from ($A.121$) we obtain:

$$\frac{\partial^2 \ln |\mathbf{X}|}{\partial \operatorname{vec}[\mathbf{X}] \, \partial \operatorname{vec}[\mathbf{X}]'} = -\mathbf{K}_{NN} \left((\mathbf{X}')^{-1} \otimes \mathbf{X}^{-1} \right). \tag{A.129}$$

B

Functional Analysis

In this appendix we provide a very loose review of linear functional analysis. Due to the extension of the topic and the scope of the book, this presentation relies on intuition more than mathematical rigor. In order to support intuition we present the subject as a generalization to the infinite-dimensional world of calculus of the familiar formalism and concepts of linear algebra. For this reason we parallel as closely as possible the discussion in Appendix A. For a more rigorous discussion the reader is referred to references such as Smirnov (1964), Reed and Simon (1980), Rudin (1991), and Whittaker and Watson (1996).

B.1 Vector space

The natural environment of linear functional analysis are infinite-dimensional spaces of functions that are a direct extension of the finite-dimensional Euclidean space discussed in Appendix A.1.

The main difference (and analogy) between the Euclidean space and a vector space of functions is that the discrete integer index n of the Euclidean vectors becomes a continuous index $\mathbf{x}$. Furthermore, we also let the value of the vector be complex. Therefore we define an element of the yet to be defined vector space by extending (A.2) as follows:

$$v : \mathbf{x} \in \mathbb{R}^N \to v(\mathbf{x}) \in \mathbb{C}. \tag{B.1}$$

Notice that we denote as v the function, to be compared with the boldface notation $\mathbf{v}$ in (A.2), which denotes a vector; on the other hand we denote as $v(\mathbf{x})$ the specific value of that function in $\mathbf{x}$, to be compared with the entry of the vector v_n in (A.2).

We represent the analogy between (B.1) and (A.2) graphically in Figure B.1, which parallels Figure A.1.

The set of functions (B.1) is a vector space, since the following operations are properly defined on its elements.

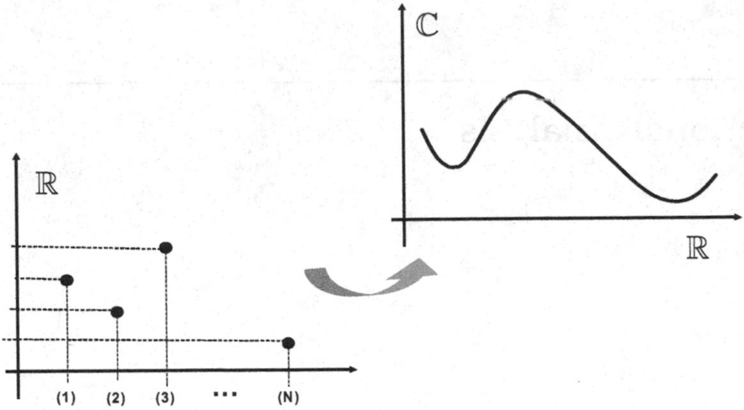

Fig. B.1. From linear algebra to functional analysis

The *sum* of two functions is defined point-wise as follows:

$$[u+v](\mathbf{x}) \equiv u(\mathbf{x}) + v(\mathbf{x}). \tag{B.2}$$

This is the infinite-dimensional version of the "parallelogram rule" of a Euclidean space, compare with (*A*.3).

The *multiplication by a scalar* is defined point-wise as follows:

$$[\alpha v](\mathbf{x}) \equiv \alpha v(\mathbf{x}), \tag{B.3}$$

compare with (*A*.4).

Combining sums and multiplications by a scalar we obtain *linear combinations* of functions.

We have seen the striking resemblance of the definitions introduced above with those introduced in Appendix A.1. With the further remark that finite sums become integrals in this infinite-dimensional world, we can obtain almost all the results we need by simply changing the notation in the results for linear algebra. We summarize the main notational analogies between linear algebra and linear functional analysis in the following table:

	linear algebra	**functional analysis**
index/dimension	$n \in \{1, \ldots, N\}$	$\mathbf{x} \in \mathbb{R}^N$
element	$\mathbf{v} : n \to v_n \in \mathbb{R}$	$v : \mathbf{x} \to v(\mathbf{x}) \in \mathbb{C}$
sum	$\sum_{n=1}^{N} [\cdot]$	$\int_{\mathbb{R}^N} [\cdot] \, d\mathbf{x}$

(B.4)

A set of functions are *linearly independent* if the parallelotope they generate is non-degenerate, i.e. if no function can be expressed as a linear combination of the others.

Using the analogies of Table B.4 we can generalize the definition of inner product given in $(A.5)$ and endow our space of functions with the following *inner product*:

$$\langle u, v \rangle \equiv \int_{\mathbb{R}^N} u(\mathbf{x}) \overline{v(\mathbf{x})} d\mathbf{x}, \tag{B.5}$$

where $\bar{\cdot}$ denotes the *conjugate transpose*:

$$\overline{a + ib} \equiv a - ib, \qquad a, b \in \mathbb{R}, \quad i \equiv \sqrt{-1}. \tag{B.6}$$

By means of an inner product we can define orthogonality. Similarly to $(A.11)$ a pair of functions (u, v) are *orthogonal* if:

$$\langle u, v \rangle = 0. \tag{B.7}$$

Orthogonal functions are in particular linearly independent, since the parallelotope they generate is not skewed, and thus non-degenerate. For a geometrical interpretation refer to the Euclidean case in Figure A.2.

As in the finite-dimensional setting of linear algebra, the inner product allows us to define the *norm* of a function, i.e. its "length", by means of the *Pythagorean theorem*. Using the analogies of Table B.4 the definition of norm given in $(A.6)$ becomes:

$$\|v\| \equiv \sqrt{\langle v, v \rangle} = \sqrt{\int_{\mathbb{R}^N} |v(\mathbf{x})|^2 \, d\mathbf{x}}. \tag{B.8}$$

When defined, this is a norm, since it satisfies the properties $(A.7)$.

Nevertheless, unlike in the finite-dimensional setting of linear algebra, for most functions the integrals $(B.5)$ and $(B.8)$ are not defined. Therefore, at this stage we restrict our space to the set of vectors with finite length:

$$L_2(\mathbb{R}^N) \equiv \left\{ v \text{ such that } \int_{\mathbb{R}^N} |v(\mathbf{x})|^2 \, d\mathbf{x} < \infty \right\}. \tag{B.9}$$

This set is clearly a restriction of the original set of functions $(B.1)$. Furthermore, we extend this set to include in a natural way a set of *generalized functions*, namely elements that behave like functions inside an integral, but are not functions as $(B.1)$ in the common sense of the word. This way the space $(B.9)$ becomes a *complete vector space*.[1]

[1] This extension can be understood intuitively as follows. Consider the set of numbers:

$$\mathcal{S} \in \left\{ \frac{1}{x}, x \in \mathbb{R} \right\}. \tag{B.10}$$

This set is the real axis deprived of the zero. Adding the zero element to this set makes it a *complete* set, which is a much richer object.

A complete vector space where the norm is defined in terms of an inner product as in $(B.8)$ is called a Hilbert space. Therefore the space of functions $L_2(\mathbb{R}^N)$ is a Hilbert space: this is the closest infinite-dimensional generalization of a finite-dimensional Euclidean vector space. In Table B.11 we summarize how the properties of the Hilbert space $L_2(\mathbb{R}^N)$ compare to the properties of the Euclidean space $\mathbb{R}^N$.

	linear algebra	**functional analysis**
space	Euclid $\mathbb{R}^N$	Hilbert $L_2(\mathbb{R}^N)$
inner product	$\langle \mathbf{u}, \mathbf{v} \rangle \equiv \sum_{n=1}^N u_n v_n$	$\langle u, v \rangle \equiv \int_{\mathbb{R}^N} u(\mathbf{x}) \overline{v(\mathbf{x})} d\mathbf{x}$
norm (length)	$\Vert\mathbf{v}\Vert \equiv \sqrt{\sum_{n=1}^N v_n^2}$	$\Vert v \Vert \equiv \sqrt{\int_{\mathbb{R}^N} \vert v(\mathbf{x})\vert^2 d\mathbf{x}}$

(B.11)

We conclude this section mentioning a more general vector space of functions. Indeed, instead of $(B.8)$ we can define a norm as follows:

$$\Vert v \Vert_p \equiv \left(\int_{\mathbb{R}^N} |v(\mathbf{x})|^p d\mathbf{x} \right)^{\frac{1}{p}}, \tag{B.12}$$

where $1 \leq p < \infty$. Notice that $(B.8)$ corresponds to the particular case $p \equiv 2$ in $(B.12)$. It can be proved that $(B.12)$ is also a norm, as it satisfies the properties $(A.7)$. This norm is defined on the following space of functions:

$$L_p(\mathbb{R}^N) \equiv \left\{ v \text{ such that } \int_{\mathbb{R}^N} |v(\mathbf{x})|^p d\mathbf{x} < \infty \right\}. \tag{B.13}$$

Unlike $(B.8)$, in the general case $p \neq 2$ this norm is not induced by an inner product. A complete normed space without inner product is called a *Banach space*. Therefore the spaces $L_p(\mathbb{R}^N)$ are Banach spaces.

B.2 Basis

A basis is a set of linearly independent elements of a vector space that can generate any vector in that space by linear combinations. According to Table B.4, the discrete integer index n of the Euclidean vectors becomes a continuous index $\mathbf{y} \in \mathbb{R}^N$. Therefore the definition $(A.13)$ of a basis for a Euclidean space is generalized to the Hilbert space $L_2(\mathbb{R}^N)$ as follows. A basis for $L_2(\mathbb{R}^N)$ is a set of linearly independent functions indexed by $\mathbf{y}$:

$$e^{(\mathbf{y})}, \quad \mathbf{y} \in \mathbb{R}^N, \tag{B.14}$$

such that any function v of $L_2(\mathbb{R}^N)$ can be expressed as a linear combination:

$$v = \int_{\mathbb{R}^N} \alpha(\mathbf{y})\, e^{(\mathbf{y})} d\mathbf{y}. \tag{B.15}$$

In analogy with (A.16), the *canonical basis* of $L_2(\mathbb{R}^N)$ is the set of functions $\delta^{(\mathbf{y})}$ indexed by $\mathbf{y}$ such that the inner product of a generic element of this basis $\delta^{(\mathbf{y})}$ with a generic function v in $L_2(\mathbb{R}^N)$ yields the "$\mathbf{y}$-th entry", i.e. the value of the function at that point:

$$\left\langle v, \delta^{(\mathbf{y})} \right\rangle \equiv v(\mathbf{y}). \tag{B.16}$$

The generic element $\delta^{(\mathbf{x})}$ of this basis is called the *Dirac delta centered in* $\mathbf{x}$.

We notice that the Dirac delta is not a standard function: it is a generalized function, since it only makes sense within an integral. Indeed, no regular function can possibly satisfy (B.16), since from (B.5) for all functions v this hypothetical regular function should satisfy the following equality:

$$\int_{\mathbb{R}^N} v(\mathbf{x})\, \overline{\delta^{(\mathbf{y})}(\mathbf{x})} d\mathbf{x} = v(\mathbf{y}). \tag{B.17}$$

In order for (B.17) to be true, $\delta^{(\mathbf{y})}(\mathbf{x})$ should be zero for all values of $\mathbf{x}$, except for $\mathbf{x} \equiv \mathbf{y}$. In this case the above integral would be zero, no matter the value of $\delta^{(\mathbf{y})}$ in $\mathbf{x} \equiv \mathbf{y}$.

Therefore, the Dirac delta is not a standard function: instead, it is a limit case of standard functions. Define an approximation of the Dirac delta in terms of the *Gauss exponential function* as follows:

$$\delta_\epsilon^{(\mathbf{y})}(\mathbf{x}) \equiv \frac{1}{(2\pi)^{\frac{N}{2}} \epsilon^N} e^{-\frac{1}{2\epsilon^2}(\mathbf{x}-\mathbf{y})'(\mathbf{x}-\mathbf{y})}. \tag{B.18}$$

This is a bell-shaped, smooth function that reaches its peak in $\mathbf{x} \equiv \mathbf{y}$ and whose width is of the order of ϵ. We plot in Figure B.2 this function for different values of ϵ. As the width ϵ approaches zero the bell becomes taller and thinner around the peak $\mathbf{y}$. Intuitively, as $\epsilon \to 0$ the function $\delta_\epsilon^{(\mathbf{y})}$ becomes zero everywhere, except at the point $\mathbf{y}$ where its value becomes infinite. This profile generalizes the finite-dimensional canonical basis (A.15).

Furthermore, the integral of the approximate Dirac delta function (B.18) over the whole real axis is one, since it is a specific case of the multivariate normal probability density function (2.156). Thus the inner product of $\delta_\epsilon^{(\mathbf{y})}$ with another function v is a weighted average of the values of v, where the most weight is given to the points in a neighborhood of $\mathbf{y}$ of radius ϵ. Therefore:

$$\left\langle v, \delta_\epsilon^{(\mathbf{y})} \right\rangle \approx v(\mathbf{y}). \tag{B.19}$$

In the limit $\epsilon \to 0$ this approximation becomes the equality (B.16).

We might be puzzled that the elements of the basis of the space $L_2(\mathbb{R}^N)$, which is a set of functions, is not a function. In reality, this is not a problem:

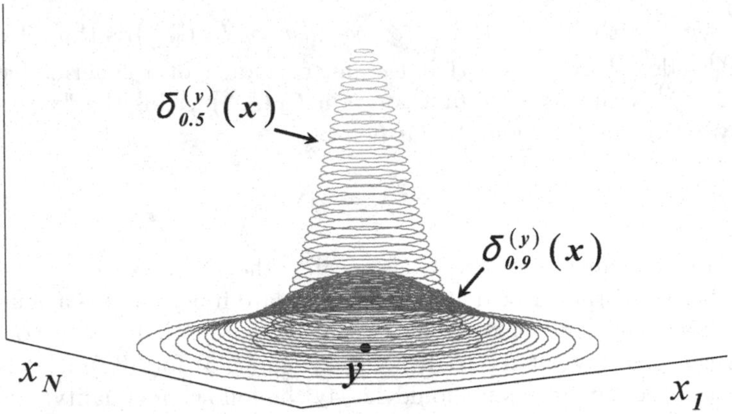

Fig. B.2. Approximation of the Dirac delta with Gaussian exponentials

we recall that in its definition we extended the space $L_2(\mathbb{R}^N)$ to include all the natural limit operations, in such a way to make it complete, see $(B.10)$.

We summarize in the table below the analogies between the basis in the finite-dimensional Euclidean vector space $\mathbb{R}^N$ and in the infinite-dimensional Hilbert space $L_2(\mathbb{R}^N)$ respectively.

	linear algebra	**functional analysis**
basis	$\{\mathbf{e}^{(n)}\}_{n\in\{1,\ldots,N\}}$	$\{e^{(\mathbf{y})}\}_{\mathbf{y}\in\mathbb{R}^N}$
canonical basis	$\langle \mathbf{v}, \boldsymbol{\delta}^{(n)} \rangle = v_n$	$\langle v, \delta^{(\mathbf{y})} \rangle = v(\mathbf{y})$

(B.20)

As an application, consider a random a variable $\mathbf{X}$ that takes on specific values $\mathbf{x}_1, \mathbf{x}_2, \ldots$ with finite probabilities $p_{\mathbf{x}_1}, p_{\mathbf{x}_2}, \ldots$ respectively. The variable $\mathbf{X}$ has a *discrete distribution*. In this situation no regular probability density function $f_{\mathbf{X}}$ can satisfy (2.4). Indeed, if this were the case, the following equality would hold:

$$p_{\mathbf{x}_i} = \mathbb{P}\{\mathbf{X} = x_i\} = \int_{\{\mathbf{x}_i\}} f_{\mathbf{X}}(\mathbf{x})\, d\mathbf{x}. \tag{B.21}$$

Nevertheless, the integral of any regular function on the singleton $\{\mathbf{x}_i\}$ is null. On the other hand, if we express the probability density function $f_{\mathbf{X}}$ as a generalized function this problem does not exist. Indeed, if we express $f_{\mathbf{X}}$ in terms of the Dirac delta $(B.16)$ as follows:

$$f\mathbf{x} = \sum_i p_{\mathbf{x}_i} \delta^{(\mathbf{x}_i)}, \tag{B.22}$$

then this generalized function satisfies ($B.21$).

In particular, consider the case of a discrete random variable that can only take on one specific value $\widetilde{\mathbf{x}}$ with associated probability $p_{\widetilde{\mathbf{x}}} \equiv 1$: this is not a random variable, as the outcome of its measurement is known with certainty. Instead, it is a constant vector $\widetilde{\mathbf{x}}$. The formalism of generalized functions allows us to treat constants as special cases of a random variable. The visualization of the probability density function of this "not-too-random" variable in terms of the regularized Dirac delta is a bell-shaped function centered around $\widetilde{\mathbf{x}}$ that spikes to infinity as the approximation becomes exact.

B.3 Linear operators

Consider in analogy with ($A.17$) a transformation A that maps functions v of the Hilbert space $L_2(\mathbb{R}^N)$ into functions that might belong to the same Hilbert space, or to some other space of functions F:

$$A: v \in L_2(\mathbb{R}^N) \mapsto u \equiv A[v] \in F. \tag{B.23}$$

In the context of functional analysis, such transformations are called *functionals* or *operators* and generalize the finite-dimensional concept of function.

In analogy with ($A.18$), a functional A is called a *linear operator* if it preserves the sum and the multiplication by a scalar:

$$\begin{aligned} A[u+v] &= A[u] + A[v] \\ A[\alpha v] &= \alpha A[v]. \end{aligned} \tag{B.24}$$

Geometrically, this means that infinite-dimensional parallelotopes are mapped into infinite-dimensional parallelotopes, as represented in Figure A.3.

For example, consider the *differentiation operator*:

$$\mathcal{D}_n[v](\mathbf{x}) \equiv \frac{\partial v(\mathbf{x})}{\partial x_n}. \tag{B.25}$$

This operator is defined on a subset of smooth functions in $L_2(\mathbb{R}^N)$. It is easy to check that the differentiation operator is linear.

The inverse A^{-1} of a linear operator A is the functional that applied either before or after the linear operator A cancels the effect of the operator A. In other words, in analogy with ($A.20$), for all functions v the inverse functional A^{-1} satisfies:

$$(A^{-1} \circ A)[v] = v = (A \circ A^{-1})[v]. \tag{B.26}$$

As in the finite-dimensional case, in general the inverse transformation is not defined. If it is defined, it is linear.

For example, consider the *integration operator*, defined as follows:

$$\mathcal{I}_n [v] (\mathbf{x}) \equiv \int_{-\infty}^{x_n} v (x_1, \ldots, z_n, \ldots x_N) \, dz_n \tag{B.27}$$

The *fundamental theorem of calculus* states that the integration operator is the inverse of the differentiation operator (B.25). It is easy to check that the integration operator is linear.

B.3.1 Kernel representations

In Appendix A.3.1 we saw that in the finite-dimensional case every linear transformation A admits a matrix representation A_{mn}. Therefore we expect that every linear operator on $L_2 (\mathbb{R}^N)$ be expressible in terms of a continuous version of a matrix. By means of the notational analogies of Table B.4 and Table B.11 this "continuous" matrix must be an integral *kernel*, i.e. a function $A (\mathbf{y}, \mathbf{x})$ such that

$$A [v] (\mathbf{y}) \equiv \int_{\mathbb{R}^N} A (\mathbf{y}, \mathbf{x}) \, v (\mathbf{x}) \, d\mathbf{x}, \tag{B.28}$$

which parallels (A.25). Such a representation does not exist in general. The operators that admit a kernel representation are called *Hilbert-Schmidt operators*. Nevertheless, we can always find and use an approximate kernel, which becomes exact in a limit sense.

For example the kernel of the differentiation operator (B.25) is not defined (we consider the one-dimensional case for simplicity). Nevertheless, the following kernel is well defined:

$$D^{\epsilon} (y, x) \equiv \frac{1}{\epsilon} \left(\delta_{\epsilon}^{(y+\epsilon)} (x) - \delta_{\epsilon}^{(y)} (x) \right), \tag{B.29}$$

where $\delta_{\epsilon}^{(y)}$ is the one-dimensional approximate Dirac delta (B.18). In the limit $\epsilon \longrightarrow 0$ we obtain:

$$\lim_{\epsilon \longrightarrow 0} \int_{\mathbb{R}} D^{\epsilon} (y, x) \, v (x) \, dx = [\mathcal{D} v] (y) . \tag{B.30}$$

Therefore (B.29) is the kernel representation of the differentiation operator in a limit sense.

B.3.2 Unitary operators

Unitary operators are the generalization of rotations to the infinite-dimensional world of functional analysis. Therefore, in analogy with (A.30), an operator U

is unitary if it does not alter the length, i.e. the norm (*B*.8), of any function in $L_2(\mathbb{R}^N)$:

$$\|U[v]\| = \|v\|. \tag{B.31}$$

For example, it is immediate to check that the *reflection operator* defined below is unitary:

$$\text{Refl}[v](\mathbf{x}) \equiv v(-\mathbf{x}). \tag{B.32}$$

Similarly the *shift operator* defined below is unitary:

$$\text{Shift}_{\mathbf{a}}[v](\mathbf{x}) \equiv v(\mathbf{x}-\mathbf{a}). \tag{B.33}$$

The most notable application of unitary operators is the *Fourier transform*. This transformation is defined in terms of its kernel representation as follows:

$$\mathcal{F}[v](\mathbf{y}) \equiv \int_{\mathbb{R}^N} e^{i\mathbf{y}'\mathbf{x}} v(\mathbf{x})\, d\mathbf{x}. \tag{B.34}$$

We prove below that for all functions in $L_2(\mathbb{R}^N)$ the following result holds:

$$\|\mathcal{F}[v]\| = (2\pi)^{\frac{N}{2}} \|v\|. \tag{B.35}$$

Therefore the Fourier transform is a (rescaled) unitary operator.

For example, consider the normal probability density function (2.156):

$$f^{\text{N}}_{\boldsymbol{\mu},\boldsymbol{\Sigma}}(\mathbf{x}) \equiv \frac{1}{(2\pi)^{\frac{N}{2}} |\boldsymbol{\Sigma}|^{\frac{1}{2}}} e^{-\frac{1}{2}(\boldsymbol{\mu}-\mathbf{x})'\boldsymbol{\Sigma}^{-1}(\boldsymbol{\mu}-\mathbf{x})}. \tag{B.36}$$

From (2.14), the Fourier transform of the normal pdf is the characteristic function of the normal distribution. From (2.157) it reads:

$$\mathcal{F}\left[f^{\text{N}}_{\boldsymbol{\mu},\boldsymbol{\Sigma}}\right](\mathbf{y}) = e^{i\boldsymbol{\mu}'\mathbf{y}-\frac{1}{2}\mathbf{y}'\boldsymbol{\Sigma}\mathbf{y}}. \tag{B.37}$$

In particular from (*B*.37) and (*B*.18), i.e. the fact that in the limit $\boldsymbol{\Sigma} \to \mathbf{0}$ the normal density $f^{\text{N}}_{\mathbf{x},\boldsymbol{\Sigma}}$ becomes the Dirac delta $\delta^{(\mathbf{x})}$, we obtain the following notable result:

$$\mathcal{F}\left[\delta^{(\mathbf{x})}\right] = \exp(i\mathbf{x}'\cdot). \tag{B.38}$$

In a Euclidean space rotations are invertible and the inverse is a rotation. Similarly, a unitary operator is always invertible and the inverse is a unitary operator. Furthermore, in a Euclidean space the representation of the inverse rotation is the transpose matrix, see (*A*.31). Similarly, it is possible to prove that the kernel representation of the inverse of a unitary transformation is the complex conjugate of the kernel representation of the unitary operator. In formulas:

$$U^{-1}[v](\mathbf{x}) = \int_{\mathbb{R}^N} \overline{U(\mathbf{y},\mathbf{x})} v(\mathbf{y})\, d\mathbf{y}. \tag{B.39}$$

By this argument, the *inverse Fourier transform* is defined in terms of its kernel representation as follows:

$$\mathcal{F}^{-1}[v](\mathbf{x}) \equiv (2\pi)^N \int_{\mathbb{R}^N} e^{-i\mathbf{x}'\mathbf{y}} v(\mathbf{y})\, d\mathbf{y}, \tag{B.40}$$

where the factor $(2\pi)^N$ appears because the Fourier transform is a rescaled unitary transformation.

In particular, inverting $(B.38)$ and substituting $v(\mathbf{y}) \equiv e^{i\mathbf{z}'\mathbf{y}}$ in $(B.40)$ we obtain the following useful identity:

$$\delta^{(\mathbf{z})}(\mathbf{x}) = \mathcal{F}^{-1}[\exp(i\mathbf{z}'\cdot)](\mathbf{x}) = (2\pi)^N \int_{\mathbb{R}^N} e^{i(\mathbf{z}-\mathbf{x})'\mathbf{y}} d\mathbf{y}. \tag{B.41}$$

Using this identity we can show that the Fourier transform is a rescaled unitary transformation. Indeed:

$$\begin{aligned} \|\mathcal{F}[v]\|^2 &\equiv \int_{\mathbb{R}^N} \left[\int_{\mathbb{R}^N} e^{i\mathbf{y}'\mathbf{x}} v(\mathbf{x})\, d\mathbf{x}\right] \overline{\left[\int_{\mathbb{R}^N} e^{i\mathbf{y}'\mathbf{z}} v(\mathbf{z})\, d\mathbf{z}\right]} d\mathbf{y} \\ &= \int_{\mathbb{R}^N}\int_{\mathbb{R}^N} \left(\int_{\mathbb{R}^N} e^{(i\mathbf{y}'(\mathbf{x}-\mathbf{z}))} d\mathbf{y}\right) v(\mathbf{x}) \overline{v(\mathbf{z})} d\mathbf{x} d\mathbf{z} \\ &= (2\pi)^N \int_{\mathbb{R}^N}\int_{\mathbb{R}^N} \delta^{(\mathbf{z})}(\mathbf{x})\, v(\mathbf{x}) \overline{v(\mathbf{z})} d\mathbf{x} d\mathbf{z} = (2\pi)^N \|v\|^2 . \end{aligned} \tag{B.42}$$

B.4 Regularization

In the Hilbert space of functions $L_2(\mathbb{R}^N)$ it is possible to define another operation that turns out very useful in applications.

The *convolution* of two functions u and v in this space is defined as follows:

$$[u * v](\mathbf{x}) \equiv \int_{\mathbb{R}^N} u(\mathbf{y})\, v(\mathbf{x}-\mathbf{y})\, d\mathbf{y}. \tag{B.43}$$

The convolution shares many of the features of the multiplication between numbers. Indeed it is commutative, associative and distributive:

$$\begin{aligned} u * v &= v * u \\ (u * v) * z &= u * (v * z) \\ (u + v) * z &= u * z + v * z. \end{aligned} \tag{B.44}$$

Furthermore, the Fourier transform $(B.34)$ of the convolution of two functions is the product of the Fourier transforms of the two functions:

$$\mathcal{F}[u * v] = \mathcal{F}[u]\,\mathcal{F}[v]. \tag{B.45}$$

This follows from the series of identities:

$$\begin{aligned}\mathcal{F}[u*v](\mathbf{y}) &\equiv \int_{\mathbb{R}^N} e^{i\mathbf{y}'\mathbf{x}}[u*v](\mathbf{x})\,d\mathbf{x} \\ &= \int_{\mathbb{R}^N} u(\mathbf{z})\left[\int_{\mathbb{R}^N} e^{i\mathbf{y}'\mathbf{x}} v(\mathbf{x}-\mathbf{z})\,d\mathbf{x}\right] d\mathbf{z} \\ &= \int_{\mathbb{R}^N} u(\mathbf{z})\left[e^{i\mathbf{y}'\mathbf{z}}\mathcal{F}[v](\mathbf{y})\right] d\mathbf{z} \\ &= \mathcal{F}[v](\mathbf{y})\,\mathcal{F}[u](\mathbf{y}).\end{aligned} \tag{B.46}$$

An important application of the convolution stems from the immediate result that the Dirac delta (B.16) centered in zero is the neutral element of the convolution:

$$\left[\delta^{(\mathbf{0})} * v\right] = v. \tag{B.47}$$

By approximating the Dirac delta with the smooth function $\delta_\epsilon^{(\mathbf{0})}$ defined in (B.18), we obtain from (B.47) an approximate expression for a generic function v, which we call the *regularization* of v with *bandwidth* ϵ:

$$v_\epsilon \equiv \left[\delta_\epsilon^{(\mathbf{0})} * v\right] \approx v. \tag{B.48}$$

From (B.43), the regularization of v reads explicitly as follows:

$$v_\epsilon(\mathbf{x}) \equiv \frac{1}{(2\pi)^{\frac{N}{2}}\epsilon^N}\int_{\mathbb{R}^N} e^{-\frac{(\mathbf{y}-\mathbf{x})'(\mathbf{y}-\mathbf{x})}{2\epsilon^2}} v(\mathbf{y})\,d\mathbf{y}. \tag{B.49}$$

Due to the bell-shaped profile of the Gaussian exponential in the above integral, the regularized function $v_\epsilon(\mathbf{x})$ is a *moving average* of $v(\mathbf{x})$ with its surrounding values: the effect of the surrounding values fades away as their distance from $\mathbf{x}$ increases. The size of the "important" points that determine the moving average is determined by the bandwidth ϵ of the bell-shaped Gaussian exponential.

The regularization (B.49) becomes exact as the bandwidth ϵ tends to zero: indeed, in the limit where ϵ tends to zero, the Gaussian exponential tends to the Dirac delta, and we recover (B.47). Furthermore, the regularized function v_ϵ is smooth: indeed, since the Gaussian exponential is smooth, the right hand side of (B.49) can be derived infinite times with respect to $\mathbf{x}$.

To become more acquainted with the regularization technique we use it to compute the derivative of the Heaviside function $H^{(\mathbf{y})}$ defined in (B.73). The partial derivative of the Heaviside function along any coordinate is zero everywhere, except in $\mathbf{y}$, where the limit that defines the partial derivative diverges to infinity. This behavior resembles that of the Dirac delta $\delta^{(\mathbf{y})}$, so we are led to conjecture that the combined partial derivatives of the Heaviside function are the Dirac delta:

$$(\mathcal{D}_1 \circ \cdots \circ \mathcal{D}_N)\left[H^{(\mathbf{y})}\right] = \delta^{(\mathbf{y})}. \tag{B.50}$$

We can verify this conjecture with the newly introduced operations. First we notice that in general the convolution of a function with the Heaviside function is the combined integral ($B.27$) of that function:

$$\left(v * H^{(\mathbf{y})}\right)(\mathbf{x}) \equiv \int_{\mathbb{R}^N} v(\mathbf{z}) H^{(\mathbf{y})}(\mathbf{x}-\mathbf{z})\, d\mathbf{z} \tag{B.51}$$
$$= \int_{-\infty}^{x_1-y_1} \cdots \int_{-\infty}^{x_N-y_N} v(\mathbf{z})\, d\mathbf{z} = (\mathcal{I}_1 \circ \cdots \circ \mathcal{I}_N)[v](\mathbf{x}-\mathbf{y}).$$

Applying this result to the regularization ($B.48$) of the Heaviside function and recalling from ($B.27$) that the integration is the inverse of the differentiation, we obtain:

$$(\mathcal{D}_1 \circ \cdots \circ \mathcal{D}_N)\left[H_\epsilon^{(\mathbf{y})}\right] \equiv (\mathcal{D}_1 \circ \cdots \circ \mathcal{D}_N)\left[H^{(\mathbf{y})} * \delta_\epsilon^{(\mathbf{0})}\right] \tag{B.52}$$
$$= (\mathcal{D}_1 \circ \cdots \circ \mathcal{D}_N)\left[(\mathcal{I}_1 \circ \cdots \circ \mathcal{I}_N)\left[\delta_\epsilon^{(\mathbf{y})}\right]\right] = \delta_\epsilon^{(\mathbf{y})}.$$

Taking the limit $\epsilon \to 0$ we obtain the proof of the conjecture ($B.50$).

Using ($B.50$) we can compute the cumulative distribution function of a discrete distribution, whose probability density function is ($B.22$). Indeed, from the definition of cumulative distribution function (2.10) we obtain:

$$F_{\mathbf{X}} \equiv (\mathcal{I}_1 \circ \cdots \circ \mathcal{I}_N)\left[\sum_{\mathbf{x}_i} p_{\mathbf{x}_i} \delta^{(\mathbf{x}_i)}\right] = \sum_{\mathbf{x}_i} p_{\mathbf{x}_i} H^{(\mathbf{x}_i)}, \tag{B.53}$$

where we used the fact that the integration operator is linear.

An important application of the regularization technique concerns the probability density function $f_{\mathbf{X}}$ of a generic random variable $\mathbf{X}$. Indeed, consider the regularization ($B.49$) of $f_{\mathbf{X}}$, which can be a very irregular function, or even a generalized function:

$$f_{\mathbf{X};\epsilon}(\mathbf{x}) \equiv \frac{1}{(2\pi)^{\frac{N}{2}} \epsilon^N} \int_{\mathbb{R}^N} e^{-\frac{(\mathbf{y}-\mathbf{x})'(\mathbf{y}-\mathbf{x})}{2\epsilon^2}} f(\mathbf{y})\, d\mathbf{y}. \tag{B.54}$$

It is immediate to check that $f_{\mathbf{X};\epsilon}$ is strictly positive everywhere and integrates to one over the entire domain: therefore it is a probability density function.

Furthermore, we notice that in general the probability density function $f_{\mathbf{X}}$ of a random variable $\mathbf{X}$ is not univocally defined. Indeed, from its very definition (2.4) the probability density function only makes sense within an integral. For instance, if we change its value at one specific point, the ensuing altered probability density is completely equivalent to the original one. More precisely, a probability density function is an equivalence class of functions that are identical *almost everywhere*, i.e. they are equal to each other except possibly on a set of zero probability (such as one point). The regularization technique ($B.54$) provides a univocally defined, smooth and positive probability density function.

Therefore, whenever needed, we can replace the original probability density function $f_{\mathbf{X}}$ with its regularized version $f_{\mathbf{X};\epsilon}$, which is smooth and approximates the original probability density function $f_{\mathbf{X}}$ to any degree of accuracy. If necessary, we can eventually consider the limit $\epsilon \rightarrow 0$ in the final solution to our problem, in order to recover an exact answer that does not depend on the bandwidth ϵ. Nevertheless, from a more "philosophical" point of view, in most cases we do not need to consider the limit $\epsilon \rightarrow 0$ in the final solution. Indeed, in most applications it is impossible to distinguish between a statistical model based on the original probability density function $f_{\mathbf{X}}$ and one based on the regularized probability density function $f_{\mathbf{X};\epsilon}$, provided that the bandwidth ϵ is small enough. Therefore it becomes questionable which of the two probability density functions is the "real" and which is the "approximate" model.

B.5 Expectation operator

Consider a random variable $\mathbf{X}$, whose probability density function is $f_{\mathbf{X}}$. Consider a new random variable Y defined in terms of a generic function g of the original variable $\mathbf{X}$:

$$Y \equiv g(\mathbf{X}). \tag{B.55}$$

We recall that the set of functions of the random variable $\mathbf{X}$ is a vector space, since sum and multiplication by a scalar are defined in a natural way as in ($B.2$) and ($B.3$) respectively.

To get a rough idea of the possible outcomes of the random variable Y defined in ($B.55$) it is intuitive to weigh each possible outcome by its respective probability. This way we are led to the definition of the *expectation operator* associated with the distribution $f_{\mathbf{X}}$. This operator associates with any function of a random variable the probability-weighted average of all its possible outcomes:

$$\mathrm{E}\{g(\mathbf{X})\} \equiv \mathrm{E}_{\mathbf{X}}\{g\} \equiv \int_{\mathbb{R}^N} g(\mathbf{x}) f_{\mathbf{X}}(\mathbf{x})\, d\mathbf{x}. \tag{B.56}$$

To simplify the notation we might at times drop the symbol $\mathbf{X}$. From this definition it is immediate to check that the expectation operator is linear, i.e. it satisfies ($B.24$).

The expectation operator endows the functions of $\mathbf{X}$ with a norm, and thus with the structure of Banach space. Indeed, consider an arbitrary positive number p. We define the p-norm of g as follows:

$$\|g\|_{\mathbf{X};p} \equiv \left(\mathrm{E}_{\mathbf{X}}\{|g|^p\}\right)^{\frac{1}{p}}. \tag{B.57}$$

When defined, it is possible to check that this is indeed a norm, as it satisfies the properties ($A.7$). In order to guarantee that the norm is defined, we restrict the generic space of functions of $\mathbf{X}$ to the following subspace:

$$L^p_{\mathbf{X}} \equiv \{g \text{ such that } \mathrm{E}_{\mathbf{X}} \{|g|^p\} < \infty\} . \tag{B.58}$$

Given the norm, we can define the distance $\|g - h\|_{\mathbf{X};p}$ between two generic functions g and h in $L^p_{\mathbf{X}}$.

The space $L^2_{\mathbf{X}}$ is somewhat special, as it can also be endowed with the following inner product:

$$\langle g, h \rangle_{\mathbf{X}} \equiv \mathrm{E}\{g\overline{h}\} . \tag{B.59}$$

It is easy to check that in $L^2_{\mathbf{X}}$ the norm is induced by the inner product:

$$\|\cdot\|_{\mathbf{X};2} = \sqrt{\langle \cdot, \cdot \rangle_{\mathbf{X}}} . \tag{B.60}$$

Therefore, $L^2_{\mathbf{X}}$ is a Hilbert space and in addition to the properties ($A.7$) also the Cauchy-Schwartz inequality ($A.8$) is satisfied:

$$|\langle g, h \rangle_{\mathbf{X}}| \leq \|g\|_{\mathbf{X};2} \|h\|_{\mathbf{X};2} . \tag{B.61}$$

As in ($A.9$)-($A.10$) the equality in this expression holds if and only if $g \equiv \alpha h$ almost everywhere for some scalar α. If the scalar α is positive:

$$\langle g, h \rangle_{\mathbf{X}} = \|g\|_{\mathbf{X};2} \|h\|_{\mathbf{X};2} ; \tag{B.62}$$

if the scalar α is negative:

$$\langle g, h \rangle_{\mathbf{X}} = - \|g\|_{\mathbf{X};2} \|h\|_{\mathbf{X};2} . \tag{B.63}$$

It is easy to check that the operator $\langle \cdot, \cdot \rangle_{\mathbf{X}}$ is, like all inner products, *symmetric* and *bilinear*. Explicitly, this means that for all functions g and h in $L^2_{\mathbf{X}}$:

$$\langle g, h \rangle_{\mathbf{X}} = \langle h, g \rangle_{\mathbf{X}} ; \tag{B.64}$$

and that for any function g in $L^2_{\mathbf{X}}$ the application $\langle g, \cdot \rangle_{\mathbf{X}}$ is linear. This implies in particular that the inner product of a linear combination of functions with itself can be expressed as follows:

$$\left\langle \sum_{m=1}^{M} \alpha_m g_m, \sum_{m=1}^{M} \alpha_m g_m \right\rangle_{\mathbf{X}} = \boldsymbol{\alpha}' \mathbf{S} \boldsymbol{\alpha}, \tag{B.65}$$

where $\mathbf{S}$ is an $M \times M$ matrix:

$$S_{mn} \equiv \langle g_m, g_n \rangle_{\mathbf{X}} . \tag{B.66}$$

It is easy to check that this matrix is symmetric, i.e. it satisfies ($A.51$), and positive, i.e. it satisfies ($A.52$).

In particular, if we consider the functions

$$g_m(\mathbf{X}) \equiv X_m - \mathrm{E}\{X_m\} \tag{B.67}$$

the matrix (B.66) becomes the covariance matrix (2.67):

$$\begin{aligned} S_{mn} &\equiv \langle X_m - \mathrm{E}\{X_m\}, X_n - \mathrm{E}\{X_n\}\rangle_{\mathbf{X}} \\ &= \mathrm{Cov}\{X_m, X_n\}. \end{aligned} \tag{B.68}$$

The Cauchy-Schwartz inequality (B.61) in this context reads:

$$|\mathrm{Cov}\{X_m, X_n\}| \leq \mathrm{Sd}\{X_m\}\,\mathrm{Sd}\{X_n\}. \tag{B.69}$$

In particular, from (B.62) and the affine equivariance of the expected value (2.56) we obtain:

$$\mathrm{Cov}\{X_m, X_n\} = \mathrm{Sd}\{X_m\}\,\mathrm{Sd}\{X_n\} \Leftrightarrow X_m = a + bX_n, \tag{B.70}$$

where a is a scalar and b is a positive scalar. Similarly, from (B.63) and the affine equivariance of the expected value (2.56) we obtain:

$$\mathrm{Cov}\{X_m, X_n\} = -\,\mathrm{Sd}\{X_m\}\,\mathrm{Sd}\{X_n\} \Leftrightarrow X_m = a - bX_n, \tag{B.71}$$

where a is a scalar and b is a positive scalar. These properties allow us to define the correlation matrix.

B.6 Some special functions

We conclude with a list of special functions that recur throughout the text. See Abramowitz and Stegun (1974) and `mathworld.com` for more information.

The *indicator function* of a set $\mathcal{S} \in \mathbb{R}^N$ is defined as follows:

$$\mathbb{I}_{\mathcal{S}}(\mathbf{x}) \equiv \begin{cases} 1 \text{ if } \mathbf{x} \in \mathcal{S} \\ 0 \text{ if } \mathbf{x} \notin \mathcal{S}. \end{cases} \tag{B.72}$$

The *Heaviside function* $H^{(\mathbf{y})}$ is a step function:

$$H^{(\mathbf{y})}(\mathbf{x}) \equiv \begin{cases} 1 \text{ where } x_1 \geq y_1, \ldots, x_N \geq y_N \\ 0 \text{ otherwise.} \end{cases} \tag{B.73}$$

We can define equivalently the Heaviside function in terms of the indicator function (B.72) as follows:

$$H^{(\mathbf{y})} \equiv \mathbb{I}_{[y_1,+\infty)\times\cdots[y_N,+\infty)}. \tag{B.74}$$

The *error function* is defined as the integral of the Gaussian exponential:

$$\mathrm{erf}(x) \equiv \frac{2}{\sqrt{\pi}} \int_0^x e^{-u^2} du. \tag{B.75}$$

The error function is odd:

$$\operatorname{erf}(-x) = -\operatorname{erf}(x). \tag{B.76}$$

Furthermore, the error function is normalized in such a way that:

$$\operatorname{erf}(\infty) = 1. \tag{B.77}$$

This implies the following relation for the *complementary error function*:

$$\operatorname{erfc}(x) \equiv \frac{2}{\sqrt{\pi}} \int_x^{+\infty} e^{-u^2} du = 1 - \operatorname{erf}(x). \tag{B.78}$$

The *factorial* is a function defined only on integer values:

$$n! \equiv 1 \times 2 \times 3 \times \cdots \times (n-1) \times n. \tag{B.79}$$

The *gamma function* is defined by the following integral:

$$\Gamma(a) \equiv \int_0^{+\infty} u^{a-1} \exp(-u)\, du. \tag{B.80}$$

The gamma function is an extension to the complex and real numbers of the factorial. Indeed it is easy to check from the definition (B.80) that the following identity holds:

$$\Gamma(n) = (n-1)!. \tag{B.81}$$

For half-integer arguments, it can be proved that the following identity holds:

$$\Gamma\left(\frac{n}{2}\right) = \frac{(n-2)(n-4)\cdots n_0\sqrt{\pi}}{2^{\frac{n-1}{2}}}, \tag{B.82}$$

where $n_0 \equiv 1$ if n is odd and $n_0 \equiv 2$ if n is even.

The *lower incomplete gamma function* is defined as follows:

$$\gamma(x;a) \equiv \int_0^x u^{a-1} e^{-u} du. \tag{B.83}$$

The *upper incomplete gamma function* is defined as follows:

$$\Gamma(x;a) \equiv \int_x^{+\infty} u^{a-1} e^{-u} du. \tag{B.84}$$

The *lower regularized gamma function* is defined as follows:

$$P(x;a) \equiv \frac{\gamma(x;a)}{\Gamma(a)}. \tag{B.85}$$

The *upper regularized gamma function* is defined as follows:

$$Q(x;a) \equiv \frac{\Gamma(x;a)}{\Gamma(a)}. \tag{B.86}$$

The regularized gamma functions satisfy:

$$P(x;a) + Q(x;a) = 1. \tag{B.87}$$

The *beta function* is defined by the following integral:

$$B(a,b) \equiv \int_0^1 u^{a-1} (1-u)^{b-1} \, du. \tag{B.88}$$

The beta function is related to the gamma function through this identity:

$$B(a,b) = \frac{\Gamma(a)\,\Gamma(b)}{\Gamma(a+b)}. \tag{B.89}$$

The *incomplete beta function* is defined by the following integral:

$$B(x;a,b) \equiv \int_0^x u^{a-1} (1-u)^{b-1} \, du. \tag{B.90}$$

The *regularized beta function* is a normalized version of the incomplete beta function:

$$I(x;a,b) \equiv \frac{B(x;a,b)}{B(a,b)}. \tag{B.91}$$

Therefore the regularized beta function satisfies

$$I(0;a,b) = 0, \qquad I(1;a,b) = 1. \tag{B.92}$$

The *Bessel functions of first, second, and third kind* are solutions to the following differential equation:

$$x^2 \frac{d^2 w}{dx^2} + x \frac{dw}{dx} + \left(x^2 - \nu^2\right) w = 0.$$

In particular, the Bessel function of the second kind admits the following integral representation, see Abramowitz and Stegun (1974) p. 360:

$$\begin{aligned} Y_\nu(x) = &\frac{1}{\pi} \int_0^\pi \sin\left(x \sin(\theta) - \nu\theta\right) d\theta \\ &-\frac{1}{\pi} \int_0^{+\infty} \left(e^{\nu u} + e^{-\nu u} \cos(\nu\pi)\right) e^{x \sinh(u)} du. \end{aligned} \tag{B.93}$$

References

Abramowitz, M., and I. A. Stegun, 1974, *Handbook of Mathematical Functions with Formulas, Graphs, and Mathematical Tables* (Dover).

Acerbi, C., 2002, Spectral measures of risk: A coherent representation of subjective risk aversion, *Journal of Banking and Finance* 26, 1505–1518.

———, and D. Tasche, 2002, On the coherence of expected shortfall, *Journal of Banking and Finance* 26, 1487–1503.

Aitchison, J., and I. R. Dunsmore, 1975, *Statistical Prediction Analysis* (Cambridge University Press).

Alexander, C., 1998, Volatility and correlation: Measurement, models and applications, in C. Alexander, ed.: *Risk Management and Analysis, I* . pp. 125–171 (Wiley).

———, and A. Dimitriu, 2002, The cointegration alpha: Enhanced index tracking and long-short equity market neutral strategies, *ISMA Finance Discussion Paper No. 2002-08.*

Amerio, A., G. Fusai, and A. Vulcano, 2002, Pricing of implied volatility derivatives, *Working Paper.*

Anderson, H. M., C. W. J. Granger, and A. D. Hall, 1990, Treasury bill yield curves and cointegration, *University of California, San Diego Discussion Paper 90-24.*

Anderson, T. W., 1984, *An Introduction to Multivariate Statistical Analysis* (Wiley) 2nd edn.

Artzner, P., F. Delbaen, J. M. Eber, and D. Heath, 1997, Thinking coherently, *Risk Magazine* 10, 68–71.

———, 1999, Coherent measures of risk, *Mathematical Finance* 9, 203–228.

Balkema, A. A., and L. De Haan, 1974, Residual life time at great age, *Annals of Probability* 2, 792–804.

Bawa, V. S., S. J. Brown, and R. W. Klein, 1979, *Estimation Risk and Optimal Porfolio Choice* (North Holland).

Ben-Tal, A., and A. Nemirovski, 1995, Optimal design of engineering structures, *Optima* pp. 4–9.

———, 2001, *Lectures on modern convex optimization: analysis, algorithms, and engineering applications* (Society for Industrial and Applied Mathematics).

Berger, J. O., 1985, *Statistical Decision Theory and Bayesian Analysis* (Springer) 2nd edn.

A. Meucci, *Risk and Asset Allocation,* Springer Finance,

Bertsimas, D., G. J. Lauprete, and A. Samarov, 2004, Shortfall as a risk measure: Properties, optimization and applications, *Journal of Economic Dynamics and Control* 28, 1353–1381.

Best, M. J., and R. R. Grauer, 1991, On the sensitivity of mean-variance-efficient portfolios to changes in asset means: Some analytical and computational results, *Review of Financial Studies* 4, 315–342.

Bilmes, J. A., 1998, A gentle tutorial of the EM algorithm and its application to parameter estimation for Gaussian mixture and hidden markov models, *Working Paper.*

Bjork, T., 1998, *Arbitrage Theory in Continous Time* (Oxford University Press).

Black, F., 1995, Interest rates as options, *Journal of Finance* 50, 1371–1376.

———, and R. Litterman, 1990, Asset allocation: combining investor views with market equilibrium, *Goldman Sachs Fixed Income Research.*

———, 1992, Global portfolio optimization, *Financial Analyst Journal.*

Black, F., and M. S. Scholes, 1973, The pricing of options and corporate liabilities, *Journal of Political Economy* 81, 637–654.

Bollerslev, T., 1986, Generalized autoregressive conditional heteroskesdasticity, *Journal of Econometrics* 31, 307–327.

Bordley, R., and M. LiCalzi, 2000, Decision analysis using targets instead of utility functions, *Decisions in Economics and Finance* 23, 53–74.

Box, G. E. P., and G. M. Jenkins, 1976, *Time Series Analysis: Forecasting and Control, Revised Edition* (Holden-Day).

Boyd, S., and L. Vandenberghe, 2004, *Convex Optimization* (Cambridge University Press).

Brace, A., B. Goldys, J. Van der Hoek, and R. Womersley, 2002, Market model of stochastic implied volatility with application to the BGM model, *Working Paper.*

Brigo, D., and F. Mercurio, 2001, *Interest Rate Models* (Springer).

Brillinger, D. R., 2001, *Time Series: Data Analysis and Theory* (Society for Industrial and Applied Mathematics, Classics in Applied Mathematics).

Britten-Jones, M., 1999, The sampling error in estimates of mean-variance efficient portfolio weights, *Journal of Finance* 54, 655–671.

Burnham, K. P., and D. Anderson, 2002, *Model Selection and Multi-Model Inference* (Springer).

Campbell, J. Y., A. W. Lo, and A. C. MacKinlay, 1997, *The Econometrics of Financial Markets* (Princeton University Press).

Campbell, J. Y., and L. M. Viceira, 2002, *Strategic Asset Allocation* (Oxford University Press).

Campbell, N. A., 1980, Robust procedures in multivariate analysis I: Robust covariance estimation, *Applied Statistics* 29, 231–237.

Casella, G., and R. L. Berger, 2001, *Statistical Inference* (Brooks Cole) 2nd edn.

Castagnoli, E., and M. LiCalzi, 1996, Expected utility without utility, *Theory and Decision* 41, 281–301.

Ceria, S., and R. A. Stubbs, 2004, Incorporating estimation errors into portfolio selection: Robust efficient frontiers, *Axioma Inc. Technical Report.*

Chopra, V., and W. T. Ziemba, 1993, The effects of errors in means, variances, and covariances on optimal portfolio choice, *Journal of Portfolio Management* pp. 6–11.

Connor, G., and R. A. Korajczyk, 1993, A test for the number of factors in an approximate factor model, *Journal of Finance* 48, 1263–1292.

———, 1995, The arbitrage pricing theory and multifactor models of asset returns, in R. A. Jarrow, V. Maksimovic, and W. T. Ziemba, ed.: *Finance* . pp. 87–144 (North-Holland).

Corielli, F., and A. Meucci, 2004, Linear models for style analysis, *Statistical Methods and Applications* 13, 105–129.

Cornish, E. A., and R. A. Fisher, 1937, Moments and cumulants in the specification of distributions, *Extrait de la Revue de l'Institute International de Statistique* 4, 1–14.

Crouhy, M., D. Galai, and R. Mark, 1998, The new 1998 regulatory framework for capital adequacy: "standardized approach" versus "internal models", in C. Alexander, ed.: *Risk Management and Analysis, I* . pp. 1–37 (Wiley).

Cuppens, R., 1975, *Decomposition of Multivariate Probabilities* (Academic Press).

Dantzig, G., 1998, *Linear Programming and Extensions* (Princeton University Press).

David, F. N., and D. E. Barton, 1962, *Combinatorial Chance* (Griffin).

David, H. A., 1981, *Order Statistics* (Wiley) 2nd edn.

De Santis, G., and S. Foresi, 2002, Robust optimization, *Goldman Sachs Technical Report.*

Dempster, A. P., M. N. Laird, and D. B. Rubin, 1977, Maximum likelihood from incomplete data via the EM algorithm, *Journal of the Royal Statistical Society* 39, 1–22.

Dickey, J. M., 1967, Matric-variate generalizations of the multivariate t distribution and the inverted multivariate t distribution, *Annals of Mathematical Statistics* 38, 511–518.

El Ghaoui, L., and H. Lebret, 1997, Robust solutions to least-squares problems with uncertain data, *SIAM Journal on Matrix Analysis and Applications* 18, 1035–1064.

Embrechts, P., McNeil A., and D. Straumann, 2002, Correlation and dependence in risk management: Properties and pitfalls, *Risk Management: Value at Risk and Beyond, Cambridge University Press.*

Embrechts, P., C. Klueppelberg, and T. Mikosch, 1997, *Modelling Extremal Events* (Springer).

Engle, R. F., 1982, Autoregressive conditional heteroscedasticity with estimates of the variance of United Kingdom inflation, *Econometrica* 50, 987–1007.

Evans, S. N., and P. B. Stark, 1996, Shrinkage estimators, Skorokhod's problem, and stochastic integration by parts, *Annals of Statistics* 24, 809–815.

———, 2002, Inverse problems as statistics, *Inverse Problems* 18, R55–R97.

Fabozzi, F. J., ed., 2005, *The Handbook of Fixed Income Securities* (McGraw-Hill) 7th edn.

Fama, E. F., and K. R. French, 1992, The cross-section of expected stock returns, *Journal of Finance* 47, 427–465.

———, 1993, Common risk factors in the returns on stocks and bonds, *Journal of Financial Economics* 33, 3–56.

Fang, K. T., S. Kotz, and K. W. Ng, 1990, *Symmetric Multivariate and Related Distributions* (CRC Press).

Fang, K. T., and Y. T. Zhang, 1990, *Generalized Multivariate Analysis* (Springer).

Fengler, M. R., W. Haerdle, and P. Schmidt, 2003, The analyis of implied volatilities, *Working Paper.*

Ferson, W. E., and A. F. Siegel, 2001, The efficient use of conditioning information in portfolios, *Journal of Finance* 56, 967–982.

Feuerverger, A., and A. C. Wong, 2000, Computation of value at risk for nonlinear portfolios, *Journal of Risk* 3, 37–55.

Fischer, T., 2003, Risk capital allocation by coherent risk measures based on one-sided moments, *Insurance: Mathematics and Economics* 32, 135–146.

Forbes, K. J., and R. Rigobon, 2002, No contagion, only interdependence: measuring stock market co-movements, *Journal of Finance* 57, 2223–2261.

Frittelli, M., and E. Rosazza Gianin, 2002, Putting order in risk measures, *Journal of Banking and Finance* 26, 1473–1486.

Frost, P. A., and J. E. Savarino, 1988, For better performance: Constrain portfolio weights, *Journal of Portfolio Management* 15, 29–34.

Fusai, G., and A. Meucci, 2002, A dynamic factor model for bond portfolio allocation, *Working Paper.*

———, 2003, Assessing views, *Risk Magazine* 16, S18–S21.

Geweke, J., 1999, Using simulation methods for Bayesian econometric models: Inference, development and communication, *Econometric Reviews* 18, 1–126.

Goel, P.M., and A. Zellner, 1986, *Bayesian Inference and Decision Techniques: Essays in Honor of Bruno De Finetti* ., vol. 6 of *Studies in Bayesian Econometrics and Statistics* (Elsevier Science).

Goldfarb, D., and G. Iyengar, 2003, Robust portfolio selection problems, *Mathematics of Operations Research* 28, 1–38.

Gollier, C., 2001, *The Economics of Risk and Time* (MIT press).

Gourieroux, C., J. P. Laurent, and O. Scaillet, 2000, Sensitivity analysis of values at risk, *Journal of Empirical Finance* 7, 225–245.

Graham, R. L., D. E. Knuth, and O. Patashnik, 1994, *Concrete Mathematics: A Foundation for Computer Science* (Addison-Wesley) 2nd edn.

Green, R. C., and B. Hollifield, 1992, When will mean-variance efficient portfolios be well diversified?, *Journal of Finance* 47, 1785–1809.

Greene, W. H., 1999, *Econometric Analysis* (Prentice Hall).

Grinold, R. C., 1996, Domestic grapes from imported wine, *Journal of Portfolio Management* 26, 29–40.

———, and K. K. Easton, 1998, Attribution of performance and holdings, in W. T. Ziemba, and J. M. Mulvey, ed.: *Worldwide Asset and Liability Modeling* . pp. 87–113 (Cambridge University Press).

Grinold, R. C., and R. Kahn, 1999, *Active Portfolio Management. A Quantitative Approach for Producing Superior Returns and Controlling Risk* (McGraw-Hill) 2nd edn.

Guthoff, A., A. Pfingsten, and J. Wolf, 1997, On the compatibility of value at risk, other risk concepts and expected utility maximization, *Diskussionsbeitrag 97-01, Westfaelische Wilhelms-Universitaet Muenster.*

Haan, W. J., and A. T. Levin, 1996, A practitioner's guide to robust covariance matrix estimation, *NBER technical Working Paper.*

Haerdle, W., and L. Simar, 2003, *Applied Multivariate Statistical Analysis* (www.quantlet.com/mdstat/scripts/mva/htmlbook).

Halldorsson, B. V., and R. H. Tutuncu, 2003, An interior-point method for a class of saddle-point problems, *Journal of Optimization Theory and Applications* 116, 559–590.

Hallerbach, W., 2003, Decomposing portfolio value-at-risk: A general analysis, *Journal of Risk* 5, 1–18.

Hamilton, J. D., 1994, *Time Series Analysis* (Princeton University Press).

Hampel, F. R., 1973, Robust estimation: A condensed partial survey, *Zeitschrift fuer Wahrscheinlichkeitstheorie und Verwandte Gebiete* 27, 87–104.

———, E. M. Ronchetti, P. J. Rousseeuw, and W. A. Stahel, 1986, *Robust Statistics, the Approach Based on Influence Functions* (Wiley).

Harvey, A. C., 1981, *The Econometric Analysis of Time Series* (Wiley).

He, G., and R. Litterman, 2002, The intuition behind Black-Litterman model portfolios, *ssrn.com*.

Huber, P. J., 1964, Robust estimation for a location parameter, *Annals of Mathematical Statistics* 35, 73–101.

———, 1981, *Robust Statistics* (Wiley).

Hull, J. C., 2002, *Options, Futures and Other Derivatives* (Prentice Hall) 5th edn.

Ingersoll, E. J., 1987, *Theory of Financial Decision Making* (Rowman and Littlefield).

Jagannathan, R., and T. Ma, 2003, Risk reduction in large portfolios: Why imposing the wrong constraints helps, *Journal of Finance* 58, 1651–1683.

James, J., and N. Webber, 2000, *Interest Rate Modelling* (Wiley).

Jaschke, S. R., 2002, The Cornish-Fisher expansion in the context of delta-gamma-normal approximations, *Journal of Risk* 4.

Jobson, J. D., and B. Korkie, 1980, Estimation for Markowitz efficient portfolios, *Journal of the American Statistical Association* 75, 544–554.

———, 1981, Putting Markowitz theory to work, *Journal of Portfolio Management* pp. 70–74.

Jorion, P., 1986, Bayes-Stein estimation for portfolio analysis, *Journal of Financial and Quantitative Analysis* 21, 279–291.

———, 1992, Portfolio optimization in practice, *Financial Analyst Journal* pp. 68–74.

———, 1996, Risk 2: Measuring the risk in value-at-risk, *Financial Analysts Journal* pp. 47–56.

Kahneman, D., and A. Tversky, 1979, Prospect theory: An analysis of decision under risk, *Econometrica* 47, 263–291.

Kennedy, D. P., 1997, Characterizing and filtering Gaussian models of the term structure of interest rates, *Mathematical Finance* 7, 107–118.

Kotz, S., N. Balakrishnan, and N. L. Johnson, 1994, *Continuous Univariate Distributions* (Wiley) 2nd edn.

———, 2000, *Continuous Multivariate Distributions: Models and Applications* (Wiley).

Kotz, S., and S. Nadarajah, 2004, *Multivariate T Distributions and Their Applications* (Cambridge University Press).

Kusuoka, S., 2001, On law invariant coherent risk measures, *Advances in Mathematical Economics* 3, 83–95.

Lang, S., 1997, *Introduction to Linear Algebra* (Springer) 2nd edn.

Ledoit, O., and M. Wolf, 2003, Improved estimation of the covariance matrix of stock returns with an application to portfolio selection, *Journal of Empirical Finance* 10, 603–621.

———, 2004, A well-conditioned estimator for large-dimensional covariance matrices, *Journal of Multivariate Analysis* 88, 365–411.

Lehmann, E. L., and G. Casella, 1998, *Theory of Point Estimation* (Springer) 2nd edn.

Leibowitz, M. L., L. N. Bader, and S. Kogelman, 1996, *Return Targets and Shortfall Risks* (McGraw-Hill).

Levy, H., 1998, *Stochastic Dominance: Investment Decision Making under Uncertainty* (Kluwer Academic Publishers).

LiCalzi, M., and A. Sorato, 2003, The Pearson system of utility functions, *Working Paper.*

Lindskog, F., A. McNeil, and U. Schmock, 2003, Kendall's tau for elliptical distributions, in G. Bol, G. Nakhaeizadeh, S. T. Rachev, and T. Ridder, ed.: *Credit Risk Measurement, Evaluation and Management* . pp. 149–156 (Physica-Verlag).

Lintner, J., 1965, The valuation of risky assets and the selection of risky investments in stock portfolios and capital budgets, *Review of Economics and Statistics* 47, 13–37.

Litterman, R., 1996, Hot spots and hedges, *Goldman Sachs and Co., Risk Management Series.*

———, and Goldman Sachs Asset Management Quantitative Resources Group, 2003, *Modern Investment Management* (Wiley).

Litterman, R., and J. Scheinkman, 1991, Common factors affecting bond returns, *Journal of Fixed Income* 1, 54–61.

Litterman, R., and K. Winkelmann, 1998, Estimating covariance matrices, *Goldman Sachs, Risk Management Series.*

Little, R. J. A., and D. B. Rubin, 1987, *Statistical Analysis with Missing Data* (Wiley).

Lo, A. W., and A. C. MacKinlay, 2002, *A Non-Random Walk Down Wall Street* (Princeton University Press).

Lobo, M., L. Vandenberghe, S. Boyd, and H. Lebret, 1998, Applications of second-order cone programming, *Linear Algebra and its Applications, Special Issue on Linear Algebra in Control, Signals and Image Processing* 284, 193–228.

Longstaff, F. A., P. Santa-Clara, and E. S. Schwartz, 2001, The relative valuation of caps and swaptions: Theory and empirical evidence, *Journal of Finance* 56, 2067–2109.

Loretan, M., and W. B. English, 2000, Evaluating correlation breakdowns during periods of market volatility, *Board of Governors of the Federal Reserve System International Finance Working Paper.*

Luenberger, D. G., 1998, *Investment Science* (Oxford University Press).

Magnus, J. R., and H. Neudecker, 1979, The commutation matrix: Some properties and applications, *Annals of Statistics* 7, 381–394.

———, 1999, *Matrix Differential Calculus with Applications in Statistics and Econometrics, Revised Edition* (Wiley).

Mardia, K. V., 1970, Measures of multivariate skewness and kurtosis with applications, *Biometrika* 57, 519–530.

———, J. T. Kent, and J. M. Bibby, 1979, *Multivariate Analysis* (Academic Press).

Markowitz, H. M., 1991, *Portfolio Selection: Efficient Diversification of Investments* (Blackwell) 2nd edn.

———, and N. Usmen, 2003, Resampled frontiers versus diffuse Bayes: an experiment, *Journal of Investment Management* 1, 9–25.

Maronna, R. A., 1976, Robust M-estimators of multivariate location and scatter, *Annals of Statistics* 1, 51–67.

Merton, R. C., 1969, Lifetime portfolio selection under uncertainty: The continuous case, *Review of Economical Statistics* 51, 247–257.

———, 1992, *Continuous-Time Finance* (Blackwell).

Meucci, A., 2001, Common pitfalls in mean-variance asset allocation, *Wilmott Technical Article.*

———, 2004, Broadening horizons, *Risk Magazine* 17, 98–101.
———, 2005, Robust Bayesian asset allocation, *Working Paper ssrn.com.*
Michaud, R. O., 1998, *Efficient Asset Management: A Practical Guide to Stock Portfolio Optimization and Asset Allocation* (Harvard Business School Press).
Minka, T. P., 2003, Old and new matrix algebra useful for statistics, *Working Paper.*
Mood, A. M., F. A. Graybill, and D. C. Boes, 1974, *Introduction to the Theory of Statistics* (McGraw-Hill) 3rd edn.
Morrison, D. F., 2002, *Multivariate Statistical Methods* (Duxbury Press).
Nelsen, R. B., 1999, *An Introduction to Copulas* (Springer).
Nesterov, Y., and A. Nemirovski, 1995, *Interor-Point Polynomial Algorithms in Convex Programming* (Society for Industrial and Applied Mathematics).
NRS, 1988-1992, *Numerical Recipes in C: The Art of Scientific Computing* (Cambridge University Press).
O'Hagan, A., 1994, *Kendall's Advanced Theory of Statistics: Bayesian Inference, Vol 2B* (Edward Arnold).
Oksendal, B., 1998, *Stochastic Differential Equations, an Introduction with Applications* (Springer) 5th edn.
Papoulis, A., 1984, *Probability, Random Variables, and Stochastic Processes* (McGraw-Hill) 2nd edn.
Parzen, E., K. Tanabe, and G. Kitagawa, 1998, *Selected Papers of Hirotugu Akaike* (Springer).
Pastor, L., 2000, Portfolio selection and asset pricing models, *Journal of Finance* 55, 179–223.
———, and R. F. Stambaugh, 2002, Investing in equity mutual funds, *Journal of Financial Economics* 63, 351–380.
Pearson, K., 1895, Memoir on skew variation in homogenous material, *Philosophical Transactions of the Royal Society* 186, 343–414.
Perret-Gentil, C., and M. P. Victoria-Feser, 2003, Robust mean-variance portfolio selection, *Cahiers du Departement d'Econometrie, University of Geneva.*
Pickands, J., 1975, Statistical inference using extreme order statistics, *Annals of Statistics* 3, 119–131.
Poston, W. L., E. J. Wegman, C. E. Priebe, and J. L. Solka, 1997, A deterministic method for robust estimation of multivariate location and shape, *Journal of Computational and Graphical Statistics* 6, 300–313.
Press, S. J., 1982, *Applied Multivariate Analysis* (Krieger) 2nd edn.
Priestley, M. B., 1981, *Spectral Analysis and Time Series* (Academic Press).
Quenouille, M. H., 1956, Notes on bias in estimation, *Biometrika* 43, 353–360.
Raiffa, H., and R. Schlaifer, 2000, *Applied Statistical Decision Theory* (Wiley).
Rau-Bredow, H., 2002, Value at risk, expected shortfall, and marginal risk contribution, *Working Paper.*
Rebonato, R., 1998, *Interest-Rate Option Models: Understanding, Analyzing and Using Models for Exotic Interest-Rate Options* (Wiley) 2nd edn.
Reed, M., and B. Simon, 1980, *Methods of Modern Mathematical Physics - Vol I* (Academic Press).
Roll, R., 1992, A mean-variance analysis of tracking error, *Journal of Portfolio Management* pp. 13–22.
Rose, C., and M. D. Smith, 2002, *Mathematical Statistics with Mathematica* (Springer).
Ross, S., 1976, The arbitrage theory of capital asset pricing, *Journal of Economic Theory* 13, 341–360.

Rousseeuw, P. J., and A. M. Leroy, 1987, *Robust Regression and Outlier Detection* (Wiley).

Rousseeuw, P. J., and K. VanDriessen, 1999, A fast algorithm for the minimum covariance determinant estimator, *Journal of the American Statistical Association* 41, 212–223.

Rudin, W., 1976, *Principles of Mathematical Analysis* (McGraw-Hill) 3rd edn.

———, 1991, *Functional Analysis* (McGraw-Hill) 2nd edn.

Scherer, B., 2002, Portfolio resampling: Review and critique, *Financial Analyst Journal* 58, 98–109.

Schoenbucher, P. J., 1999, A market model for stochastic implied volatility, *Philosofical Transactions of the Royal Society* 357, 2071–2092.

Searle, S. R., 1982, *Matrix Algebra Useful for Statistics* (Wiley).

Sharpe, W. F., 1964, Capital asset prices: A theory of market equilibrium under conditions of risk, *Journal of Finance* 19, 425–442.

———, 1974, Imputing expected returns from portfolio composition, *Journal of Financial and Quantitative Analysis* pp. 463–472.

Shirayaev, A. N., 1989, *Probability* (Springer) 2nd edn.

Smirnov, V. I., 1964, *A Course of Higher Mathematics - Vol V. Integration and Functional Analysis* (Pergamon Press).

———, 1970, *Linear Algebra and Group Theory* (Dover).

Stambaugh, R. F., 1997, Analyzing investments whose histories differ in length, *Journal of Financial Economics* 45, 285–331.

Stein, C., 1955, Inadmissibility of the usual estimator for the mean of a multivariate normal distribution, *Proceedings of the 3rd Berkeley Symposium on Probability and Statistics.*

———, 1975, Estimation of a covariance matrix, *Rietz Lecture, 39th Annual Meeting IMS.*

Stock, J. H., and M. W. Watson, 1988, Testing for common trends, *Journal of the American Statistical Association* 83, 1097–1107.

Sutradhar, B. C., 1986, On the characteristic function of the multivariate Student t distribution, *Canadian Journal of Statistics* 14, 329–337.

———, 1988, Author's revision, *Canadian Journal of Statistics* 16, 323.

Tasche, D., 1999, Risk contributions and performance measurement, *Working Paper, Technische Universitaet Muenchen.*

———, 2002, Expected shortfall and beyond, *Journal of Banking and Finance* 26, 1519–1533.

Thorin, O., 1977, On the infinite divisibility of the lognormal distribution, *Scandinavian Actuarial Journal* pp. 121–148.

Tukey, J.W., 1958, Bias and confidence in not-quite large samples, *Annals of Mathematical Statistics* 29, 614.

———, 1977, *Exploratory Data Analysis* (Addison-Wesley).

Varian, R. H., 1992, *Microeconomic Analysis* (Norton) 3rd edn.

Watson, G. S., 1984, *Statistics on Spheres* (Wiley).

Whittaker, E. T., and G. N. Watson, 1996, *A Course of Modern Analysis* (Cambridge University Press) 4th edn.

Wilmott, P., 1998, *Derivatives* (Wiley).

Wilson, T., 1994, Plugging the gap, *Risk Magazine* 7, 74–80.

Yamai, Y., and T. Yoshiba, 2002, Comparative analyses of expected shortfall and value-at-risk (2): Expected utility maximization and tail risk, *Monetary and Economic Studies.*

Zellner, A., and V. K. Chetty, 1965, Prediction and decision problems in regression models from the Bayesian point of view, *Journal of the American Statistical Association* 60, 608–616.

List of Figures

A. Meucci, *Risk and Asset Allocation*, Springer Finance,

Notation

Generic

www.2.4	Technical appendix to Chapter 2, Section 4, at symmys.com	
$a \approx b$	a is approximately equal to b	
$a \equiv b$	a is defined as b	
K	number of factors in factor model	p. 132
N	number of securities / dimension of market invariants	p. 101
S	dimension of market parameters	p. 186
T	length of time series (also investment decision date)	Fig. 3.11

Time

t	generic time	
τ	time distance to investment horizon	Fig. 3.11
$\widetilde{\tau}$	estimation interval for the market invariants	Fig. 3.11
T	investment decision date (also length of time series)	Fig. 3.11

General distribution theory

$\mathbb{P}$	probability	(2.3)
$\mathbf{X}$ (X)	generic random variable (univariate)	p. 34
$\mathbf{x}$ (x)	realized value of the r.v. $\mathbf{X}$ (X)	p. 35
$\mathbf{X} \sim \mathrm{D}$	the distribution of the r.v. $\mathbf{X}$ is D	
$\mathbf{X} \stackrel{d}{=} \mathbf{Y}$	the distributions of the r.v. $\mathbf{X}$ and $\mathbf{Y}$ are the same	
$\mathbf{X}_B \vert \mathbf{x}_A$	conditional distribution of the r.v. $\mathbf{X}_B$ for given $\mathbf{x}_A$	p. 45
$X_{r:T}$	r-th order statistics in a sample of T i.i.d. r.v. X_t	(2.247)
$f_{\mathbf{X}}$	probability density function of the r.v. $\mathbf{X}$	(2.4)
$F_{\mathbf{X}}$	cumulative distribution function of the r.v. $\mathbf{X}$	(2.9)

A. Meucci, *Risk and Asset Allocation*, Springer Finance,

$\phi_{\mathbf{X}}$	characteristic function of the r.v. $\mathbf{X}$	(2.13)
Q_X	quantile of the r.v. X (univariate)	(1.17)

Summary statistics

CM_k^X	central moments of the univariate r.v. X	(1.48)
$\mathrm{CM}_{n_1 \cdots n_k}^{\mathbf{X}}$	central moments of the multivariate r.v. $\mathbf{X}$	(2.92)
$\mathrm{Cor}\{\mathbf{X}\}$	correlation matrix of the r.v. $\mathbf{X}$	(2.133)
$\mathrm{Cov}\{\mathbf{X}\}$	covariance matrix of the r.v. $\mathbf{X}$	(2.67)
$\mathrm{E}\{\mathbf{X}\}$	expected value of the r.v. $\mathbf{X}$	(2.54)
$\mathrm{Ku}\{X\}$	kurtosis of the univariate r.v. X	(1.51)
$\mathrm{MAD}\{X\}$	mean absolute deviation of the univariate r.v. X	(1.41)
$\mathrm{MDis}\{\mathbf{X}\}$	modal dispersion of the r.v. $\mathbf{X}$	(2.65)
$\mathrm{Ma}(\mathbf{x}, \boldsymbol{\mu}, \boldsymbol{\Sigma})$	Mahalanobis distance of $\mathbf{x}$ from $\boldsymbol{\mu}$ through the metric induced by $\boldsymbol{\Sigma}$	(2.61)
$\mathrm{Med}\{X\}$	median of the univariate r.v. X	(1.26)
$\mathrm{Mod}\{\mathbf{X}\}$	mode of the r.v. $\mathbf{X}$	(2.52)
$\mathrm{Ran}\{X\}$	range of the univariate r.v. X	(1.37)
RM_k^X	raw moments of the univariate r.v. X	(1.47)
$\mathrm{RM}_{n_1 \cdots n_k}^{\mathbf{X}}$	raw moments of the multivariate r.v. $\mathbf{X}$	(2.91)
$\mathrm{Sd}\{X\}$	standard deviation of the univariate r.v. X	(1.42)
$\mathrm{Sk}\{X\}$	skewness of the univariate r.v. X	(1.49)
$\mathrm{SW}\{X_m, X_n\}$	Schweizer and Wolff measure of dependence	(2.103)
$\mathrm{Var}\{X\}$	variance of the univariate r.v. X	(1.43)
Z_X	z-score of the univariate r.v. X	(1.35)
$\rho\{X_m, X_n\}$	Spearman's rho	(2.130)
$\tau\{X_m, X_n\}$	Kendall's tau	(2.128)

Specific distributions

$\mathrm{U}([a, b])$	uniform distribution on the interval $[a, b]$	(1.54)
$\mathrm{U}(\mathcal{S})$	uniform distribution on the set $\mathcal{S}$	(2.144)
$\mathrm{N}(\mu, \sigma^2)$	univariate normal distribution with expected value μ and variance σ^2	(1.66)
$\mathrm{N}(\boldsymbol{\mu}, \boldsymbol{\Sigma})$	multivariate normal distribution with expected value $\boldsymbol{\mu}$ and covariance matrix $\boldsymbol{\Sigma}$	(2.155)
$\mathrm{N}(\mathbf{M}, \boldsymbol{\Sigma}, \mathbf{S})$	matrix-variate normal distribution with expected value $\mathbf{M}$, column covariance $\boldsymbol{\Sigma}$ and row covariance $\mathbf{S}$	(2.181)
$\mathrm{St}(\nu, \mu, \sigma^2)$	univariate Student t distribution with ν degrees of freedom, location μ and dispersion σ	(1.85)
$\mathrm{St}(\nu, \boldsymbol{\mu}, \boldsymbol{\Sigma})$	multivariate Student t distribution with ν degrees of freedom, location $\boldsymbol{\mu}$ and scatter matrix $\boldsymbol{\Sigma}$	(2.187)

$\mathrm{St}(\nu, \mathbf{M}, \mathbf{\Sigma}, \mathbf{S})$	matrix-variate Student t distribution with with ν degrees of freedom, expected value $\mathbf{M}$, column scatter matrix $\mathbf{\Sigma}$ and row scatter matrix $\mathbf{S}$	(2.198)
$\mathrm{Ca}(\mu, \sigma^2)$	univariate Cauchy distribution with location μ and dispersion σ	(1.78)
$\mathrm{Ca}(\boldsymbol{\mu}, \mathbf{\Sigma})$	multivariate Cauchy distribution with location $\boldsymbol{\mu}$ and scatter matrix $\mathbf{\Sigma}$	(2.208)
$\mathrm{LogN}(\mu, \sigma^2)$	univariate lognormal distribution: $\ln(X)$ has expected value μ and variance σ^2	(1.94)
$\mathrm{LogN}(\boldsymbol{\mu}, \mathbf{\Sigma})$	multivariate lognormal distribution: $\ln(\mathbf{X})$ has expected value $\boldsymbol{\mu}$ covariance matrix $\mathbf{\Sigma}$	(2.217)
$\mathrm{Ga}(\nu, \sigma^2)$	(central) gamma distribution with ν degrees of freedom and scale parameter σ^2	(1.108)
χ^2_ν	chi-square distribution with ν degrees of freedom	(1.109)
$\mathrm{W}(\nu, \mathbf{\Sigma})$	Wishart distribution with ν degrees of freedom and scale parameter $\mathbf{\Sigma}$	(2.223)
$\mathrm{IW}(\nu, \mathbf{\Psi})$	inverse Wishart distribution with ν degrees of freedom and inverse scale parameter $\mathbf{\Psi}$	(2.232)
$\mathrm{El}(\boldsymbol{\mu}, \mathbf{\Sigma}, g)$	Elliptical distribution with location $\boldsymbol{\mu}$, scatter matrix $\mathbf{\Sigma}$ and pdf generator g	(2.268)
$\mathrm{SS}(\alpha, \boldsymbol{\mu}, m_{\mathbf{\Sigma}})$	Symmetric stable distribution with thickness α, location $\boldsymbol{\mu}$ and elliptical measure $m_{\mathbf{\Sigma}}$	(2.285)
$\mathrm{Em}(i_T)$	Empirical distribution determined by the observations i_T	(2.239)

Market variables

$\mathbf{M}$	market vector: affine transformation of security prices at the investment horizon	(5.11)
$\mathbf{P}_t\ (\mathbf{P}_{T+\tau})$	prices of the market securities at time t (at the investment horizon)	p. 101
$\mathcal{T}$	transaction costs	p. 305
$C_{t,\tau}$	compounded return over an interval τ that becomes known at time t	(3.11)
$H_{t,\tau}$	total return over an interval τ that becomes known at time t	(3.9)
$L_{t,\tau}$	linear return over an interval τ that becomes known at time t	(3.10)
Y_t^v	yield to maturity v at time t	(3.30)
$Z_t^{(E)}$	value at time t of a zero-coupon bond that matures at time E	p. 109
$\sigma_t^{(K,E)}$	implied percentage volatility for strike K and expiry E at time t	(3.40)
Conv	convexity of a forward swap	(3.255)
PVBP	present value of a basis point in a forward swap	(3.254)
RD	roll-down in a forward swap	(3.253)

Market estimation and modeling

$\mathbf{X}_{t,\tau}$	market invariants relative to time interval τ	p. 103
$\mathbf{X}_t$	market invariants relative to normalized time interval ($\tau \equiv 1$ in suitable units)	p. 171
i_T	market information available at time T, typically the time series of the market invariants $\mathbf{x}_1, \ldots, \mathbf{x}_T$	(4.8)
I_T	market information before the realization, typically a set of market invariants $\mathbf{X}_1, \ldots, \mathbf{X}_T$	(4.14)
R^2	generalized r-square	(3.116)
$\mathbf{B}$	factor loadings	(3.117)
$\mathbf{F}_t$ ($\mathbf{F}$)	factors	(3.117)
$\mathbf{U}_t$ ($\mathbf{U}$)	residuals	(3.117)
$\mathbf{E}$	orthogonal matrix of eigenvectors in principal component decomposition	(A.62)
$\mathbf{\Lambda}$	diagonal matrix of decreasing eigenvalues in principal component decomposition	(A.65)
$\widehat{\mathbf{G}}$	generic estimator	(4.9)
$\widehat{\boldsymbol{\theta}}$	parametric estimator	(7.2)
$\widehat{\boldsymbol{\theta}}_{\mathrm{ce}}$	Bayesian classical-equivalent estimator	(7.5)
Bias	bias of estimator	(4.25)
Err	error of estimator	(4.23)
Inef	inefficiency of estimator	(4.26)
Loss	loss of estimator	(4.19)
CN	condition number	(4.115)
IF	influence function	(4.185)
SC	sensitive curve	(4.166)

Allocations

$\boldsymbol{\alpha}$	generic allocation	p. 239
$\boldsymbol{\alpha}_{\mathrm{B}}$	Bayesian allocation	(9.9)
$\boldsymbol{\alpha}_{\mathrm{BL}}$	Black-Litterman allocation	(9.32)
$\boldsymbol{\alpha}_{\mathrm{ce}}$	classical-equivalent allocation	(9.13)
$\boldsymbol{\alpha}_{MV}$	global minimum variance portfolio	(6.99)
$\boldsymbol{\alpha}_{\mathrm{p}}$	prior allocation	(8.64)
$\boldsymbol{\alpha}_{\mathrm{r}}$	robust allocation	(9.110)
$\boldsymbol{\alpha}_{\mathrm{rB}}$	robust Bayesian allocation	(9.133)
$\boldsymbol{\alpha}_{\mathrm{rs}}$	resampled allocation	(9.86)
$\boldsymbol{\alpha}_{\mathrm{s}}$	sample-based allocation	(8.81)
$\boldsymbol{\alpha}_{SR}$	maximum Sharpe ratio portfolio	(6.100)

Investor's preferences / profile

A	Arrow-Pratt absolute risk-aversion	(5.121)
$\mathcal{C}$	investor's constraints	p. 307
$\mathcal{C}^+$	cost of constraint violation	p. 393
CE	certainty-equivalent	(5.93)
Coh	coherent index of satisfaction	p. 288
CVaR	conditional value at risk	(5.208)
EOP	expected overperformance	(6.178)
ES	expected shortfall	(5.207)
IR	information ratio	(6.180)
OC	opportunity cost	(8.16)
Q	quantile-based index of satisfaction	(5.159)
RP	risk-premium	(5.122)
$\mathcal{S}$	index of satisfaction	(5.48)
Spc	spectral index of satisfaction	p. 290
SR	Sharpe ratio	(5.51)
TE	tracking error	(6.179)
VaR	value at risk	(5.158)
Ψ	investor's objective	p. 239

Linear algebra and geometry

$\mathbb{R}^N$	Euclidean N-dimensional vector space	p. 465
$\mathcal{E}_{\mathbf{m},\mathbf{S}}$	ellipsoid centered in $\mathbf{m}$ with shape parameter $\mathbf{S}$ and unit radius factor	$(A.76)$
$\mathbf{v}$	row vector	$(A.1)$
v_n, $[\mathbf{v}]_n$, $v^{(n)}$	n-th entry of the vector $\mathbf{v}$	
$\|\mathbf{v}\|$	norm of $\mathbf{v}$	$(A.7)$
$\boldsymbol{\delta}^{(n)}$	canonical basis in $\mathbb{R}^N$	$(A.15)$
$\mathbf{A}$	matrix	
$\mathbf{A}'$	transpose of the matrix $\mathbf{A}$	
$\|\mathbf{A}\|$	determinant of the square matrix $\mathbf{A}$	$(A.39)$
$\operatorname{tr}(\mathbf{A})$	trace of the square matrix $\mathbf{A}$	$(A.43)$
$\mathbf{S} \succeq \mathbf{0}$	$\mathbf{S}$ positive and symmetric	
$\operatorname{diag}(d_1,\ldots,d_N)$	square matrix: all entries null, except the diagonal, which is $(d_1,\ldots,d_N)$	
$\mathbf{1}$, $\mathbf{1}_N$	N-dimensional vector of ones	
$\mathbf{I}$, $\mathbf{I}_N$	$N \times N$ identity matrix	$(A.27)$
$\otimes$	Kronecker product	$(A.96)$
vec	operator that stacks the columns of a matrix	$(A.104)$
vech	operator that stacks the columns of a symmetric matrix, skipping the redundant entries	$(A.113)$
$\mathbf{D}_N$	duplication matrix	$(A.113)$
$\mathbf{K}_{NK}$	commutation matrix	$(A.108)$

Functional analysis

$\mathcal{D}$ ($\mathcal{D}_n$)	differentiation operator (multivariate, in n-th coordinate)	(B.25)
$\mathcal{F}[v]$	Fourier transform of the function v	(B.34)
$\mathcal{F}^{-1}[v]$	inverse Fourier transform of the function v	(B.40)
$\mathcal{I}$ ($\mathcal{I}_n$)	integration operator (multivariate, in n-th coordinate)	(B.27)
$L_2(\mathbb{R}^N)$	set of generalized functions on $\mathbb{R}^N$ with integrable square absolute value	(B.9)
$\|v\|$	L_2-norm of the function v	(B.8)
$\|g\|_{\mathbf{X};p}$	p-norm induced by the distribution of the r.v. $\mathbf{X}$	(B.57)

Special functions

$\delta^{(\mathbf{x})}$	Dirac delta centered in $\mathbf{x}$	(B.16)
$\delta_\epsilon^{(\mathbf{y})}$	regularized Dirac delta function centered in $\mathbf{x}$ with bandwidth ϵ	(B.18)
$\mathbb{I}_{\mathcal{S}}$	indicator function of the set $\mathcal{S}$	(B.72)
$H^{(\mathbf{x})}$	Heaviside function with step in $\mathbf{x}$	(B.74)
erf	error function	(B.75)
Γ	gamma function	(B.80)
B	beta function	(B.88)
I	regularized beta function	(B.91)

Index